POINT PROCESSES AND THEIR STATISTICAL INFERENCE

PROBABILITY: PURE AND APPLIED

A Series of Textbooks and Reference Books

Editor

MARCEL NEUTS

University of Arizona
Tucson, Arizona

1. Introductory Probability Theory, *by Janos Galambos*
2. Point Processes and Their Statistical Inference, *by Alan F. Karr*

Other Volumes in Preparation

POINT PROCESSES AND THEIR STATISTICAL INFERENCE

ALAN F. KARR
Department of Mathematical Sciences
The Johns Hopkins University
Baltimore, Maryland

MARCEL DEKKER, INC. New York and Basel

Library of Congress Cataloging-in-Publication Data

Karr, Alan F., [date]
Point processes and their statistical inference.

(Probability, pure and applied ; 2)
Includes bibliographies and index.
1. Point processes. I. Title. II. Series.
QA274.42.K37 1986 519.2 85-31127
ISBN 0-8247-7513-9

MARCEL DEKKER, INC.
270 Madison Avenue, New York, New York 10016

Current printing (last digit):
10 9 8 7 6 5 4 3 2 1

PRINTED IN THE UNITED STATES OF AMERICA

To thank my wife *Marilyn* and my daughters *Betsy* and *Cathy* is impossible. They cheerfully tolerated (or was it enjoyed?) my being away from home for much of 1983 and treated with unflagging good humor my mental absence on numerous other occasions. Not being mathematicians, they will probably never read the book past this point, nevertheless their influence permeates every page. That without them the book would not exist is trite but true; in gratitude I dedicate it lovingly to them.

Preface

Inference for stochastic processes seems to have become a distinct subfield of probability and statistics. No longer viewed as falling between probability and statistics but part of neither, it is now recognized as spanning both. Questions of inference are current important concerns—not only in classical senses but also in state estimation—for the traditionally important stochastic processes (Markov chains, diffusions, and point processes).

For point processes, progress has been especially rapid because of simple qualitative structure, the wealth of tractable but broad special cases, and, above all, persistent, stimulating contact with applications. This contact provides the impetus for the solution of problems at the same time as, in consonance with most of the history of mathematics, it confirms that important physical problems lead to the most challenging and interesting mathematical questions. Yet as recently as the mid-1970s inference for point processes was more a collection of disparate techniques than a coherent field of scientific investigation. The introduction of martingale methods for statistical estimation, hypothesis testing, and state estimation revolutionized the then inchoate field, at once expanding the family of models amenable to meaningful analysis and unifying, albeit conceptually more than practically, the entire subject. Developments since then have been numerous and diverse, but without engendering fragmentation.

The goal of this book is to present a unified (but necessarily, of course, incomplete) description of inference for point processes before the field becomes so broad that such a goal is hopeless. By depicting the probabilistic and statistical heart of the subject, as well as several of its boundaries, I have endeavored to convey simultaneously the unity and diversity whose interplay and tension I find so alluring.

Although conceived primarily as a research monograph, the book is nevertheless suitable as a text for graduate-level courses and seminars. The style of presentation is meant to be accessible to probabilists and statisticians alike, but (reflecting my own development as a probabilist who became—and remains—very interested in statistics) more background is presumed in probability than in statistics. Therefore, most probabilistic ideas are introduced without explanation while comparably difficult statistical concepts are usually described briefly at least. To attempt to read the book without a semester graduate course in probability (at the level, for example, of Chung, 1975), including an introduction to discrete time martingales, and a corresponding course in mathematical statistics (e.g., from Bickel and Doksum, 1977) would be unrealistic and ineffective, but this background should suffice. No deep prior knowledge of point processes is presumed; even such basic processes as Poisson processes are treated in some detail.

The most difficult decision during writing was how to treat continuous time martingale theory. A full development was deemed beyond the scope of the book; instead I have summarized relevant major results in Appendix B, and throughout most of the text have imposed assumptions sufficiently strong to render arcane theoretical details unnecessary. The slight loss of generality is more than offset by the gain in clarity and cogency; one is able to concentrate on the really important questions. A reader familiar with discrete time martingales can understand the key ideas by analogy.

I have resisted, in most cases, the temptation to generalize beyond recognition. In Freeman Dyson's beautiful metaphor I am more part of Rutherford's Manchester than Plato's Athens, in spirit a diversifier and a seeker of unity in the small. But unfettered diversity is chaos and a book without unifying ideas merely a dictionary. There are important themes: maximum likelihood, empirical averages (in several guises) as estimators, martingale representations, and, above all, exploitation of special structure (whether that structure be broadly or narrowly prescribed) that do pervade the entire book, although sometimes in the minor mode and often in variations. This somewhat schizophrenic but, I think, inescapable, viewpoint surfaces not only in the large scale structure of the book, but also in the structure of individual chapters and sections. Chapters 1–5 concern broad subjects while Chapters 6–10 investigate in more detail correspondingly narrower topics. In Chapters 6–9 rather stringent structural assumptions are stipulated but once the context is established we impose few additional restrictions; hence our problems and methods are distinctly "nonparametric."

The book can be used in a variety of advanced graduate-level courses or research seminars. Chapters 1 and 2 could, perhaps with supplementation, be used for a one-quarter or one-semester introduction to point processes. Together with Chapters 4 and 5, they present the two main lines of theory and inference: distributional theory/empirical inference and intensity theory/martingale inference; with judicious omissions they could be covered in one semester, but a more leisurely pace is also feasible. Chapter 3 (which was surprisingly difficult to write) is meant less for detailed study than as guide and appetizer for what follows. To omit it seems a mistake; to devour it is unnecessary. Each of Chapters 6–10 presents diversity in the form of inference for an important special class of point processes; they are largely independent of one another and are probably best covered selectively as interests of instructor and students dictate. Any of them could form the basis for a single set of lectures. The accompanying figure shows the chapter interrelationships.

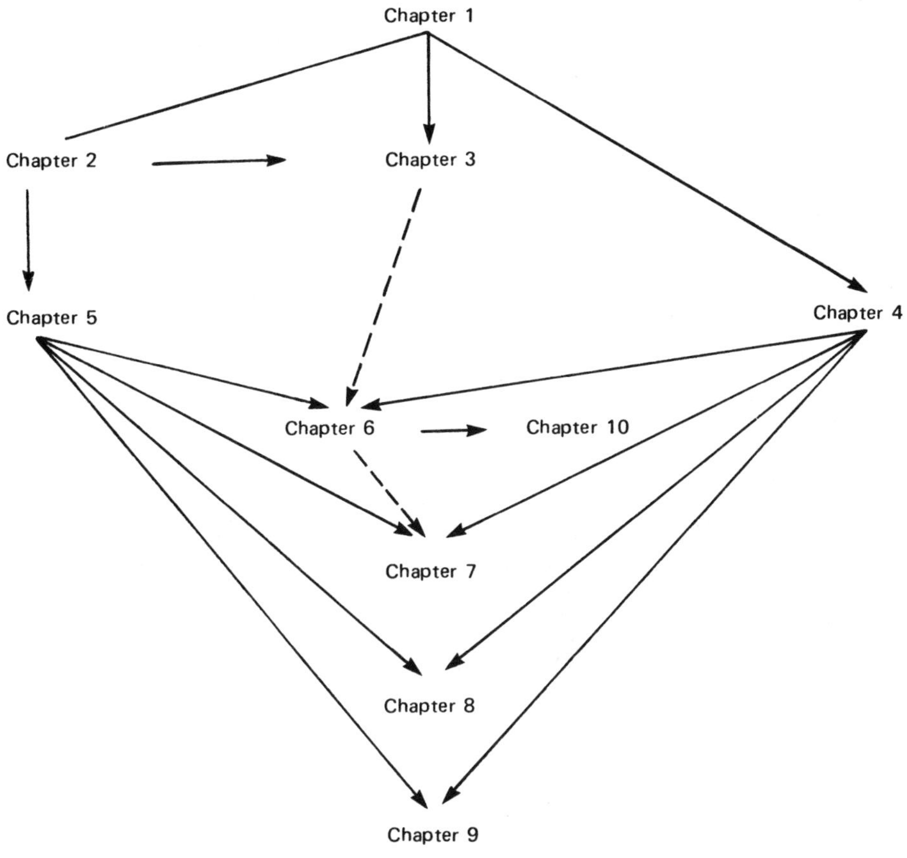

Every chapter contains a number of exercises, whose purpose is to extend and enrich the main body of material. No essential points have been relegated to exercises, but a reading of the book that ignores them is incomplete. They are important agents of diversity as well as the principal means whereby applications are presented. The level of difficulty varies: some are computations, some are improvements pertaining to special cases of results in the text, still others are theorems taken from the literature that are not central to our development but are of importance and interest notwithstanding.

An author should also, I think, say what a book is not. This is not a book about data analysis, nor does it pretend to present applications in depth or breadth even though the text and exercises taken together do provide a reasonable sample. Bayesian ideas are crucial to state estimation, but Bayesian inference per se is absent. There is much more emphasis on estimation than on hypothesis testing; the argument that every estimator implicitly defines a test statistic is somewhat disingenuous because our estimators are often estimating infinite-dimensional objects and their distributional properties are typically too poorly understood to permit analysis of power, let alone optimality. Among important classes of point processes clustered point processes have received, relatively, the shortest shrift; such choices are inevitable, and this is one class of processes for which alternative treatments do exist. Cox processes may seem overrepresented, but I cannot remain dispassionate on a subject which has been a personal special interest for so long.

Much of the book was written during an itinerant year in 1983. M. R. Leadbetter and S. Cambanis of the University of North Carolina, where I spent a sabbatical semester, were especially hospitable, as were, for shorter periods, P. Franken of Humboldt University and E. Arjas of the University of Oulu. Extremely helpful readings of the manuscript were generously provided by R. Gill (Amsterdam), S. Johansen (Copenhagen), O. Kallenberg (Goteborg), R. Leadbetter (Chapel Hill) and J. Smith (Washington, D.C.); I hope they can recognize how valuable their comments were. Errors that remain, of course, are mine. I also wish to thank R. Serfling of The Johns Hopkins University for his constant encouragement and friendship, and the Air Force Office of Scientific Research for their support of my research, leading to a number of results that appear here for the first time.

Alan F. Karr

Contents

1

Point Processes: Distribution Theory

The origins of our subject are assuredly ancient, dating from speculation concerning the distribution of stars in a sky once thought spherical and from records of floods, earthquakes, and other natural events exhibiting no apparent periodic structure. By contrast, its modern mathematical history is not lengthy. It stems from the work of Palm and later Khinchin on problems of teletraffic theory and work, notably by Gibbs, on statistical mechanics, and from increased understanding of the most fundamental of point processes, the Poisson process. Since that work was done, there has developed on the one hand a general, powerful, and elegant corpus of mathematical theory having both a life of its own and important consequences in other branches of probability, and on the other hand a plethora of applications to such diverse fields as biology, geography, meteorology, physics, operations research, and engineering. Following the best mathematical tradition, the two aspects have nurtured one another and, despite some recent signs of divergence, remain in healthy contact.

A point process is a model of indistinguishable points distributed randomly in some space. Points may represent times of events, locations of objects, or even, as elements of function spaces, paths followed by a stochastic system. Here we emphasize point processes on general (well-behaved topological) spaces. Thus events or objects may have secondary characteristics associated with them, but are viewed simply as points in a larger space; for example, a point of mass u located at $x \in \mathbf{R}^3$ becomes a single point (x,u) of $\mathbf{R}^3 \times \mathbf{R}_+$. Marked point processes, then, are mathematically nothing more than point processes on product spaces and can be treated using point process methods.

As random processes, point processes are subject to *statistical inference,* one principal theme of the book. For example, the probability law of a point process N may be unknown, and need to be determined from independent, identically distributed (i.i.d.) realizations or from just one realization (which might be impossible). We emphasize the case that the law of N is entirely unknown or of known structure but with an unknown infinite-dimensional "parameter." Two general approaches to inference will be developed. In Chapter 4 we treat empirical estimators of objects such as the Laplace functional that determine the law of N and from which additional quantities of interest, for example mean and covariance measures, can be calculated. The strength of these methods is generality in regard to the underlying space and the structure of the point process, but they require multiple realizations. Our second broad class of techniques, the martingale methods presented in Chapter 5, apply only to point processes on $\mathbf{R}_+$ and only if certain structural assumptions are imposed, but within this context, their power is extraordinary.

Alternatively, a point process may be stipulated to admit particular qualitative structure, for example that of a Poisson or Cox process, rendering the estimation problem somewhat simpler in form but not necessarily easier in substance since typically the unknown "parameter" is infinite-dimensional. In such instances the central problem is to devise methods that exploit the special structure. Other classes of point processes treated here include renewal processes (Chapter 8) and stationary point processes (Chapter 9).

Our second salient theme is *state estimation* for point processes that are only partially observed. The observations are represented by a σ-algebra $\mathcal{H}$ ("observability" means measurability with respect to $\mathcal{H}$) and the objective is to reconstruct, with minimum error, unobservable functionals of N [i.e., random variables $X \in \sigma(N)$ not measurable with respect to $\mathcal{H}$]; it is required that the law of N be known. Usually, observations correspond to complete knowledge of N over a subset A of the underlying space. Realization-by-realization reconstruction of unobserved aspects of N entails estimation of random variables, leading to the term "state estimation," as opposed to "parameter" or "law" estimation. Throughout we employ the nearly universal criterion of mean squared error, with respect to which the optimal state estimator of the random variable X—with minimum mean squared error—is, of course, the conditional expectation $\hat{X} = E[X \mid \mathcal{H}]$: provided that $E[X^2] < \infty$, $E[(\hat{X} - X)^2] \leq E[(Y - X)^2]$ for all $Y \in L^2(\mathcal{H})$. Were one to stop here, the problem of state estimation would lack independent content. The key issues involve computation of optimal state estimators as explicit functions (nonlinear in general) of the observed data and, especially, recursive representation of state estimators.

A recursive representation enables one to update a state estimator as additional data becomes available, rather than compute from scratch a revised estimator that incorporates both old and new observations. We illustrate in stylized form. Suppose that observations $\mathcal{H}_1$ have been obtained previously and that $\hat{X}_1 = E[X \mid \mathcal{H}_1]$ has been calculated (by whatever method). At a subsequent time further observations $\mathcal{H}_2$ are obtained. A recursive representation avoids computing the revised estimator $\hat{X}_{12} = E[X \mid \mathcal{H}_1 \vee \mathcal{H}_2]$ directly, by substituting an expression of the generic form $\hat{X}_{12} - \hat{X}_1 = F(\hat{X}_1, Y_2)$, where $Y_2 \in \mathcal{H}_2$ is a suitable random element and F is some function. Two things are accomplished here: only the increment $\hat{X}_{12} - \hat{X}_1$ in the estimator is computed ($\hat{X}_1$ is known already), and contributions of the old and new observations are separated. The ultimate in recursive representations are stochastic differential equations for state estimators for processes on $\mathbf{R}_+$, as discussed in Chapters 5, 6, and 7. There the linear order structure of the real line is essential because it determines a natural sequence in which observations arise.

Finally, the problems of statistical inference and state estimation may be concatenated. Suppose that there are given i.i.d. copies N_i of a point process N with unknown probability law $\mathcal{L}$ and that $N_1, \ldots, N_n$ have been observed previously. Suppose that N_{n+1} has just been observed over the set A (mathematically, the observations are the σ-algebra $\mathcal{F}_A^{N_{n+1}}$) and that it is desired to perform state estimation for unobserved parts of it. Without knowledge of $\mathcal{L}$ the "true" state estimator, $\hat{X}_{n+1} = E[X_{n+1} \mid \mathcal{F}_A^{N_{n+1}}]$ cannot be computed. If, however, one knew the functional form $\hat{X}_{n+1} = G(\mathcal{L}, N_{n+1}|_A)$, where $N_{n+1}|_A$ is the restriction of N_{n+1} to A, then a "pseudo-state estimator" $\hat{X}^*_{n+1}$ can be formed by invoking the *principle of separation*, which dictates that one utilize $N_1, \ldots, N_n$ (and in practice N_{n+1} as well) to form an estimator $\hat{\mathcal{L}}$ of $\mathcal{L}$, then take as the psuedo-state estimator, $\hat{X}^*_{n+1} = G(\hat{\mathcal{L}}, N_{n+1}|_A)$. The key to the problem is to impose assumptions that allow computation of G without knowledge of $\mathcal{L}$. Mixed Poisson processes admit the most satisfactory solution, as described in Chapter 7; other cases are examined in Chapters 6, 8, and 10.

More generally, one may envision state estimation for random variables that are not $\mathcal{F}^N$-measurable but are related to N nevertheless. The prime example is Cox processes, which have specified structural and distributional properties conditional on a directing random measure that is unobservable in all important applications. The random variable X may then be a functional of the directing measure. Questions and issues outlined above remain germane.

The remainder of this chapter introduces point processes and their distributional theory. Chapter 2 treats the martingale theory of point processes on $\mathbf{R}$. Chapter 3 formulates in some detail the kinds of problems of statistical inference and state estimation treated in this book and by way of illustration examines the most important point processes, ordinary Poisson processes on $\mathbf{R}_+$. Chapters 1 and 4, and 2 and 5, form parallel pairs. Chapter 4 addresses distribution-based inference using empirical estimators, with focus on large-sample properties, corresponding in this case to observation of independent realizations. Chapter 5 presents an analogously broad description of intensity-based inference for point processes on the line. Important special classes of point processes—Poisson processes, Cox processes, renewal processes, and stationary point processes—are the subjects of Chapters 6–9, which are unified by our exploiting special structure as effectively as possible in order to derive particularly precise results. A slightly tangential—but nonetheless interesting—topic is introduced in Chapter 10: inference for stochastic processes sampled according to a point process.

1.1. RANDOM MEASURES AND POINT PROCESSES

Let E be a locally compact Hausdorff space whose topology has a countable base (hereafter abbreviated LCCB) with Borel σ-algebra $\mathscr{E}$. As in Appendix A, let $\mathscr{B}$ denote the ring of bounded Borel sets, and recall that a measure μ on E is a Radon measure if $\mu(A) < \infty$ for every $A \in \mathscr{B}$. With $\mathbf{M}$ the set of Radon measures, elements of the subsets $\mathbf{M}_p = \{\mu \in \mathbf{M} : \mu(A) \in \mathbf{N}$ for all $A \in \mathscr{B}\}$, $\mathbf{M}_s = \{\mu \in \mathbf{M}_p : \mu(\{x\}) \leq 1$ for all $x \in E\}$, $\mathbf{M}_a = \{\mu \in \mathbf{M} : \mu$ is purely atomic$\}$, and $\mathbf{M}_d = \{\mu \in \mathbf{M} : \mu$ is diffuse$\}$ are point measures, simple point measures, purely atomic measures, and diffuse measures, respectively. On $\mathbf{M}$ we define the σ-algebra $\mathcal{M}$ given by (A.2): $\mathcal{M}$ is generated by the coordinate mappings $\mu \to \mu(f) = \int f\, d\mu$, where f ranges over the set C_K of continuous functions on E whose support is compact. By $\mathcal{M}_p$, $\mathcal{M}_s$, $\mathcal{M}_a$, $\mathcal{M}_d$ we denote the trace σ-algebras on $\mathbf{M}_p$, $\mathbf{M}_s$, $\mathbf{M}_a$, $\mathbf{M}_d$. Convergence in $\mathbf{M}$ is vague convergence: $\mu_n \to \mu$ if and only if $\mu_n(f) \to \mu(f)$ for every $f \in C_K$. The resultant Borel σ-algebra is $\mathcal{M}$. For additional details and results, see Appendix A.

We come now to the principal definitions.

Definition 1.1. Let $(\Omega, \mathscr{F}, P)$ be a probability space.

a) A *random measure* on E is a measurable mapping M of $(\Omega, \mathscr{F})$ into $(\mathbf{M}, \mathcal{M})$;

b) A *point process* on E is a measurable mapping N of $(\Omega, \mathscr{F})$ into $(\mathbf{M}_p, \mathcal{M}_p)$.

For a random measure M and $A \in \mathscr{E}$, the random variable $M(A)$ is termed the measure or mass of A and the integral $M(f) = \int_E f(x) M(dx)$, a well-defined random variable for f either nonnegative and $\mathscr{E}$-measurable or in C_K, is called the integral of f with respect to M. A random measure is interpreted as a random distribution of mass or charge (but only of one sign) in E.

A point process N is a random distribution of indistinguishable points in E; $N(A)$ is the number of points in A. There exists a representation

$$N = \sum_{i=1}^{K} \varepsilon_{X_i}, \tag{1.1}$$

where ε_K is the point mass at x [$\varepsilon_X(A)$ is 1 or zero according as $x \in A$ or not], where K is a random variable with values in $\bar{\mathbf{N}} = \{0,1, \ldots, \infty\}$, and where the X_i are measurable mappings of $(\Omega,\mathcal{F})$ into $(E,\mathcal{E})$, termed the points or atoms of N. They need not be distinct and the order of their indices is not unique. We employ (1.1) throughout; in particular, we then have $N(f) = \Sigma_{i=1}^{K} f(X_i)$.

When $E = \mathbf{R}_+$ the X_i are customarily replaced by T_i chosen so that $0 \le T_1 \le T_2 \le \ldots$ and we adopt the usual terminology: the T_i are times of events or *arrival times*, their differences $U_i = T_i - T_{i-1}$ are *interarrival times*, and $N(t) = N([0,t])$ is the number of arrivals in the time interval $[0,t]$. The fundamental relation between the *counting process* $N(t)$ and arrival time sequence T_n is that for each n and t

$$\{N(t) \ge n\} = \{T_n \le t\}; \tag{1.2}$$

thus each process contains information sufficient to reconstruct the other.

A point process N is *simple* if $P\{N \in \mathbf{M}_s\} = 1$. In this case the points X_i in (1.1) are distinct [almost surely (a.s.)] and on $\mathbf{R}_+$ we may take $0 \le T_1 < T_2 < \cdots$, with the interpretation that there are no simultaneous arrivals. For simple point processes on $\mathbf{R}$, especially stationary point processes, the most convenient representation is

$$N = \sum_{i=-K_1}^{K_2} \varepsilon_{T_i},$$

where K_1, K_2 are random variables taking values in $\bar{\mathbf{N}}$ and $\cdots < T_{-1} < T_0 \le 0 < T_1 < T_2 < \cdots$.

Similarly, a random measure M is *purely atomic* if $P\{M \in \mathbf{M}_a\} = 1$, in which case (1.1) generalizes [see (A.2)] to become

$$M = \sum_{i=1}^{K} U_i \varepsilon_{X_i} \tag{1.3}$$

with K and the X_i as in (1.1) and the U_i nonnegative random variables. For many purposes one can identify M with the point process

$$N = \sum_{i=1}^{K} \varepsilon_{(X_i, U_i)} \tag{1.4}$$

on $E \times \mathbf{R}_+$; we use this device several times to extend point process results and techniques to atomic random measures (see, e.g., Section 6.5).

Finally, a random measure M is *diffuse* if $P\{M \in \mathbf{M}_d\} = 1$.

Associated with a random measure M are σ-algebras representing various forms of observation of M. The most important correspond to complete knowledge of M over a set A, namely $\mathcal{F}^M(A) = \sigma\{M(B) : B \in \mathscr{E}, B \subset A\}$; for simplicity, we put $\mathcal{F}^M = \mathcal{F}^M(E)$, which represents complete observation of M. In Chapters 3 and 6 we also consider integral data corresponding to σ-algebras $\mathcal{F}^M(g_1, \ldots, g_m)$, which depict knowledge of M limited to the integrals $M(g_1), \ldots, M(g_m)$. When $E = \mathbf{R}_+$ we write $\mathcal{F}_t^M$ for $\mathcal{F}^M([0,t])$. In this case particularly, but also to some extent in general, we refer to the $\mathcal{F}^M(\cdot)$ individually and *en masse* as the *internal history* of M. Chapters 2 and 5 examine histories $\mathcal{G}(\cdot)$ such that $\mathcal{F}^M(A) \subset \mathcal{G}(A)$ for each A, in which case M is said to be *adapted* to the $\mathcal{G}(A)$.

We now introduce several important classes of point processes. Of these, Poisson processes are truly fundamental, while Cox processes, because of their role in connection with state estimation and because of their being on the one hand rather general, yet on the other hand closely linked to Poisson processes, play nearly an equal role.

Definition 1.2. Let μ be an element of $\mathbf{M}$. A point process N on E is a *Poisson process* with mean measure μ if

a) Whenever $A_1, \ldots, A_k \in \mathscr{E}$ are disjoint the random variables $N(A_1), \ldots, N(A_k)$ are independent;

b) For each $A \in \mathscr{E}$ and $k \geq 0$,

$$P\{N(A) = k\} = \frac{e^{-\mu(A)}\mu(A)^k}{k!} \tag{1.5}$$

[By convention $N(A) = \infty$ a.s. if $\mu(A) = \infty$.]

Thus N has *independent increments* and for each A the random number of points in A has a Poisson distribution with mean $\mu(A)$. (The term "mean measure" is introduced formally in Section 1.2.) Each of these properties nearly implies the other (see Section 1.5 and the chapter notes for details). The Poisson processes earliest to be understood are the ordinary (or homogeneous) Poisson processes on $\mathbf{R}_+$, for which $\mu(A) = \lambda|A|$, where $|A|$ is the Lebesgue measure of A and $\lambda > 0$ is known as the *rate* or *intensity*, terminology justified in Sections 1.8 and 3.5. Other Poisson processes on $\mathbf{R}_+$ are sometimes called nonhomogeneous Poisson processes. Inference for Poisson processes is the topic of Chapter 6.

A Cox process is a Poisson process with the mean measure made random.

Definition 1.3. Let M be a random measure on E and let N be a point process on E, both defined over the same probability space. Then N is a *Cox process directed by* M if conditional on M, N is a Poisson process with mean measure M:

a) Whenever $A_1, \ldots, A_k$ are disjoint the random variables $N(A_1), \ldots, N(A_k)$ are conditionally independent given $\mathcal{F}^M$;

b) For each A and each $k \geq 0$,

$$P\{N(A) = k|\mathcal{F}^M\} = \frac{e^{-M(A)}M(A)^k}{k!} \tag{1.6}$$

We refer to (N,M) as a *Cox pair*; M is the *directing measure* of N. (The terms "doubly stochastic Poisson process" and "Poisson process in a random environment" are synonymous with "Cox process.")

The most elementary Cox processes are the Poisson processes ($M \equiv \mu$ is deterministic). Next are the *mixed Poisson processes*, with directing measures $M = Y\nu$, where Y is a nonnegative random variable and ν is a fixed element of $\mathbf{M}$. A mixed ordinary Poisson process on $\mathbf{R}_+$ has rate $\lambda(\omega)$ that is random but not a function of time. Other specific classes of Cox processes are discussed in later chapters. It is shown below that the mapping of probability laws of random measures into laws of Cox processes directed by them is a bijection; this property is useful in establishing theoretical results as well as in applications (see, e.g., Exercise 1.7). The principal characterization of Cox processes as limits of thinned point processes appears in Section 1.5, along with the associated invariance property. Cox processes also present the challenging state estimation problem of reconstructing the directing measure from observation of the Cox process, which we solve in generality in Chapter 7.

Although empirical processes are not examined in complete detail in this book, several results do appear in Chapter 4, and in any case they are in some ways the most basic point processes, especially since Poisson processes can be constructed as mixed empirical processes (Exercise 1.1); this construction is central to some of our inference procedures.

Definition 1.4. Given i.i.d. random elements $X_1, \ldots, X_n$ of E, the point process

$$N = \sum_{i=1}^{n} \varepsilon_{X_i} \tag{1.7}$$

is an (n-sample) *empirical process* on E.

Thus N represents the positions of n points in E chosen independently and with the same distribution. In a mixed empirical process the number of points also becomes random.

Definition 1.5. Let $X_1, X_2, \ldots$ be i.i.d. random elements of E and let K be a nonnegative integer-valued random variable independent of the X_i. The point process

$$N = \sum_{i=1}^{K} \varepsilon_{X_i} \tag{1.8}$$

is a *mixed empirical process*.

On the real line, point processes can be defined via the interarrival times [the "interval" approach, as opposed to the "counts" approach based on the counting process $(N(t))$]; the simplest structural assumption is that the interarrival times be i.i.d.

Definition 1.6. A point process $N = \Sigma_{n=1}^{\infty} \varepsilon_{T_n}$ is a *renewal process* with interarrival distribution F (a probability on $\mathbf{R}_+$) if $T_0 = 0$ and if the interarrival times $U_i = T_i - T_{i-1}$ are independent and each has distribution F.

Renewal processes are significant because many stochastic processes contain embedded renewal processes, whose points typically are times of events at which the underlying process (called a regenerative process) renews itself probabilistically. Important properties of renewal processes are of two types: those associated with the renewal equation and asymptotic properties following from the renewal theorem. Some of each type are discussed in Chapter 8.

The remaining class of point processes accorded chapter-length treatment (in Chapter 9) is stationary point processes. Some machinery is needed to formulate the most usable definition, which is postponed to Section 1.8, but for completeness here is a provisional version.

Definition 1.7. A point process $N = \Sigma\, \varepsilon_{X_i}$ on $E = \mathbf{R}^d$ is *stationary* if for each x the translated point process $N_{(x)} = \Sigma\, \varepsilon_{X_i - x}$ has the same law as N.

Additional classes—marked point processes, cluster processes, infinitely divisible point processes, and symmetrically distributed point processess—are introduced elsewhere in the chapter.

1.2. DISTRIBUTIONAL DESCRIPTORS AND UNIQUENESS

The distribution of a point process (more generally, of a random measure) can be specified in alternative forms, some of which—particularly the Laplace functional—are relatively more amenable to explicit calculation, computational manipulation, and derivation of characterization and convergence properties. Incomplete but useful descriptions in the form of first-, second-, and higher-order moment measures are also available. Definitions and results in this section are formulated for point processes, but we indicate the extent to which they generalize to random measures. We begin with objects that uniquely determine (possibly with mild additional assumptions) the distribution of a point process.

Definition 1.8. Given a point process N defined over $(\Omega,\mathcal{F},P)$, the *distribution* of N is the probability measure $\mathcal{L}_N = PN^{-1}$on $(\mathbf{M}_p,\mathcal{M}_p)$.

Definition 1.9. Point processes N_1, N_2 (not necessarily defined over the same probability space) are *identically distributed* if $\mathcal{L}_{N_1} = \mathcal{L}_{N_2}$.

Equality in distribution is written as $N_1 \stackrel{d}{=} N_2$.

Definition 1.10. The *Laplace functional* of a point process N is the mapping defined by

$$L_N(f) = E[e^{-N(f)}] = E[\exp(-\Sigma f(X_i))], \tag{1.9}$$

where f is nonnegative and $\mathcal{E}$-measurable.

Definition 1.11. The *zero-probability functional* of N is the mapping $z_N: \mathcal{E} \to [0,1]$ defined by

$$z_N(A) = P\{N(A) = 0\} \tag{1.10}$$

All four definitions extend verbatim to random measures, except that in Definition 1.8 $(\mathbf{M}_p,\mathcal{M}_p)$ is replaced by $(\mathbf{M},\mathcal{M})$; however, the zero-probability functional is useful mainly for (simple) point processes. The Laplace functional is evidently a generalized Laplace transform and is shown below to possess the conventional properties of a transform: uniqueness, computational tractability in the sense of its being explicitly computable in special cases and for transformations of point processes (see Section 1.5), computability of moments from it, and determination of convergence in distribution (Section 1.3). It is a fundamental tool in this book.

The main uniqueness theorem can now be given.

Theorem 1.12. For point processes N_1, N_2 the following assertions are equivalent:

a) $N_1 \stackrel{d}{=} N_2$;
b) $(N_1(A_1), \dots, N_1(A_k)) \stackrel{d}{=} (N_2(A_1), \dots, N_2(A_k))$ for all k and all $A_1, \dots, A_k \in \mathcal{B}$;
c) $L_{N_1}(f) = L_{N_2}(f)$ for all $f \in p\mathcal{E}$.

If in addition N_1 and N_2 are simple, then each of a) – c) is equivalent to

d) $z_{N_1}(A) = z_{N_2}(A)$ for all $A \in \mathscr{E}$.

Proof: (Sketch). Equivalence of a) and b) is standard. If b) holds and $f = \sum_{i=1}^{k} a_i 1_{A_i}$ is a simple, nonnegative function, then

$$\begin{aligned} L_{N_1}(f) &= E\left[\exp\left\{-\sum_{i=1}^{k} a_i N_1(A_i)\right\}\right] \\ &= E\left[\exp\left\{-\sum_{i=1}^{k} a_i N_2(A_i)\right\}\right] = L_{N_2}(f); \end{aligned} \quad (1.11)$$

extension to general f is by approximation. Conversely, if c) holds, then in particular (1.11) does, from which b) follows by the uniqueness theorem for Laplace transforms. That a) entails d) is evident; the reverse implication comes from the property that the family of events of the form $\{\mu : \mu(A) = 0\}$ is a π-system on $\mathbf{M}_s$ that generates $\mathcal{M}_s$, so that in the simple case d) implies a) by the monotone class theorem. □

Equivalence of a), b), and c) remains valid for random measures. Although d) has no direct counterpart, certain results derived by means of Cox processes exist for diffuse random measures; for details of these and several formally weaker versions of Theorem 1.12, see the exercises and chapter notes.

Moments of a point process are described in the following manner.

Definition 1.13. The *mean measure* or *first moment measure* of a point process N is the measure μ_N on E given by $\mu_N(A) = E[N(A)]$; N is *integrable* if $\mu_N \in \mathbf{M}$. More generally, for each k let N^k be the point process $N^k(dx_1, \ldots, dx_k) = N(dx_1) \cdots N(dx_k)$; then the *moment measure of order k* is the measure $\mu_N^k = E[N^k]$.

Definition 1.14. The *covariance measure* of N is the signed measure

$$\rho_N(A \times B) = \mathrm{Cov}(N(A), N(B)) = \mu_N^2(A \times B) - \mu_N(A)\mu_N(B) \quad (1.12)$$

on $E \times E$. [To avoid complications we use the covariance measure only when the second moment measure belongs to $\mathbf{M}(E^2)$.]

Moment and covariance measures for a random measure are defined exactly as in Definitions 1.13 and 1.14.

For many purposes (e.g., definition of Palm distributions), ordinary moment measures of point processes are unsatisfactory because the products N^k contain points on various diagonals. Without contributing information not already present, such points can engender "spurious" singularity of moment measures. Their removal is effected by replacing N^k by the point process

$$N^{(k)}(dx_1, \dots, dx_k) = N(dx_1)(N - \varepsilon_{x_1})(dx_2) \cdots \left(N - \sum_{i=1}^{k-1} \varepsilon_{x_i}\right)(dx_k),$$

which—provided that N is simple—consists only of k-tuples of points whose coordinates are distinct. In terms of the representation (1.1) we have

$$N^k = \sum_{i_1} \cdots \sum_{i_\mu} \varepsilon_{(X_{i_1}, \dots, X_{i_k})},$$

but

$$N^{(k)} = \sum_{\substack{i_1 \\ i_1 \neq \cdots \neq i_k}} \cdots \sum_{i_k} \varepsilon_{(X_{i_1}, \dots, X_{i_k})}$$

We refer to the measures $\mu_N^{(k)} = E[N^{(k)}]$ as *factorial moment measures* of N.

Moment measures can be computed from the Laplace functional:

$$\mu_N(f) = -\frac{d}{d\alpha} L_N(\alpha f)|_{\alpha=0},$$

while from

$$E[N(f)^2] = \frac{d^2}{d\alpha^2} L_N(\alpha f)|_{\alpha=0}$$

we obtain

$$E[N(f)N(g)] = \frac{1}{2}\{E[N(f+g)^2] - E[N(f)^2] - E[N(g)^2]\},$$

from which the covariance measure can be calculated. Since in addition $z_N(A) = \lim_{t\to\infty} L_N(t1_A)$, the Laplace functional truly is the fundamentally useful descriptor of $\mathcal{L}_N$.

Here are the key examples.

Example 1.15. (Poisson processes). Let N be a Poisson process with mean measure μ. Then

$$L_N(f) = \exp[-\int_E (1 - e^{-f})\, d\mu]; \tag{1.13}$$

consequently (or directly from the definition)

$$z_N(A) = e^{-\mu(A)}, \tag{1.14}$$

$$\mu_N(A) = \mu(A), \tag{1.15}$$

and

$$\rho_N(A \times B) = \mu(A \cap B) \tag{1.16}$$

[That $N(A)$, $N(B)$ can only be positively correlated follows from the independent increments property of N.] Furthermore, N is simple if and only if μ is diffuse. The difference between ordinary and factorial moment measures can be inferred from the formulas

$$\mu_N^2(f) = E[\sum f(X_i,X_j)] = \int f(x,x)\mu(dx) + \iint f(x,y)\mu(dx)\mu(dy),$$

and

$$\mu_N^{(2)}(f) = E\left[\sum_{i \neq j} f(X_i,X_j)\right] = \iint f(x,y)\mu(dx)\mu(dy);$$

the latter measure is absolutely continuous with respect to $\mu \times \mu$ but the former is not. □

Example 1.16. (Cox processes). Let N be a Cox process directed by the random measure M. Then

$$L_N(f) = L_M(1 - e^{-f}), \tag{1.17}$$

$$z_N(A) = L_M(1_A), \tag{1.18}$$

$$\mu_N = \mu_M, \tag{1.19}$$

$$\rho_N(A \times B) = \rho_M(A \times B) + \mu_M(A \cap B) \tag{1.20}$$

The Cox process N is simple if and only if the directing measure M is diffuse. □

To derive (1.17), for example, one conditions on M and applies (1.13):

$$L_N(f) = E[E[e^{-N(f)}|M]] = E[\exp(-\int (1 - e^{-f})\, dM)] = L_M(1 - e^{-f})$$

Application of (1.14) – (1.16) to deduce (1.18) – (1.20) is analogous.

A useful consequence reduces some random measure results (particularly for diffuse random measures) to properties of point processes (see Exercise 1.7 for an application).

Proposition 1.17. Let N be a Cox process directed by M. Then $\mathcal{L}_N$ and $\mathcal{L}_M$ determine each other uniquely. □

We conclude the section by discussing independence for point processes. Point processes N_1 and N_2 defined over the same probability space are *independent* if the σ-algebras $\mathcal{F}^{N_1}$, $\mathcal{F}^{N_2}$ are independent in the usual sense. Theorem 1.12 furnishes sufficient conditions.

Proposition 1.18. For point processes N_1, N_2 the following are equivalent:

a) N_1, N_2 are independent;

b) $E[\exp(-N_1(f)-N_2(g))] = L_{N_1}(f)L_{N_2}(g)$ for all f, g.

If N_1, N_2 are simple these are equivalent to

c) $P\{N_1(A)=0, N_2(B)=0\} = z_{N_1}(A)z_{N_2}(B)$ for all A, B. □

Neither of the naive conditions $E[\exp\{-N_1(f)-N_2(f)\}] = L_{N_1}(f)L_{N_2}(f)$ for all f nor $P\{N_1(A)=0, N_2(A)=0\} = z_{N_1}(A)z_{N_2}(A)$ for all A is sufficient for independence; they fail even for Poisson processes on a singleton set.

The preceding discussion extends to more than two point processes as well as to general random measures.

1.3. CONVERGENCE IN DISTRIBUTION

Several important classes of point processes, especially Cox processes and infinitely divisible point processes, admit characterizations as the class of all limits in distribution of sequences of point processes satisfying some sort of rarefaction condition; see Sections 1.4 and 1.5 for details. Here we formulate convergence in distribution for point processes and derive criteria for it. Recall that convergence in $\mathbf{M}_p$ is vague convergence (see Appendix A): $\mu_n \to \mu$ if and only if $\mu_n(f) \to \mu(f)$ for all $f \in C_K$. The vague topology is metrizable in such a way that $\mathbf{M}_p$ becomes a complete, separable metric space (Theorem A.9), and one can then speak of weak convergence of probability measures on $(\mathbf{M}_p, \mathcal{M}_p)$, since $\mathcal{M}_p$ is the Borel σ-algebra engendered by the vague topology. This leads to the following definition.

Definition 1.19. Let $N, N_1, N_2, \ldots$ be point processes on E. We say that (N_n) *converges in distribution* to N, written $N_n \xrightarrow{d} N$, provided that $\mathcal{L}_{N_n} \to \mathcal{L}_N$ weakly as probability measures on $(\mathbf{M}_p, \mathcal{M}_p)$, that is, that

$$E[H(N_n)] \to E[H(N)] \tag{1.21}$$

for every bounded, continuous function $H: \mathbf{M}_p \to \mathbf{R}$.

Of course, these point processes need not be defined on the same probability space, but for notational simplicity we treat them as if they were.

In proving convergence in distribution the crucial step is usually to establish relative compactness, which by Prohorov's theorem (see Billingsley, 1968) is equivalent to *tightness*; a sequence (P_n) of probabilities on a (complete, separable) metric space is tight if for every $\varepsilon > 0$ there is a compact set K such that $P_n(K) > 1 - \varepsilon$ for every n. According to Theorem A.10, a subset Γ of $\mathbf{M}_p$ is relatively compact if and only if

$$\sup\{|\mu(f)| : \mu \in \Gamma\} < \infty \tag{1.22}$$

for every $f \in C_K$ or, equivalently, if

$$\sup\{\mu(A) : \mu \in \Gamma\} < \infty \tag{1.23}$$

for each $A \in \mathcal{B}$. Using these characterizations we are able to derive the fundamental and highly useful property that a sequence of point processes is tight if and only if all of its one-dimensional distributions are tight.

Lemma 1.20. For a sequence (N_n) of point processes the following are equivalent:

a) (N_n) is tight [i.e., $(\mathcal{L}_{N_n})$ is tight];
b) $(N_n(f))$ is tight for every $f \in C_K$;
c) $(N_n(A))$ is tight for every $A \in \mathcal{B}$.

Proof: (Sketch). Equivalence of b) and c) is clear, and they are deduced from a) via (1.22) and (1.23). To show that b) entails a), let (f_k) be a sequence of nonnegative functions in C_K with $f_k \uparrow 1$; then given $\varepsilon > 0$ for each k there is c_k such that $P\{N_n(f_k) > c_k\} < \varepsilon/2^k$ for all n. The set $\Gamma = \cap_k\{\mu : \mu(f_k) \leq c_k\}$ is relatively compact in $\mathbf{M}_p$ and (with $\bar{\Gamma}$ the closure of Γ) $P\{N_n \notin \bar{\Gamma}\} \leq \Sigma_k P\{N_n(f_k) > c_k\} \leq \varepsilon$ for all n, and consequently a) holds. □

There ensues the major theorem on convergence in distribution for point processes.

Theorem 1.21. Let (N_n), N be point processes on E. Then the following assertions are equivalent:

a) $N_n \xrightarrow{d} N$;
b) $(N_n(A_1), \ldots, N_n(A_k)) \xrightarrow{d} (N(A_1), \ldots, N(A_k))$ for all k and all $A_1, \ldots, A_k \in \mathcal{B}$ such that $P\{N(\partial A_i) = 0\} = 1$ for each i;
c) $N_n(f) \xrightarrow{d} L_N(f)$ for all $f \in C_K$;
d) $L_{N_n}(f) \to L_N(f)$ for all nonnegative functions $f \in C_K$.

Proof: (Sketch). Equivalence of c) and d) is essentially immediate, while the implications a) $\Rightarrow$ b) and a) $\Rightarrow$ c) are consequences of the continuous mapping theorem (Billingsley, 1968). If c) is fulfilled, then (N_n) is tight by Lemma 1.20, so that every subsequence $(N_{n'})$ admits a further subsequence $(N_{n''})$ converging in distribution to some point process $\tilde{N}$. By the implication a) $\Rightarrow$ c) and Theorem 1.12, we have that $\tilde{N} \stackrel{d}{=} N$; hence $N_{n''} \stackrel{d}{\to} N$. Since convergence in distribution is metrizable, $N_n \stackrel{d}{\to} N$. That b) entails a) is shown in a similar manner. □

For simple point processes the zero-probability functional yields another criterion for convergence in distribution. However, convergence of zero-probability functionals does not by itself imply tightness; additional assumptions are necessary.

Proposition 1.22. Let (N_n), N be point processes with N simple and assume that (N_n) is tight. Then $N_n \stackrel{d}{\to} N$ if and only if $z_{N_n}(A) \to z_N(A)$ for all $A \in \mathcal{B}$ with $P\{N(\partial A) = 0\} = 1$. □

The proof is a straightforward modification of that of Theorem 1.21. No improvement results from assuming that the N_n are also simple. Tightness of (N_n) can be established using Lemma 1.20 or other criteria stated next.

Proposition 1.23. A sequence (N_n) of point processes is tight if either of the following conditions is satisfied:

a) For each $A \in \mathcal{B}$ and each m there is a finite partition $A = \Sigma_{j=1}^{k} A_{mj}$ such that

$$\lim_{m\to\infty} \overline{\lim_{n\to\infty}} \sum_j P\{N_n(A_{mj}) \geq 2\} = 0;$$

b) For each $A \in \mathcal{B}$ with $P\{N(\partial A) = 0\} = 1$,

$$\overline{\lim_{n\to\infty}} E[N_n(A)] \leq E[N(A)] < \infty \quad \square$$

The definition of convergence in distribution and Theorem 1.21 extend to random measures with no substantive modification. Using Proposition 1.17 one can reduce random measures versions of Theorem 1.21 to point process versions, and can obtain specialized results applicable when the limit is diffuse. For example (compare Exercise 1.7), if (M_n) is tight and M is diffuse, then $M_n \stackrel{d}{\to} M$ if and only if there is $t \in (0,\infty)$ such that $E[e^{-tM_n(A)}] \to E[e^{-tM(A)}]$ for each $A \in \mathcal{B}$. Sufficiency of this condition (necessity is apparent) is demonstrated by constructing Cox processes N_n, N directed by tM_n, tM, respectively. Then together with (1.18), the condition implies that $z_{N_n}(A) \to z_N(A)$ for all $A \in \mathcal{B}$, and since tightness of (M_n) is equivalent to that of (N_n) [by (1.22) and its random measure analog], Proposition 1.22 yields $N_n \stackrel{d}{\to} N$. Finally, (1.17) and the random measure version of Theorem 1.21 provide the desired conclusion.

Similar reasoning shows that the mapping associating the law of the directing measure and that of the Cox process is not only a bijection (Proposition 1.17) but also continuous in both directions.

For point processes on **R** or $\mathbf{R}_+$ specialized results hold. Specifically, let $N_n = \Sigma\, \varepsilon_{T_{ni}}$ and $N = \Sigma \varepsilon_{T_i}$ be point processes on $\mathbf{R}_+$ with $0 \le T_{n1} \le T_{n2} \le \cdots$ for each n (and similarly for the T_i), and let (U_{ni}), (U_i) be the associated interarrival time sequences. Then the following statements are equivalent:

a) $N_n \xrightarrow{d} N$;

b) $(T_{n1}, \dots, T_{nk}) \xrightarrow{d} (T_1, \dots, T_k)$ for each k;

c) $(U_{n1}, \dots, U_{nk}) \xrightarrow{d} (U_1, \dots, U_k)$ for each k;

An analogous statement holds concerning point processes on **R**.

We conclude with two examples; more are given in the exercises.

Example 1.24. (Poisson processes). Using equivalence of a) and d) in Theorem 1.21, one sees using (1.13) that for Poisson processes $N, N_1, N_2, \dots$ with respective mean measures $\mu, \mu_1, \mu_2, \dots, N_n \xrightarrow{d} N$ if and only if $\mu_n \to \mu$ (vaguely). □

Example 1.25. (Renewal processes). The law of a renewal process depends continuously on the interarrival distribution in the sense of weak convergence on $\mathbf{R}_+$ of the latter. Indeed, more can be shown: if (N_n) is a sequence of renewal processes and $N_n \xrightarrow{d} N$, then N is a renewal process. □

1.4 MARKED POINT PROCESSES AND CLUSTER PROCESSES

Marked point processes are a fundamental tool in the study of thinned and compound point processes, Palm distributions, stationary point processes, and cluster processes, as well as in applications (e.g., to queueing and reliability theory). We introduce them in this section, along with several important constructions.

Definition 1.26. Let $N = \Sigma\, \varepsilon_{X_i}$ be a point process on E and let E' be a second LCCB space. A *marked point process* with underlying process N is any point process

$$\bar{N} = \sum \varepsilon_{(X_i, Z_i)} \tag{1.24}$$

on $E \times E'$. The random element Z_i of E' is the *mark* associated to X_i. □

In one sense a marked point process is merely a point process on the product space $E \times E'$, so that there is little theory of marked point processes per se, but interpretation and usage show that the concept does have a life of its own. One can think of the X_i as primary qualities of the objects under study and the marks Z_i as secondary attributes. Often the distinction is that the stochastic structure of the marks is simpler than that of the X_i or dependent on the X_i, as for processes with position-dependent marks. As an example, the X_i may be locations of unit charge ions, each marked by $Z_i = 1$ or $Z_i = -1$ according as its charge is positive or negative, or the arrival time of a customer at a queueing system may be marked by the service time or the total waiting time. Classical compound Poisson processes on $\mathbf{R}$ (see Section 1.5) constitute another example.

More generally, a purely atomic random measure $M = \Sigma\, U_i \varepsilon_{X_i}$ on E may be viewed as the marked point process $N = \Sigma\ \varepsilon_{(X_i,\, U_i)}$ on $E \times \mathbf{R}_+$; hence point process methods and results are applicable more widely than is initially apparent. See, for example, the discussion of infinitely divisible random measures with independent increments in this section, as well as the exercises.

In the most elementary form of marking the marks are simply independent of the point process N.

Example 1.27. (Independent marking). Let $N = \Sigma\, \varepsilon_{X_i}$ be a point process on E and let Z_i be i.i.d. random elements of E' such that N and (Z_i) are independent. Then the marked point process $\bar{N}$ of (1.24) is said to be obtained from N by *independent marking*. For f on $E \times E'$,

$$L_{\bar{N}}(f) = E[E[e^{-\Sigma f(X_i,\, Z_i)}|N]] = L_N(H),$$

where $H(x) = -\log\,(\int_{E'} e^{-f(x,z)} \rho(dz))$, with ρ the distribution of the Z_i. For $E' = \mathbf{R}_+$, the associated purely atomic random measure $M = \Sigma\, U_i \varepsilon_{X_i}$ is studied in Section 1.5 under the slight misnomer "compound point process." □

More general but equally tractable is the case that the marks are conditionally independent given N.

Example 1.28. (Position-dependent marking). Given a transition probability K from E to E', we may construct the mark sequence so that the Z_i are conditionally independent given N with $P\{Z_i \in B|N\} = K(X_i, B)$, so that in addition the distribution of Z_i depends only on X_i. We then have $L_{\bar{N}}(f) = L_N(H)$, for $H(x) = -\log\,(\int_{E'} e^{-f(x,z)} K(x, dz))$. □

For an example from hydrology, see Exercise 1.12.

In study of Palm distributions for stationary point processes (Section 1.8) one deals with marked point processes in which each point X_i of the underlying process N is marked by the translate of N by X_i:

$$\bar{N} = \sum_i \varepsilon_{(X_i, \Sigma_j \varepsilon_{X_j - X_i})}$$

on $\mathbf{R}^d \times \mathbf{M}_p$. In particular the mean measure of $\bar{N}$ is the product of Lebesgue measure and the very important Palm measure of N.

Finally, we view clustered point processes as marked point processes.

Definition 1.29. Let $N = \Sigma\, \varepsilon_{X_i}$ be a point process on E and let $E' = \mathbf{M}_p(\widetilde{E})$, where $\widetilde{E}$ is also LCCB. Let $\bar{N} = \Sigma\, \varepsilon_{(X_i, \widetilde{N}_i)}$ be a marked point process with underlying process N and mark space E'. Then provided that it is locally finite, the point process

$$\widetilde{N} = \sum_i \widetilde{N}_i \tag{1.25}$$

on $\widetilde{E}$ is a *clustered point process*.

When $\widetilde{E} = E$ the interpretation below is particularly precise, but it makes sense in other cases as well. Think of the X_i as primary points or *cluster centers*, to each of which is associated a *cluster* of secondary points or *cluster members* in $\widetilde{E}$ represented by the point process $\widetilde{N}_i$. The clustered point process $\widetilde{N}$ is composed of all the cluster members. Although the term "cluster" is picturesque and suggestive of intuitive notions of clustering, it should not be taken too literally. Even when $\widetilde{E} = E$ the points of $\widetilde{N}_i$ need not be near each other or X_i; thus, here as elsewhere, one must take care not to draw unwarranted conclusions based on intuitive connotations of terms.

Among clustered point processes the most important are the Poisson cluster processes.

Definition 1.30. Let $\widetilde{N}$ be a clustered point process with underlying marked point process $\bar{N} = \Sigma\, \varepsilon_{(X_i, \widetilde{N}_i)}$. Then $\widetilde{N}$ is a *Poisson cluster process* on $\widetilde{E}$ if

a) $N = \Sigma\, \varepsilon_{X_i}$ is a Poisson process on E;
b) $\bar{N}$ is obtained from N by position-dependent marking (Example 1.28).

That is, there is a transition kernel K from E to $E' = \mathbf{M}_p(\widetilde{E})$ such that the $\widetilde{N}_i$ are conditionally independent given N, with $P\{\widetilde{N}_i \in \Gamma | N\} = K(X_i, \Gamma)$, $\Gamma \in \mathcal{M}_p(\widetilde{E})$. The Laplace functional $L_{\widetilde{N}}$ is given by $L_{\widetilde{N}}(f) = L_{\bar{N}}(H)$, where $H(x,v) = v(f)$, and consequently,

$$L_{\widetilde{N}}(f) = \exp\left[-\int \mu(dx) \int K(x,dv)(1 - e^{-v(f)})\right], \tag{1.26}$$

where μ is the mean measure of N. This expression appears again in Section 1.5.

Every Poisson process is a Poisson cluster process: with $\tilde{E} = E$ and $K(x, \cdot) = \varepsilon_{\varepsilon_x}$ it follows that $\tilde{N}_i = \varepsilon_{X_i}$ and hence that $\tilde{N} = N$. For a Poisson cluster process $\tilde{N}$ both the marked point process $\bar{N}$ and the point process $\Sigma\, \varepsilon_{X_i}$ are Poisson (Exercise 1.11); therefore, Poisson processes provide the key to study of Poisson cluster processes. The question then arises: How broad is the class of Poisson cluster processes? Theorem 1.32 contains the answer that a point process is a Poisson cluster process if and only if it is infinitely divisible, which we now define.

Definition 1.31. A point process N on E is *infinitely divisible* if for every n there exist i.i.d. point processes $N_1, \ldots, N_n$ such that

$$N \stackrel{d}{=} \sum_{i=1}^{n} N_i \tag{1.27}$$

[There is no difficulty in confirming that the finite sum on the right-hand side of (1.27), formally defined in Section 1.5 as the superposition of $N_1, \ldots, N_n$, defines a point process.]

If N is Poisson with mean measure μ, then taking $N_1, \ldots, N_n$ to be Poisson with mean μ/n gives

$$E\left[\exp\left(-\sum_{i=1}^{n} N_i(f)\right)\right] = [\exp(-n^{-1}\int (1 - e^{-f})d\mu)]^n = L_N(f);$$

consequently, N is infinitely divisible.

The class of infinitely divisible point processes carries a rich, elegant theory extending the classical theory of infinitely divisible probability measures on $\mathbf{R}_+$ (or that of increasing Lévy processes on $\mathbf{R}_+$). In Section 1.5 we develop results analogous to those valid in the classical case; however, the following characterization of infinitely divisible point processes as Poisson cluster processes is special to the point process setting. For compatibility with Definition 1.30 we use the notation employed there.

Theorem 1.32. Assume that $\tilde{E}$ is unbounded and let $\tilde{N}$ be a point process on $\tilde{E}$ with $E[\tilde{N}(K)] < \infty$ for each compact set K. Then the following statements are equivalent:

a) $\tilde{N}$ is infinitely divisible;
b) $\tilde{N}$ is a Poisson cluster process (one can take $E = \tilde{E}$ in Definition 1.30).

Proof: (Partial). We defer the proof that a) implies b) to Section 1.5. For the converse, as observed above the marked point process

$\bar{N} = \Sigma\, \varepsilon_{(X_i, \tilde{N}_i)}$ is Poisson and hence for each n can be represented as

$$\bar{N} \overset{d}{=} \sum_{j=1}^{n} \sum_{i} \varepsilon_{(X_{ji}, \tilde{N}_{ji})},$$

where the point processes $\bar{N}_j = \Sigma_i\, \varepsilon_{(X_{ji}, \tilde{N}_{ji})}$, $j = 1, \dots, n$, are i.i.d. Poisson processes. Then

$$\tilde{N} \overset{d}{=} \sum_{j=1}^{n} \sum_{i} \tilde{N}_{ji}, \tag{1.28}$$

which shows not only that $\tilde{N}$ is infinitely divisible but also that the components may be taken to be Poisson cluster processes. (The other implication shows that they must be taken so.) □

Additional discussion of infinitely divisible point processes appears in Section 1.5. Infinite divisibility for random measures is defined in precise analogy with Definition 1.30, and versions of the results here and elsewhere in the chapter are valid. One must however, take care to maintain the distinction between point process infinite divisibility and random measure infinite divisibility since there exist point processes (e.g., nonzero deterministic point processes) that are not infinitely divisible as point processes but are infinitely divisible as random measures. We use the term "infinitely divisible point process" *only* in the sense of Definition 1.30.

One class of infinitely divisible random measures is particularly amenable to analysis using point process methods, that of infinitely divisible random measures with independent increments.

Definition 1.33. A random measure M has *independent increments* if $M(A_1), \dots, M(A_k)$ are independent whenever $A_1, \dots, A_k \in \mathcal{B}$ are disjoint.

Concerning random measures with independent increments, the following is the principal characterization/decomposition.

Theorem 1.34. A random measure M with independent increments admits a decomposition

$$M = \mu_0 + \sum_{j=1}^{k} W_j\, \varepsilon_{x_j} + \sum_{i} U_i\, \varepsilon_{X_i}, \tag{1.29}$$

where

i) μ_0 is a fixed element of $\mathbf{M}$ (the deterministic part of M);
ii) $0 \le k \le \infty$, (x_j) is a sequence in E without accumulation points, and the W_j are independent random variables with $P\{W_j > 0\} > 0$ for each j (the x_j are *fixed atoms of* M);
iii) The marked point process $\bar{N} = \Sigma\, \varepsilon_{(X_i, U_i)}$ is a Poisson process on $E \times (0, \infty)$ and is independent of (W_j).

It follows that the random measure $\Sigma\, U_i\, \varepsilon_{X_i}$ is infinitely divisible with independent increments.

Proof: (Sketch). Let M' be the component of M remaining after all fixed atoms are removed. Given $A \in \mathcal{B}$, let (A_{nj}) be a null array of partitions of A (i.e., $A = \Sigma_j A_{nj}$ for each n, the partitions are successive refinements, and $\max_j \operatorname{diam} A_{nj} \to 0$). Since M' has independent increments but no fixed atoms, the triangular array $M'(A_{nj})$ is a null array of random variables [independent in j for each n and satisfying $\max_j P\{M'(A_{nj}) > \varepsilon\} \to 0$ for every $\varepsilon > 0$]. Moreover, $M'(A) = \Sigma_j M'(A_{nj})$ for each n; hence by the classical theory of infinitely divisible distributions on $\mathbf{R}_+$ (see, e.g., Chung, 1974) $M'(A)$ is infinitely divisible. The independent increments property of M' implies, then, that all finite-dimensional distributions—of random vectors $(M'(A_1), \ldots, M'(A_k))$—are infinitely divisible. Consequently, M' is itself infinitely divisible. We now remove from M' any deterministic component, after which we need only show that the residual random measure M'' has the purely atomic form indicated by (1.29). By the random measure version of Theorem 1.32, M'' can be represented as $M'' = \Sigma\, M_i$, where $\hat{M} = \Sigma\, \varepsilon_{M_i}$ is a Poisson process on $\mathbf{M}$, whose mean measure we denote by η. It follows that

$$L_{M''}(f) = \exp[-\int (1 - e^{-\mu(f)})\, \eta(d\mu)], \tag{1.30}$$

a version of the important expression (1.43) in Section 1.5. The key consequence of (1.30) is that the independent increments property of M'' entails that η is concentrated on the set of measures $\{u\varepsilon_x : x \in E, u > 0\}$ (the "degenerate" measures). Indeed, if $A \cap B = \varnothing$, then from $L_{M''}(1_{A \cup B}) = L_{M''}(1_A) L_{M''}(1_B)$ and (1.30), we infer that

$$\int \{(1 - e^{-\mu(A \cup B)}) - [(1 - e^{-\mu(A)}) + (1 - e^{-\mu(B)})]\}\, \eta(d\mu) = 0,$$

which yields $\mu(A) \wedge \mu(B) = 0$ almost everywhere with respect to η. By induction, for disjoint $A_1, \ldots, A_\ell$, almost everywhere with respect to η at most one of $\mu(A), \ldots, \mu(A_\ell)$ is nonzero. Thus η is concentrated on the degenerate measures. Using the obvious identification $u\, \varepsilon_x \leftrightarrow (x,u)$, (1.30) becomes

$$L_{M''}(f) = \exp[-\int (1 - e^{-uf(x)})\, \gamma\, (dx,du)] \tag{1.30a}$$

for some measure γ on $E \times (0,\infty)$, so (1.29) holds as asserted. □

We conclude with a familiar special case.

Example 1.35. (Poisson processes). By definition a Poisson process has independent increments and we have already noted its infinite divisibility. Conversely, let N be a simple point process without fixed atoms that simultaneously is infinitely divisible and has independent increments. Then in the representation (1.29) the U_i must all be one and hence N is Poisson. □

1.5 TRANSFORMATIONS OF POINT PROCESSES

Two important operations, marking and clustering, were discussed in Section 1.4. In this section we introduce additional ways in which point processes are transformed or combined to produce further point processes. We characterize infinitely divisible point processes in one more way, as limits of superpositions of null arrays of point processes. Related "limit" characterizations of Poisson and Cox processes are also described.

The simplest transformation is mapping.

Definition 1.36. Let $N = \Sigma\, \varepsilon_{X_i}$ be a point process on E and let E' be a second LCCB space. Given a measurable mapping $f: E \to E'$, provided that it is locally finite, the point process $Nf^{-1} = \Sigma\, \varepsilon_{f(X_i)}$ is the *mapping* of N by f.

In Section 1.8 and Chapter 9 we consider in detail the case that $E = \mathbf{R}^d$ for some $d \geq 1$ and f is a translation: $f(y) = y - x$ for fixed $x \in E$. Point processes with translation-invariant distributional properties are termed stationary. A second special case of mapping, used in Section 1.4 and again in Section 6.3, is projection, whereby one restricts attention to a single component of a point process on a product space. Since mapping amounts to a change of variable, it follows by standard results that $L_{Nf^{-1}}(g) = L_N(g \circ f)$, from which other computations (e.g., pertaining to the mean measure and the zero-probability functional) ensue. If N is Poisson with mean μ and if μf^{-1} belongs to $\mathbf{M}(E')$, then Nf^{-1} is Poisson with mean μf^{-1}. Applications of this property are given in Exercise 1.19.

Our next class of transformations involves additional randomness and yields compound point processes (which are in general not point processes at all but rather, purely atomic random measures; however, the terminology is well established and we use it despite its not being precisely correct), and in particular, thinned point processes.

Definition 1.37. Let $N = \Sigma\, \varepsilon_{X_i}$ be a point process on E and let $\bar{N} = \Sigma\, \varepsilon_{(X_i, U_i)}$ be a marked point process with mark space $\mathbf{R}_+$, obtained from N by position-dependent marking using a transition kernel K (Example

1.28). Then the purely atomic random measure

$$M = \sum U_i \varepsilon_{X_i} \tag{1.31}$$

is a *K-compound* of N, or *compound point process* determined by N and K.

In the special case that $K(x, \cdot) \equiv \rho(\cdot)$ for a probability ρ on $\mathbf{R}_+$, the U_i are i.i.d. and independent of N and we call M a ρ-compound of N, or *classical* compound point process. Classical compound Poisson processes on $\mathbf{R}_+$ are the best known examples; they have independent increments in general and independent and stationary increments if N is a homogeneous Poisson process. Theorem 1.34 demonstrates that except for fixed atoms and a deterministic part, every random measure with independent increments is of the form (1.31), with the marked point of process $\bar{N}$ a Poisson process on $E \times \mathbf{R}_+$.

For study of a compound point process $M = \Sigma\, U_i \varepsilon_{X_i}$ it is often convenient to deal instead with the marked point process $\bar{N} = \Sigma\, \varepsilon_{(X_i, U_i)}$ from which it is constructed, for then point process theories and computational methods can be brought to bear. However, in inference one must take care if the U_i can assume the value zero, because then $\bar{N}$ cannot be reconstructed from observation of M alone. (This is true in particular for thinned point processes defined momentarily, but they *are* point processes and can be dealt with directly.) Of course, if $U_i > 0$ for each i, then observation of M is equivalent to that of $\bar{N}$ and the two may be used interchangeably for inference purposes.

When the U_i take only the values zero and one, the compound point process $M = \Sigma\, U_i \varepsilon_{X_i}$ is a point process N' satisfying $N'(A) \leq N(A)$ for every A; such a process we call a *thinning* of N. In the setting of Definition 1.37, the mark space is $\{0,1\}$ and we may identify the kernel K with the function $p(x) = K(x,\{1\})$ from E into $[0,1]$.

Definition 1.38. Let $N = \Sigma\, \varepsilon_{X_i}$ be a point process on E and let $p: E \rightarrow [0,1]$ be measurable. Construct the marked point process $\bar{N} = \Sigma\, \varepsilon_{(X_i, U_i)}$ with mark space $\{0,1\}$ using position-dependent marking with the kernel $K(x,\{1\}) = 1 - K(x,\{0\}) = p(x)$. Then the point process

$$N' = \sum U_i \varepsilon_{X_i} \tag{1.32}$$

is a p-thinning of N, or *thinned point process*.

The interpretation: starting with N one forms N' by for each i retaining the point X_i of N—in the same location—with probability $p(X_i)$ and deleting it with the complementary probability $1-p(X_i)$. Different points are retained or deleted independently. It is important to remember that p represents *retention* probabilities, not deletion probabilities.

The principal computational properties, established by conditioning, are

$$L_{N'}(f) = L_N(-\log\{1-p+pe^{-f}\}), \tag{1.33}$$

$$z_{N'}(A) = L_N(1_A \cdot \{-\log(1-p)\}), \tag{1.34}$$

and

$$\mu_{N'}(dx) = p(x)\mu_N(dx) \tag{1.35}$$

If $p(x)>0$ for all x, then by (1.33) the distributions $\mathcal{L}_N$ and $\mathcal{L}_{N'}$ determine each other uniquely. For p a constant function, the U_i are i.i.d. Bernoulli random variables independent of N and we call N' a *classical* p-thinning of N.

In the context of thinning the most important point processes are the Cox processes, whose fundamental role we now explain. If N is a Cox process directed by the random measure M and if N' is the p-thinning of N, then from (1.17) and (1.33),

$$L_{N'}(f) = L_N(-\log\{1-p+pe^{-f}\}) = L_M(p(1-e^{-f})) = L_{pM}(1-e^{-f}), \tag{1.36}$$

where $pM(dx) = p(x)M(dx)$; consequently, N' is a Cox process directed by pM. The family of Cox processes is thus closed under thinning, but more is true: Cox processes are characterized by their being p-thinnings for every p.

Theorem 1.39. For a point process N on E the following statements are equivalent:

a) N is a Cox process, directed by some random measure M;
b) For every function $p\colon E \to (0,1]$ there is a point process N_p such that $N \stackrel{d}{=} N'_p$, where N'_p is the p-thinning of N_p.

Proof: To show that a) implies b), it suffices to take N_p to be the Cox process directed by $(1/p)M$ and to apply (1.36). Conversely, in particular for each n, N has the same distribution as the $p_n \equiv 1/n$-thinning of some N_n, and that b) implies a) will follow once Theorem 1.40 is proved. □

Nearly every invariance property is a simplified manifestation of a deeper limiting property, as is true in the situation at hand.

Theorem 1.40. Let (N_n) be a sequence of point processes on E and let (p_n) be a sequence of functions from E to $(0,1]$ such that $p_n \to 0$ uniformly.

For each n, let N'_n be the p_n-thinning of N_n. Then the following assertions are equivalent:

a) There is a point process N' such that $N'_n \xrightarrow{d} N'$;

b) There is a random measure M such that $p_n N_n \xrightarrow{d} M$.

When a) and b) hold, N' has the distribution of a Cox process directed by M.

Proof: In view of the convergence assumption on (p_n), for each f

$$\begin{aligned} L_{N'_n}(f) &= L_{N_n}(-\log\{1-p_n+p_n e^{-f}\}) \\ &\sim L_{N_n}(p_n(1-e^{-f})) = L_{p_n N_n}(1-e^{-f}) \end{aligned} \tag{1.37}$$

Since (N'_n) converges if and only if $\lim L_{N'_n}(f)$ exists for all f (Theorem 1.21) and similarly for $(p_n N_n)$, equivalence of a) and b) is established. Taking limits in (1.37) yields $L_{N'}(f) = L_M(1-e^{-f})$, which confirms the last statement. □

The Cox processes, therefore, are those point processes that arise as limits in distribution of sequences of point processes subjected to successively more severe thinning. Another class of point processes characterized as a "set of all limits under a rarefaction mechanism" is the infinitely divisible point processes. Before presenting this characterization, we formalize—by investing with a special name—the operation of addition of point processes.

Definition 1.41. Given point processes $N_1, \ldots, N_k$ defined on the same space E and over the same probability space, their *superposition* is the point process

$$N(A) = \sum_{i=1}^{k} N_i(A) \tag{1.38}$$

Superposition is of particular interest when the components N_i are independent, for then Laplace functionals multiply:

$$L_N = \prod_{i=1}^{k} L_{N_i}, \tag{1.39}$$

which is not only useful for establishing various results but also confirms the analogy between Laplace functionals and Laplace transforms.

Example 1.42. (Poisson processes). Let N_1, N_2 be independent Poisson processes with mean measures μ_1, μ_2. Then by (1.39)

$$L_{N_1+N_2}(f) = L_{N_1}(f) L_{N_2}(f) = \exp[-\int (1-e^{-f})\, d(\mu_1+\mu_2)]$$

and hence by Theorem 1.12 the superposition is Poisson with mean $\mu_1 + \mu_2$. For a partial converse, see Exercise 1.18. □

In addition to their being characterized as Poisson cluster processes, infinitely divisible point processes are all limits in distribution of superpositions of uniformly sparse point processes. Formulation of this entails one final bit of terminology.

Definition 1.43. Let $\{N_{nk} : n \geq 1, 1 \leq k \leq k_n\}$ be a triangular array of point processes, all on the same space E; then (N_{nk}) is a *null array* provided that

a) For each n, $N_{n1}, \ldots, N_{nk_n}$ are independent;
b) For every $B \in \mathcal{B}$,

$$\lim_{n\to\infty} \max_{k\leq k_n} P\{N_{nk}(B) \geq 1\} = 0 \tag{1.40}$$

Thus we arrive at one final characterization of infinitely divisible point processes.

Theorem 1.44. For (N_{nk}) a null array of point processes on E, the following statements are equivalent:

a) There is a point process N such that

$$\sum_{k=1}^{k_n} N_{nk} \xrightarrow{d} N; \tag{1.41}$$

b) There is a measure λ on $\mathbf{M}_p \setminus \{0\}$ satisfying

$$\int_{\mathbf{M}_p} (1 - e^{-\mu(B)})\lambda(d\mu) < \infty$$

for all $B \in \mathcal{B}$, such that

$$\sum_{k=1}^{k_n} [1 - L_{N_{nk}}(f)] \to \int_{\mathbf{M}_p} (1 - e^{-\mu(f)})\lambda(d\mu) \tag{1.42}$$

for all $f \in C_K^+$ (the set of nonnegative functions in C_K).

When a) and b) hold the limit N is necessarily infinitely divisible with

$$L_N(f) = \exp[-\int_{\mathbf{M}_p} (1 - e^{-\mu(f)})\lambda(d\mu)] \tag{1.43}$$

Proof: (Sketch). If a) holds, then $\Sigma_k \log L_{N_{nk}}(f) \to \log L_N(f)$ for each f and together with the null array assumption this implies that

$$\sum_k [1 - L_{N_{nk}}(f)] \to -\log L_N(f) \tag{1.44}$$

for $f \in C_K^+$. Let

$$\lambda = \lim_{n \to \infty} \sum_k \mathcal{L}_{N_{nk}}, \tag{1.45}$$

where $\mathcal{L}_{N_{nk}}$ is the probability law of N_{nk} and the limit is in the sense of vague convergence of measures on $\mathbf{M}_p$. Existence of λ is the incompletely resolved point in this argument. Combined, (1.44) and (1.45) yield

$$-\log L_N(f) = \lim_{n \to \infty} \sum_k \int (1 - e^{-\mu(f)}) \mathcal{L}_{N_{nk}}(d\mu) = \int (1 - e^{-\mu(f)}) \lambda(d\mu)$$

which demonstrates both (1.42) and (1.43).

Conversely, that (1.43) defines the Laplace functional of a point process N (in fact, a Poisson cluster process) is straightforward. Under the null array assumption, (1.42) implies that for $f \in C_K$,

$$\sum_k \log L_{N_{nk}}(f) \sim -\sum_k [1 - L_{N_{nk}}(f)] \to -\int (1 - e^{-\mu(f)}) \lambda(d\mu) = \log L_N(f);$$

consequently, a) holds. □

Since every infinitely divisible point process is the limit of the null array of Definition 1.31, its Laplace functional has the form (1.43). Furthermore, it is evident that a point process with Laplace functional (1.43) is a Poisson cluster process. These two observations constitute the proof that a) entails b) in Theorem 1.32.

To recapitulate, we now have three characterizations of infinitely divisible point processes: as Poisson cluster processes, as limits of row superpositions of null arrays, and as having Laplace functionals of the form (1.43). The latter two are analogous to classical results for infinitely divisible probability measures on $\mathbf{R}_+$, to which, indeed, the point process results reduce when E consists of but one point. Note also that (1.26) is of the form (1.43).

Theorem 1.44 admits random measure analogs (see Kallenberg, 1983, for details). The null array limit characterization holds provided that the definition is phrased as $\lim_n \max_j P\{M_{nj}(B) > \varepsilon\} = 0$ for every $\varepsilon > 0$. In the representation (1.43) λ becomes a measure on $\mathbf{M} \setminus \{0\}$ and a deterministic component may be present. Sharper forms are available when the limit has independent increments. For point processes, such a limit is a Poisson process; this convergence is treated at some length in the next section.

1.6. APPROXIMATION OF POINT PROCESSES

Because Poisson and Cox processes are the fundamental classes of point processes, it is of interest to investigate whether and how well a point process or sequence of point processes, arising, for example, as row superpositions of a triangular array, can be approximated by a Poisson or Cox process. Limit theorems, of course, provide some information but it is incomplete unless rates of convergence are also available. In this section we describe a paradigm for approximation of point processes, based on approximation error as the key concept, and carry out the steps of the paradigm for the most important special case: approximation of Bernoulli point processes and their superpositions by Poisson processes.

The main elements of the paradigm are:

1) A *metric* between distributions of point processes (which depends on an underlying metric on $\mathbf{M}_p$);
2) An *inequality* that provides an upper bound on the distance between two point processes (typically, one specified in advance and the other a well-chosen approximator);
3) Use of the inequality to derive *limit theorems with rates of convergence*.

Additional questions such as optimality of a given approximation within a set of alternatives are also of interest.

Throughout the section we assume that E is a *compact* metric space. The vague (weak) topology on $\mathbf{M}_p$ is metrizable by the *Prohorov metric*

$$d(\mu_1,\mu_2) = \inf\{\varepsilon > 0 : \mu_1(A) \leq \mu_2(A^\varepsilon) + \varepsilon \text{ for all } A \in \mathscr{E}\} \wedge 1, \quad (1.46)$$

where $A^\varepsilon = \{x : d^*(x,A) < \varepsilon\}$, with d^* the metric on E, and $d^*(x,A) = \inf\{d^*(x,y) : y \in A\}$. For point processes N_1 and N_2 the metric d leads to two distances between $\mathscr{L}_{N_1}$ and $\mathscr{L}_{N_2}$:

a) The *variation distance*

$$\alpha(\mathscr{L}_{N_1},\mathscr{L}_{N_2}) = \sup\{|P\{N_1 \in \Gamma\} - P\{N_2 \in \Gamma\}| : \Gamma \in \mathcal{M}_p\}; \quad (1.47)$$

b) The *Prohorov distance*

$$\rho(\mathscr{L}_{N_1},\mathscr{L}_{N_2}) = \inf\{\varepsilon > 0 : P\{N_1 \in \Gamma\} \leq P\{N_2 \in \Gamma^\varepsilon\} + \varepsilon \text{ for all } \Gamma \in \mathcal{M}_p\}, \quad (1.48)$$

where $\Gamma^\varepsilon = \{\mu : d(\mu,\Gamma) < \varepsilon\}$.

Evidently, $\rho(\mathscr{L}_{N_1},\mathscr{L}_{N_2}) \leq \alpha(\mathscr{L}_{N_1},\mathscr{L}_{N_2})$; we concentrate on bounds for the latter. The Prohorov distance provides a metric for convergence in distribution: $N_n \xrightarrow{d} N$ if and only if $\rho(\mathscr{L}_{N_n},\mathscr{L}_N) \to 0$ (see Billingsley, 1968).

A significant technique for dealing with the variation distance is coupling. A *coupling* of point processes with laws $\mathscr{L}_{N_1}$, $\mathscr{L}_{N_2}$ is a construction of N_1 and N_2 on the same probability space. Given a coupling (N_1,N_2) it follows that

$$\alpha(\mathscr{L}_{N_1},\mathscr{L}_{N_2}) \leq P\{N_1 \neq N_2\} = P\{d(N_1,N_2) > 0\} \quad (1.49)$$

and that

$$\rho(\mathcal{L}_{N_1},\mathcal{L}_{N_2}) \leq \inf\{\varepsilon > 0 : P\{d(N_1,N_2) \geq \varepsilon\} < \varepsilon\}; \tag{1.50}$$

(see Strassen, 1965, for details). A *maximal coupling* achieves equality in (1.49) or (1.50) and hence is useful for analysis of the distance between $\mathcal{L}_{N_1}$ and $\mathcal{L}_{N_2}$. It is known that maximal couplings exist for every choice of $\mathcal{L}_{N_1}$ and $\mathcal{L}_{N_2}$, in both the α- and ρ-senses, and that in general the two maximal couplings (neither is unique) cannot be realized simultaneously. Unfortunately, relatively few maximal couplings are known explicitly.

One way to carry out the paradigm, then, is as follows. Starting with the point process N to be approximated, for each candidate approximation N' construct a coupling (N,N')—maximal, if possible—and use either (1.49) or (1.50) to derive a bound on the distance between $\mathcal{L}_N$ and $\mathcal{L}_{N'}$. Then determine the optimal approximation N^*. To illustrate, we consider the problem of approximating Bernoulli point processes and their superpositions by Poisson processes; as a consequence of our inequalities, we obtain the Poisson limit theorem of Grigelionis (1963) (see also Franken, 1963).

Given $p \in (0,1]$ and a probability F on E, a *Bernoulli point process* with parameters (p,F) is a point process $N = U\varepsilon_X$, where X is a random element of E with distribution F and U is a Bernoulli random variable independent of X with $p = P\{U = 1\}(= 1 - P\{U = 0\})$. Thus with probability $1 - p$, N has no points at all, while with probability p, N has a single point with distribution F. For sufficiently small p the optimal Poisson approximation to N, in the sense of minimizing the α-distance between N and a Poisson process, is given by the next result.

Theorem 1.45. Let N be a Bernoulli point process on E with parameters (p,F) and let N^* be a Poisson process with mean measure

$$\mu^* = [-\log(1-p)]F \tag{1.51}$$

Then

$$\alpha(\mathcal{L}_N,\mathcal{L}_{N^*}) = p + (1-p)\log(1-p) \tag{1.52}$$

If $p < 1 - 1/e$, then for any Poisson process N' with mean measure $\mu' \neq \mu^*$,

$$\alpha(\mathcal{L}_N,\mathcal{L}_{N^*}) \leq \alpha(\mathcal{L}_N,\mathcal{L}_{N'}); \tag{1.53}$$

consequently, for $p < 1/e$, N^* is the unique optimal Poisson approximation to N.

Proof: (Sketch). Construct a maximal coupling (N,N^*) as follows.

Starting with N^*, let N be the point process obtained by randomly choosing one atom of N^*. [See Kallenberg, 1983; more precisely, $N = 0$ on $\{N^* = 0\}$ and on $\{N^*(E) = k\}$, $k \geq 1$, assuming that on this event N^* has the representation $N^* = \Sigma_{i=1}^k \varepsilon_{X_i}$, where X_i are i.i.d. with distribution F, put $N = \Sigma_{i=1}^k V_i \varepsilon_{X_i}$, where $V = (V_1, \dots, V_k)$ is independent of N with $P\{V = (1,0,\dots,0)\} = \dots = P\{V = (0,\dots,0,1)\} = 1/k$.] Then by conditional uniformity of N^*—the key to the construction (see Exercise 1.1)—N is indeed Bernoulli with parameters (p,F); moreover,

$$P\{N \neq N^*\} = P\{N^*(E) \geq 2\} = p + (1-p)\log(1-p)$$

Maximality is apparent: for any coupling of N and N^*, $P\{N \neq N^*\} \geq P\{N^*(E) \geq 2\}$ because N is Bernoulli. Hence (1.52) holds.

To prove optimality, suppose that N' is Poisson with mean measure written as $\mu' = \lambda G$, where $\lambda > 0$ and G is a probability on E. Then first of all, since N is Bernoulli, for the maximal coupling (N,N') (we do not construct it explicitly)

$$\alpha(\mathcal{L}_N, \mathcal{L}_{N'}) = P\{N \neq N'\} \geqslant P\{N'(E) \geq 2\} = 1 - e^{-\lambda} - \lambda e^{-\lambda};$$

the latter function increases in λ and hence we may restrict attention to values satisfying $\lambda \leq \lambda^* = -\log(1-p)$. With $\lambda \leq \lambda^*$ and N' Poisson with mean measure λF, the maximal coupling of N and N' satisfies $P\{N \neq N'\} = p - \lambda e^{-\lambda}$; consequently, with λ fixed, the optimal choice of G is the obvious choice $G = F$, since for the maximal coupling of N and a Poisson process N'' with mean measure λG,

$$\begin{aligned} P\{N \neq N''\} &= P\{N''(E) \geq 2\} + P\{N'' = 0,\ N \neq 0\} + P\{N''(E) = 1,\ N \neq N''\} \\ &= 1 - e^{-\lambda} - \lambda e^{-\lambda} + P\{N'' = 0,\ N \neq 0\} \\ &\quad + P\{N''(E) = 1,\ N \neq N''\} \\ &\geq p - \lambda e^{-\lambda} = P\{N \neq N'\}, \end{aligned}$$

with the inequality strict unless $G = F$. The problem is thus reduced to a one-dimensional minimization over λ, solution of which is easily seen to be $\lambda = \lambda^*$. □

Approximation of the superposition of independent Bernoulli point processes is then straightforward, but the following result does not yield an optimal approximation even if E consists of only one point (in this case Poisson approximations for point processes reduce to results for Bernoulli and binomial random variables).

Proposition 1.46. Let $N_1, \dots, N_k$ be independent Bernoulli point processes with parameters $(p_1,F_1), \dots, (p_k,F_k)$, let $N = \Sigma_{i=1}^k N_i$, and let N'

be Poisson with mean measure $\mu = \Sigma_{i=1}^{k} [-\log(1 - p_i)]F_i$. Then

$$\alpha(\mathcal{L}_N, \mathcal{L}_{N'}) \leq 1 - \prod_{i=1}^{k} [(1 - p_i)(1 - \log(1 - p_i))] \quad \square \tag{1.54}$$

From Proposition 1.46 one can deduce the Poisson limit theorem of Grigelionis (1963). However, a somewhat stronger version can be obtained from the next result, which embeds a triangular array of Bernoulli point processes in a single Poisson process.

Let $\{N_{ni} : n \geq 1, 1 \leq i \leq k_n\}$ be a triangular array of Bernoulli point processes with respective parameters (p_{ni}, F_{ni}) (for each n the N_{ni} are independent as i varies) and let $N_n = \Sigma_{i=1}^{k_n} N_{ni}$ be the nth row superposition.

Theorem 1.47. Assume that there exists a finite measure μ and E such that

$$-\sum_{i=1}^{k_n} [\log(1 - p_{ni})]F_{ni} \ll \mu \tag{1.55}$$

for each n, and let r_n be the Radon-Nikodym derivative. Then there exists a probability space on which are defined a Poisson process N' with mean measure μ and a version of the triangular array (N_{ni}) such that for each n,

$$P\{N_n \neq N'\} \leq 1 - \exp\left(-\int |1 - r_n|\, d\mu\right) \prod_{i=1}^{k_n} [(1 - p_{ni})[1 - \log(1 - p_{ni})]] \tag{1.56}$$

Consequently, if

$$\lim_{n\to\infty} \max_i p_{ni} = 0 \tag{1.57}$$

and

$$\lim_{n\to\infty} \int |1 - r_n|\, d\mu = 0, \tag{1.58}$$

then as $n \to \infty$

$$\alpha(\mathcal{L}_{N_n}, \mathcal{L}_{N'}) = O(\max p_{ni} + \int |1 - r_n|\, d\mu) \tag{1.59}$$

Proof: (Sketch). Let N' be Poisson with mean μ and by repeated thinning decompose N' as

$$N' = \widetilde{N}_n + \sum_{i=1}^{k_n} \widetilde{N}_{ni}$$

where $\hat{N}_n$ and the $\hat{N}_{ni}$ are independent Poisson processes, $\hat{N}_{ni}$ has mean measure $[(r_n(x) \wedge 1)/r_n(x)][-\log(1-p_{ni})]F_{ni}(dx)$, and $\hat{N}_n$ has mean measure $[1 - r_n(x) \wedge 1]\mu(dx)$. Also, let $\bar{N}_n$ be Poisson with mean measure $[r_n(x) - 1]^+\mu(dx)$, independent of N' and decomposed into independent components $\bar{N}_n = \Sigma_{i=1}^{k_n} \bar{N}_{ni}$, where $\bar{N}_{ni}$ is Poisson with mean measure $\{1 - [r_n(x) \wedge 1]/r_n(x)\}[-\log(1-p_{ni})]F_{ni}(dx)$. For each n and i, $N'_{ni} = \hat{N}_{ni}$ is Poisson with mean $[-\log(1-p_{ni})]F_{ni}$, and to it, couple N_{ni} as in Theorem 1.45, preserving independence of the N_{ni} in the process. Then

$$P\{N_n = N'\} \geq P\{\hat{N}_n = 0,\ \bar{N}_n = 0,\ N_{ni} = N'_{ni},\ i = 1, \ldots, k_n\}$$
$$= P\{\hat{N}_n = 0\}P\{\bar{N}_n = 0\} \sum_{i=1}^{k_n} P\{N_{ni} = N'_{ni}\}$$

Evaluation of the last expression using (1.52) yields (1.59). □

We illustrate with two examples.

Example 1.48. (Point process analog of the classical Poisson limit theorem for binomial distributions). Suppose that for each n the N_{ni}, $i = 1, \ldots, n$, are i.i.d. with parameters $(\lambda/n, F)$, where $\lambda > 0$ and F is a fixed probability on E. Taking $\mu = \lambda F$ in Theorem 1.47 leads to the convergence rate $\alpha(\mathcal{L}_{N_n}, \mathcal{L}_{N'}) = O(n^{-1})$; observe that it applies to the α-distance rather than just the ρ-distance. □

Example 1.49. (Empirical processes). Let $X_1, X_2, \ldots$ be i.i.d., strictly positive random variables with distribution function F admitting positive, continuous density f on $[0,\varepsilon]$ for some $\varepsilon > 0$. Let $E = [0,t]$, where $t > 0$ is fixed, and put $N_{ni} = 1(X_i \leq t/n)\varepsilon_{nX_i}$, $1 \leq i \leq n$. Then N_{ni} is Bernoulli with parameters $(F(t/n), F_n)$, where $F_n(A) = F(n^{-1}A)/F(t/n)$ for $A \subset E$. Choosing $\mu(dx) = \lambda\, dx$, where $\lambda = f(0)$, in Theorem 1.47—so that N' is an ordinary Poisson process with rate λ— and performing the necessary computations yields the estimate

$$\alpha(\mathcal{L}_{N_n}, \mathcal{L}_{N'}) = O(n^{-1}); \tag{1.60}$$

for details, see Exercise 1.21. □

An alternative methodology for approximation of point processes on $\mathbf{R}_+$, which has particular application to Poisson approximation, is based on martingales and compensators and is described in Section 2.3. On $\mathbf{R}_+$ this method is able to approximate more general point processes than the coupling method can (thus far, at least), but its lack of applicability on general spaces can be a severe restriction. By constrast, the methods in this section are sufficiently broad to yield infinitely divisible approximations to quite arbitrary point processes, as we now demonstrate.

Let N be a point process on a compact space E and put $p = P\{N \neq 0\}$, $G(\Gamma) = P\{N \in \Gamma | N \neq 0\}$. Rather than approximate N directly, one can approximate the Bernoulli point process $\tilde{M} = 1(N \neq 0)\varepsilon_N$ on the space $\mathbf{M}_p$. Let $\tilde{N}^* = \Sigma\, \varepsilon_{N_j}$ be a Poisson process on $\mathbf{M}_p$, with mean $[-\log(1-p)]G$, such that (1.52) holds; then $N^* = \Sigma\, N_j$ is an infinitely divisible point process and we obtain the following approximation.

Proposition 1.50. Given N, there exists an infinitely divisible point process N^* such that

$$P\{N \neq N^*\} = p + (1-p)\log(1-p), \tag{1.61}$$

where $p = P\{N \neq 0\}$. □

1.7. PALM DISTRIBUTIONS

As will emerge in later sections, Palm distributions are a powerful tool for inference for point processes, especially Cox processes (Chapter 7) and stationary point processes (Chapter 9). In addition, structural and conditioning properties are often best or most easily described using Palm distributions. Heuristically, a (reduced) Palm distribution of a simple point process N is the probability law, conditional on the presence of points of N at prescribed locations, of the *remainder* of N, but since the event in question ordinarily has probability zero, an indirect approach is needed.

Let N be a *simple* point process on E and assume that for each $k \geq 1$ the kth-order moment measure $\mu_N^k(B) = E[N^k(B)]$ belongs to $\mathbf{M}(E^k)$; the same is then true of the factorial moment measure $\mu_N^{(k)} = E[N^{(k)}]$, where, as in Section 1.2,

$$N^{(k)}(dx_1, \ldots, dx_k) = N(dx_1)(N - \varepsilon_{x_1})(dx_2) \cdots \left(N - \sum_{i=1}^{k-1} \varepsilon_{x_i}\right)(dx_k)$$

For each k, let $\mathbf{M}_s(k) = \{\mu \in \mathbf{M}_s : \mu(E) = k\}$; note that $\mathbf{M}_s^* = \cup_k \mathbf{M}_s(k)$ is the set of finite elements of $\mathbf{M}_s$.

We begin with the fundamental definitions.

Definition 1.51. For each $k \geq 0$ the k-variate *Campbell measure* of N is the measure $C_N^{(k)}$ on $\mathbf{M}_s(k) \times \mathbf{M}_s$ defined by

$$C_N^{(k)}(\Lambda \times \Gamma) = E\left[\int_{E^k} 1\left(\sum_{i=1}^{k} \varepsilon_{x_i} \in \Lambda\right) 1\left(N - \sum_{i=1}^{k} \varepsilon_{x_i} \in \Gamma\right) N^{(k)}(dx)\right] \tag{1.62}$$

The *compound Campbell measure* of N is the measure

$$C_N(\Lambda \times \Gamma) = \sum_{k=0}^{\infty} \frac{1}{k!} C_N^{(k)}(\Lambda \times \Gamma) \tag{1.63}$$

For $k=0$ we take $C_N^{(0)}(\Lambda \times \Gamma) = 1(0 \in \Lambda)P\{N \in \Gamma\}$; note also that

$$\mu_N^{(k)}(f) = \int \mu(f) C_N^{(k)}(d\mu \times \mathbf{M}_s) \tag{1.64}$$

We further introduce the *compound factorial moment measure* [the name is justified by (1.64)]

$$K_N(\Lambda) = C_N(\Lambda \times \mathbf{M}_s)$$

Evidently, $C_N(\Lambda \times \Gamma) \leq K_N(\Lambda)$ for each Λ and Γ; consequently, standard (or, at least, known) results in measure theory imply the existence of a disintegration

$$C_N(d\mu, d\nu) = K_N(d\mu) Q_N(\mu, d\nu), \tag{1.65}$$

where Q_N is a transition probability from $\mathbf{M}_s^*$ to $\mathbf{M}_s$.

Definition 1.52. The probability distributions $Q_N(\mu, \cdot)$, $\mu \in \mathbf{M}_s^*$, are the *reduced Palm distributions* of N.

Note that the reduced Palm distribution $Q_N(\mu, \cdot)$ is defined only for μ a *finite* element of $\mathbf{M}_s$. For $\mu \in \mathbf{M}_s(k)$ we have

$$\begin{aligned} Q_N(\mu, \Gamma) &= \frac{C_N(d\mu \times \Gamma)}{K_N(d\mu)} \\ &= \frac{C_N(d\mu \times \Gamma)}{C_N(d\mu \times \mathbf{M}_s)} \\ &= \frac{E[1(N - \mu \in \Gamma)(1/k!) \int 1(\Sigma_{i=1}^k \varepsilon_{x_i} \in d\mu) N^{(k)}(dx)]}{E[(1/k!) \int 1(\Sigma_{i=1}^k \varepsilon_{x_i} \in d\mu) N^{(k)}(dx)]} \end{aligned}$$

leading to the crucial interpretation

$$Q_N(\mu, \Gamma) = P\{N - \mu \in \Gamma | N - \mu \geq 0\}, \tag{1.66}$$

of $Q_N(\mu, \cdot)$ as the law of $N - \mu$ conditional on the event that N has a point at every point of μ. "Reduced" in Definition 1.52 refers to our having subtracted μ from N.

There are various ways in which (1.66) can be made precise in terms of limits; here is one. Let (A_{nj}) be a null array of partitions of E [i.e., max diam$(A_{nj}) \to 0$ (see Theorem 1.34)] and for $\mu \in \mathbf{M}_s^*$ and each n let $A_n(\mu) = \cup\{A_{nj} : \mu(A_{nj}) > 0\}$. If μ has k (distinct) points, then eventually $A_n(\mu)$ consists of k disjoint components $B_{1n}(\mu), \ldots, B_{nk}(\mu)$, each containing exactly one point of μ.

Theorem 1.53. For almost every μ with respect to K_N,

$$Q_N(\mu,\cdot)=\lim P\{N-1_{A_n(\mu)}N\in(\cdot)|N(B_{nj}(\mu))=1, j=1,\dots,\mu(E)\} \quad (1.67)$$

in the sense of weak convergence of probability measures, where $1_{A_n(\mu)}N(dx)=1(x\in A_n(\mu))N(dx)$ is the restriction of N to $A_n(\mu)$.

Proof: (Sketch). To simplify the notation, we assume that $\mu\in\mathbf{M}_s(1)$; to streamline the argument, we suppose that the mapping $\mu\to Q_N(\mu,\cdot)$ is continuous. Finally, let $\tilde{A}_n(\mu)=\{\varepsilon_y : y\in A_n(\mu)\}$ and let $\Lambda_n(\mu)=\{\nu:\nu(A_n(\mu))=0\}$. Then

$$\begin{aligned}
&P\{N-1_{A_n(\mu)}N\in(\cdot)|N(B_{nj}(\mu))=1, j=1,\dots,\mu(E)\}\\
&\quad=\frac{P\{N-1_{A_n(\mu)}N\in(\cdot), N(B_{nj}(\mu))=1, j=1,\dots,\mu(E)\}}{P\{N(B_{nj}(\mu))=1, j=1,\dots,\mu(E)\}}\\
&\quad=\frac{E[\int_{A_n(\mu)}1(N-\varepsilon_x\in\Gamma\cap\Lambda_n(\mu))N(dx)]}{E[\int_{A_n(\mu)}1(N-\varepsilon_x\in\Lambda_n(\mu))N(dx)]}\\
&\quad=\frac{C_N^{(1)}\{\tilde{A}_n(\mu)\times[\Gamma\cap\Lambda_n(\mu)]\}}{C_N^{(1)}[\tilde{A}_n(\mu)\times\Lambda_n(\mu)]}\\
&\quad=\frac{\int_{\tilde{A}_n(\mu)}K_N(d\nu)Q_N(\nu,\Gamma\cap\Lambda_n(\mu))}{\int_{\tilde{A}_n(\mu)}K_N(d\nu)Q_N(\nu,\Lambda_n(\mu))}
\end{aligned}$$

[by (1.65)]

$$\cong\frac{Q_N(\mu,\Gamma\cap\Lambda_n(\mu))}{Q_N(\mu,\Lambda_n(\mu))}$$

by the continuity assumption. But furthermore, since $A_n(\mu)\downarrow\operatorname{supp}\mu$ and since N is simple,

$$\begin{aligned}
\int_{\mathbf{M}_s(1)}K_N(d\nu)Q_N(\nu,\Lambda_n(\mu))&=C_N^{(1)}(\mathbf{M}_s(1)\times\Lambda_n(\mu))\\
&=E\Big[\int 1(N-\varepsilon_x\in\Lambda_n(\mu))N(dx)\Big]\\
&\to E[N(E)]=K_N(\mathbf{M}_s(1)),
\end{aligned}$$

where the last equality is by (1.64), it follows that $Q_N(\cdot,\Lambda_n(\mu))\to 1$ a.e. (almost everywhere), which suffices to complete the proof. □

A point process N_μ with probability law $Q_N(\mu,\cdot)$ is termed a *reduced Palm process* of N; introduction of these processes, defined on an unspecified probability space, simplifies the presentation materially.

Once again we illustrate with Poisson and Cox processes. The form of the reduced Palm distributions for a Poisson process is yet another manifestation of their independent increments.

Example 1.54. (Poisson processes). Let N be a Poisson process on E with mean measure μ. Then for almost every ν the reduced Palm process N_ν is Poisson with mean μ. Indeed, if $\nu = \Sigma_{i=1}^k \varepsilon_{x_i}$, then for disjoint neighborhoods V_i of x_i, f a continuous function, and $U = (\cup_{i=1}^k V_i)^c$,

$$\frac{E[\Pi_{i=1}^k N(V_i) e^{\{-N(f)+\Sigma f(x_i)\}}]}{E[\Pi_{i=1}^k N(V_i)]}$$

$$= E[\exp[-\int_U f\,dN] \prod_{i=1}^k \frac{N(V_i)\exp[-\int_{V_i} f\,dN + f(x_i)]}{\mu(V_i)}]$$

$$= \exp[-\int_U (1-e^{-f})\,d\mu] \prod_{i=1}^k \frac{\exp[-\int_{V_i}(1-e^{-f})\,d\mu][\int_{V_i} e^{-f}\,d\mu]e^{f(x_i)}}{\mu(V_i)}$$

$$\to \exp[-\int (1-e^{-f})\,d\mu]$$

from which the assertion follows immediately. In fact, this property is characteristic of Poisson processes; see Proposition 1.57 for a related characterization of mixed Poisson processes. □

With appropriate modifications—mainly no longer subtracting $\Sigma_{i=1}^k \varepsilon_{x_i}$—one can extend the concept of Palm distributions to random measures; however, the conditioning interpretation of (1.66) and Theorem 1.53 does not carry over. The compound Campbell measure of (1.63) is replaced by the measure

$$C_M(\Lambda \times \Gamma) = \sum_{k=0}^{\infty} \frac{1}{k!} E\left[\int 1\left(\sum_{i=1}^{k} \varepsilon_{x_i} \in \Lambda\right) M^k(dx) 1(M \in \Gamma)\right] \tag{1.63a}$$

on $\mathbf{M}_s \times \mathbf{M}$, which then can be disintegrated as

$$C_M(d\mu, d\eta) = K_M(d\mu) Q_M(\mu, d\eta), \tag{1.65a}$$

where

$$K_M(\Lambda) = \sum_{k=0}^{\infty} \frac{1}{k!} E\left[\int 1\left(\sum_{i=1}^{k} \varepsilon_{x_i} \in \Lambda\right) M^k(dx)\right]$$

The probability measures $Q_M(\mu,\cdot)$, $\mu \in \mathbf{M}_s$, are called *unreduced Palm distributions* of M; we also define *Palm processes* M_μ, with respective probability laws $Q_M(\mu, \cdot)$.

Within this context the following property holds for Cox processes; it is used in Section 7.2 for analysis of the state estimation problem of reconstruction of the directing measure of a Cox pair from observations of the Cox process alone. Theorem 4.27 provides further confirmation of the central role of Palm distributions in state estimation problems.

Proposition 1.55. Let M be a random measure on E and let N be a Cox process directed by M. Then $K_N = K_M$ and for almost every $\mu \in \mathbf{M}_p$ with respect to K_N, the reduced Palm process N_μ has the distribution of a Cox process directed by the Palm process M_μ.

Proof: For each k and positive functions f and g, we have

$$E[\exp[-M(1-e^{-f})]\int e^{-\Sigma g(x_i)}M^k(dx)] = E[e^{-N(f)}\int e^{-\Sigma g(x_i)}e^{\Sigma f(x_i)}N^{(k)}(dx)]$$

because N is a Cox process directed by M. It follows that $K_N = K_M$ and hence that

$$\int K_M(d\mu)e^{-\mu(g)}\int Q_M(\mu,d\eta)\exp[-\eta(1-e^{-f})]$$
$$= \int K_N(d\mu)e^{-\mu(g)}\int Q_N(\mu,d\nu)e^{-\nu(f)}$$
$$= \int K_M(d\mu)e^{-\mu(g)}\int Q_N(\mu,d\nu)e^{-\nu(f)}$$

This implies that for fixed f, $L_{M_\mu}(1-e^{-f}) = L_{N_\mu}(f)$ almost everywhere with respect to K_M, and the proposition follows by standard reasoning and Theorem 1.12. □

No analog of (1.66) holds for random measures and unreduced Palm distributions, although we do have that for $\mu = \Sigma_{i=1}^k \varepsilon_{x_i}$,

$$Q_M(\mu,\Gamma) = \frac{E[M^k(dx)1(M \in \Gamma)]}{E[M^k(dx)]}$$

If M happens to be a simple point process, we encounter a notational inconsistency: the unreduced Palm distributions Q_M are not the reduced Palm distributions Q_N of Definition 1.51. This is resolved as follows.

Definition 1.56. Given a simple point process N, the Palm distributions from (1.65a) are termed *unreduced Palm distributions* of N and denoted by $Q_N^*(\mu,\cdot)$.

The interpretation, evidently, is that

$$Q_N^*(\mu,\cdot)=P\{N\in(\cdot)|N-\mu\geq 0\} \tag{1.66a}$$

Only in connection with Palm measures for stationary point processes do unreduced Palm distributions play an important role; otherwise, for point processes they are something of a historical artifact.

Further applications of Palm distributions pertain to characterization and continuity, as well as to conditioning properties discussed at length in Section 2.6 and to inference problems treated in later chapters. Rather than attempt complete coverage we give only one final characterization—for mixed Poisson processes—related to those given already for Poisson and Cox processes; its proof is omitted.

Proposition 1.57. Let N be a point process on E with infinite mean measure μ_N. Then the following are equivalent:

a) $\mathcal{L}_{N_\mu}$ does not depend on μ (a.e. K_N);

b) N is a mixed Poisson process. □

1.8 STATIONARY POINT PROCESSES

Stationarity for point processes means distributional invariance under translations; among other properties a stationary point process admits a single intensity parameter, which under minimal ergodicity assumptions can be reconstructed from almost every realization. Of course for stationarity to make sense the underlying space must have group structure; for simplicity we assume throughout the section that E is a closed additive subgroup of $\mathbf{R}^d$ for some $d\geq 1$. The main cases of interest, of course, are $E=\mathbf{R}^d$ and $E=\mathbf{Z}^d$.

For $x\in E$ let τ_x denote the translation operator $\tau_x y=y-x$; for μ a measure on E we then have $\mu\tau_x^{-1}(A)=\mu(A+x)$. Let λ be a Haar measure for E, which we fix throughout the discussion; for $E=\mathbf{R}^d$ we take λ to be Lebesgue measure.

Let N be a point process on E defined over a probability space $(\Omega,\mathcal{F},P)$ rich enough that one can define on it operators $(\theta_x)_{x\in E}$ satisfying

a) $(x,\omega)\rightarrow\theta_x\omega$ is jointly measurable;

b) $\theta_0=I$, the identity on E;

c) $\theta_{x+y}=\theta_x\theta_y$ for each x, y.

This is not really a restriction: on the canonical space $\Omega=\mathbf{M}_p$, one can take $\theta_x\mu=\mu\tau_x^{-1}$. We then put

$$N_{(x)}=N\tau_x^{-1} \tag{1.68}$$

For $N=\Sigma\,\varepsilon_{X_i}$, $N_{(x)}=\Sigma\,\varepsilon_{X_i-x}$, that is, each point of N is translated by x.

Here is the key definition.

Definition 1.58. The point process N is *stationary* with respect to (θ_x) if

a) (θ_x) is measure-preserving for $P: P\theta_x^{-1} = P$ for all $x \in E$;
b) for each x,

$$N_{(x)} = N \circ \theta_x \tag{1.69}$$

In some contexts, statistical inference in particular, stationarity (along with Palm measures and ergodicity) is more a property of the probability P than of N, but in this section we adopt the latter viewpoint. A somewhat broader concept, *stationarity in law*, which requires only that $N_{(x)} \stackrel{d}{=} N$ for each x, is also in use. Obviously, stationarity in the sense of Definition 1.58 implies stationarity in law; conversely, a point process stationary in law admits a canonical version that is stationary. We therefore restrict ourselves to the more convenient and powerful form.

The following extreme examples exhibit very different behaviors.

Example 1.59. (Poisson processes). Let $\Omega = \mathbf{M}_p$ and $\mathcal{F} = \mathcal{M}_p$, with $N(\omega) = \omega$. If under P, N is Poisson with mean measure λ, then N is stationary with respect to operators $\theta_x\omega = \omega\tau_x^{-1}$. We then need only verify Definition 1.58a); for it,

$$\begin{aligned} E[e^{-N\circ\theta_x(f)}] &= E[e^{-N(f\circ\tau_x)}] \\ &= \exp[-\int (1 - e^{-f\circ\tau_x})\,d\lambda] \\ &= \exp[-\int (1 - e^{-f})\,d\lambda] = E[e^{-N(f)}], \end{aligned}$$

where the third equality is by translation invariance of λ. □

Example 1.60. (Lattice process). Let $E = \mathbf{R}$, $\Omega = [0,1]$, $\mathcal{F} = \mathcal{B}(\Omega)$, and let P be the uniform distribution, so that $X(\omega) = \omega$ is uniformly distributed. Let $N = \Sigma_{n=-\infty}^{\infty} \varepsilon_{X+n}$; that is, N is a random translation of the integer lattice. Then N is stationary with respect to (θ_x), where $\theta_x(\omega) = \{\omega - x\}$, the fractional part of $\omega - x$ (see Exercise 1.26). □

If N is stationary, then the first moment measure μ_N is translation invariant and hence a multiple of λ.

Definition 1.61. Let N be a stationary point process. The constant $\nu_N \in [0,\infty]$ satisfying $\mu_N = \nu_N\lambda$ is the *intensity* of N.

Corresponding simplifications of higher-order moment measures are derived in Proposition 9.4.

Note that the intensity depends on the choice of λ. For Examples 1.59 and 1.60, the intensity is 1. Definition 1.61 is a macro-level definition of intensity. At least two other heuristic concepts of "intensity" can be formulated: a *local limit* $\lim P\{N(V_n) = 1\}/\lambda(V_n)$ with $V_n \downarrow \{0\}$ and a *global limit* $\lim N(A_n)/\lambda(A_n)$ as $A_n \uparrow E$. We shall see that in nice cases the three notions coincide. The micro-level version is approached through concepts of regularity and orderliness and results known (originally for $E = \mathbf{R}$) as Khinchin-Koroljuk theorems (see Corollary 1.66). To treat the global limit version requires an ergodic theory for stationary point processes, given in Theorem 1.68 and extended in Theorem 9.2. Both versions are best explained via Palm measures.

Let N be a stationary point process on E with finite intensity ν_N. Since stationarity implies that for each univariate *unreduced* Palm process $N^*_{\varepsilon_x}$,

$$N^*_{\varepsilon_x}\tau_x^{-1} \stackrel{d}{=} N^*_{\varepsilon_o}, \tag{1.70}$$

it suffices to study the Palm distribution $Q^*_N(\varepsilon_0, \cdot)$. We do so, but slightly indirectly, in terms of the Palm measure P^*.

Definition 1.62. Let N be a stationary point process on E with finite intensity ν_N. The finite measure

$$P^* = \nu_N Q^*_N(\varepsilon_0, \cdot) \tag{1.71}$$

on Ω is the *Palm measure* of N (or, depending on emphasis, of P). □

Even though, as seen momentarily, P^* need not be a probability, integrals with respect to it are written as "expectations" E^*. From Section 1.7, the *Palm distribution* $P^*(\cdot)/P^*(\Omega)$ is interpreted as the probability law of N conditional on there being a point at the origin.

The following characterization is used repeatedly.

Proposition 1.63. Let N be stationary with respect to (θ_x). Then the Palm measure is the unique measure P^* on Ω satisfying

$$E\Big[\int G(\theta_x\omega, x)N(\omega, dx)\Big] = E^*\Big[\int G(\omega, x)\lambda(dx)\Big] \tag{1.72}$$

for all bounded, measurable functions $G: \Omega \times E \to \mathbf{R}$.

Proof: It suffices to establish (1.72) for G of the form $G(\omega, x) = 1(N(\omega) \in \Gamma)1(x \in A)$, where $\Gamma \in \mathcal{M}_p$, $A \in \mathcal{E}$; in this case,

$$E^*\left[\int G(\omega,x)\lambda(dx)\right] = \lambda(A)P^*\{N \in \Gamma\}$$
$$= \nu_N \lambda(A) Q_N(0,\Gamma)$$

[by (1.71), where $Q_N(x,\cdot) = Q_N(\varepsilon_x,\cdot)$]

$$= \nu_N \int_A \lambda(dx) Q_N(x, \{\mu : \mu\tau_x^{-1} \in \Gamma\})$$

[by (1.70)]

$$= \int_A \mu_N(dx) \int Q_N(x,d\mu) 1(\mu\tau_x^{-1} \in \Gamma)$$
$$= \int\int C_N^1(dx,d\mu) 1(\mu\tau_x^{-1} \in \Gamma) 1(x \in A)$$
$$= E[\int_A 1(N_{(x)} \in \Gamma) N(dx)] = E[\int G(\theta_x\omega,x) N(dx)]$$

where the next-to-last equality is by an analog of (1.62) for unreduced Palm distributions. □

In particular, as also apparent from (1.71), $\nu = P^*(\Omega)$.

Palm distributions for Poisson processes were determined in Section 1.7; the following example continues Example 1.60.

Example 1.64. (Lattice process). For the stationary point process $N = \Sigma\, \varepsilon_{X+n}$ of Example 1.60, $P^* = \varepsilon_0$. To see this, apply (1.72) with $G(\omega,x) = e^{-\alpha\omega} 1(x \in [0,1])$, where $\alpha > 0$, to obtain

$$\int_\Omega P^*(d\omega) e^{-\alpha\omega} = E[\int_{[0,1]} e^{-\alpha\theta_x(\omega)} N(dx)] = E[\exp(-\alpha\theta_{X(\omega)}\omega)] = 1$$

Note that $P^* = \varepsilon_0$ is consistent with the conditioning interpretation of Palm distributions given in (1.66a). □

For the stationary case that interpretation can be given in alternative forms.

Theorem 1.65. Let N be a stationary point process with finite intensity and Palm measure P^*.

a) If E is discrete (i.e., countable), then $P^* << P$ with

$$\frac{dP^*}{dP}(\omega) = \frac{N(\omega,\{0\})}{\lambda(\{0\})}; \tag{1.73}$$

b) If E is nondiscrete and Z is a positive random variable such that $x \to Z(\theta_x\omega)$ is continuous for each ω and if (V_n) is a sequence of neighborhoods of $0 \in E$ such that $\bar{V}_n \downarrow \{0\}$, then

$$E^*[Z] = \lim_{n\to\infty} \frac{E[ZN(V_n)]}{\lambda(V_n)} \tag{1.74}$$

Furthermore, if N is simple, then

$$\frac{E^*[Z]}{P^*(\Omega)} = \lim_{n\to\infty} E[Z|N(V_n) > 0] \tag{1.75}$$

Proof: a) This part is straightforward, and is omitted.

b) Taking $G(\omega,x) = Z(\theta_{-x}\omega)1(x \in V_n)$ in (1.72) gives

$$E[ZN(V_n)] = E^*[\int_{V_n} Z(\theta_{-x}\omega)\lambda(dx)],$$

from which (1.74) follows by continuity of Z and the dominated convergence theorem. To establish (1.75) we show first that

$$\lim_{n\to\infty} E[N(V_n)1(N(V_n) \geq 2)] = 0 \tag{1.76}$$

Indeed, yet another application of (1.72) verifies that

$$\begin{aligned}\lambda(V_n)^{-1}E[N(V_n)1(N(V_n) \geq 2)] &= \lambda(V_n)^{-1}E^*[\int_{V_n} 1(N(V_n - x) \geq 2)\lambda(dx)] \\ &\leq P^*\{N(V_n - V_n) \geq 2\} \\ &\to P^*\{N(\{0\}) \geq 2\} = 0,\end{aligned}$$

where the last equality holds because N is simple. Therefore,

$$\lim \lambda(V_n)^{-1}\left|E[ZN(V_n)] - E[Z1(N(V_n) > 0)]\right| \to 0; \tag{1.77}$$

for $Z \equiv 1$ we obtain

$$\lim \lambda(V_n)^{-1}P\{N(V_n) > 0\} = P^*(\Omega) \tag{1.78}$$

Finally, (1.75) follows by combining (1.74), (1.77), and (1.78). □

Within the proof of Theorem 1.65—especially in (1.78)—is the microlevel interpretation of the intensity.

Corollary 1.66. (Khinchin-Koroljuk theorem). Let N be a simple, stationary point process with finite intensity ν_N and let (V_n) be as in Theorem 1.65b. Then

$$\lim \lambda(V_n)^{-1}P\{N(V_n) > 0\} = \lim \lambda(V_n)^{-1}P\{N(V_n) = 1\} = \nu_N \quad \square \tag{1.79}$$

In addition, (1.72) can be viewed in terms of marked point processes. Let $N = \Sigma\, \varepsilon_{X_i}$ be stationary and note that for each i the translate $N_{(X_i)} = N\tau_{X_i}^{-1} = \Sigma_j\, \varepsilon_{X_j - X_i}$ has a point at the origin. Form the marked point process, with mark space $\mathbf{M}_p$, in which each point X_i is marked by the translate $N_{(X_i)}$. For $G(\omega,x) = H(N(\omega),x)$, with $H: \mathbf{M}_p \times E \to \mathbf{R}$, $\int H(N,x)N(dx) = \int_{\mathbf{M}_p \times E} H\, d\bar{N}$, and hence (1.72) states that the mean measure of $\bar{N}$ is the product measure $\lambda \times P^*$. This observation is crucial to nonparametric estimation techniques for stationary point processes that we develop in Chapter 9.

Before presenting the global limit interpretation of the intensity we examine ergodicity for stationary point processes. Suppose that N is stationary with respect to (θ_x).

Definition 1.67. An event $\Gamma \in \mathcal{F}$ is *invariant* for (θ_x) if $\theta_x^{-1}(\Gamma) = \Gamma$ for all x. The point process N (more properly, the probability P) is *ergodic* if $P(\Gamma)$ is zero or one for every invariant event Γ.

Denote by $\mathcal{I}$ the σ-algebra of invariant events.

The ergodic theorem leads to the global limit interpretation of the intensity. For it (unlike earlier results) the condition $\lambda(E) = \infty$ is essential.

Theorem 1.68. Let P^* be the Palm measure of the stationary point process N. Then

a) $P^* << P$ on $\mathcal{I}$ and if Y is the $\mathcal{I}$-measurable derivative dP^*/dP, then

$$E[N(f)|\mathcal{I}] = Y\lambda(f) \tag{1.80}$$

for all $f \in C_K^+$.

b) If $f \in C_K^+$, then provided that $\lambda(E) = \infty$,

$$\lim_{x\to\infty} (\lambda\{y: 0 \le y \le x\})^{-1} \int_{\{y:0\le y\le x\}} N\tau_y^{-1}(f)\lambda(dy) = Y\lambda(f) \tag{1.81}$$

almost surely with respect to P and hence almost everywhere with respect to P^*. [In (1.81), inequalities and the limit are meant componentwise.]

Proof: (Sketch). a) Given $\Gamma \in \mathcal{I}$ and $f \ge 0$, (1.72) implies that

$$E[E[N(f)|\mathcal{I}];\Gamma] = E[N(f);\Gamma] = P^*(\Gamma)\lambda(f)$$

Choosing f_0 with $0 < \lambda(f_0) < \infty$ and putting $Y = E[N(f_0)|\mathcal{I}]/\lambda(f_0)$ thus gives a version of dP^*/dP on $\mathcal{I}$, and then (1.80) is a direct computation.

b) Existence of the limit and its being equal to $E[N(f)|\mathcal{I}]$ ensue from general results in ergodic theory (see Dunford and Schwartz, 1958; Matthes et al. 1978; and Theorem 9.2). Identification of the limit is effected using (1.80). □

In particular, if N is ergodic the limit in (1.81) is equal to $P^*(\Gamma)\lambda(f) = \nu_N\lambda(f)$, which provides the promised third interpretation of intensity.

EXERCISES

1.1. a) Let N be a Poisson process with mean measure μ. Establish *conditional uniformity* of N: if $0 < \mu(A) < \infty$, then for each k, $P\{N_A \in \Gamma | N(A) = k\} = P\{\widetilde{N}_k \in \Gamma\}$, where $\widetilde{N}_k$ is the k-sample empirical process on A induced by the probability $\rho(B) = \mu(B \cap A)/\mu(A)$. (Given that N has k points in A, their locations are independent and identically distributed according to ρ.)
b) Show that if $\mu(E) < \infty$, then the Poisson process with mean measure μ is a mixed empirical process $N = \Sigma_{i=1}^{K} \varepsilon_{X_i}$, where the X_i are i.i.d. with distribution $\rho(B) = \mu(B)/\mu(E)$ as above and the random variable K is independent of the X_i and Poisson distributed with mean $\mu(E)$.

1.2. Let N be a Poisson process with mean measure μ. Prove that N is simple if and only if μ is diffuse.

1.3. Let N be a mixed Poisson process on E with diffuse mean measure μ. Prove that N is *symmetrically distributed* with respect to μ: whenever $A_1, \ldots, A_k \in \mathcal{B}$ are disjoint with $\mu(A_1) = \cdots = \mu(A_k) < \infty$, the random variables $N(A_1), \ldots, N(A_k)$ are interchangeable.

1.4. Let N be a Cox process on $\mathbf{R}_+$ directed by the diffuse random measure M, and let $N_t = N([0,t])$, with M_t defined analogously. Assume that $M_\infty = \infty$ almost surely, and let $A_s = \inf\{t : M_t > s\}$, $s \in \mathbf{R}_+$, and define a point process $\bar{N}$ by $\bar{N}_s = N(A_s)$. Prove that $\bar{N}$ is an ordinary Poisson process with rate 1.

1.5. Let $N = \Sigma_{i=1}^{\infty} \varepsilon_{T_i}$ be a renewal process on $\mathbf{R}_+$ with interarrival distribution F satisfying $F(0) = 0$ (which implies that N is simple).
a) Let $\widetilde{N}_t = \Sigma_{i=0}^{\infty} 1(T_i \le t)$, where $T_0 = 0$. Show that the mean measure of $\widetilde{N}$ is the *renewal measure* $R = \sum_{n=0}^{\infty} F^n$, where F^n is the n-fold convolution of F with itself and $F^0 = \varepsilon_0$.
b) Recalling that $N_t = \Sigma_{i=1}^{\infty} 1(T_i \le t)$, show that for $f \ge 0$ the function

$$g(t) = L_N(f(t + \cdot)) = E\left[\exp\left(-\sum_{i=1}^{\infty} f(t + T_i)\right)\right]$$

satisfies the equation $g(t) = \int_0^\infty g(t + x)e^{-f(t+x)}F(dx)$.
c) Given $A \in \mathcal{B}$, derive a similar equation for $z_N(A + t)$.

1.6. Let N be a simple point process on E such that $N(A)$ is Poisson distributed for every $A \in \mathcal{B}$. Prove that N is a Poisson process.

1.7. Let M_1, M_2 be diffuse random measures on E. Prove that if there exists $t \in (0,\infty)$ such that $E[e^{-tM_1(A)}] = E[e^{-tM_2(A)}]$ for all $A \in \mathcal{E}$, then $M_1 \overset{d}{=} M_2$. [*Hint*: Use Cox processes.]

1.8. Let N be a point process on E. Prove that the zero-probability functional z_n is *completely monotone* in the following sense. For a mapping Ψ of $\mathcal{B}$ into $\mathbf{R}$ and $A \in \mathcal{B}$ define $\Delta_A \Psi(B) = \Psi(A \cup B) - \Psi(B)$ and define iterates $\Delta_{A_1} \cdots \Delta_{A_n} \Psi$ by recursion. Then Ψ is completely monotone if $(-1)^n \Delta_{A_1} \ldots \Delta_{A_n} \cdot \Psi(B) \geqslant 0$ for all choices of n, $A_1, \ldots, A_n$, and B.

1.9. Let N be the n-sample empirical process engendered by a probability ρ on E. Prove that $\mathcal{L}^N$ depends continuously on ρ. Formulate and prove an analogous assertion for mixed empirical processes.

1.10. Let (N_n) be a sequence of point processes on E, all with mean measure $\mu \in \mathbf{M}$. Prove that (N_n) is tight.

1.11. Let $N = \Sigma\, \varepsilon_{X_i}$ be a Poisson process on E with mean measure μ and let K be a transition probability from $(E,\mathcal{E})$ to an LCCB space $(E',\mathcal{E}')$.

a) Prove that (the underlying probability space can if necessary be enlarged in such a manner that) there exist random elements Y_i of E' that are conditionally independent given N with $P\{Y_i \in B|N\} = K(X_i,B)$, $B \in \mathcal{E}'$, for each i.

b) Prove that $\bar{N} = \Sigma\, \varepsilon_{(X_i,Y_i)}$ is a Poisson process on $E \times E'$ with mean measure $\mu(dx)K(x,dy)$.

1.12. This exercise applies the construction in Exercise 1.11 to a problem in hydrology. Let $E = E' = \mathbf{R}_+$, let N represent time of flood peak exceedances in a river, and let Y_i be the height above the base flow of the flood peak occurring at time T_i. (An "exceedance" is a local maximum in the flow; between successive exceedances the flow must drop below the base level.) Assume that $N = \Sigma\, \varepsilon_{T_i}$ and (Y_i) satisfy the hypotheses of Exercise 1.11. For $A \in \mathcal{B}(\mathbf{R}_+)$ let $M(A) = \sup\{Y_i : T_i \in A\}$ be the maximum flood peak occurring in the time set A.

a) Prove that for each A and t, $P\{M(A) \leq t\} = \exp[-\int_A \mu(dx)K(x,(t,\infty))]$.

b) Prove that the stochastic process $M_t = M([0,t])$ is a (nonhomogeneous) strong Markov process.

1.13. Let N be a Cox process directed by the random measure M. Prove that if M is infinitely divisible as a random measure, then N is infinitely divisible as a point process.

1.14. Verify (1.33) – (1.35).

1.15. Let $M = \Sigma\, U_i \varepsilon_{X_i}$ be the classical ρ-compound of a point process $N = \Sigma\, \varepsilon_{X_i}$ on E (Definition 1.37).

a) Calculate the Laplace functional and mean measure of M.

b) Prove that $\mathcal{L}_M$ depends continuously on $\mathcal{L}_N$ and ρ.

c) Calculate the Laplace functional, mean measure, and zero-probability functional of the marked point process $\bar{N} = \Sigma\, \varepsilon_{(X_i,U_i)}$.

1.16. a) Let N be a Poisson process on E with mean measure μ and let N' be a p-thinning of N, where $p: E \to [0,1]$ (Definition 1.38). Prove that N' and $N'' = N - N'$ are independent Poisson processes, and determine their mean measures.

b) Let N be Poisson and let N' be a point process on E satisfying $N'(A) \leq N(A)$ for each A. Prove that if N' and $N'' = N - N'$ are independent, then N' is a p-thinning of N, with $p(x) = (d\mu_{N'}/d\mu_N)(x)$.

1.17. Let N_1, N_2 be independent point processes on E. Prove that the superposition $N_1 + N_2$ is simple if and only if N_1 and N_2 are simple and $A_1 \cap A_2 = \emptyset$, where A_i is the set of fixed atoms of N_i [i.e., points x such that $P\{N_i(\{x\}) > 0\} > 0$].

1.18. a) Let N_1, N_2 be independent point processes on E whose superposition $N_1 + N_2$ is a Poisson process. Prove that N_1 and N_2 are Poisson. [*Hint*: Use Raikov's theorem and Exercise 1.6.]

b) Construct an example of Poisson processes N_1, N_2 that are *not* independent but for which $N_1 + N_2$ is Poisson notwithstanding.

1.19. This exercise describes a Poisson model of light traffic on an infinite one-dimensional road. Let X_i be the position in $\mathbf{R}$ at $t = 0$ of the ith vehicle and let V_i be its velocity, which is random but does not vary over time. Assume that $N = \Sigma\, \varepsilon_{X_i}$ is a homogeneous Poisson process with rate 1 and that the V_i are independent of N and i.i.d. with distribution G satisfying $G(\{0\}) = 0$. ("Light traffic" means that the assumption that the velocities V_i be independent is plausible.)

a) Prove that $\bar{N} = \Sigma\, \varepsilon_{(X_i, V_i)}$ is Poisson on $\mathbf{R}^2$ and determine its mean measure.

b) Prove that for each $t \in \mathbf{R}$, the point process $N^t = \Sigma\, \varepsilon_{X_i + tV_i}$, which gives vehicle positions at time t, is Poisson with rate 1.

c) For $y \in \mathbf{R}$ let $T_i(y) = (y - X_i)/V_i$ be the time (possibly negative) at which vehicle i passes the point y. Then $N^+(y) = \Sigma\, 1(V_i > 0)\varepsilon_{T_i(y)}$ gives times at which vehicles moving to the right pass y, and $N^-(y) = \Sigma\, 1(V_i < 0)\varepsilon_{T_i(y)}$ describes left-moving vehicles. Prove that for each y, $N^+(y)$ and $N^-(y)$ are independent Poisson processes and calculate their mean measures.

1.20. This exercise presents some properties of the queueing system $M/G/\infty$. Let $E = \mathbf{R}$ and let $N = \Sigma\, \varepsilon_{T_i}$ be a homogeneous Poisson process with rate 1; points of N are arrival times to a queueing system with infinitely many servers. (Taking $E = \mathbf{R}$ removes transient effects.) Let Y_i be the service time of the customer arriving at time T_i; there is no wait for service, so this customer departs at time $T_i + Y_i$. Assume that the Y_i are independent of N and i.i.d. with distribution G.

a) Prove that the departure process $N^* = \Sigma\, \varepsilon_{T_i + Y_i}$ is Poisson with rate 1. (That this is true regardless of G implies that observation of N^* alone yields no information concerning G.)

b) Suppose that N is an arbitrary point process but that the Y_i and N^* are as above and that N^* is a Poisson process with rate 1. Prove or refute the assertion that N must also be Poisson.

1.21. Provide the details needed to verify (1.60).

1.22. Let N be a point process on E and put $p = P\{N \neq 0\}$, $F(A) = P\{N(A) = 1 | N(E) = 1\}$. Let N' be Poisson with mean measure $\mu = [-\log(1-p)]F$. Prove that there exists a coupling (N, N') satisfying $P\{N \neq N'\} = \max\{P\{N(E) \geq 2\}, P\{N'(E) \geq 2\}\}$. (This result can be used in conjunction with results in Section 1.6 to provide Poisson approximations to non-Bernoulli point processes.)

1.23. Let N be the k-sample empirical process engendered by a probability ρ on E. Calculate the reduced Palm distributions of N.

1.24. Let M be the random measure $M(dx) = Y\,dx$ on $\mathbf{R}$, where Y is a positive random variable and dx is Lebesgue measure. Determine the Palm distributions of M.

1.25. Let N be a simple point process on E with Palm processes N_μ, $\mu \in \mathbf{M}_s$. Prove that for μ, $\nu \in \mathbf{M}_s$, $N_{\mu+\nu} = (N_\mu)_\nu$.

1.26. Establish the assertions in Example 1.60.

1.27. Prove (1.70).

1.28. Let N be a simple, stationary point process on $\mathbf{R}$. Prove that $P\{N = 0\} + P\{N(-\infty,0) = N(0,\infty) = \infty\} = 1$.

1.29. Let N be a stationary point process on $\mathbf{R}$ with finite intensity and let P^* be its Palm measure. Define *Palm functions* $p_k^*(t) = P^*\{N(0,t) = k\}$.

a) Prove that the *Palm-Khinchin equations*

$$P\{N(0,t) > k\} = \int_0^t p_k^*(s)\,ds, \qquad k \geq 0,$$

are fulfilled. [*Hint*: Verify and use the identity

$$1(N(0,t) > k) = \int_{(0,t)} 1(N(x,t) = k)N(dx)]$$

b) Show that if N is a homogeneous Poisson process the Palm-Khinchin equations are the Kolmogorov forward equations for the Markov counting process $N_t = N(0,t]$.

1.30. Show that the continuity assumption in Theorem 1.65b cannot be suppressed. [*Hint*: Use the stationary point process of Example 1.60.]

1.31. This exercise concerns stationary renewal processes. Let F be a probability distribution on $(0,\infty)$ with finite mean m and let G be the distribution $G(t) = m^{-1}\int_0^t [1 - F(u)]\,du$. Let $N = \Sigma_{i=-\infty}^{\infty} \varepsilon_{T_i}$ be the point process constructed as follows. (Recall that $U_i = T_i - T_{i-1}$ are the interarrival times.)

i) $\cdots < T_0 \leq 0 < T_1 < \cdots$;
ii) $(-T_0, U_1)$ and U_i, $i \neq 1$, are independent;
iii) The U_i, $i \neq 1$, have distribution F;
iv) $(-T_0, U_1)$ has distribution $m^{-1}1(u < v)\,du\,F(dv)$.

The point process N is the *stationary renewal process* induced by F.

a) Prove that N is stationary.
b) Prove that $-T_0$ and T_1 both have distribution G.
c) Prove that U_1 has distribution $H(t) = m^{-1}\int_0^t vF(dv)$.
d) Prove that $P^*(\Omega) = m^{-1}$, where P^* is the Palm measure for N.
e) Prove that under the Palm distribution $Q_N^*(\varepsilon_0, \cdot) = P^*(\cdot)/P^*(\Omega)$, the U_i are i.i.d. with distribution F.

1.32. Let $N = \Sigma_{n=-\infty}^{\infty} \varepsilon_{T_n}$ be a simple, stationary point process on $\mathbf{R}$ with finite intensity and Palm measure P^*.

a) Prove that $P^*\{T_0 \neq 0\} = 0$.
b) Prove that N is simple if and only if $P^*\{N(\{0\}) \neq 1\} = 0$.
c) Let $F^*(\cdot) = P^*\{T_1 \in (\cdot)\}$. Prove that $\int vF^*(dv) = 1$.
d) Prove that $P\{-T_0 \in du,\ U_1 \in dv\} = 1(0 < u < v)\, du\, F^*(dv)$.

1.33. Let N be a Cox process on $\mathbf{R}^d$ with directing random measure $M(dx) = X(x)\lambda(dx)$, where $(X(x))$ is a positive, measurable stochastic process parametrized by E and λ is Lebesgue measure. (In Chapter 7, X is termed the directing intensity of N.) Prove that N is stationary if and only if X is strictly stationary.

NOTES

General sources on the distributional—as opposed to martingale—theory of point processes are Cox and Isham (1980), at a relatively elementary level; Kallenberg (1983), measure-theoretic and complete, albeit often terse; Matthes et al. (1978), voluminous but very difficult reading; and Snyder (1975), which, though restricted to point processes on **R**, contains extensive consideration of engineering applications.

Section 1.1

The material here is standard. A proof of (1.1) is given in Kallenberg (1983). Poisson processes are in many senses the most fundamental point processes. On the line the history is lengthy, but that of Poisson processes on general spaces is briefer; two early sources are Doob (1953) and Dobrushin (1956). Cox processes are first treated in Lundberg (1940) and Cox (1955); Grandell (1976) and Snyder (1975) are more modern expositions focusing on state estimation, a topic treated in detail in Section 7.2. Empirical processes, whose role is fundamental, are the subject of Gaenssler (1984) (see also Kallenberg, 1983). Feller (1971) is the definitive reference on much of renewal theory but does not include recent developments; see Chapter 8 for additional sources as well as a summary of key components of the theory.

Section 1.2

Uniqueness results are not given here in the sharpest form: the class of sets of functions for which equalities in Theorem 1.12, for example, must be verified, need not be as large as stated. Although the added generality is not purely formal, we do not need it. "Best" versions of these and related results, including a proof of Proposition 1.17, are in Kallenberg (1983). A characteristic functional for random measures, with definition and properties analogous to those of the Laplace functional, is analyzed in von Waldenfels (1968).

Section 1.3

Kallenberg (1983) and Matthes et al. (1978) contain more detailed and complete presentations of convergence in distribution for point processes and (in the former) random measures. Lemma 1.20 is exceedingly important: it renders tightness (typically the most difficult step in proving convergence in distribution) entirely routine. Proposition 1.23 is proved in Kallenberg (1983).

Section 1.4

Marked point processes are introduced in Matthes (1963a); see Matthes et al. (1978) and Franken et al. (1981) for expository treatments. The former contains detailed discussion of clustered point processes. Matthes (1963a) first examined infinitely divisible point processes; P. M. Lee (1967) is another early reference. Despite its being difficult, Matthes et al. (1978) is *the* reference on infinitely divisible point processes. The Poisson cluster representation in Theorem 1.32 is due in stages to P. M. Lee (1967), Jagers (1972), and Matthes et al. (1978). Random measures with independent increments were first investigated, under the rubric "completely random measures," by Kingman (1967).

Section 1.5

Thinned and compound Poisson processes have been studied at least since the 1940s. Jagers and Lindvall (1974) and, especially, Kallenberg (1975a) formalized the terminology and established characterizations via limits of rarefied processes (an analog of Theorem 1.40 holds for compound point processes).The two characterizations of Cox processes (Theorems 1.39 and 1.40) are due to Mecke (1968) and Kallenberg (1975b), respectively. Theorem 1.44 comes originally from Kerstan and Matthes (1964a).

Section 1.6

Coupling ideas are not new, of course, even in the context of point processes, but the Poisson/Bernoulli coupling appears here for the first time, as does the embedding scheme in Theorem 1.47; they were derived jointly with R. J. Serfling and are based on conditional uniformity of Poisson processes together with coupling techniques for Poisson and Bernoulli random variables introduced by Serfling (1978). Poisson approximations of empirical processes are discussed in Gaenssler (1984). An important alternative methodology, based on martingales and stochastic intensities, for Poisson approximation of rather arbitrary point processes on **R** is discussed in Section 2.3.

Section 1.7

Ideas identifiable as pertaining to what are now termed Palm distributions are indeed found in Palm (1943), but Khinchin (1960) is the first rigorous and systematic investigation; Ryll-Nardzewski (1961) is another principal early work. Extensions to more general spaces and processes were accomplished by Jagers (1973) and Papangelou (1974b). A complete, up-to-date development is presented in Kallenberg (1983), where Palm distributions are linked with concepts from statistical physics such as Gibbs distributions (see Section 2.6 for some further discussion). As mentioned in the text, except in the context of stationary point processes (see

Definition 1.62), unreduced Palm distributions for point processes are no longer of serious interest. Proposition 1.55 is given for $\mu = \varepsilon_x$ in the unreduced case in Grandell (1976) and for general μ in the reduced case by Karr (1985a). The mixed Poisson characterization of Proposition 1.57 comes from Kallenberg (1983).

Section 1.8

The earliest general treatment of stationary point processes is Khinchin (1960); Matthes et al. (1978), and Neveu (1977) are recent expository presentations. Much of the general theory (for spaces other than **R**) was given first in Matthes (1963a), Kerstan and Matthes (1964a), and Franken et al. (1965), Proposition 1.63 is due to Ryll-Nardzewski (1961) and Slivnjak (1962). Orderliness and regularity are examined in Leadbetter (1968, 1972a), Daley and Vere-Jones (1972), and Daley (1974). The result now known as Dobrushin's lemma—that a stationary, orderly point process [in the sense that $t^{-1}P\{N(0,t) > 2\} \to 0$ as $t \to 0$, a weakened form of (1.78)] is simple—is attributed to Dobrushin by Volkonskii (1960). Koroljuk's theorem, which identifies the limit in (1.79), is so-named in Khinchin (1960), but seems never to have been published under Koroljuk's name. Khinchin's contribution to Corollary 1.66 is existence of the limit. Elaboration of ergodicity for stationary point processes can be found in Matthes et al. (1978) and Neveu (1977); see also Chapter 9, which contains (in Theorem 9.2) a rather broad generalization of Theorem 1.68.

Exercises

1.1. A systematic analysis of conditional uniformity occupies several sections of Matthes et al. (1978). Conditional uniformity characterizes mixed Poisson processes among Cox processes.

1.3. Kallenberg (1975b) shows that this interchangeability property also characterizes mixed Poisson processes and mixed empirical processes. His definition of "symmetrically distributed" is different from ours but equivalent.

1.4. See Serfozo (1977) for general discussion of random time changes for random measures.

1.6. This result is due to Rényi (1967).

1.7. Kallenberg (1983) employes this device systematically; see also the discussion following Proposition 1.23.

1.8. For the converse (that, given minor additional assumptions, a completely monotone mapping Ψ with $\Psi(\varnothing) = 1$ is the zero-probability functional of some point process), see Kurtz (1974).

1.12. The model and results are due to Karr (1976a).

1.13. The converse is false even when E is a singleton.

1.16. See Karr (1985b) for an extension to arbitrary point processes.

1.17. Although intuitive this property is nontrivial (see Franken et al, 1981, for a proof).

1.18. Çinlar (1972) presents this and many related results. In connection with b), see Jacod (1975a).

1.19. The model is examined in Rényi (1964a) and Solomon and Wang (1972). Limit versions of these invariance properties are proved in Breiman (1963) and Stone (1968a).

1.20. This queue, of course, is classical; (see, e.g., Franken et al, 1981, for expository treatment). M. Brown (1970) and Thédeen (1964, 1967a) consider related properties and the extent to which they characterize Poisson processes (see also exercise 6.24). More on departure processes, which have been an important motivation for study of non-Poisson point processes, is given in Brémaud (1981) and Franken et al. (1981).

1.21. See Dudley (1972), Gaenssler (1984), and Whitt (1973).

1.22. This is an unpublished result of Karr/Serfling.

1.28. Like Exercise 1.17, this property is intuitive but not trivial; it is due to Ryll-Nardzewski (1961). A more general zero-infinity law is given in Proposition 9.1.

1.29. The equations are due to Khinchin (1960); the suggested proof, from Neveu (1977), typifies the elegance of the development there. For additional material, see Cox and Isham (1980), Matthes et al. (1978), and Neveu (1977).

1.31. See Neveu (1977) for more on stationary renewal processes. Further aspects, including inference, are treated in Chapter 9.

1.32. Note the analogy to Exercise 1.31.

1.33. See also Theorem 9.18.

2

Point Processes: Intensity Theory

The compensator of a point process N on $\mathbf{R}_+$ is a random measure A interpretible as the local conditional mean of N given its strict past (together perhaps with some additional information): informally, $A(dt) = E[N(dt)|\mathcal{F}^N([0,t))]$. The compensator is ordinarily smoother than N and in nice cases $A_t = \int_0^t \lambda_u \, du$ for some random process λ, the *stochastic intensity* of N. In this chapter we develop the theory of compensators and stochastic intensities for point processes on $\mathbf{R}_+$. As will emerge, this theory is linked intimately to the modern theory of martingales, because the difference between a point process and its compensator is a (mean zero) martingale, to which various interpretations can be assigned. In information terms it is an innovation process conveying information about N that is not expressible from the strict past. In statistical settings it is, by contrast, a noise process (akin, by virtue of its being orthogonal to the past, to the residuals in a regression problem) representing aspects that somehow cannot be estimated. A rather general theory of inference for point processes in which the statistical model is specified via the stochastic intensity is described in Chapter 5. In this setting martingales serve a fundamental role in computations of means, variances, and covariances, and as the basis for important central limit theorems. Applications to specific classes of point processes appear in Chapters 7, 8, and 9.

For point processes on general spaces the whole topic of conditioning is *much* more difficult, principally because of the absence of a linear order defining a natural way in which the process is observed. Notwithstanding, useful results do exist; the objects of central importance are the Palm distributions introduced in Section 1.7. In addition, we treat the theory of exvisible projections, which can be developed along lines parallel to the development of compensators.

The latter topic constitutes Section 2.6; the remainder is organized as follows. Section 2.1, after setting notation and terminology, portrays rather precisely the structure of the internal history of a marked point process on $\mathbf{R}_+$. Relevant stochastic Stieltjes integrals are treated briefly in Section 2.2. In Section 2.3 we define compensators of point processes and marked point processes and establish important properties. These include not only characterizations and explicit constructions but also applications to convergence in distribution and approximation of point processes. When the compensator is absolutely continuous, the derivative is a random process called the stochastic intensity; this special, especially important, and (fortunately) rather frequent case is the topic of Section 2.4. Only one result appears in Section 2.5: every $\mathcal{F}^N$-martingale is the integral of a predictable process with respect to the innovation martingale $M = N - A$; this theorem confirms rather convincingly the innovation interpretation of M and is a key result in the book.

2.1 THE INTERNAL HISTORY OF A MARKED POINT PROCESS

Let $N = \Sigma_{n=1}^{\infty} \varepsilon_{(T_n, Z_n)}$ be a marked point process on $\mathbf{R}_+$ with mark space $(E, \mathcal{E})$. The T_n are times of occurrences of events: the nth event occurs at T_n and has mark $Z_n \in E$. The interarrival times are denoted by $U_n = T_n - T_{n-1}$, with $U_1 = T_1$. Throughout the chapter we assume that the point process $\Sigma_{n=1}^{\infty} \varepsilon_{T_n}$ on $\mathbf{R}_+$ is simple. For $B \in \mathcal{E}$ let $N_t(B) = \Sigma_{n=1}^{\infty} 1(T_n \leq t, Z_n \in B)$, with $N_t = N_t(E)$. Here and elsewhere, when convenient we switch freely between measure and distribution function notation. As in Section 1.1, define

$$\mathcal{F}_t^N = \sigma(N_s(B) : 0 \leq s \leq t, B \in \mathcal{E}) \tag{2.1}$$

and further define

$$\widetilde{\mathcal{F}}_t^N = \mathcal{F}_t^N \vee \mathcal{G}, \tag{2.2}$$

where $\mathcal{G}$ is the family of all P-null sets engendered by $\mathcal{F}_\infty^N$. Our principal goal in this section is to describe the structure of $\mathcal{F}^N = (\mathcal{F}_t^N)$, virtually all of which is inherited by $(\widetilde{\mathcal{F}}_t^N)$; the former is the real object of interest, but sometimes it is necessary for technical reasons to work with the latter. We term each the *internal history* of N.

We digress briefly to recall some ideas from the *"théorie générale des processus."* Although phrased in terms of $\mathcal{F}^N$ for concreteness, they apply equally to any history. There are two key concepts, stopping times and various forms of measurability for stochastic processes.

A random variable T taking values in $[0,\infty]$ is a *stopping time* if $\{T \leq t\} \in \mathcal{F}_t^N$ for every t; that is, whether T has occurred before t can be determined by observation of N over $[0,t]$. Associated with the stopping time T are the σ-algebra $\mathcal{F}_T^N = \{\Lambda \in \mathcal{F}_\infty^N : \Lambda \cap \{T \leq t\} \in \mathcal{F}_t^N \text{ for all } t\}$, which represents evolution of N until the random time T (see Proposition 2.3 for a more precise statement) and the σ-algebra $\mathcal{F}_{T-}^N = \sigma(\Lambda \cap \{T > t\} : t \geq 0, \Lambda \in \mathcal{F}_t^N) \vee \mathcal{F}_0^N$. Note that only generators of $\mathcal{F}_{T-}^N$ are specified. The latter comprises the *strict past* of N at T in much the same way that $\mathcal{F}_{t-}^N = \mathcal{F}^N([0,t))$ is the strict past at the deterministic time t. We have $\mathcal{F}_{T-}^N \subset \mathcal{F}_T^N$ and $T \in \mathcal{F}_{T-}^N$. (Here and throughout we write that a random variable belongs to a σ-algebra to mean that it is measurable.)

Suppose now that $X = (X_t)$ is a random process. The least restrictive measurability relationship between X and $\mathcal{F}^N$ is that as a mapping on $\mathbf{R}_+ \times \Omega$, X be *measurable* with respect to $\mathcal{B}(\mathbf{R}_+) \times \mathcal{F}_\infty^N$ and that X be *adapted* to $\mathcal{F}^N$ in the sense that $X_t \in \mathcal{F}_t^N$ for each t. These minimal assumptions are too weak in most instances; they fail, for example, to imply that the process $Y_t = \int_0^t X_u du$ (let alone more complicated processes) even be adapted. A stronger condition resolves this difficulty: X is *progressive* (sometimes, *progressively measurable*) if for every t the mapping $(u,\omega) \to X_u(\omega)$ on $[0,t] \times \Omega$ is measurable with respect to $\mathcal{B}([0,t]) \times \mathcal{F}_t^N$. Evidently, if X is progressive, then so is the process $\int_0^t X_u \, du$. Every left- or right-continuous process is progressive (Exercise 1.2).

Predictability is a fundamental concept. In applications such as state estimation and martingale inference it is essential to distinguish processes whose value for every t is determined by $\mathcal{F}_{t-}^N$ from those for which less-than-certain prediction of the current state from the strict past is possible. After some terminology this leads to the class of predictable processes. The *predictable σ-algebra* $\mathcal{P}$ generated by N is the σ-algebra on $\mathbf{R}_+ \times \Omega$ generated by the π-system of sets

$$(s,t] \times \Lambda, \tag{2.3}$$

where $s < t$ and $\Lambda \in \mathcal{F}_s^N$. Other characterizations include:

a) $\mathcal{P} = \sigma((s,\infty) \times \Lambda : s \geq 0, \Lambda \in \mathcal{F}_s^N)$;
b) $\mathcal{P} = \sigma(\{(t,\omega) : t \leq T(\omega)\} : T$ is a bounded stopping time);
c) $\mathcal{P} = \sigma(X : X$ is adapted and left-continuous);
d) $\mathcal{P} = \sigma(X : X$ is adapted and continuous).

A process X is *predictable* if $X \in \mathcal{P}$ as a function on $\mathbf{R}_+ \times \Omega$; every predictable process is progressive. Justification of the term resides in the property that if X is predictable and T is a finite stopping time, then X_T is $\mathcal{F}_{T-}^N$-measurable. (If X is progressive, then $X_T \in \mathcal{F}_T^N$ but no stronger statement holds in general.) For practical purposes predictable may be taken to be synonymous with *left-continuous* and adapted.

We now describe the structure of the internal history $\mathcal{F}^N$. The heuristic description is simple: the accumulating information represented by $\mathcal{F}^N$ changes only at the arrival times T_n, and whereas $T_n \in \mathcal{F}_{T_n-}^N$, the mark Z_n

belongs to $\mathcal{F}^N_{T_n}$ but not to $\mathcal{F}^N_{T_n-}$. Finally, Z_n is the *only* information in $\mathcal{F}^N_{T_n}$ not already in $\mathcal{F}^N_{T_n-}$.

Here are the most elementary properties.

Proposition 2.1. a) For each t,

$$\mathcal{F}^N_t = \sigma(1(T_n \le s, Z_n \in B) : n \ge 1,\ 0 \le s \le t,\ B \in \mathcal{E}) \tag{2.4}$$

b) The history $\mathcal{F}^N$ is right-continuous: for every t,

$$\mathcal{F}^N_t = \bigcap_{\varepsilon > 0} \mathcal{F}^N_{t+\varepsilon} \quad \square \tag{2.5}$$

We next note properties of $\mathcal{F}^N$-stopping times.

Proposition 2.2. a) For each n, T_n is a stopping time of $\mathcal{F}^N$;
b) For each n,

$$\mathcal{F}^N_{T_n} = \sigma((T_1,Z_1), \dots, (T_n,Z_n)), \tag{2.6}$$

while

$$\mathcal{F}^N_{T_n-} = \sigma((T_1,Z_1), \dots, (T_{n-1},Z_{n-1}),T_n) \tag{2.7}$$

The proof is deferred until after the next result, which generalizes (2.4) and is a key property.

Proposition 2.3. If T is a stopping time of $\mathcal{F}^N$, then

$$\mathcal{F}^N_T = \sigma(N_{s \wedge T}(B) : s \ge 0,\ B \in \mathcal{E}) \tag{2.8}$$

Proof: (Sketch). That $\supset$ holds in (2.8) is clear, and we may prove the opposite inclusion under the restriction that T assumes only countably many values $0 \le t_0 < t_1 < \cdots \le \infty$. Let t be fixed and let $\Lambda \in \mathcal{F}^N_t$ be partitioned as $\Lambda = \Sigma_{k=0}^{\infty} \Lambda \cap \{T = t_k\}$; since $\Lambda_k = \Lambda \cap \{T = t_k\}$ belongs to $\mathcal{F}^N_{t_k}$ there exist a countable dense subset V of $\mathbf{R}_+$ and a function f_k such that

$$1_{\Lambda_k} = f_k(N_t(B) : t \in V \cap [0,t_k],\ B \in \mathcal{E}) \tag{2.9}$$

But $t \wedge T = t$ on Λ_k whenever $t \le t_k$, and this converts (2.9) to $1_{\Lambda_k} = f_k\,(N_{t \wedge T}(B) : t \in V \cap [0,t_k],\ B \in \mathcal{E})$. $\square$

The proof of (2.6) proceeds as follows. To prove $\subset$, it suffices by (2.8) to observe that $N_{t \wedge T_n}(B) \in \sigma((T_1,Z_1), \dots, (T_n,Z_n))$ for all n, t, and B, which is obvious, as is $\supset$ in (2.6). For (2.7), $\supset$ holds because $T_n \in \mathcal{F}^N_{T_n}$ and for $\subset$ by the monotone class theorem we need only show that generators of $\mathcal{F}^N_{T_n}$ of the form $\{N_s(B) = k, t < T_n\}$ ($s \leqslant t$, $B \in \mathcal{E}$) belong to $\sigma((T_1, Z_1), \ldots, (T_{n-1}, Z_{n-1}), T_n)$, which follows immediately from

$$\{N_s(B) = k, t < T_n\} = \{t < T_n\} \cap \left\{\left[\sum_{i=1}^{n} 1(T_i \leq s, Z_i \in B)\right] = k\right\}$$

Two additional results elucidate further properties of stopping times.

Proposition 2.4. Given a finite stopping time T of $\mathcal{F}^N$, for each n,

$$\mathcal{F}_T^N \cap \{T_n \leq T < T_{n+1}\} = \mathcal{F}_{T_n}^N \cap \{ \leqslant \mathrm{T} < \mathrm{T}_{n+1}\} \qquad (2.10)$$

Proof: For each t, B and k,

$$\{N_{t \wedge T}(B) = k\} \cap \{T_n \leq T < T_{n+1}\} = \{N_{t \wedge T_n}(B) = k\} \cap \{T_n \leq T < T_{n+1}\},$$

and (2.10) follows by two applications of (2.8) and the monotone class theorem. □

Proposition 2.5. Let T be a stopping time of $\mathcal{F}^N$. For each n there is a random variable $U_n \in \mathcal{F}_{T_n}^N$ such that on $\{T_n \leq T\}$,

$$T \wedge T_{n+1} = (T_n + U_n) \wedge T_{n+1} \qquad (2.11)$$

Proof: Propositions 2.2 and 2.4 entail existence of functions g_n satisfying $T1(T_n \leq T < T_{n+1}) = g_n(T_1, Z_1, \ldots, T_n, Z_n)1(T_n \leq T < T_{n+1})$, so we may take $U_n = [g_n(T_1, Z_1, \ldots, T_n, Z_n) - T_n]^+$. □

As will be seen in the proofs of Theorems 2.18 and 2.34, the point of (2.11) is that $U_n \in \mathcal{F}_{T_n}^N$.

We conclude the section with two constructions of predictable processes with particular properties.

Proposition 2.6. For each n let Y_n be a $\mathcal{B}(\mathbf{R}_+) \times \mathcal{F}_{T_n}^N$-measurable mapping on $\mathbf{R}_+ \times \Omega$. Then the process $X_t = Y_n(t)$, $t \in (T_n, T_{n+1}]$, is predictable.

Proof: The processes $1(T_n < t \leq T_{n+1})$ are predictable, so X may be treated "piecewise." By the monotone class theorem we may suppose that $Y_n(t, \omega) = 1(t \in V)1(\omega \in \Lambda)$ for $V \in \mathcal{B}(\mathbf{R}_+)$ and $\Lambda \in \mathcal{F}_{T_n}^N$, in which case

$$Y_n(t)1(T_n < t \leq T_{n+1}) = 1(t \in V)1(\widetilde{T}_n < t \leq \widetilde{T}_{n+1}), \qquad (2.12)$$

where $\widetilde{T}_n = T_n$ on Λ and is infinite on Λ^c; the right-hand side of (2.12) is evidently predictable. □

Proposition 2.7. Given $\Lambda \in \mathcal{F}_{T_n}^N$, there is a predictable process $X \geq 0$ such that $X_t = 0$ unless $t \in (T_{n-1}, T_n]$ and such that $X_{T_n} = 1_\Lambda$. □

2.2 STOCHASTIC INTEGRATION

Here we summarize some ideas associated with stochastic integration that are needed in the book, but with no pretense of completeness. In particular, all integrals are simply stochastic Stieltjes integrals, because the integrators are either increasing processes (e.g., point processes) or differences of increasing processes (with, therefore, paths of bounded variation over compact intervals). Specifically, the latter are usually innovation martingales. A central idea is the intimate relationship among martingales, predictable processes, and integrals with respect to martingales.

Suppose that $(\Omega,\mathcal{F},P)$ is a probability space, that $Y=(Y_t)$ is a measurable stochastic process with sample paths of locally bounded variation, and that X is a measurable process whose sample paths are locally bounded. Then the process $X*Y$ defined by

$$(X*Y)_t = \int_0^t X_s\,dY_s \tag{2.13}$$

is called the *stochastic Stieltjes integral* of X with respect to Y. In (2.13) and elsewhere the interval of integration is $(0,t]$; that the right-hand endpoint is included is crucial. For each t and ω the integral in (2.13) exists as an ordinary Lebesgue-Stieltjes integral and for each t, $(X*Y)_t$ is a random variable. However, as hinted in Section 2.1, further measurability typically is needed; usually, we assume that X is predictable and Y is progressive, in which case $X*Y$ is progressive.

The definition (2.13) remains meaningful if Y is an increasing process [i.e., $t \to Y_t(\omega)$ is increasing for each ω] and X is nonnegative. In particular, if $N=\Sigma\,\varepsilon_{T_n}$ is a point process on $\mathbf{R}_+$, then for X measurable and either locally bounded or nonnegative, the stochastic integral $X*N$ exists and for each t,

$$(X*N)_t = \sum_{n=1}^{\infty} 1(T_n \le t)X_{T_n} \tag{2.14}$$

In view of (2.13) properties of $X*Y$ mirror corresponding properties of Lebesgue-Stieltjes integrals and do not require special mention. For the sake of completeness, however, we do include two formulas for integration by parts, to be used later in the chapter.

Proposition 2.8. Let a and b be functions of locally bounded variation on $\mathbf{R}_+$, each right-continuous with left-hand limits. Then for each t,

$$a(t)b(t) - a(0)b(0) = \int_0^t a(s^-)\,db(s) + \int_0^t b(s)\,da(s) \tag{2.15}$$

and

$$a(t)b(t) - a(0)b(0) = \int_0^t a(s^-)\,db(s) + \int_0^t b(s^-)\,da(s) + \sum_{s\le t} \Delta a(s)\,\Delta b(s), \tag{2.16}$$

where $a(s^-)$ is the left-hand limit of a at s and $\Delta a(s) = a(s) - a(s^-)$. □

Suppose now that $\mathcal{H}$ is a history on $(\Omega,\mathcal{F},P)$ and that M is a martingale with *locally integrable variation*: for each $t>0$,

$$E[\int_0^t d|M|] < \infty, \tag{2.17}$$

where $|M|$ is the total variation process associated with M: $|M|_t$ is the variation of the path $s \to M_s$ over $[0,t]$. One of the truly important results in stochastic integration is that the integral of a predictable process C with respect to M is itself a martingale. Heuristically, the argument is simple but revealing. In differential form the martingale M is characterized by the property that $E[dM_t|\mathcal{H}_t] = 0$ for each t. It follows that

$$E[d(C*M)_t|\mathcal{H}_t] = E[C_{t+}\,dM_t|\mathcal{H}_t] = C_{t+}E[dM_t|\mathcal{H}_t] = 0; \tag{2.18}$$

note the crucial role of the predictability assumption in enabling one to bring C_{t+} outside the conditional expectation.

Here are a more formal statement and more rigorous proof.

Theorem 2.9. Let M be a martingale fulfilling (2.17) and let C be a predictable process satisfying

$$E[\int_0^t |C|d|M|] < \infty \tag{2.19}$$

for each t. Then the stochastic integral $C*M$ is a martingale.

Proof: Suppose that C is a "primitive" predictable process of the form $C(u,\omega) = 1(s<u\le t)1(\omega\in\Lambda)$, where $s<t$ and $\Lambda\in\mathcal{H}_s$ [see (2.3)]. Then $C*M$ exists, is adapted, and evidently satisfies $E[|(C*M)_t|]<\infty$ for each t. Moreover, for $u<v$ and $\Gamma\in\mathcal{H}_u$,

$$E[\{(C*M)_v - (C*M)_u\};\Gamma] = E[(M_{v\wedge t} - M_{u\wedge t});\Lambda\cap\Gamma] = 0,$$

which confirms that $C*M$ is a martingale. An application of the monotone class theorem completes the proof. □

2.3 COMPENSATORS

We begin with a point process $N = \sum_{n=1}^{\infty} \varepsilon_{T_n}$ on $\mathbf{R}_+$, with counting process $N_t = \sum_{n=1}^{\infty} 1(T_n \leq t)$, and a history $\mathcal{H} = (\mathcal{H}_t)$ such that the following hold.

Assumptions 2.10. a) $\mathcal{H}$ satisfies "*les conditions habituelles du théorie générale des processus*" [i.e., "the usual conditions" (see Dellacherie, 1972)].

b) $\mathcal{F}_t^N \subset \mathcal{H}_t$ for each t (N is adapted to $\mathcal{H}$).

c) $E[N_t] < \infty$ for each t.

The first condition is required for technical reasons, the second is obviously needed, and the third is for simplicity; it permits us to avoid local martingales. Moreover, it is fulfilled in many, if not most, important applications. When it fails one "localizes" the results given below (see Appendix B); the standard choice of localizing times is the arrival times T_n.

The notion of the compensator of a point process is the fundamental new idea of the section.

Definition 2.11. Let N be a point process on $\mathbf{R}_+$ satisfying Assumptions 2.10. The *compensator* of N with respect to $\mathcal{H}$ is the unique random measure A on $\mathbf{R}_+$ such that

a) The process (A_t) is $\mathcal{H}$-predictable;

b) For every nonnegative $\mathcal{H}$-predictable process C,

$$E\left[\int_0^{\infty} C\, dN\right] = E\left[\int_0^{\infty} C\, dA\right] \tag{2.20}$$

Existence and uniqueness of the compensator can be proved, as in Dellacherie (1972), using projection methods. One first defines the *predictable projection* $(\widetilde{X}_t)$ of an adapted, positive process (X_t) to satisfy $E[\widetilde{X}_T] = E[X_T]$ for every (finite) predictable stopping time T (see Exercise 2.2); roughly speaking, this means that $\widetilde{X}_t = E[X_t | \mathcal{H}_{t-}]$ for suitably chosen versions of these conditional expectations. This definition is equivalent to

$$E\left[\int_0^{\infty} \widetilde{X}\, dB\right] = E\left[\int_0^{\infty} X\, dB\right] \tag{2.21}$$

for every predictable random measure B [in the sense of Definition 2.11a)]. In this context, (2.20) becomes the adjoint of the operator (2.21).

Alternatively, the counting process (N_t) is a submartingale with respect to $\mathcal{H}$ and the compensator can be realized as the predictable increasing process in the Doob-Meyer decomposition of N (see Theorems 2.14 and B.6 for further details).

The compensator depends not only on $\mathcal{H}$ but also—which is crucial in martingale inference—on the probability P; however, we suppress these dependences whenever possible.

Here is the simplest example.

Example 2.12. (Poisson processes). Let N be a Poisson process on $\mathbf{R}_+$ with diffuse mean measure μ and let $\mathcal{H} = \mathcal{F}^N$, the completed internal history. Then the compensator A is deterministic and equal to μ. Indeed, the process $t \to \mu_t$ is trivially predictable and for $C(u,\omega) = 1(s < u \leq t)1(\omega \in \Lambda)$ [here $s < t$ and $\Lambda \in \mathcal{F}_s^N$; it suffices by (2.3) and the monotone class theorem to consider only such processes], the independent increments character of N implies that

$$E[\int_0^\infty C\,dN] = E[(N_t - N_s);\Lambda] = (\mu_t - \mu_s)P\{\Lambda\} = E[\int_0^\infty C\,d\mu] \quad \square$$

The next example is more complicated, but shows clearly the role of Cox processes.

Example 2.13. (Cox processes). Let N be a Cox process directed by the diffuse random measure M and let $\mathcal{H}_t = \mathcal{F}_\infty^M \vee \mathcal{F}_t^N$, suitably completed. Then the compensator of N is the directing measure M (see Exercise 2.10). However, for $\mathcal{H}_t = \mathcal{F}_t^N$ the compensator is in general different. $\square$

The following characterization of the compensator is essential to the *dynamic* viewpoint of point processes that accompanies the "martingale approach," as well as in statistical inference. It states that the difference $M = N - A$ between a point process and its compensator is a martingale, but of a schizophrenic sort. In some contexts it is a noise process for which estimation is declared impossible, but in others it contains all relevant information about the development of N. Together with Theorem 2.21 and Proposition 2.23, Theorem 2.14 forms the basis for variance and covariance computations, as well as for important central limit theorems.

Theorem 2.14. Given a point process N satisfying Assumptions 2.10 and a predictable random measure M, the following are equivalent:

a) A is the compensator of N;

b) The process $M_t = N_t - A_t$ is a (mean zero) martingale.

Proof: For C a generator of the predictable σ-algebra of the form (2.3), Definition 2.11 yields

$$E[(N_t - N_s);\Lambda] = E[(A_t - A_s);\Lambda], \tag{2.22}$$

which by Assumption 2.10c) transforms to

$$E[(N_t - A_t);\Lambda] = E[(N_s - A_s);\Lambda],$$

and hence b) holds if a) does. Conversely, if $N - A$ is a martingale, then A is integrable, so (2.22) obtains, and (2.20) follows by the monotone class theorem. □

In infinitesimal form (2.22) becomes the heuristic expression

$$dA_t = E[dN_t|\mathcal{H}_{t-}] \tag{2.22a}$$

The time increment dt must be thought of as *forward* in time from t. Thus $dM_t = dN_t - dA_t$ is that portion of dN_t that cannot be foreseen from observation of N over $[0,t)$ and whatever other information is part of the strict past $\mathcal{H}_{t-}$. This motivates the following terminology.

Definition 2.15. The martingale $M = N - A$ is the *innovation process* of N.

The innovation interpretation is consistent with previous examples. Thus, since a Poisson process has independent increments, for $\mathcal{H}_t = \mathcal{F}_t^N$ the innovation martingale contains everything other than trivially predictable mean values that could have been predicted even in total absence of observations. At another extreme is the "degenerate" (but good for counterexamples) case that $\mathcal{H}_t = \mathcal{F}_\infty^N$ for all t: there is no innovation at all—everything about N is revealed at $t = 0$—and $A \equiv N$. At neither of these extremes lie processes for which all innovation occurs at a single random time, as illustrated by the following example.

Example 2.16. (Lattice process). Let $N = \Sigma_{n=0}^{\infty} \varepsilon_{X+n}$ be (the restriction to $\mathbf{R}_+$ of) the stationary point process of Example 1.60 and suppose that $\mathcal{H}_t = \mathcal{F}_t^N$. Then once $T_1 = X$, which is uniformly distributed on $[0,1]$, has been observed, it is known exactly where all remaining points lie. Using Theorem 2.18,

$$A_t = \int_0^t (1-x)^{-1}\,dx$$

for $t \in [0,T_1]$, while for $n \geq 1$ and $t \in (T_n, T_{n+1}]$,

$$A_t = \int_0^{T_1} (1-x)^{-1}\,dx + (n-1) + 1(t \geq T_n + 1) \quad \square \tag{2.23}$$

Explicit calculation of compensators is possible (in principle, at least) using either of the next two results. The first is more general (and related to the construction of conditional intensities in Section 2.6) and correspondingly less useful in applications where there is special structure. The second posits a specific form for $\mathcal{H}$, but it is the "typical" form, so the result is broadly applicable after all.

Theorem 2.17. Let N be a point process on $\mathbf{R}_+$ satisfying Assumptions 2.10 and let A be the compensator. Then for each t, almost surely,

$$A_t = \lim_{n\to\infty} \sum_{k=0}^{2^n-1} E\left[N\left(\frac{(k+1)t}{2^n}\right) - N\left(\frac{kt}{2^n}\right) \mid \mathcal{H}_{kt/2^n}\right] \quad \square$$

Theorem 2.18. Let N be a point process on $\mathbf{R}_+$ satisfying Assumptions 2.10 with respect to the history

$$\mathcal{H}_t = \mathcal{H}_0 \vee \mathcal{F}_t^N \tag{2.24}$$

(suitably completed). For each n, let $F_n(\omega,du)$ be a regular conditional distribution of $U_{n+1} = T_{n+1}T_n$ given $\mathcal{H}_{T_n}: F_n(\omega,du) = P\{U_{n+1} \in du|\mathcal{H}_{T_n}\}(\omega)$. Then on $(T_n,T_{n+1}]$,

$$A_t = A_{T_n} + \int_0^{t-T_n} \frac{F_n(dx)}{F_n[x,\infty)} \tag{2.25}$$

Proof: (Sketch). It will do to show that for each n and each finite stopping time T, $E[N_{T\wedge T_n}] = E[A_{T\wedge T_n}]$, with A given by (2.25). By (2.24) and Proposition 2.5 there is $V_n \in \mathcal{H}_{T_n}$ (this is crucial!) satisfying (2.11). The rest of the proof is computational. On the one hand,

$$E\left[\sum_{j=0}^{n-1}\left[\int_0^{v_j\wedge u_{j+1}} \frac{F_j(dx)}{F_j[x,\infty)}\right] 1(T \geqslant T_j)\right]$$

$$= \sum_{j=0}^{n-1} E\left[E\left[\int_0^{v_j\wedge u_{j+1}} \frac{F_j(dx)}{F_j[x,\ \infty)} \mid \mathcal{H}_{T_j}\right] 1(T \geq T_j)\right]$$

$$= \sum_{j=0}^{n-1} E[\int_0^{V_j} F_j(dx) 1(T\geq T_j)],$$

but on the other hand,

$$E[N_{T\wedge T_n}] = E\left[\sum_{j=0}^{n-1} (N_{T\wedge T_{j+1}} - N_{T\wedge T_j}) 1(T\geq T_j)\right]$$

$$= E\left[\sum_{j=0}^{n-1} 1(V_j \geq U_{j+1}) 1(T\geq T_j)\right] = \sum_{j=0}^{n-1} E[\int_0^{V_j} F_j(du) 1(T\geq T_j)] \quad \square$$

The assumption (2.24) limits information that is accumulated incrementally to arise from the point process alone. Other information, as occurs in the Cox process example following Definition 2.11, must be present at time zero and cannot increase over time. In applications, of course, this may not be reasonable physically.

A consequence of Theorem 2.18 is the following uniqueness property.

Theorem 2.19. Let N be a point process defined over $(\Omega, \mathcal{F})$ and let P and P' be probabilities with respect to which N has $\mathcal{H}$-compensators A and A', respectively, and assume that (2.24) is fulfilled. Then provided that $P = P'$ on $\mathcal{H}_0$, $P = P'$ on $\mathcal{H}_\infty$.

Proof: Theorem 2.18 shows that the conditional distributions $F_n(du) = P\{U_{n+1} \in du | \mathcal{H}_{T_n}\}$ and $F_n'(du) = P'\{U_{n+1} \in du | \mathcal{H}_{T_n}\}$ coincide for every n and from this the conclusion follows by Propositions 2.2 and 2.3. □

We elaborate on one interpretation of (2.25). Let T be a positive random variable whose distribution function F admits density $f = F'$. The function

$$h(t) = \frac{f(t)}{1 - F(t)} = \frac{P\{T \leq t + dt | T > t\}}{dt} \tag{2.26}$$

is known in reliability and survival analysis as the *hazard function* of F. Comparison of (2.25) and (2.26) identifies the compensator as a (cumulative) *conditional hazard function*. See also Theorem 2.31 and associated discussion.

Here is one further example.

Example 2.20. (Renewal processes). Let $N = \Sigma_{n=1}^{\infty} \varepsilon_{T_n}$ be a renewal process (Definition 1.6) with interarrival distribution F having density f. Then $F_n(\omega, \cdot) = F(\cdot)$ for all n and ω, and (2.25) becomes

$$A_t = -\sum_{n=1}^{\infty} \log[1 - F(t \wedge T_n)] \tag{2.27}$$

See Section 8.1 for additional discussion and interpretations. □

The innovation martingale M has associated with it a predictable variation process $\langle M \rangle$ (see Appendix B for fundamental properties). The predictable variation and, more generally, predictable covariation processes arising from marked point processes not only play an important computational role, but are also the basis for central limit theorems for martingale estimators. We now calculate $\langle M \rangle$.

Theorem 2.21. Let N be a point process with $\mathcal{H}$-compensator A and assume that the innovation martingale $M = N - A$ is square integrable. Let $\Delta A_t = A_t - A_{t-}$. Then

$$<M>_t = \int_0^t (1 - \Delta A_s)\, dA_s \tag{2.28}$$

The integration-by-parts formula (2.16) implies that

$$M_t^2 = 2\int_0^t M_{s-}\, dM_s + \sum_{s \leq t} (\Delta M_s)^2$$

Expanding the summation and using the identity $(\Delta N)^2 = \Delta N$ yields

$$M_t^2 = \int_0^t (2M_{s-} + 1 - 2\,\Delta A_s)\, dM_s + \int_0^t (1 - \Delta A_s)\, dA_s,$$

where we have also used the relationship $M = N - A$. The preceding formula exhibits M^2 as the sum of a martingale (Theorem 2.9) and a predictable increasing process, and thus constitutes a Doob-Meyer decomposition of the submartingale M^2. However, that decomposition is unique (Theorem B.8), and hence (2.28) holds. □

In particular, if A is continuous in t almost surely (especially if A is absolutely continuous as in Section 2.4), then $\Delta A = 0$ and $<M> = A$. The differential interpretation $d<M> = dA$ together with (2.22a) then yields

$$E[dN_t - E[dN_t|\mathcal{H}_{t-}])^2|\mathcal{H}_{t-}] = dA_t = E[dN_t|\,\mathcal{H}_{t-}]; \tag{2.29}$$

that is, a point process with a continuous compensator is locally and conditionally a Poisson process in the sense that the mean and variance are equal.

Now let $N = \Sigma\, \varepsilon_{(T_n, Z_n)}$ be a marked point process with mark space $(E, \mathcal{E})$. For each $B \in \mathcal{E}$ the counting process $N_t(B) = \Sigma\, 1(T_n \leq t, Z_n \in B)$ admits $\mathcal{H}$-compensator $A_t(B)$. But one can also regard $N_t(\cdot)$ as a stochastic process parametrized by $\mathbf{R}_+$ and taking values in $\mathbf{M}_p(E)$, and it is desirable, then, that for each t, $A_t(B)$ be a random measure in B, which can be accomplished using a construction identical to that used to produce regular conditional probabilities, and yields the following result.

Theorem 2.22. Let N be a marked point process such that the point process $\Sigma\, \varepsilon_{T_n}$ satisfies Assumptions 2.10. Then there exists a unique random measure A on $\mathbf{R}_+ \times E$ such that

a) for each B the process $A_t(B) = A([0,t\,] \times B)$ is predictable;

b) for each nonnegative predictable process C (i.e., defined on $\mathbf{R}_+ \times E$ and measurable with respect to $\mathcal{P} \times \mathcal{E}$)

$$E[\int_{\mathbf{R}_+ \times E} C\,dN] = E[\int_{\mathbf{R}_+ \times E} C\,dA] \quad \square \tag{2.30}$$

Predictably, the random measure A is called the *compensator* of N. For each B, $M_t(B) = N_t(B) - A_t(B)$ is a martingale; moreover, $M_t(\cdot) = N_t(\cdot) - A_t(\cdot)$ is a measure-valued (but the values need not be positive measures) martingale, which we term the *innovation process*.

For the case that the innovation martingales $M(B_1)$ and $M(B_2)$ associated with disjoint subsets B_1 and B_2 of E are square integrable, the following result calculates the predictable covariation process. The most useful form arises when there exists a stochastic intensity (Proposition 2.32) or, more generally, if either $A_t(B_1)$ or $A_t(B_2)$ is merely continuous in t, for then these martingales are orthogonal.

Proposition 2.23. Let N be a marked point process with compensator A and let B_1, B_2 be disjoint sets in $\mathcal{E}$ for which the martingales $M(B_1)$ and $M(B_2)$ are square integrable. Then their predictable covariation is

$$<M(B_1),M(B_2)>_t = -\int_0^t \Delta A_s(B_1)\,\Delta A_s(B_2)\,ds \tag{2.31}$$

Proof: On the one hand, application of Theorem 2.21 to the point process $N_t(B_1, \cup B_2)$, which has compensator $A_t(B_1 \cup B_2) = A_t(B_1) + A_t(B_2)$, gives

$$<M(B_1 \cup B_2)>_t = \int_0^t (1 - \Delta A_s(B_1) - \Delta A_s(B_2))[d\,A_s(B_1) + dA_s(B_2)],$$

while on the other hand

$$\begin{aligned} <M(B_1 \cup B_2)>_t &= <M(B_1)>_t + <M(B_2)>_t + 2<M(B_1),M(B_2)>_t \\ &= \int_0^t (1 - \Delta A_s(B_1))\,dA_s(B_1) + \int_0^t (1 - \Delta A_s(B_2))\,dA_2(B_2) \\ &\quad + 2<M(B_1),M(B_2)>_t \end{aligned}$$

by another appeal to Theorem 2.21. The remainder of the proof is computational. □

The "negative correlation" represented by (2.31) has an obvious physical basis: the processes $N(B_1)$ and $N(B_2)$ cannot jump simultaneously.

Hence to the extent that a jump is deemed relatively more likely to be in $N(B_1)$, it must be relatively less so in $N(B_2)$.

By combining (2.28) and (2.31) and using the inner product property of predictable covariation, for general $B_1, B_2 \in \mathscr{E}$,

$$\begin{aligned}<M(B_1),M(B_2)>_t = &\int_0^t [1 - \Delta A(B_1 \cap B_2)]\, dA\,(B_1 \cap B_2)\\ &- \int_0^t \Delta A(B_1 \cap B_2)\, \Delta A(B_1\, \Delta B_2)\, ds\\ &- \int_0^t \Delta A(B_1 \backslash B_2)\, \Delta A(B_2 \backslash B_1)\, ds \end{aligned} \tag{2.32}$$

We illustrate a simple special case.

Example 2.24. (Position-dependent marking). Let $\widetilde{N}$ be a point process on $\mathbf{R}_+$ from which N is constructed by position-dependent marking (Example 1.28), using a kernel $K: \mathbf{R}_+ \to E$. Then the $\mathscr{F}^N$-compensator of N is

$$A(dt,dx) = \widetilde{A}\,(dt)K(t,dx), \tag{2.33}$$

where $\widetilde{A}$ is the $\mathscr{F}^{\tilde{N}}$-compensator of $\tilde{N}$. □

When (2.24) holds Theorem 2.18 admits a rather obvious extension for marked point processes. We do not state it formally, but see Theorem 2.30 for the corresponding result when there exists a stochastic intensity.

Another application of compensators is to convergence in distribution for point processes, as exemplified by the following Poisson limit theorem.

Theorem 2.25. Let $N^1, N^2, \ldots$ be simple point processes on $\mathbf{R}_+$ having compensators $A^1, A^2, \ldots$ for their respective internal histories, and let N be a point process with deterministic, continuous compensator A (i.e., N is a Poisson process with mean measure A). If for each t,

$$A_t^n \xrightarrow{d} A_t, \tag{2.34}$$

then $N^n \xrightarrow{d} N$.

Proof: (Sketch). We prove only the special case that there exists a constant C such that

$$A_\infty^n = \lim_{t\to\infty} A_t^n \le C \tag{2.35}$$

almost surely for each n. Details on removal of (2.35) are given in Kabanov et al. (1980a).

By Theorem 1.21 it suffices to show that for $f \geq 0$ a simple function on $\mathbf{R}_+$ with compact support,

$$\lim_{n\to\infty} E[\exp[-N^n(f) + \int_0^\infty (1 - e^{-f})\, dA]] = 1 \tag{2.36}$$

The key point is each process

$$Z_t^n = \exp[-\int_0^t f\, dN^n + \int_0^t (1 - e^{-f})\, dA^n]$$

is (under (2.35)) a square integrable martingale, in consequence of which

$$E[\exp[-\int_0^t f\, dN^n + \int_0^t (1 - e^{-f})\, dA^n]] = 1 \tag{2.37}$$

A heuristic argument is the following. For small values of Δt

$$\begin{aligned} &E[\exp[-\int_t^{t+\Delta t} f\, dN^n + \int_t^{t+\Delta t} (1 - e^{-f})\, dA^n] | \mathcal{F}_t^{N_n}] \\ &\cong E[(1 + [e^{-f(t)} - 1]\, \Delta N_t^n)(1 + [1 - e^{-f(t)}]\, dA_t^n) | \mathcal{F}_t^{N_n}] \\ &= 1 + (e^{-f(t)} - 1) E[\Delta N_t^n - dA_t^n | \mathcal{F}_t^{N_n}] - (e^{-f(t)} - 1)^2 E[\Delta N_t^n dA_t^n | \mathcal{F}_t^{N_n}] \\ &= 1 - (e^{-f(t)} - 1)^2 (dA_t^n)^2 \cong 1, \end{aligned}$$

where (2.22a) is used at the next-to-last step. Hence by (2.37)

$$\begin{aligned} &|E[\exp[-N^n(f) + \int_0^\infty (1 - e^{-f})\, dA]] - 1| \\ &= |E[\exp[-N^n(f) + \int_0^\infty (1 - e^{-f})\, dA] - \exp[-N^n(f) + \int_0^\infty (1 - e^{-f})\, dA^n]]| \\ &\leq E[|\exp[\int_0^\infty (1 - e^{-f})\, dA] - \exp[\int_0^\infty (1 - e^{-f})\, dA^n]|], \end{aligned}$$

and this last quantity converges to zero by (2.34) and the assumption that f is simple; therefore, (2.36) holds. □

A variant yields an inequality for the total variation distance between a point process and a Poisson process. [See Section 1.6 for definitions and details; given random elements X and Y of the same space, we write $\alpha(X,Y)$ for the variation distance between their laws $\mathcal{L}_X$ and $\mathcal{L}_Y$.]

Theorem 2.26. Let N be a point process on $[0,t]$ with $\mathcal{F}^N$-

compensator A and let N^* be a Poisson process with mean measure μ. Then

$$\alpha(\mathcal{L}_N,\mathcal{L}_{N^*}) \leq E[\|A-\mu\|] + E\left[\sum_{s\leqslant t} (\Delta A_s)^2\right], \tag{2.38}$$

where $\|A-\mu\|$ is the total variation of the signed measure $A-\mu$ over $[0,t]$.

Proof: (Partial). We prove (2.38) under the stipulation that the compensator A be continuous, so that the second term on the right-hand side is zero, and we further assume that $A_\infty = \infty$ a.s. It follows (Exercise 2.11) that if $\widetilde{A}_s = \inf\{t: A_t \geq s\}$, then $\widetilde{N}_s = N(A_s)$ is a stationary Poisson process with rate 1. Hence for $t \geq 0$ and X_λ having a Poisson distribution with mean λ,

$$\begin{aligned}\alpha(N_t,X_\lambda) &\leq P\{N_t \neq X_\lambda\}\\ &\leq E[|N_t - \widetilde{N}_\lambda|]\\ &= E\left[\int_0^\infty \{1(\widetilde{A}_\lambda < s \leq t) + 1(t < s \leq \widetilde{A}_\lambda)\}\, dN_s\right]\\ &= E\left[\int_0^\infty \{1(\widetilde{A}_\lambda < s \leq t) + 1(t < s \leq \widetilde{A}_\lambda)\}\, dA_s\right]\end{aligned}$$

(the integrand is predictable)

$$= E[|A(\widetilde{A}_\lambda) - A_t|] = E[|A_t - \lambda|],$$

the last step by continuity of A. Suppose now that $0 = t_0 < t_1 < \cdots < t_k$ and that $X_{\mu(t_j)-\mu(t_{j-1})}$ are independent, Poisson distributed random variables with the indicated means. The preceding computation implies that

$$\begin{aligned}&\alpha(\{N_{t_j} - N_{t_{j-1}}\},\{X_{\mu(t_j)-\mu(t_{j-1})}\})\\ &\quad\leq E\left[\sum_{j=1}^k \left|A_{t_j} - A_{t_{j-1}} - (\mu(t_j) - \mu(t_{j-1}))\right|\right] \leq E[\|A-\mu\|],\end{aligned}$$

and this gives (2.38). □

We conclude the section by comparing the bound (2.38) with that implied by Proposition 1.46.

Example 2.27. Let $N = \Sigma_{i=1}^k N_i$, where the N_i are i.i.d. Bernoulli processes on $[0,1]$ with parameters (p,F) and suppose that F admits density f. Then the compensator of N is

$$dA_u = (k - N_{u-})\frac{pf(u)}{1-pF(u)}du \tag{2.39}$$

Suppose that N^* is Poisson with mean $\mu = kpF$. Then (2.38) yields

$$\begin{aligned}\alpha(N,N^*) &\leq \int_0^1 E\left[|(k - N_{u-})\frac{pf(u)}{1 - pF(u)} - kpf(u)|\right] du \\ &= p\int_0^1 E\left[|\frac{k - N_{u-}}{1 - pF(u)} - 1|\right] f(u)\,du \\ &= p\int_0^1 E[|N_u - kpf(u)|]\frac{f(u)}{1 - pF(u)}\,du \\ &\leq 2kp^2\int_0^1 f(u)\,dF(u) = kp^2\end{aligned}$$

By contrast, for N^{**} Poisson with mean $\mu = k[-\log(1-p)]F$, Proposition 1.46 yields $\alpha(N,N^{**}) \leq kp^2/2$. Application of (2.38) to estimate $\alpha(N,N^{**})$ is very difficult computationally. □

2.4 STOCHASTIC INTENSITIES

Let N be a point process on $\mathbf{R}_+$ with compensator A. In many important applications (and all of Chapter 5) A admits a (jointly measurable) density with respect to Lebesgue measure (the density is a random process in general), and results in Section 2.3 simplify, especially Theorem 2.21 and—for marked point processes—Proposition 2.23. In addition, the interpretation of the compensator as integrated conditional hazard function becomes manifestly clear. In the following discussion the history $\mathcal{H}$ is fixed; we require that Assumptions 2.10 be satisfied.

Definition 2.28. Let N be a simple point process on $\mathbf{R}_+$. A positive, predictable process $\lambda = (\lambda_t)$ such that

$$A_t = \int_0^t \lambda_s\,ds \tag{2.40}$$

is the compensator of N is called the *stochastic intensity* of N.

It follows from the discussion relating to Definition 2.11 that if it exists, the stochastic intensity is unique. The heuristic interpretation [compare (2.22a)] is that for each t,

$$\lambda_t\,dt = E[\Delta N_t|\mathcal{H}_{t-}] = P\{\Delta N_t = 1|\mathcal{H}_{t-}\} = P\{\Delta N_t > 0|\mathcal{H}_{t-}\} \tag{2.41}$$

(Recall that N is simple.) Further, there is a persuasive analogy between stochastic intensities and hazard functions of ordinary random variables, one version of which is given in Theorem 2.30 and another in Theorem 2.31 (see also Exercise 2.13).

For a Poisson process N with mean measure $\mu(dt) = \alpha_t\,dt$ (α is often termed the *intensity* function of N) the $\mathcal{F}^N$-stochastic intensity exists and is deterministic and equal to α. If N is a Cox process directed by the absolutely continuous random measure $M(dt) = X_t\,dt$ and if $\mathcal{H}_t = \mathcal{F}_t^N \vee \mathcal{F}_\infty^X$, then (X_t) is the $\mathcal{H}$-stochastic intensity of N and is called the *directing intensity*.

Not every point process admits a stochastic intensity. A deterministic point process is its own compensator and hence has no stochastic intensity. A somewhat less degenerate example is a renewal process with discrete interarrival distribution, which by (2.27) does not possess a stochastic intensity even with respect to its internal history. Indeed, existence of a stochastic intensity depends as much on the history as on N; the smaller $\mathcal{H}$, the better the chance that a stochastic intensity exists. By making $\mathcal{H}$ large enough that $\mathcal{F}_\infty^N \subset \mathcal{H}_0$, one forces N to be its own compensator. Conversely, if N admits $\mathcal{H}$-stochastic intensity λ and if $\mathcal{F}_t^N \subset \mathcal{G}_t \subset \mathcal{H}_t$ for each t, then N admits $\mathcal{G}$-stochastic intensity $\tilde{\lambda}$ satisfying

$$\tilde{\lambda}_t = E[\lambda_t | \mathcal{G}_t] \tag{2.42}$$

almost surely for each t (Exercise 2.16). For the Cox process example, the $\mathcal{F}^N$-stochastic intensity $\tilde{X}$ satisfies

$$\tilde{X}_t = E[X_t | \mathcal{F}_t^N] \tag{2.43}$$

In Chapters 5 and 7—as in (2.42) and (2.43)—we examine state estimation problems that are really calculations of stochastic intensities.

Existence and exact nature of a stochastic intensity depend also on the probability P, but not in an entirely arbitrary manner (see Theorem 2.31).

In view of these remarks, one should not expect "simple" sufficient conditions for existence of a stochastic intensity, and there do not seem to be any, even for the internal history. Theorem 2.30 does provide one condition; we give it after extending the concept of stochastic intensity to marked point processes.

Suppose now that $N = \Sigma\, \varepsilon_{(T_n, Z_n)}$ is a marked point process with mark space E.

Definition 2.29. A stochastic process $(\lambda_t(B): t \geq 0, B \in \mathcal{E})$ is the *stochastic intensity* of N provided that

a) For each t, $B \to \lambda_t(B)$ is a random measure on E;
b) For each B, the process $(\lambda_t(B))$ is the stochastic intensity of the counting process $N_t(B) = \Sigma\, 1(T_n \leq t, Z_n \in B)$.

The never-explicitly-stated extension of Theorem 2.18 to marked point processes has the following analog.

Theorem 2.30. Let N be a marked point process and suppose that Assumptions 2.10 are satisfied for

$$\mathcal{H}_t = \mathcal{H}_0 \vee \mathcal{F}_t^N \tag{2.44}$$

Suppose that for each n the conditional distribution $F_n(du,dx) = P\{U_{n+1} \in du, Z_{n+1} \in dx | \mathcal{H}_{T_n}\}$ has the form

$$F_n(du,dx) = f_n(u,dx)\, du, \tag{2.45}$$

where $f_n(u,dx)$ is a measurable, measure-valued stochastic process. Then the process given by

$$\lambda_t(B) = \frac{f_n(t - T_n, B)}{F_n([t - T_n, \infty), E)} \tag{2.46}$$

for $t \in (T_n, T_{n+1}]$ defines a stochastic intensity of N. □

In particular, if N is an ordinary point process and (2.44) holds, then provided that $P\{U_{n+1} \in du | \mathcal{H}_{T_n}\} = f_n(u)\, du$, N admits stochastic intensity

$$\lambda_t = \frac{f_n(t - T_n)}{\int_{t-T_n}^{\infty} f_n(u) du} \tag{2.47}$$

on $(T_n, T_{n+1}]$. See Exercise 2.14 for the special form of (2.47) when N is a renewal process.

Comparison of (2.41) and (2.26) makes manifest the interpretation of the stochastic intensity as a conditional hazard function. In further pursuit of this analogy, consider a random variable T with hazard function $h = f/(1 - F)$. Viewing T as a sample of size 1 from the distribution function F, we write the likelihood function $\ell(T,F) = f(T)$ in terms of the hazard function:

$$\ell(T,F) = \exp[-\int_0^t h(s)\, ds + \log h(T)] \tag{2.48}$$

A similar representation holds for point processes, with the stochastic intensity playing the role of the hazard function in (2.48). For simplicity we formulate it only for point processes without marks.

Theorem 2.31. Let $(\Omega, \mathcal{F}, P)$ be a probability space over which is defined a simple point process N having $\mathcal{F}^N$-stochastic intensity λ. Suppose that P_0 is a probability measure on $(\Omega, \mathcal{F})$ with respect to which N is a stationary Poisson process with rate 1. Then for each t, $P \ll P_0$ on $\mathcal{F}_t^N$, with derivative (*likelihood ratio*)

$$\frac{dP}{dP_0}(N) = \exp[\int_0^t (1 - \lambda_s)\, ds + \int_0^t \log \lambda_s\, dN_s] \tag{2.49}$$

Conversely, if P_0 is as above and P is a probability measure on $(\Omega,\mathcal{F})$ satisfying $P \ll P_0$, then there exists a predictable process λ such that N has stochastic intensity λ with respect to P.

Proof: For $u \le t$ define

$$Z_u = \exp[\int_0^u (1-\lambda_s)\,ds + \int_0^u \log\lambda_s\,dN_s];$$

then it is easily verified using the independent increments property of Poisson processes that $(Z_u)_{u\le t}$ is a martingale over $(\Omega,\mathcal{F},P_0)$. By (2.20) and Theorem 2.19 it suffices to show that if P' is the probability defined by $dP' = Z_t\,dP_0$, then for each bounded, predictable process C,

$$E'[\int_0^t C_s\,dN_s] = E'[\int_0^t C_s\lambda_s\,ds] \tag{2.50}$$

But

$$E'[\int_0^t C_s\,dN_s] = E_0[Z_t\int_0^t C_s\,dN_s]$$

$$= E_0[\int_0^t Z_s C_s\,dN_s]$$

(since Z is a P_0-martingale)

$$= E_0[\int_0^t Z_{s-}\lambda_s C_s\,dN_s]$$

(by direct computation, $Z_{T_n} = Z_{T_n-}\lambda_{T_n}$ for each arrival time T_n)

$$= E_0[\int_0^t Z_{s-}\lambda_s C_s\,ds]$$

(the P_0-stochastic intensity if N is identically 1)

$$= E_0[\int_0^t Z_s C_s\lambda_s\,ds]$$

(almost surely, $Z_{s-} = Z_s$ for almost every $s \in [0,t]$)

$$= E_0[Z_t\int_0^t C_s\lambda_s\,ds] = E'[\int_0^t C_s\lambda_s\,ds],$$

and therefore (2.50) holds.

The proof of the converse is deferred to Section 2.5. □

Note that the not-yet-proved part of Theorem 2.31 provides—albeit indirectly—a sufficient condition for existence of a stochastic intensity.

With the aid of (2.49) we can interpret a point process as a dynamic, uncountable set of independent Bernoulli trials, one for each time t. Taking λ_t as the success probability for the trial at t [see (2.41)] we conclude that—for observation over the time interval $[0,t]$—the "conditional" probability of successes at times $T_1, \dots, T_{N_t}$ is

$$\prod_{k=1}^{N_t} \lambda_{T_k} = \exp[-\int_0^t (-\log \lambda)\, dN], \tag{2.51}$$

while by (2.10) and (2.47) the "conditional" probability of no successes in the interval (T_{j-1}, T_j) is

$$1 - F_{j-1}(T_j - T_{j-1}) = \exp\left(-\int_{T_{j-1}}^{T_j} \lambda_s\, ds\right) \tag{2.52}$$

Multiplying the terms in (2.51) and (2.52), along with a term (similar to the latter) corresponding to the probability of no successes in (T_{N_t}, t), gives that the probability of successes at and only at the points $T_1, \dots, T_{N_t}$ is

$$\exp\left(-\int_0^t \lambda_s\, ds + \int_0^t \log \lambda_s\, dN_s\right) \tag{2.53}$$

For the Poisson process, $\lambda \equiv 1$ and (2.53) reduces to e^{-t}; taking the quotient of these two probabilities yields the likelihood ratio (2.49) and also a heuristic proof of Theorem 2.31. Note the resemblance between (2.48) and (2.49), confirmation of the claim that the stochastic intensity is a conditional hazard function.

We continue by noting explicitly the simplified forms of Theorem 2.21 and Proposition 2.23 that obtain for point processes with a stochastic intensity.

Proposition 2.32. a) Let N be a point process on $\mathbf{R}_+$ with stochastic intensity λ and assume that the innovation martingale $M_t = N_t - \int_0^t \lambda_s\, ds$ is square integrable. Then for each t,

$$\langle M \rangle_t = \int_0^t \lambda_s\, ds \tag{2.54}$$

b) Let N be a marked point process on $\mathbf{R}_+$ with mark space E and stochastic intensity $\lambda_t(B)$, and let B_1, B_2 be disjoint sets for which the innovation martingales $M_t(B_i)$, $i = 1,2$, are square integrable. Then $M(B_1)$ and $M(B_2)$ are orthogonal:

$$\langle M(B_1), M(B_2) \rangle \equiv 0 \quad \square \tag{2.55}$$

In particular, (2.55) implies that the process $M_t(B_1)M_t(B_2)$ is a martingale. This property will be crucial in Chapter 5 for calculation of covariances.

We conclude the section with an example whose inference will be examined in detail in Chapter 5.

Example 2.33. (System of Markov processes). Let S be a finite set and let $(X_t(1)),\dots,(X_t(K))$ be be K independent Markov processes, each with state space S and generator A. Suppose that we can observe only so-called "macro data," that is, how many processes are in each state at each time, but nothing more. We may represent the observations as a marked point process in the following manner. Let $E = \{(i,j) : i, j \in S, i \neq j\}$ and for $(i,j) \in E$ define the counting process

$$N_t(i,j) = \sum_{k=1}^{K} \sum_{s \leqslant t} 1(X_{s-}(k) = i, X_s(k) = j);$$

$N_t(i,j)$ is the observed number of jumps from i to j among the K processes during the time interval $[0,t]$.

For each i and t, let $Y_t(i) = \sum_{k=1}^{K} 1(X_{t-}(k) = i)$ be the number of processes in state i at time $t-$, and let $\mathcal{H}_t = \bigvee_{k=1}^{K} \mathcal{F}_t^{X(k)}$ with, as usual, appropriate completion. Then Assumptions 2.10 are fulfilled and (Exercise 2.19) N has $\mathcal{H}$-stochastic intensity

$$\lambda_t(i,j) = A(i,j) Y_t(i) \tag{2.56}$$

(Intuitively, this is clear from the interpretation of off-diagonal elements in the generator as transition intensities.) Furthermore, each process

$$M_t(i,j) = N_t(i,j) - A(i,j) \int_0^t Y_s(i)\, ds \tag{2.57}$$

is a square integrable martingale over every bounded time interval [for $K = 1$, (2.57) reduces to *Dynkin's formula*] and for $(i,j) \neq (i',j')$ these martingales are orthogonal. □

2.5 REPRESENTATION OF POINT PROCESS MARTINGALES

Let N be a point process on $\mathbf{R}_+$ having compensator A with respect to the completed internal history $\widetilde{\mathcal{F}}^N$ given by (2.2). According to Theorem 2.9, if H is a bounded predictable process and $M = N - A$ is the $\widetilde{\mathcal{F}}^N$-innovation martingale, then the process

$$\widetilde{M}_t = \int_0^t H_s dM_s \tag{2.58}$$

is also a martingale. This section is devoted to the converse assertion that every (suitably regular) $\widetilde{\mathcal{F}}^N$-martingale has the form (2.58). One consequent interpretation is that the innovation martingale M does indeed contain all information concerning the evolution of N. The only theorem in the section is formulated for marked point processes but, for simplicity and clarity, proved only for point processes without marks.

Theorem 2.34. Let $N=\Sigma\,\varepsilon_{(T_n,Z_n)}$ be a marked point process on $\mathbf{R}_+$ with mark space E and let M be the innovation martingale with respect to the internal history $\widetilde{\mathcal{F}}^N$. Let Assumptions 2.10 be satisfied and let $(\widetilde{M}_t)$ be a uniformly integrable $\widetilde{\mathcal{F}}^N$-martingale. Then there exists a process $\{H(t,x): t\geq 0, x\in E\}$, jointly measurable in (ω,t,x) and predictable in (ω,t) for each fixed x, such that

$$\widetilde{M}_t=\widetilde{M}_0+\int_{(0,t]\times E} H(s,x)\,dM_s(dx) \tag{2.59}$$

for every t.

Proof: (Sketch). For a point process without marks, (2.59) reduces to

$$\widetilde{M}_t=\widetilde{M}_0+\int_0^t H_s\,dM_s, \tag{2.60}$$

where $(H_t)_{t\geq 0}$ is predictable. For the remainder of the proof we assume that the compensator (A_t) is continuous.

Given this assumption, construction of the process H is accomplished as follows. First, for each n let $F_n(du)=P\{U_{n+1}\in du|\widetilde{\mathcal{F}}^N_{T_n}\}$ [we assume that these are regular conditional distributions (see Theorem 2.18)]. Second, by Proposition 2.2 for each n there exists a function $V_n(u,\omega)$ that is $\mathcal{B}(\mathbf{R}_+)\times\widetilde{\mathcal{F}}^N_{T_n}$-measurable and such that

$$\widetilde{M}_{T_{n+1}}(\omega)=V_n(U_{n+1}(\omega),\omega) \tag{2.61}$$

Finally, as a consequence of Proposition 2.4, for each n and t there is $W_n(t)\in\widetilde{\mathcal{F}}^N_{T_n}$ satisfying

$$\widetilde{M}_t 1(T_n<t<T_{n+1})=W_n(t)1(T_n<t<T_{n+1}) \tag{2.62}$$

We will show that H_t for $t\in(T_n,T_{n+1}]$ is given by

$$H_t=V_n(t-T_n)-\int_{t-T_n}^{\infty}\frac{V_n(u)F_n(du)}{F_n(t-T_n,\infty)} \tag{2.63}$$

[It follows from Theorem 2.18 and the continuity assumption that the distributions $F_n(\cdot)$ are continuous.]

To confirm the representation (2.60) we first observe that since M is a uniformly integrable martingale the optional sampling theorem implies that for $A\in\widetilde{\mathcal{F}}^N_{T_n}$,

$$E[\widetilde{M}_t 1(T_n\leq t<T_{n+1})1_A]=E[\widetilde{M}_{T_{n+1}}1(T_n\leq t<T_{n+1})1_A]; \tag{2.64}$$

we next evaluate each side of (2.64). Since $W_n(t)1(T_n \le t) \in \widetilde{\mathcal{F}}^N_{T_n}$, (2.62) implies that

$$E[\widetilde{M}_t 1(T_n \le t < T_{n+1})1_A] = E[1(U_{n+1} > t - T_n)W_n(t)1(T_n \le t)1_A]$$
$$= E[F_n(t - T_n, \infty)W_n(t)1(T_n \le t)1_A]$$

Similarly, because $1(T_n \le t)1_A \in \widetilde{\mathcal{F}}^N_{T_n}$,

$$E[\widetilde{M}_{T_{n+1}} 1(T_n \le t < T_{n+1})1_A] = E[V_n(U_{n+1})1(U_{n+1} > t - T_n)1(T_n \le t)1_A]$$
$$= E[(\int_{t-T_n}^{\infty} V_n(u)F_n(du))1(T_n \le t)1_A],$$

where we have also used (2.61).

It follows that for fixed n and t,

$$\widetilde{M}_t 1(T_n \le t < T_{n+1}) = \int_{t-T_n}^{\infty} \frac{V_n(u)F_n(du)}{F_n(t - T_n, \infty)} 1(T_n \le t) \tag{2.65}$$

almost surely, and the P-null set can be made independent of t (and thence of n) by the continuity assumption. Brute force differentiation of the right-hand side of (2.65) produces the following, valid for $t \in (T_n, T_{n+1})$:

$$dM_t = \frac{-V_n(t - T_n)F_n(dt - T_n)F_n(t - T_n, \infty)}{F_n(t - T_n, \infty)^2}$$
$$+ \int_{t-T_n}^{\infty} \frac{V_n(u)F_n(du)F_n(dt - T_n)}{F_n(t - T_n, \infty)^2}$$
$$= -H_t\, dA_t,$$

where the last equality holds by Theorem 2.18. Consequently, (2.60) holds on (T_n, T_{n+1}). Taking limits in (2.65) as $t \uparrow T_{n+1}$ yields

$$\widetilde{M}_{T_{n+1}} - \widetilde{M}_{(T_{n+1})-} = V_n(U_{n+1}) - \int_{U_{n+1}}^{\infty} \frac{V_n(u)F_n(du)}{F_n(U_{n+1}, \infty)},$$

and therefore (2.60) holds at T_{n+1}.

Now it remains only to observe that the process H is predictable by virtue of Proposition 2.6. □

We can now complete the proof of Theorem 2.31.

Proof: (Of the converse of Theorem 2.31). We need to show that if $P \ll P_0$, then there is a predictable process λ that is the $(P, \mathcal{F}^N)$-stochastic intensity of N. Define the P_0-martingale $Z_t = E_0[dP/dP_0 | \widetilde{\mathcal{F}}^N_t]$. By Theorem 2.34—applied to Z over the probability space $(\Omega, \mathcal{F}, P_0)$, over which the stochastic intensity of N is identically 1—there is a predictable process H

such that

$$Z_t = 1 + \int_0^t H_s(dN_s - ds) \tag{2.66}$$

Given a bounded, predictable process C vanishing outside $[0,t]$,

$$E\Big[\int_0^t C_s\,dN_s\Big] = E_0\Big[Z_t \int_0^t C_s\,dN_s\Big]$$

$$= E_0\Big[\int_0^t Z_s C_s\,dN_s\Big]$$

$$= E_0\Big[\int_0^t (Z_{s-} + H_s) C_s\,dN_s\Big]$$

by (2.66), $Z_{T_n} = Z_{T_n-} + H_{T_n}$ at each arrival time T_n of N]

$$= E_0\Big[\int_0^t (Z_{s-} + H_s) C_s\,ds\Big]$$

$$= E_0\left[\int_0^t \left(1 + \frac{H_s}{Z_{s-}}\right) 1(Z_{s-} > 0) Z_{s-} C_s\,ds\right]$$

[as a positive martingale, Z is trapped at 0 if ever it reaches 0, in which case, by (2.66), H becomes 0 as well]

$$= E_0\left[Z_t \int_0^t \left(1 + \frac{H_s}{Z_{s-}}\right) 1(Z_{s-} > 0) C_s\,ds\right]$$

$$= E\left[\int_0^t \left(1 + \frac{H_s}{Z_{s-}}\right) 1(Z_{s-} > 0) C_s\,ds\right]$$

and hence we may take $\lambda_t = (1 + H_t/Z_{t-})1(Z_{t-} > 0)$. □

Among other consequences of Theorem 2.34 is that it defines the class of allowable error processes in the context of martingale estimation (Chapter 5).

2.6 CONDITIONING IN GENERAL SPACES

The foregoing exposition makes clear the fundamental role of the linear order structure of $\mathbf{R}_+$ in the theory of conditioning for point processes given partial observations. Such ideas as the past of a point process, the strict past, stopping time, predictable process, and innovation seem linked inextricably to point processes on $\mathbf{R}_+$ being observed in the natural order of increasing time. Even in higher-dimensional Euclidean spaces (where there are at least intrinsic partial orders), let alone in general LCCB spaces, there seems little prospect of a full-fledged analog of the compensator/stochastic intensity approach. The central difficulty is that the theory of martingales is in many senses inherently one-dimensional.

Nevertheless, the situation is not hopeless: there do exist analogs of some concepts (notably, predictability) discussed in preceding sections. In this section we describe two approaches to conditioning for point processes on general spaces. The first, based on *exvisibility* (a spatial version of predictability), uses projection methods and mimics fairly closely some ideas involving compensators that were presented in Section 2.3.

By contrast the second approach is a general theory of conditional distributions $P\{N_B \in (\cdot)|\mathcal{F}^N(B^c)\}$, where B is a (bounded) subset of the underlying space E and N_B is the restriction of N to B: $N_B(A) = N(A \cap B)$, broad enough to range from analogs of compensators at one extreme to Palm distributions at the other. The heart of this theory is a suitably formulated extension of the statement in Proposition 2.5 that for a point process on $\mathbf{R}_+$ the known information changes only at the arrival times. For a point process N on a general space this becomes: if $B \subset A$ and $N(A \backslash B) = 0$, then $\mathcal{F}^N(A^c)$ and $\mathcal{F}^N(B^c)$ contain the same information pertaining to N_B. Since $\mathcal{F}^N(A^c) \subset \mathcal{F}^N(B^c)$ the latter contains more information in principle, but only for realizations of N for which $A \backslash B$ contains at least one point of N. A precise formulation is embodied in the following consistency property.

Lemma 2.35. Let N be a point process on E and suppose that $B \subset A$. Then on $\{N(A\backslash B) = 0\}$,

$$P\{N_B \in (\cdot)|\mathcal{F}^N(B^c)\} = \frac{P\{N_B \in (\cdot), N(A\backslash B) = 0|\mathcal{F}^N(A^c)\}}{P\{N(A\backslash B) = 0|\mathcal{F}^N(A^c)\}} \tag{2.67}$$

Proof: For $\Lambda \in \mathcal{F}^N(B^c)$ of the form $\Lambda = \{N_{A\backslash B} \in \Gamma_1, N_{A^c} \in \Gamma_2\}$ and $\Delta \in \mathcal{F}^N(B)$,

$$\begin{aligned}
&P\{N_B \in \Delta, N(A\backslash B) = 0, \Lambda\} = 1(0 \in \Gamma_1)P\{N_B \in \Delta, N(A\backslash B) = 0,\ N_{A^c} \in \Gamma_2\}\\
&= 1(0 \in \Gamma_1)E[P\{N_B \in \Delta, N(A\backslash B) = 0|\mathcal{F}^N(A^c)\}; \{N_{A^c} \in \Gamma_2\}]\\
&= 1(0 \in \Gamma_1)E\left[\frac{P\{N_B \in \Delta, N(A\backslash B) = 0|\mathcal{F}^N(A^c)\}}{P\{N(A\backslash B) = 0|\mathcal{F}^N(A^c)\}}\, 1(N(A\backslash B) = 0)1(N_{A^c} \in \Gamma_2)\right]\\
&= E\left[\frac{P\{N_B \in \Delta, N(A\backslash B) = 0|\mathcal{F}^N(A^c)\}}{P\{N(A\backslash B) = 0|\mathcal{F}^N(A^c)\}}\, 1(N(A\backslash B) = 0);\Lambda\right]
\end{aligned}$$

and (2.67) holds by the monotone class theorem. □

It follows that given sets B_1, B_2,

$$P\{N_{B_1 \cap B_2} \in (\cdot)|\mathcal{F}^N(B_1^c)\} = P\{N_{B_1 \cap B_2} \in (\cdot)|\mathcal{F}^N(B_2^c)\} \qquad (2.68)$$

on $N(B_1 \Delta B_2) = 0$. That is, for two sets containing the same points of N, one makes the same prediction concerning the behavior of N on their complements. As seen below and as suggested by (2.67) and (2.68), all of the conditional distributions of N can be realized from a single kernel, the *Gibbs kernel*. Moreover, also as suggested by (2.67), the Gibbs kernel can be expressed in terms of the (reduced) Palm distributions $Q(\mu,dv)$ of N (see Section 1.7). Intuitively, suppose that B were fixed in (2.67) and that we could take A to be the random set $B \cup (\text{supp } N)^c$, where supp N = $\{x : N(\{x\}) = 1\}$ is the *support* of N. Then $N(A \backslash B) = 0$ and (2.67) becomes

$$P\{N_B \in (\cdot)|\mathcal{F}^N(B^c)\} = P\{N_B \in (\cdot)|\mathcal{F}^N(B^c \cap \text{supp } N)\} = Q(N_{B^c}, \cdot) \qquad (2.69)$$

However, we are jumping ahead. Before providing details of this general theory of conditioning we will explicate the recently developed theory of exvisibility along with the older idea of the conditional intensity of a point process; this theory recognizably parallels that in Sections 2.3 – 2.5.

Let N be a simple point process on E with mean measure belonging to M. Exvisibility ("visibility from the outside") is analogous to predictability but is not a generalization since the two need not coincide for point processes on $\mathbf{R}_+$ (Exercise 2.29).

Definition 2.36. a) The *exvisible σ-algebra* engendered by N is the σ-algebra $\mathcal{V}$ on $E \times \Omega$ generated by sets

$$B \times \Lambda, \qquad (2.70)$$

where $B \in \mathcal{B}$ (i.e., B is a bounded Borel set) and $\Lambda \in \mathcal{F}^N(B^c)$.

b) A stochastic process X parametrized by E is *exvisible* with respect to N if the mapping $(y,\omega) \rightarrow X_y(\omega)$ is $\mathcal{V}$-measurable.

c) A random measure M on E is *exvisible* with respect to N if M is $\mathcal{F}^N$-measurable and if the process $X_y = M(\{y\})$ is exvisible.

Roughly speaking, a process is exvisible if for each y its value at y is determined by the restriction of N to $\{y\}^c$, and a random measure M is exvisible if for each y, knowledge of N on $\{y\}^c$ determines whether there is an atom of M at y and, if so, its size. In particular, every diffuse random measure is exvisible (diffuseness is the spatial analog of continuity). The point process N may be exvisible with respect to itself: on $\mathbf{R}_+$ the random translate $N = \Sigma\, \varepsilon_{X+n}$ of Example 1.60 is exvisible; or N may fail to be exvisible: a Poisson process (because of independent increments) is not exvisible. Exvisibility differs from predictability in at least two respects.

First, Definition 2.36 is formulated only for the internal history of N, and this restriction is crucial to some arguments below. Second and more important, predictability is one-sided, dealing with prediction of dN_t from $\mathcal{F}^N_{t-}$, but exvisibility on $\mathbf{R}_+$ is *two-sided*, and concerns estimation of dN_t from $\mathcal{F}^N(\{t\}^c)$. Thus a point process may be exvisible without being predictable and the exvisible projection and compensator need not be the same.

Nonetheless, the two theories are parallel in many senses. For example, as for compensators, a projection theorem is one basic result.

Theorem 2.37. Let M be a random measure measurable with respect to $\mathcal{F}^N$. Then there is a unique exvisible random measure A satisfying

$$E[\int X\,dM] = E[\int X\,dA] \tag{2.71}$$

for every positive exvisible process X. □

We term A the *exvisible projection* of M. The proof of Theorem 2.37 is analogous to that of existence of compensators: one first shows that a process $X \geq 0$ admits an exvisible projection $\widetilde{X} \geq 0$ such that $E[X_Y] = E[\widetilde{X}_Y]$ for every *exvisible random point* Y, that is, Y is a random element of E whose graph $\{(y,\omega): y = Y(\omega)\}$ belongs to $\mathcal{V}$, and then (2.71) corresponds to the adjoint of this operator.

Parallelism between the two theories extends to a "martingale" characterization of the exvisible projection.

Theorem 2.38. Let M be an $\mathcal{F}^N$-measurable random measure with $\mu_M \in \mathbf{M}$ and let A be an exvisible random measure. Then the following are equivalent:

a) A is the exvisible projection of M;

b) for each $B \in \mathcal{B}$,

$$E[M(B)|\mathcal{F}^N(B^c)] = E[A(B)|\mathcal{F}^N(B^c)] \tag{2.72}$$

Proof: By (2.70) elementary exvisible processes have the form $X_y(\omega) = 1(y \in B)1(\omega \in \Lambda)$ for $B \in \mathcal{B}$ and $\Lambda \in \mathcal{F}^N(B^c)$, and for such a process

$$E[\int X\,dM] = E[M(B);\Lambda], \tag{2.73a}$$

and

$$E[\int X\,dA] = E[A(B);\Lambda] \tag{2.73b}$$

By (2.71), a) implies equality of the left-hand sides of (2.73a) and (2.73b), yielding b); conversely, b) implies equality of the right-hand sides, which gives a) via the monotone class theorem. □

By iterated conditioning we obtain the "martingale" version.

Corollary 2.39. The exvisible projection A of M satisfies

$$E[M(C)|\mathcal{F}^N(B^c)] = E[A(C)|\mathcal{F}^N(B^c)] \tag{2.74}$$

whenever $C \subset B \in \mathcal{B}$. □

In particular, Theorem 2.38 applies to N itself, and (2.72) admits an interpretation analogous to (2.22a): if A is the exvisible projection of N, then for each y

$$A(dy) = E[N(dy)|\mathcal{F}^N(\{y\}^c)] \tag{2.72a}$$

Observe once more, however, that when $E = \mathbf{R}_+$, (2.22a) is one-sided, whereas (2.72a) is two-sided.

Historical development of the theory originated in the direction represented for compensators by Theorem 2.17, whose analog for general spaces is the following result.

Theorem 2.40. Let N be a simple, integrable point process on E and let $\mathcal{D} = (D_{nj})$ be a null array of $\mathcal{B}$-partitions of E. Then there exists a random measure A such that for each $D \in \mathcal{D}$,

$$\begin{aligned} A(D) &= \lim_{n\to\infty} \sum_{D_{nj} \subset D} P\{N(D_{nj}) = 1|\mathcal{F}^N(D_{nj}^c)\} \\ &= \lim_{n\to\infty} \sum_{D_{nj} \subset D} E[N(D_{nj})|\mathcal{F}^N(D_{nj}^c)] \end{aligned} \tag{2.75}$$

almost surely and in L^1.

Proof: (Sketch). We may presume to work with regular versions of the conditional probabilities $P\{\cdot|\mathcal{F}^N(D^c)\}$, $D \in \mathcal{D}$, which further satisfy the regularity conditions that for $C \subset D$ in $\mathcal{D}$,

$$P\{N(D \setminus C) = 0|\mathcal{F}^N(D^c)\} > 0 \tag{2.76}$$

and that on $\{N(D \setminus C) = 0\}$,

$$P\{(\cdot)|\mathcal{F}^N(C^c)\} = \frac{P\{(\cdot), N(D \setminus C) = 0|\mathcal{F}^N(D^c)\}}{P\{N(D \setminus C) = 0|\mathcal{F}^N(D^c)\}} \tag{2.77}$$

We note that (2.76) is equivalent to the condition (Σ) of Papangelou (1974b) (see the chapter notes), while (2.77) holds by Lemma 2.35.

Hence for $y \in C \subset D$,

$$P\{N(C) > 0 | \mathcal{F}^N(C^c)\} = 1 - \frac{P\{N(D) = 0 | \mathcal{F}^N(D^c)\}}{P\{N(D \setminus C) = 0 | \mathcal{F}^N(D^c)\}},$$

and the right-hand side of this expression converges as $C \downarrow \{y\}$; consequently for each y there exists the limit

$$U_y = \lim_{D \downarrow \{y\}} P\{N(D) > 0 | \mathcal{F}^N(D^c)\}$$

Moreover, as will be justified momentarily, $A^1 = \sum_{y \in E} U_y \varepsilon_y$ is a random measure on E.

The key step is to show existence of a random measure A^2 satisfying

$$A^2(\{y\})N(\{y\}) \equiv 0, \tag{2.78}$$

such that for each D,

$$A^2(\cdot \cap D) = \frac{E[N_D(\cdot)1(N(D)=1) | \mathcal{F}^N(D^c)]}{P\{N(D) = 0 | \mathcal{F}^N(D^c)\}}, \tag{2.79}$$

on $\{N(D) = 0\}$, and finally, with A_d^2 denoting the diffuse component of A^2, such that

$$A^2 - A_d^2 = \sum_y (1 - N(\{y\})) \frac{U_y}{1 - U_y} \varepsilon_y \tag{2.80}$$

In fact, one may take A^2 to be *defined* by (2.78) and (2.79), using (2.77) to ensure that the definition is consistent. Then for $y \in D$, (2.79) implies that on $\{N(D) = 0\}$,

$$\begin{aligned} A^2(\{y\}) &= \frac{P\{N(\{y\}) = N(D) = 1 | \mathcal{F}^N(D^c)\}}{P\{N(D) = 0 | \mathcal{F}^N(D^c)\}} \\ &= \frac{P\{N(\{y\}) = N(D) = 1 | \mathcal{F}^N(D^c)\}}{P\{N(D \setminus \{y\}) = 0 | \mathcal{F}^N(D^c)\}} \\ &\quad \times \frac{P\{N(D \setminus \{y\}) = 0 | \mathcal{F}^N(D^c)\}}{P\{N(D \setminus \{y\}) = 0 | \mathcal{F}^N(D^c)\} - P\{N(\{y\}) = N(D) = 1 | \mathcal{F}^N(D^c)\}} \\ &= \frac{U_y}{1 - U_y}, \end{aligned}$$

so that (2.80) holds, and hence A^2 is a well-defined (i.e., locally finite) random measure. For details of the argument needed to check that A^2 is finite *near* points of N [it vanishes *at* points of N by (2.78)], see Kallenberg (1980).

Now let

$$A = A^1 + A_d^2; \tag{2.81}$$

we will show that A satisfies (2.75). For $D \in \mathcal{D}$, (2.77) implies that on $N(D) = 0$,

$$\sum_{D_{nj} \subset D} P\{N(D_{nj}) = 1 | \mathcal{F}^N(D_{nj}^c)\} = \sum_{D_{nj} \subset D} \frac{P\{N(D_{nj}) = N(D) = 1 | \mathcal{F}^N(D^c)\}}{P\{N(D \setminus D_{nj}) = 0 | \mathcal{F}^N(D^c)\}}$$

$$\to \int_D \frac{P\{N(dy) = N(D) = 1 | \mathcal{F}^N(D^c)\}}{P\{N(D \setminus \{y\}) = 0 | \mathcal{F}^N(D^c)\}}$$

$$= A(D).$$

In general (by the monotone class theorem),

$$\sum_{D_{nj} \subset D} P\{N(D_{nj}) = 1 | \mathcal{F}^N(D_{nj}^c)\} 1(N(D_{nj}) = 0) \to \int_D (1 - N(\{y\}))A(dy)$$

$$= A(D) - \sum_{y \in D} U_y N(\{y\}) \quad (2.82)$$

But by the construction and argument in the first part of the proof,

$$\sum_{D_{nj} \subset D} P\{N(D_{nj}) = 1 | \mathcal{F}^N(D_{nj}^c)\} 1(N(D_{nj}) > 0) \to \sum_{y \in D} U_y N(\{y\}) \quad (2.83)$$

The first part of (2.75) follows at once from (2.82) – (2.83) and the second holds because N is simple and integrable. □

The random measure A of Theorem 2.40 is the *conditional intensity* of N. Under the assumptions of the theorem it is independent of $\mathcal{D}$ (see Papangelou, 1974b; Kallenberg, 1980). Exvisible projections and conditional intensities are connected by the following result.

Proposition 2.41. Let N be a point process satisfying the hypotheses of Theorem 2.40. Then the conditional intensity and the exvisible projection of N coincide.

Proof: Let A be the conditional intensity of N; we show that A satisfies the conditions of Theorem 2.38. For (2.72), if $D \in \mathcal{D}$, then by (2.75) and L^1 convergence

$$E[A(D) | \mathcal{F}^N(D^c)] = \lim E[\sum_{D_{nj} \subset D} E[N(D_{nj}) | \mathcal{F}^N(D_{nj}^c)] | \mathcal{F}^N(D^c)]$$

$$= E[N(D) | \mathcal{F}^N(D^c)],$$

which yields (2.72) by standard approximations. To show that the conditional intensity is exvisible, we use (2.81). The diffuse component A_d^2 is trivially exvisible, while A^1 is exvisible by construction. □

Thus the approach of "conditioning on the complement of a small set," which on $\mathbf{R}_+$ is represented in part by the compensator, admits an analog in general spaces. By contrast, the Palm distributions represent the opposite extreme of "conditioning on a subset of the support" of a point process. One can link these extremes. The key idea is expressed by (2.67) and (2.68): up to rescaling, all relevant information is contained in the observed atom positions. Observation over a larger set containing no additional points of the process does not—except by scaling—change the conditional distributions of the unobserved part. As noted previously, this raises the hope that conditional probabilities of the form (2.84) are all expressible from a single kernel, which itself must be expressible in terms of the Palm distributions. The heuristic expression (2.69) would then also be valid and would imply that on $\{N(B)=0\}$,

$$\frac{P\{N_B \in (\cdot)|\mathcal{F}^N(B^c)\}}{P\{N(B)=0|\mathcal{F}^N(B^c)\}} = \frac{Q(N,\cdot)}{Q(N,\{0\})} \tag{2.84}$$

Finally, (2.84) shows that any conditional probability can be calculated from the Palm distributions. We now make (2.84) precise (almost).

Theorem 2.42. Let N be a point process on E with Palm distributions $Q = Q_N$, and suppose that $E[N(E)] < \infty$. Then the random measure M on M_p defined by

$$M(dv) = \frac{[Q(N,dv)]}{Q(N,\{0\})} \tag{2.85}$$

satisfies, for each $B \in \mathcal{B}$,

$$M(dv) = \frac{P\{N_B \in dv|\mathcal{F}^N(B^c)\}}{P\{N(B)=0|\mathcal{F}^N(B^c)\}} \tag{2.86}$$

on $\{N(B)=0\} \cap \{v(B^c)=0\}$.

Proof: (Sketch). The ignored point is that the denominators in (2.85) and (2.86) are positive a.s. (even for the latter this is not a tautology). Let C be the compound Campbell measure of N [see (1.63)] and suppose that $\Lambda \in \mathcal{M}_p \cap \{\mu : \mu(B)=0\}$ and $\Gamma \in \mathcal{M}_p \cap \{\mu : \mu(B^c)=0\}$. Then, evidently,

$$E\left[\frac{P\{N_B \in \Gamma|\mathcal{F}^N(B^c)\}}{P\{N(B)=0|\mathcal{F}^N(B^c)\}} 1(N(B)=0)1(N_{B^c} \in \Lambda)\right] = P\{N_B \in \Gamma, N_{B^c} \in \Lambda\},$$

while

$$E\left[\frac{Q(N,\Gamma)}{Q(N,\{0\})} 1(N(B)=0)1(N_{B^c} \in \Lambda)\right]$$

$$= \int C(\{0\} \times d\nu)1(\nu(B) = 0)1(\nu_{B^c} \in \Lambda) \frac{Q(\nu, \Gamma)}{Q(\nu, \{0\})}$$

[since $C(\{0\}, \cdot) = Q(0, \cdot) = \mathcal{L}_N$ (see Exercise 2.31)]

$$= \int C(d\mu \times \{0\})1(\mu(B) = 0)1(\mu_{B^c} \in \Lambda) \frac{Q(\mu, \Gamma)}{Q(\mu, \{0\})}$$

(by symmetry of C)

$$= \int K_N(d\mu)Q(\mu, \{0\})1(\mu(B) = 0)1(\mu_{B^c} \in \Lambda) \frac{Q(\mu, \Gamma)}{Q(\mu, \{0\})}$$

$$= C(\{\mu : \mu(B) = 0, \mu_{B^c} \in \Lambda\} \times \Gamma)$$

By (1.62) and symmetry of C,

$$\begin{aligned} C(\{\mu : \mu(B) = 0, \mu_{B^c} \in \Lambda\} \times \Gamma) &= C(\Gamma \times \{\nu : \nu(B) = 0, \nu_{B^c} \in \Lambda\}) \\ &= P\{N_B \in \Gamma, N_{B^c} \in \Lambda\}, \end{aligned}$$

and this completes the proof. □

We formalize some additional terminology.

Definition 2.43. The kernel

$$G(\mu, d\nu) = \frac{Q(\mu, d\nu)}{Q(\mu, \{0\})} \tag{2.87}$$

is the *Gibbs kernel* of N. The random measure $M = G(N)$ is the *Gibbs measure* of N. □

If N is a Poisson process on E with finite mean measure μ, then $Q(\nu, \cdot) = \mathcal{L}_N(\cdot)$ for all ν (Section 1.7) and hence

$$G(\nu, d\eta) = e^{\mu(E)}\mathcal{L}_N(d\eta) \tag{2.88}$$

More generally, let N be Poisson and let $\tilde{N}$ be a point process such that the likelihood ratio $f = d\mathcal{L}_{\tilde{N}}/d\mathcal{L}_N$ exists and is strictly positive. Then by the independent increments and conditional uniformity properties of N,

$$\frac{P\{\tilde{N}_B \in \Gamma | \mathcal{F}^{\tilde{N}}(B^c)\}}{P\{\tilde{N}(B) = 0 | \mathcal{F}^{\tilde{N}}(B^c)\}} = \frac{E[f(N)1(N_B \in \Gamma) | \mathcal{F}^N(B^c)]}{E[f(N)1(N(B) = 0) | \mathcal{F}^N(B^c)]}$$

$$= f(\tilde{N}_{B^c})^{-1} \sum_{k=0}^{\infty} \frac{1}{k!} \int_{E^k} f\left(\tilde{N}_{B^c} + \sum_{i=1}^{k} \varepsilon_{x_i}\right) 1\left(\sum_{i=1}^{k} \varepsilon_{x_i} \in \Gamma\right) \mu^k(dx)$$

By introducing the "energy" (log-likelihood) function $U(\nu, \nu+\eta) = -\log[f(\nu+\eta)/f(\nu)]$, one obtains an explicit representation for the Gibbs kernel of $\tilde{N}$:

$$G_{\tilde{N}}(\nu,\Gamma) = \sum_{k=0}^{\infty} \frac{1}{k!} \int_{E^k} \exp\left[-U\left(\nu, \nu + \sum_{i=1}^{k} \varepsilon_{x_i}\right)\right] 1\left(\sum_{i=1}^{k} \varepsilon_{x_i} \in \Gamma\right) \mu^k(dx)$$

It remains to connect the Gibbs kernel explicitly with the "conditioning on the complement of a point" represented by exvisible projections and conditional intensities. Suppose that D is a set belonging to the null array $\mathcal{D}$ of Theorem 2.40 with $N(D) = 0$. Then with other notation as in that theorem, by (2.75) and (2.86),

$$A(D) = \lim \sum_{D_{nj} \subset D} E[N(D_{nj}) | \mathcal{F}^N(D_{nj}^c)]$$

$$= \frac{\lim \Sigma\, P\{N(D_{nj}) = 0 | \mathcal{F}^N(D_{nj}^c)\} \int Q(N,d\nu) 1(\nu(D^c) = 0)\, \nu(D_{nj})}{Q(N,\{0\})}$$

$$= Q(N,\{0\})^{-1} \lim \int Q(N,d\nu) 1(\nu(D^c) = 0) \sum_{D_{nj} \subset D} \nu(D_{nj}) P\{N(D_{nj} = 0 | \mathcal{F}^N(D_{nj}^c)\}$$

$$= Q(N,\{0\})^{-1} \int Q(N,d\nu) 1(\nu(D^c) = 0) \int_D \nu(dx)(1 - U_x)$$

For $x \in D_{nj}(x)$ [keep in mind that $N(D) = 0$]

$$1 - U_x = \lim P\{N(D_{nj}) = 0 | \mathcal{F}^N(D_{nj}(x)^c)\}$$

$$= \frac{Q(N,\{0\})}{Q(N,\{0\}) + Q(N,\{\varepsilon_x\})}$$

Consequently, on $\{N(D) = 0\}$,

$$A(D) = \int Q(N,d\nu) 1(\nu(D^c) = 0) \int_D [Q(N,\{0\}) + Q(N, \{\varepsilon_x\})]^{-1} \nu(dx) \quad (2.89)$$

Atoms of A can arise only from the random measure $A^1 = \Sigma\, U_y \varepsilon_y$ and hence the component of A not embodied in (2.89) is simply the purely atomic random measure $\tilde{A}(B) = \int_B U\, dN$.

EXERCISES

2.1. Let $(\mathcal{H}_t)$ be a history on $(\mathbf{\Omega}, \mathcal{F})$. Prove that every right- or left-continuous adapted process X is progressive.

2.2. Define a stopping time T to be *predictable* if there is a sequence of stopping times T_n such that $T_n \uparrow T$ but $T_n < T$ on $\{T > 0\}$ for each n, and *totally inaccessible* if $P\{T = S\} = 0$ for every predictable stopping time S. Prove that the arrival times $T_1, T_2, \ldots$ in a stationary Poisson process N are totally inaccessible. (The history is $\mathcal{F}^N$.)

2.3. Let $\mathcal{H}$ be a history with respect to which M is a martingale and A is an adapted increasing process. Prove that for each finite stopping time T such that $E[|M_T A_T|] < \infty$, $E[\int_0^T M_s \, dA_s] = E[M_T A_T]$. [*Hint*: Use the property that $M_s = E[M_T | \mathcal{H}_s]$ on $\{s < T\}$.]

2.4. Show by construction of an explicit example that the predictability hypothesis in Theorem 2.9 cannot be suppressed.

2.5. Let $(\mathcal{F}_t^N)$ be the internal history of a point process N on $\mathbf{R}_+$ and let M be a process that is simultaneously a martingale and predictable. Prove that almost surely $M_t = M_0$ for all $t > 0$ (i.e., M is constant over time). [*Hint*: Use Proposition 2.6.]

2.6. Let N be a Poisson process on $\mathbf{R}_+$ with diffuse mean measure μ. Prove directly that $M_t = N_t - \mu_t$ is a martingale.

2.7. Let N be a Poisson process on $\mathbf{R}_+$ with diffuse mean measure μ and let $M = N - \mu$ be the $\mathcal{F}^N$-innovation martingale.

a) Compute each of the stochastic integrals $\int_0^t N_s \, dN_s$, $\int_0^t N_{s-} \, dN_s$, $\int_0^t N_s \, dM_s$, and $\int_0^t N_{s-} \, dM_s$.

b) Determine which of these processes are martingales and interpret in terms of the discussion of predictability in Section 2.2.

2.8. Verify (2.39).

2.9. Prove that a simple point process N with continuous, deterministic compensator is necessarily a Poisson process. [*Hint*: It suffices (why?) to show that N has independent increments.]

2.10. a) Let N be a Cox process on $\mathbf{R}_+$ directed by the diffuse random measure M and let $\mathcal{H}_t = \mathcal{F}_t^N \vee \mathcal{F}_\infty^M$. Prove that the $\mathcal{H}$-compensator of N is M.

b) Conversely, let N be a point process on $\mathbf{R}_+$ adapted to $\mathcal{H}$ and suppose that there exists an increasing process A, with $A_t \in \mathcal{H}_0$ for all t, such that A is the $\mathcal{H}$-compensator of N. Prove that N is a Cox process directed by A.

2.11. Let N be a point process on $\mathbf{R}_+$ having $\mathcal{F}^N$-compensator A and assume that $P\{\lim_t A_t = \infty\} = 1$. Define $\widetilde{A}_s = \inf\{t : A_t \geq s\}$ and let $\widetilde{N}_s = N(\widetilde{A}_s)$. Prove that $\widetilde{N}$ is a stationary Poisson process with rate 1.

2.12. This exercise demonstrates how the history can affect the compensator in a nontrivial manner. Let (N_t) be a *Yule process* on $\mathbf{R}_+$, that is, a pure birth process with birth rate in state i equal to i [N is a Markov process with generator $A(i, i+1) = -A(i,i) = i$].

a) Prove that there is a random variable V such that $e^{-t} N_t \to V$ almost surely. (It is clear intuitively that the population grows exponentially.)

b) Prove that the $\mathcal{F}^N$-compensator of N is $A_t = \int_0^t N_{s-} \, ds$. (That is, N_{s-} is the $\mathcal{F}^N$-stochastic intensity of N.)

c) Show that the compensator of N with respect to $\mathcal{H}_t = \mathcal{F}_t^N \vee \sigma(V)$, where V is as in a), is $B_t = V(e^t - 1)$, and conclude with the aid of Exercise 2.10 that N is a Cox process.

2.13. Let T be a positive random variable with distribution function F admitting density f. The function $h(t) = f(t)/[1 - F(t)]$ is the *hazard function* or *failure rate function* of F. (The terms come from survival analysis and reliability.) Verify that:

a) For each t, $h(t) = \lim_{s \to 0} s^{-1} P\{T \le t + s | T > t\}$.

b) For each t, $F(t) = 1 - \exp[-\int_0^t h(u)\,du]$, and hence F is uniquely determined by h.

c) The distribution F is exponential if and only if the hazard function h is constant.

d) The expression (2.48) holds.

2.14. Let N be a renewal process with interarrival distribution F admitting hazard function h. Prove that the $\mathcal{F}^N$-stochastic intensity of N is $\lambda_t = h(V_{t-})$, where $V_t = t - T_{N_t}$, the *backward recurrence time* at t, is the time elapsed since the most recent arrival (see Chapter 8).

2.15. (Continuation of Exercise 2.14) Explain why the process $h(V_t)$ is *not* the stochastic intensity of N.

2.16. Let N be a point process with $\mathcal{H}$-stochastic intensity (λ_t) and let $\mathcal{G}$ be another history with $\mathcal{F}_t^N \subset \mathcal{G}_t \subset \mathcal{H}_t$ for each t. Prove that

a) There exists a $\mathcal{G}$-stochastic intensity $(\tilde{\lambda}_t)$;

b) For each t, $\tilde{\lambda}_t = E[\lambda_t | \mathcal{G}_t]$ almost surely.

2.17. Let $N(1)$, $N(2)$ be independent point processes on $\mathbf{R}_+$ and suppose that $N(i)$ has $\mathcal{F}^{N(i)}$-stochastic intensity $\lambda_t(i)$. Prove that with respect to $\mathcal{H}_t = \mathcal{F}_t^{N(1)} \vee \mathcal{F}_t^{N(2)}$ the superposition $N = N(1) + N(2)$ has stochastic intensity $\lambda_t = \lambda_t(1) + \lambda_t(2)$.

2.18. This exercise describes an explicit construction of a point process with a prescribed stochastic intensity. Let λ_t be a nonnegative process and let $\tilde{N} = \Sigma\, \varepsilon_{(X_i, Y_i)}$ be a stationary Poisson process on $\mathbf{R}_+^2$ with intensity 1; assume that λ and $\tilde{N}$ are independent. Define $N = \Sigma\, \varepsilon_{T_i}$ by the following recursive procedure:

i) $T_1 = \inf\{X_i : Y_i \le \lambda_0\}$ is the x-coordinate of the leftmost point of $\tilde{N}$ falling below the line $y = \lambda_0$.

ii) With $T_1, \ldots, T_n$ previously defined, $T_{n+1} = \inf\{X_i : Y_i \le \lambda_{T_n}\}$. Prove that N has stochastic intensity λ.

2.19. Verify (2.56).

WARNING! In Exercises 2.20 – 2.23 it may be difficult or impossible at this point to proceed beyond conditional probabilities or expectations yielded by Exercise 2.16. (These problems should be attacked by first calculating the stochastic intensity with respect to a larger history.) Do not attempt to proceed further now; techniques will be presented in Chapters 3, 5, and 7 for computing the required conditional expectations.

2.20. a) Let $\tilde{N}$ be a point process on $\mathbf{R}_+$ with stochastic intensity (λ_t) and lejt T be a positive random variable, independent of N, with hazard function h. Let N be the point process $N_t = \tilde{N}(T \wedge t)$. Calculate the stochastic intensity of N with respect to

i) The history $\mathcal{H}_t = \mathcal{F}_t^{\tilde{N}} \vee \sigma(T)$;

ii) The internal history $\mathcal{F}^N$.

b) Let $N(1)$, $N(2)$ be independent Poisson processes on $\mathbf{R}_+$ with rates ν_1, ν_2, respectively, and let T be a random variable independent of $N(1)$ and $N(2)$ and having an exponential distribution with parameter θ. Let N be the point process defined by $N_t = N_{T\wedge t}(1) + N_{(T-t)^+}(2)$; at the random time T, N switches from being Poisson with rate ν_1 to being Poisson with rate ν_2. Prove that N is a Cox process (see Exercise 2.10) and calculate the stochastic intensity with respect to the internal history. (This is the *disruption problem*.)

2.21. Let $\tilde{N}$ be a Poisson process on $\mathbf{R}_+$ with rate c and let (X_t) be a Markov process with state space $\{0,1\}$ and generator

$$A = \begin{bmatrix} -a & a \\ b & -b \end{bmatrix},$$

where a, $b > 0$. Assume that $\tilde{N}$ and X are independent. Calculate the stochastic intensity of the *partially observed Poisson process* $N_t = \int_0^t X_s \, d\tilde{N}_s$ with respect to its **internal history**.

2.22. Let $\tilde{N}$ be a Poisson process with rate ν, representing arrival times of particles at a Geiger counter. Each arrival locks the counter for an interval of deterministic length τ, during which other arrivals that may occur have no effect whatever.

a) Calculate the stochastic intensity of the point process N of recorded arrivals. [*Hint*: Use Theorem 2.30.]

b) Generalize to the case that the locked times are i.i.d. random variables independent of $\tilde{N}$.

2.23. This exercise examines the classical model of *censored survival data*. Let $X_1, \ldots, X_n$ be i.i.d. nonnegative random variables with absolutely continuous distribution F; these are the survival times (say, for patients undergoing some medical treatment). However, not all the X_i are observed (contact with a patient may be lost, or the study may terminate, for example); this phenomenon is called *censoring*, and may be modeled (in the simplest case) as follows. Let $Y_1, \ldots, Y_n$ be i.i.d. random variables with absolutely continuous distribution G, such that (X_i) and (Y_i) are independent; the Y_i are censoring times. The observed data for patient i are $Z_i = X_i \wedge Y_i$ and $\delta_i = 1(X_i \leq Y_i)$, so that it is known whether Z_i is a survival time or a censoring time. These data can be summarized in the marked point process $N = \sum_{i=1}^n \varepsilon_{(Z_i, \delta_i)}$. Calculate the stochastic intensity of N with respect to the internal history.

2.24. Consider a waiting facility at which objects arrive according to a point process N with $\mathcal{F}^N$-stochastic intensity λ. There they are detained and dispatched in groups. Specifically, at a fixed time τ all objects present are dispatched; however, there may be one intermediate dispatching at a stopping time $T \leq \tau$ chosen to minimize the total waiting cost over the interval $[0,\tau]$. (One can think of τ as a cycle length, so that the long-run average cost per time unit is $1/\tau$ times the cost over $[0,\tau]$.) Let $c(s)$ be the cost of storing one object for s time units; we assume c to be strictly increasing and differentiable. With no intermediate dispatch the total expected cost is $E[\int_0^\tau c(\tau - s)\, dN_s]$. The cost *saving* from a dispatch at the stopping time T is $C(T) = E[N_{T}c(\tau - T)]$, which should be maximized in order that the total cost be minimized.

a) Show that $E[\int_0^\tau c(\tau - s)\, dN_s] = E[\int_0^\tau c(\tau - s)\lambda_s\, ds]$.

b) Show that $C(T) = E[\int_0^T [c(\tau - s)\lambda_s - N_s c'(\tau - s)]\, ds]$. [*Hint*: Use integration by parts (see Proposition 2.8).]

c) Verify that if N is a Poisson process with rate ν and $c(s) \equiv s$, then $C(T) = E[\int_0^T [\nu(\tau - s) - N_s]\, ds]$.

d) Show that for the special case in c), the optimal stopping time is $T^* = \inf\{s : N_s \geq \nu(\tau - s)\}$.

2.25. Prove the Poisson convergence theorem associated with Exercise 1.24, using compensators and the following plan. Let $X_1, X_2, \ldots$ be i.i.d. positive random variables with density function f, let $\lambda = f(0)$, and for each n let $N^{(n)} = \sum_{i=1}^{n} \varepsilon_{nX_i}$. Use Exercise 2.8 to calculate the stochastic intensity $(\lambda_t^{(n)})$, and then show that for each t, $\lambda_t^{(n)} \to \lambda$ in probability. Deduce from this that the hypotheses of Theorem 2.25 are fulfilled and hence that $N^{(n)}$ converges in distribution to a Poisson process with rate λ.

2.26. Let N be a point process with stochastic intensity (λ_t) for which the innovation martingale M is square integrable. Calculate explicitly the representation (2.59) for the martingale $M_t^2 - \int_0^t \lambda_s\, ds$.

2.27. Prove that there exists no point process for which the innovation martingale is a Wiener process. [*Hint*: See Exercise 2.5.]

2.28. Let N be a Poisson process on a general space E with diffuse mean measure μ. Calculate the exvisible projection of N directly, calculate the conditional intensity using (2.75), and verify that the two are identical.

2.29. Let X be a random variable uniformly distributed on $[0,1]$ and let N be the point process $\sum_{n=1}^{\infty} \varepsilon_{X+n+(1/n)}$. Compute the compensator of N and the exvisible projection of N and show that the two differ.

2.30. Let N be a point process on a general space E with reduced Palm distributions $Q(\mu, d\nu)$. Show that $Q(0, \cdot) = \mathcal{L}_N(\cdot)$.

2.31. Let N be a point process with compound Campbell measure C given by (1.63). Prove that if $P\{N(E) < \infty\} = 1$, then C is symmetric: $C(\Lambda \times \Gamma) = C(\Gamma \times \Lambda)$ for all $\Lambda, \Gamma \in \mathcal{M}_p$.

NOTES

Section 2.1

There are by now numerous references on the measure-theoretic foundations of the theory of stochastic processes. Important earlier works are Chung and Doob (1965) and Meyer (1969); Dellacherie (1972) is the first complete treatment. More recently, Brémaud (1981), Dellacherie and Meyer (1980), Liptser and Shiryayev (1978), and Kallianpur (1980) have given expository presentations; the first of these is oriented especially to point processes and contains several of the propositions in the section.

Section 2.2

The theory of stochastic integrals with respect to martingales and semimartingales, which is founded on that of Wiener and Poisson integrals, has been developed largely by the French school of probabilists, especially Jacod, and by the Japanese, especially Itô, the founder of the theory, and their descriptions of it, even though no one would call them easy, are the best. These include Jacod (1979), Metivier and Pellaumail (1980), and Dellacherie and Meyer (1980), along with numerous papers from the Strasbourg Séminaire des Probabilités. Gill (1980b) and Liptser and Shiryayev (1978) contain developments of stochastic Stieltjes integrals specifically related to point processes and innovation martingales. The integration-by-parts formulas of Proposition 2.8, which are derived via Fubini's theorem, are standard but less widely known than they should be.

Section 2.3

Important general references are Brémaud and Jacod (1977), Dellacherie (1972), Gill (1980b), Liptser and Shiryayev (1978), and Shiryayev (1981); all but the second are devoted entirely or in significant part to point processes. In Assumption 2.10 appear some of the "*conditions habituelles*" of the French school. Although compensators can be traced at least to the "*projection dual prévisible d'un processus croissant*" of Dellacherie (1972) as well as to work of Papangelou, the key paper is Jacod (1975b), where are given in definitive form the martingale characterization of compensators (Theorems 2.14 and 2.22), the uniqueness theorem (Theorem 2.19) the form of likelihood ratios [of which Theorem 2.31 is a special case (see also Theorem 5.2)], and the representation for point process martingales (generalizing Theorem 2.34). The role of compensators in statistical inference began to emerge in Aalen (1975, 1978) and Gill (1980b) (see Chapter 5 for details). There is another important line of contributions to the theory in its formative stages. Motivated by questions of smoothing, filtering, and prediction in communications, several electrical engineers developed results that were crucial in directing the course of later research. (Nearly all of these have by now been supplanted by more general versions, but their importance is not diminished thereby.) In particular, Snyder (1972a, b, 1975) concerning state estimation for Cox processes, Boel et al. (1975) on representation of point process martingales, and van Schuppen (1977) on state estimation with point process observations are key references. The concept, interpretation, and application of innovation are due to these and other engineers, and ultimately to Kalman (1960) (see also Bucy and Kalman, 1961; Kailath, 1970).

Theorem 2.17 appears in Gill (1980b) but is more or less standard and was probably known beforehand; a shortcoming is that application of (2.23) for each t need not produce a process that is predictable in t. Theorem 2.18 is due for ordinary point processes to Dellacherie (1972). Hazard functions are treated in numerous references, especially on reliability and analysis of survival data (see, e.g., Barlow and Proschan, 1975; Kalbfleisch and Prentice, 1980; Lawless, 1982; Nelson, 1982). A good source of martingale results such as Theorem 2.21 and Proposition 2.23, along with ideas of predictable variation and predictable covariation, is Jacod (1979); some of both ideas and results date to earlier work such as Kunita and Watanabe (1967). A sketch of the theory of continuous-time martingales appears in Appendix B.

The Poisson convergence theorem (Theorem 2.25) is taken from Kabanov et al. (1980a); Theorem 2.26 is due to T. C. Brown (1983).

Section 2.4

Of the general sources noted for Section 2.3, only Brémaud (1981) and Gill (1980b) contain extended discussions of stochastic intensities; some material can be found in Brémaud and Jacod (1977) and Jacobsen (1982). Two more elementary (but sometimes oblique) treatments are Cox and Isham (1980) and Snyder (1975). Motivation for study of point processes admitting stochastic intensities stems from statistical inference (Aalen, 1975, 1978; Gill, 1980b), from queueing theory, particularly attempts to describe output processes and flows within networks of queues (see Brémaud, 1981, for discussion and references) and from communication theory. An early version of Theorem 2.31, the representation theorem for likelihood ratios, was given by Rubin (1972), but results of Jacod (1975b), valid for marked point processes with very general mark spaces, are definitive. Kailath and Segall (1975) is a related paper. In the engineering literature (see Snyder, 1975) and elsewhere, point processes with nondeterministic stochastic intensities have been termed *self-exciting*, but in the probability literature "self-exciting" usually connotes a stochastic intensity that is a deterministic function of the past (see, e.g., Proposition 7.32 and Exercise 9.22).

Section 2.5

In addition to Jacod (1975b), important papers on structure of point process martingales are Boel et al. (1975), M. H. A. Davis (1976), van Schuppen (1977), and (for multidimensional analogs) Yor (1976). Brémaud (1981) and Liptser and Shiryayev (1978) are good expository treatments.

Section 2.6

Of the conditioning concepts presented here, exvisibility and exvisible projections, whose definition and theory are due to van der Hoeven (1982), more clearly parallel the dual projection approach of Dellacherie (1972) (see also van der Hoeven, 1983). Conditional intensities were introduced by Papangelou (1974b), the source as well of two regularity conditions that are important to the theory. In the notation of the section, condition (Σ) states that for each bounded set B, $P\{N(B) = 0 | \mathcal{F}^N(B^c)\} > 0$ almost surely on the set where $P\{N(B) = 1 | \mathcal{F}^N(B^c)\} > 0$, while condition (Σ^*) states that for each $B \in \mathcal{B}$, almost surely the conditional expectation $E[N(\cdot) | \mathcal{F}^N(B^c)]$ has no atoms on B. Papangelou (1974b) shows that a point process N satisfying (Σ) and (Σ^*) satisfies a version of (2.86). Results in Kallenberg (1980) generalize those of Papangelou (1974b) by removing a square integrability hypothesis. Both Kallenberg (1980) and van der Hoeven (1982) shed further light on (Σ) and (Σ^*). The remaining ideas and results presented in the section are nearly all due to Kallenberg (1983, chaps. 12 – 14), a strikingly original work that connects conditioning for point process with various ideas in statistical mechanics. In particular, Lemma 2.35, the importance of the consequences thereof, Theorem 2.42, and Definition 2.43 appear there. A less arduous version is Kallenberg (1984).

Exercises

2.2. This is yet another manifestation of the memorylessness property of exponential distributions. See Dellacherie (1972) concerning predictability, accessibility, and total inaccessibility of stopping times.

2.5. The assumption that the history be generated by a point process is absolutely crucial; the property is not true, for example, for the internal history of a Wiener process.

2.7. Predictability of the integrand is the key to differences among these integrals.

2.9, 2.10. The characterization of Poisson processes is due to Watanabe (1964), while the Cox process analog was first given by Brémaud (1975a) (see also T. C. Brown, 1981). Cox processes are characterized by $\mathcal{H}_0$-measurability of the compensator.

2.11. See Jacod (1975b) and Brémaud and Jacod (1977). In addition, this property is related to the construction in Exercise 2.18.

2.12. Basawa and Prakasa Rao (1980) and Karlin and Taylor (1975) have more on Yule processes.

2.13. See also Section 8.2 and accompanying references.

2.14, 2.15. See Proposition 8.10, Theorem 8.17, and associated discussion.

2.16. See Brémaud and Jacod (1977).

2.18. This elegant construction was gleaned from a lecture by Brémaud.

2.19. The model appears in Aalen (1978) (see also Jacobsen, 1982); Billingsley (1961a) is the key classical reference. In van der Plas (1983) a related discrete time model is analyzed using least squares methods.

2.20. See Liptser and Shiryayev (1978). The disruption problem is analyzed further in Chapters 3, 5 (especially Example 5.30), and 7.

2.21. For additional intensity computations, see Section 6.3 and Karr (1982).

2.22. This and other counter models are treated in Feller (1971) (see also Çinlar, 1975a).

2.23. The seminal reference is Kaplan and Meier (1958), which treats estimation of F from the censored data (Z_i,δ_i), $i = 1, \ldots, n$; the ensuing literature is voluminous (see Chapter 5 for additional references).

2.24. Ross (1969) established the Poisson version. For the more general case, see Brémaud and Jacod (1977).

2.25. See references for Exercise 1.21.

2.31. See Kallenberg (1983).

3
Inference for Point Processes: Introduction

In this chapter we describe a general framework for statistical inference and state estimation for point processes. Not all of it is explored in detail in later chapters, but examples presented here illustrate many aspects. One crucial point to remember is that we interpret "inference" as a tripartite subject whose components are first, *statistical inference* in the sense of estimation and hypothesis testing (usually in a nonparametric setting); second, *state estimation*, the optimal reconstruction, realization by realization, of unobserved portions of a point process or of associated random variables or processes; and third, the problem of *combined statistical inference and state estimation*, in which attributes of the probability law of a point process required to calculate state estimators must be replaced by statistical estimators.

We pause to recall the structure of a *statistical model*, which, formulated in generality, is comprised of

1) A *sample space* $(\Omega, \mathcal{G})$ representing the set of all conceivable outcomes of a "random experiment" and a σ-algebra of events (sets of outcomes) of interest.
2) A collection of *random mechanisms* that might govern the experiment, represented as an indexed (as distinct from "parametric") family $\mathcal{P} = \{P_\alpha : \alpha \in I\}$ of candidate probability measures on $(\Omega, \mathcal{G})$. Here I is an arbitrary index set containing as few as two elements or so large that $\mathcal{P}$ is the set of all probability measures on $(\Omega, \mathcal{G})$. One element of $\mathcal{P}$, the "true" law, actually does govern the experiment.
3) *Observations* or *data* represented by a sub-σ-algebra $\mathcal{F}$ of $\mathcal{G}$. In many cases $\mathcal{F} = \sigma(X)$ for some stochastic process X. The form of

the data may be dictated exogenously by the "physics" of a particular application or may be partly or entirely under control of the statistician or experimenter.

Within this setting the central problem of statistical inference is to construct meaningful statements concerning the probability measure actually governing the experiment, or properties of it, based on the observed data. Such statements necessarily involve uncertainty, which must be controlled or at least described.

The goal of *statistical estimation* is to determine the operative probability, or functionals of it, as accurately as possible and to characterize the errors that arise. *Hypothesis testing* concerns choosing which of two values of the index α or two disjoint subsets of the index set I is more compatible (or less incompatible) with the data.

Roughly speaking, "finite sample" theory is the case that the data σ-algebra $\mathcal{F}$ is associated with observations over a compact set [since finiteness is the same as compactness in $\mathbf{N}$, this includes the case $\mathcal{F} = \sigma(X_1, \ldots, X_n)$ that is the basis of "classical" finite sample theory]. "Asymptotic" theory pertains to the case $\mathcal{F} \uparrow \mathcal{F}_\infty$ with $\mathcal{F}_\infty$ representing observation over a noncompact set and $\mathcal{F}$ observations over compact sets that increase to it. Customary usage designates as "parametric" the case that I is a subset of a finite-dimensional Euclidean space and as "nonparametric" all others, but often with a connotation that $\mathcal{P}$ is very large. The intermediate case that the "parameter" is infinite-dimensional yet $\mathcal{P}$ is severely restricted notwithstanding appears infrequently in classical statistics but is central for point processes (see, e.g., Chapter 6). Experimental design deals with collecting data suited to methods available for its analysis and making effective use of limited sampling resources. For example, a point process on $\mathbf{R}$ may be observed for a fixed length of time, until a prescribed number of arrivals occurs, or sequentially, until sufficient data are collected to make statements of prescribed accuracy.

We will not elaborate here on statistical concepts such as consistency, asymptotic normality, efficiency, sufficiency, completeness, and contiguity. Those that arise as conclusions of results are explained briefly as they appear; they and many others not mentioned above are discussed in numerous statistics texts.

Finally, a comment on the term *classical statistics* as it is used here. By it we mean the statistics of regular parametric families, especially exponential families, given i.i.d. observations. The two great principles of estimation, *maximum likelihood* and the *method of moments* (or *substitution*), and the principle of *likelihood ratios* for hypothesis testing, rule classical statistics and will be almost as predominant for us, but the more complicated nature of point processes forces greater reliance on the principle of ad hoc methods exploiting special structure, provided that such methods be demonstrated to produce informative conclusions.

3.1 FORMS OF OBSERVATION

We begin by describing different ways in which point processes may be observed. There are three principal dichotomies, not entirely independent in theory or in practice, pertinent to the form of data arising from observation of a point process N on a space E. These classifications yield the basic statistical models associated with point processes; the main of these are given explicitly below, while fundamental issues of statistical inference are presented in Section 3.2.

The three principal dichotomies are:

1) Whether the underlying space E is *compact* or *noncompact*;
2) Whether the observations are a *single realization* of N or i.i.d. *multiple copies*;
3) Whether the observation is *complete* or *partial*.

Particular kinds of partial observation will be introduced momentarily and others appear in later chapters, so with one proviso we concentrate on the first two dichotomies. The proviso is that in reality and in statements of our results, point processes can only ever really be observed over compact sets regardless of whether there is one realization or multiple copies. Consequently, estimators and test statistics must always be computed from observations over compact sets. Of course, this does not preclude one's studying asymptotic properties as the "set of observation" increases to a noncompact set. It follows that for noncompact E we are always in some sense in the partially observed case, but we prefer to think of complete observations over compact subsets of a noncompact space as a form of complete observation.

Complete Observation

Hence there are four cases of complete observation, two of which are important theoretically and in applications and are understood at least for some classes of point processes. The third is a straightforward generalization of either or both of these two, while the fourth is of evident practical interest but underdeveloped theoretically, albeit not without reason: it is very difficult, especially if one seeks results of reasonable generality.

The important cases are:

1) Observation of *multiple copies* of a point process N on a *compact* set E.

2) Observation of a *single realization* of a point process N on a *noncompact* space E. (Recall that actual observations are over compact subsets of E.)

Their importance derives primarily from their admitting—given suitable assumptions—interesting, useful, and broadly valid asymptotic results as the amount of data becomes infinite, in the former as the number of observed copies increases to infinity and in the latter as the process is observed over compact sets increasing to E.

The third case, multiple copies of a point process on a noncompact space, represents in some ways a surfeit of data. Methods from either of the two principal cases are applicable and no new difficulties arise. In view of our "axiom" that point processes are observed only over compact sets, it seems best to regard this case as an extension of the first case.

The remaining case of observation of a single realization of a point process on a compact set *is* of theoretical and practical interest, but is notoriously difficult (as are analogous "finite sample" problems in other statistical contexts), especially in terms of general techniques and results. Those that do exist deal with models in which special structure is a dominant feature.

We now give explicitly the statistical models for the two principal cases listed above, along with some variants.

Model 3.1. (i.i.d. copies of a point process on a compact space). Let N be a point process on a compact space E. The statistical model of complete observation of a sequence of i.i.d. copies of N is given as follows.

a) The sample space is

$$(\Omega, \mathcal{G}) = (\mathbf{M}_p, \mathcal{M}_p)^{\mathbf{N}} \tag{3.1}$$

Let (N_i) be the sequence of coordinate mappings; each N_i is a point process on E.

b) Let $\{\mathcal{L}_\alpha : \alpha \in I\}$ be an indexed collection of probability measures on $(\mathbf{M}_p, \mathcal{M}_p)$; these are the candidates for $\mathcal{L}_N$, the probability law of N (Definition 1.8). Then $\mathcal{P} = \{\mathcal{P}_\alpha : \alpha \in I;\}$, where for each α

$$P_\alpha = (\mathcal{L}_\alpha)^{\mathbf{N}} \tag{3.2}$$

Under P_α the N_i are i.i.d. with law $\mathcal{L}_\alpha$.

c) The data representing complete observation of $N_1, \ldots, N_n$ are

$$\mathcal{F}_n = \prod_{i=1}^{n} \mathcal{F}^{N_i}, \tag{3.3}$$

where $\mathcal{F}^N = \mathcal{F}^N(E) = \sigma(N(A) : A \in \mathcal{E})$.

Principal variations of this model are to incorporate an underlying random mechanism associated with each point process or to allow only partial observation. The former arises in connection with Cox processes (Chapter 7) and the latter in ways described later in the section.

Model 3.2. (Single realization of a point process on a noncompact space). Let N be a point process on a noncompact space E. The statistical model for complete observation of N is given as follows.

a) The sample space is

$$(\Omega, \mathcal{G}) = (\mathbf{M}_p, \mathcal{M}_p) \tag{3.4}$$

b) Let $\{\mathcal{L}_\alpha : \alpha \in I\}$ be an indexed family of candidate probability laws for N; then simply

$$\mathcal{P} = \{\mathcal{L}_\alpha : \alpha \in I\} \tag{3.5}$$

c) For A a compact subset the data representing complete observation of N over A are the σ-algebra

$$\mathcal{F}^N(A) = \sigma(N(B) : B \subset A) \tag{3.6}$$

Both of these are "one-sample" models in the sense that there is a single operative probability measure. Except for isolated instances, we do not treat two-sample problems, so we have not included them as part of the general formulations.

Partial Observation

We next list some forms of partial observation of point processes, but with no pretense of completeness; it takes little effort to concoct many more. Five forms play central roles in this book.

1) *Restriction to a subset.* This occurs when rather than N (observed over all of E) one observes only the point process

$$N_A(\cdot) = N(\cdot \cap A) \tag{3.7}$$

(i.e., the restriction of N to A). In the situation of Model 3.1 it sometimes happens that each process N_i is observed over a subset of A_i of E, so that the data σ-algebra of (3.3) becomes $\mathcal{F}_n = \prod_{i=1}^n \mathcal{F}^{N_i}(A_i)$.

2) *Observation of integral data.* Here there are specified finitely many functions $g_1, \dots, g_k$ in $C_K(E)$ (the g_j are continuous with compact support) and one observes only the integral data

$$\begin{aligned} \mathcal{F}^N(g_1, \dots, g_k) &= \sigma(N(g_1), \dots, N(g_k)) \\ &= \sigma(N(g) : g \in sp\{g_1, \dots, g_k\}), \end{aligned} \tag{3.8}$$

where "*sp*" denotes "linear space spanned by." One motivation is that observation devices inevitably introduce smoothing, which can be represented by integral data. For spatial point processes it may be possible to perform only a finite number of measurements corresponding to the g_j. Design aspects, that is, choice of the g_j, are especially important, as confirmed in Chapter 6.

3) *Observation of a mapping.* In this case instead of the point process $N = \Sigma\, \varepsilon_{X_i}$ on E, one observes the mapping (Definition 1.36) $Nf^{-1} = \Sigma\, \varepsilon_{f(X_i)}$ on a space E', where f: $E \rightarrow E'$ is measurable and possibly unknown. Of course, if f is known and one-to-one, there is no partial observation because N can be reconstructed without error from Nf^{-1}; however, Nf^{-1} might be only partially observed. In most examples f is not one-to-one. An important instance, to illustrate, is that E is a product space and f a coordinate projection. Hence a point process $N = \Sigma\, \varepsilon_{(X_i, Y_i)}$ is observable only in the form $Nf^{-1} = \Sigma\, \varepsilon_{X_i}$. This model includes such diverse effects as being unable to observe the marks in a marked point process, observation of secondary points but not cluster centers in a Poisson cluster process (Sections 1.4 and 6.5), certain observations of Poisson processes arising in emission tomography (Section 6.3), and line or transect sampling methods used in ecology and forestry. (The latter also incorporate integral data features.)

The three foregoing forms of partial observation affect only the data in the statistical model; the final two, by introduction of exogenous randomness, alter all components of the model.

4) *Observation of a thinned process.* In the heuristic sense we take this to mean that a random subset of the points of N is deleted and only the remaining points, which constitute a "thinning" of N, are observable. (The deletion operation is also called thinning.) The main thinning mechanism we examine is p-thinning in the sense of Definition 1.38. Recall that for p a function from E to $[0,1]$ the p-thinning of $N = \Sigma\, \varepsilon_{X_i}$ is the point process $N' = \Sigma\, U_i \varepsilon_{X_i}$, where the U_i are Bernoulli random variables that are conditionally independent given N, with $P\{U_i = 1 | N\} = p(X_i)$. In problems of statistical inference the thinning function p may be unknown as well as the law of N, and both might be estimated from observations of N', while for state estimation both $\mathcal{L}_N$ and p are known and one attempts to reconstruct N from complete or partial observations of N'.

By way of illustration we give explicitly the following variant of Model 3.1; it is analyzed in Section 4.5.

Model 3.3. (i.i.d. copies of a p-thinned point process on a compact space). Let N be a point process on a compact space E. The statistical model for complete observation of a sequence of i.i.d. copies of a p-thinning of N is given as follows.

a) The sample space is $(\Omega, \mathcal{G}) = (\Omega_0, \mathcal{G}_0)^{\mathbf{N}}$, where $(\Omega_0, \mathcal{G}_0) = (\mathbf{M}_p, \mathcal{M}_p)^2$, with coordinates N_i, N'_i on the ith factor space.

b) Let $\{\mathcal{L}_\alpha : \alpha \in I_0\}$ be an indexed family of candidate probability laws for N and let H denote a set of (measurable) functions $p: E \to [0,1]$, the possible thinning functions. For $\alpha \in I_0$ and $p \in H$, let $Q_{\alpha,p}$ be the joint law on $(\Omega_0, \mathcal{G}_0)$ of a point process N with law $\mathcal{L}_\alpha$ and the p-thinning N' of N. Then $\mathcal{P} = \{(Q_{\alpha,p})^{\mathbf{N}} : \alpha \in I_0, p \in H\}$.

c) The data corresponding to complete observation of $N'_1, \ldots, N'_n$ is

$$\mathcal{F}_n = \prod_{i=1}^{n} \mathcal{F}^{N'_i} \tag{3.9}$$

5) *Observation of a stochastic integral.* Here is a case where the observations may constitute a purely atomic random measure rather than a point process. Let X be a measurable, nonnegative stochastic process on E (possibly but not necessarily independent of N) and let the observations be the stochastic integral $(X * N)$ defined by

$$X * N(A) = \int_A X\, dN \tag{3.10}$$

When X assumes only the values 0 and 1, $X * N$ is a point process and, indeed, a thinning of N, although not in general a p-thinning. In several places throughout the book we study the case that N is a Poisson process on $\mathbf{R}_+$ and X is a binary Markov process independent of N. In this case the stochastic integral $X * N$ is a Cox process and a renewal process, and has received widespread application (see also Exercises 3.15 and 3.16).

The reader will have noticed that the forms of partial observation are not unambiguously distinct. For example, thinning and stochastic integrals overlap one another and also restriction to a subset [the thinning function or, in (3.10), X, could be an indicator function], and the integral data theme enters (3.10) as well.

For point processes on $\mathbf{R}_+$ there is the additional distinction of synchronous or asynchronous data. For $N = \Sigma\, \varepsilon_{T_n}$ a point process on $\mathbf{R}_+$, synchronous sampling is observation until the time T_n of the nth arrival, $n = 1,2, \ldots$, and yields *synchronous data*

$$\mathcal{F}^N_{T_n} = \sigma(T_1, \ldots, T_n) = \sigma(U_1, \ldots, U_n), \tag{3.11}$$

where $U_i = T_i - T_{i-1}$ are the interarrival times. The interval $[0,T_n]$ over which N is observed is thus random. Asynchronous sampling corresponds to observing N over fixed time intervals $[0,t]$ and gives *asynchronous data*

$$\mathcal{F}^N_t = \sigma(N_u : 0 \leq u \leq t) \tag{3.12}$$

3.2 PROBLEMS OF STATISTICAL INFERENCE

Our goal in this section is to outline a conceptual framework for statistical inference for point processes and to elucidate it with "sample" problems meant to convey the breadth and scope of point process inference, but not as a complete description. The superstructure is broad enough to include most extant theory and applications even though only part of the theory and but a small fraction of the applications are treated in detail in the remainder of the book.

The overall structure consists of five parts that will be explained in detail below:

1) Parametric estimation;
2) Nonparametric estimation;
3) Parametric tests of quantitative hypotheses;
4) Nonparametric tests of quantitative hypotheses;
5) Tests of qualitative hypotheses.

Each subdivides based on the forms of observation discussed in Section 3.1, and in particular can be interpreted as having two principal and rather different subareas corresponding to observation of a single realization of a point process and to observation of i.i.d. multiple realizations. The complete observation/partial observation classification influences methodology rather than conceptualization and so does not figure prominently in our discussion.

One can formulate further subdivision into "finite sample" results (associated with observation over compact sets) and "asymptotic" results as the set of observation increases to a noncompact set (either by increase of the set over which one point process is observed or by increase in the number of observed copies).

The parametric/nonparametric distinction for estimation and quantitative hypothesis testing is more or less as in classical statistics. *Parametric* problems are those in which the law of the point process is specified by underlying model assumptions except for finitely many (e.g., one or two) real parameters. *Nonparametric* estimation includes all other estimation problems no matter how restricted the family $\mathcal{P}$ of candidate probability laws. For example, estimation of a completely unknown mean measure is a nonparametric problem, as is estimation of the probability law $\mathcal{L}_N$ itself when $\mathcal{P}$ is, say, restricted to the Poisson processes on **R**. Even estimation of a single real parameter can be a nonparametric problem if sufficiently little is assumed about the law of the point process; an example is estimation of the intensity of a stationary point process whose law is not further restricted (see Example 3.7).

In classical statistics estimation consists of *point estimation* and *interval estimation* (more generally, construction of *confidence regions* for multidimensional parameters), linked through the latter to hypothesis testing. For point processes most estimation procedures yield point estimators, especially for nonparametric problems. In spaces of measures confidence regions are difficult even to specify, let alone construct, and the traditional requirement of compactness may be impossible to fulfill.

Nonparametric testing of quantitative hypotheses pertains to infinite-dimensional characteristics such as mean measures, Laplace functionals, zero-probability functionals, and the like, as well as to finite-dimensional characteristics when $\mathcal{P}$ is nonetheless infinite-dimensional. A test concerning the intensity of a point process assumed only to be stationary is a nonparametric test of a quantitative hypothesis. By contrast, qualitative hypotheses concern *structure* of a point process, for example whether it has independent increments or is stationary or is a Cox process. In classical statistics these are sometimes referred to as questions of "model fit." Tests of qualitative hypotheses are nearly always nonparametric because specification of the law of a point process except for finitely many parameters nearly always specifies its structure completely. Obviously, qualitative hypotheses and hypotheses concerning quantitative characteristics must be tested using rather different methods, but underlying principles such as likelihood ratio tests are common to both.

Especially for qualitative hypotheses but also for quantitative hypotheses, one encounters difficulties with specification and analysis of hypotheses. Reasonable hypotheses for point processes are often too broad. When the null hypothesis is very broad, a test constructed to have acceptably low power (probability of rejection) when it is true may then lack power under the alternative hypothesis as well, and be unlikely to reject the null hypothesis regardless of whether it holds. Even if the null hypothesis is not hopelessly broad, a broad alternative may contain too many processes that are too much like the null hypothesis process, and again the test may lack power. Severely restricting the structural assumptions, which would have the desirable effect of narrowing both null and alternative hypotheses, may be contradicted by the "physics" of the problem. As an example, it is less difficult to test whether a renewal process is a Poisson process (Example 3.14) than to test whether a point process assumed only to be stationary is Poisson (Example 3.15). Unfortunately, though, the renewal assumption may be unsupported on physical grounds, if not actually contradicted. Moreover, it is more difficult to test the hypothesis that a stationary point process is a renewal process than to test the hypothesis that it is Poisson (because the former is broader).

A related problem is computation of power, especially under the alternative hypothesis. Even for rather simple tests, an approximation of power, let alone an exact computation, may be impossible. The technique, common in classical statistics, of identifying a "worst case" representative of the alternative hypothesis, computing power for it, and then using the result as a lower bound for power elsewhere under the alternative, often is not feasible. The practical consequence is that one cannot determine, for a given test, to which kinds of deviations from the null hypothesis it is especially sensitive or insensitive. Differences among tests remain murky, and selection of a test, particularly if one anticipates that if the null hypothesis fails it is likely to do so in a known manner, is difficult.

In the examples that follow and the remainder of the book, two methodologies for statistical inference for point processes dominate:

1) Empirical inference (Chapter 4);
2) Intensity-based inference (Chapter 5);

each is the subject of a later chapter. Moreover, each has a facet reflecting the principle of maximum likelihood and another reflecting the method of moments, or substitution. We elaborate briefly.

Empirical inference is based on the notion that a quantity should be estimated by a corresponding empirical average and is most important in the multiple-copy case. In terms of likelihood, for i.i.d. samples the method is based on the role of the empirical distribution as the (unrestricted) maximum likelihood estimator of the true distribution and on the precept that the maximum likelihood estimator of a functional is the same functional of the maximum likelihood estimator. Thus one is led, for example, to empirical Laplace functionals

$$\hat{L}(f) = n^{-1} \sum_{i=1}^{n} e^{-N_i(f)} \tag{3.13}$$

analyzed in Section 4.2. Given $N_1, N_2, \ldots$ that are i.i.d. with law $\mathcal{L}$ the maximum likelihood estimator of $\mathcal{L}$ is the *empirical measure* $\hat{\mathcal{L}} = n^{-1} \sum_{i=1}^{n} \varepsilon_{N_i}$, a random, purely atomic probability distribution on $(\mathbf{M}_p, \mathcal{M}_p)$. Since $L(f) = E[\exp(-N_i(f))]$, one should therefore take $\hat{L}(f) = \int e^{-\mu(f)} \hat{\mathcal{L}}(d\mu)$, which is the same as (3.13). The substitution justification of (3.13) is simply that $L(f) = E[\exp(-N_i(f))]$. A compelling virtue of empirical methods is generality in regard to the underlying space.

Intensity-based methods apply only to point processes on $\mathbf{R}_+$, and then only if integrability assumptions are fulfilled, but within this setting they are very powerful. They are justified in likelihood terms by Theorem 2.31, which expresses the likelihood function in terms of the stochastic intensity. Thus when unknown parameters or functions appear explicitly in the stochastic intensity (Chapter 5) and if the latter is computable or of postulated form, then maximum likelihood estimation and likelihood ratio tests are feasible. The substitution basis for intensity methods resides in Theorem 2.14, which for a point process N with stochastic intensity λ asserts that the process $M_t = N_t - \int_0^t \lambda_s\,ds$ is a martingale, and in interpretation of increments in M as noise. Hence for the multiplicative intensity model of Example 5.7, where the stochastic intensity is $\lambda_t(\alpha) = \alpha_t \lambda_t$, with λ an observable (predictable) process and α an unknown deterministic function, $dN_t = \alpha_t \lambda_t$ within "innovation martingale noise," so that the indefinite integral $B_t(\alpha) = \int_0^t \alpha_s\,ds$ is estimated by the stochastic process

$$\hat{B}_t = \int_0^t \lambda_s^{-1}\,dN_s; \tag{3.14}$$

the estimation error process $(B_t(\alpha) - \hat{B}_t)$ is a martingale (for details see Chapter 5).

The principle of "ad hoc methods exploiting special structure" cannot be overlooked. Indeed, it is fundamental in Chapters 6 – 9.

Now, rather than prolong the verbal explanations, we present a series of examples intended not only to illustrate the range of statistical inference problems and methods associated with point processes, but also to elucidate the general structure that we have been attempting to describe.

Parametric Estimation

As noted previously, this case entails severely restrictive structural assumptions.

Example 3.4. (Renewal process). Let $N = \Sigma\, \varepsilon_{T_n}$ be a renewal process whose interarrival times $U_i = T_i - T_{i-1}$ have gamma density

$$f_{\alpha,\beta}(x) = \Gamma(\alpha)^{-1}\beta^{\alpha}x^{\alpha-1}e^{-\beta x}, \tag{3.15}$$

with both the shape parameter α and the scale parameter β unknown. Given synchronous data $\mathcal{F}^N_{T_n}$ [see (3.11)] the problem is classical: $U_1, \ldots, U_n$ are i.i.d. with density (3.15) and standard estimation procedures for gamma distributions apply. When both α and β are unknown, the *likelihood equations* $\partial L/\partial\alpha = 0$; $\partial L/\partial\beta = 0$, where $L(\alpha,\beta) = \Sigma_{i=1}^n \log[f_{\alpha,\beta}(U_i)]$ is the *log-likelihood function*, cannot be solved in closed form (although numerical solution is always a possibility). The conventional procedure is to use the method of moments: $f_{\alpha,\beta}$ has mean α/β and variance α/β^2. If either α or β is known, then (3.15) defines a one-parameter exponential family and a multitude of results is applicable.

For asynchronous data $\mathcal{F}^N_t$ one approach is to take as estimator of α

$$\hat{\alpha}(t) = \hat{\alpha}_{N(t)} \tag{3.16}$$

and similarly for $\hat{\beta}(t)$, where $\hat{\alpha}_n$ is the $\mathcal{F}^N_{T_n}$-estimator of α. In view of Proposition 2.4 this causes little loss of information for finite t, while asymptotic properties of the two estimators are related in a straightforward, systematic manner by the results of Serfozo (1975) (see Chapter 8 for details).

An alternative approach uses Theorem 2.31. If $h_{\alpha,\beta}$ denotes the hazard function for $f_{\alpha,\beta}$, then under $P_{\alpha,\beta}$, N has stochastic intensity $\lambda_t(\alpha,\beta) = h_{\alpha,\beta}(V_{t-})$, where $V_t = t - T_{N(t)}$ is the backward recurrence time of N at t. Using (2.49), the log-likelihood function is, up to a constant term,

$$L_t(\alpha,\beta) = H_{\alpha,\beta}(V_t) + \sum_{i=1}^{N_t} [H_{\alpha,\beta}(U_i) + \log(h_{\alpha,\beta}(U_i))], \qquad (3.17)$$

where $H_{\alpha,\beta}(t) = \int_0^t h_{\alpha,\beta}(s)\,ds$. Although not amenable to closed-form maximization, (3.17) can be handled numerically in specific applications. □

Example 3.5. (Stochastic integral). Let N be a Poisson process on $\mathbf{R}_+$ with unknown rate c and let (X_t) be a Markov process independent of N, with state space $\{0,1\}$ and *known* transition intensities a $(0 \to 1)$ and b $(1 \to 0)$ (see Exercise 2.21). The observations are the point process $N'(A) = \int_A X\,dN$, which is a Cox process and a renewal process. Techniques from these classes are available, especially for synchronous data (Exercises 3.15 and 3.16), but direct methods exist as well. Suppose that X has as initial distribution the invariant distribution $\nu = (b/(a+b),\ a/(a+b))$; then for each t, $E[N_t'] = cat/(a+b)$. The asynchronous estimators

$$\hat{c} = \frac{a+b}{a}\frac{N_t'}{t} \qquad (3.18)$$

are strongly consistent (i.e., $\hat{c} \to c$ almost surely) and asymptotically normal. The latter means that $t^{1/2}[\hat{c} - c]$ has a limiting normal distribution under P_c, which can be used to construct approximate confidence intervals and hypothesis tests for c. Note that in (3.18) dependence of the estimator on the sample size t is suppressed; for notational economy we do this whenever possible.

If X were observable in addition to N, then (3.18) would be replaced by

$$\hat{c} = \frac{N_t'}{\int_0^t X_s\,ds} = \frac{\int_0^t X\,dN}{\int_0^t X\,ds}, \qquad (3.19)$$

for which strong consistency and asymptotic normality again hold. □

Here is a multiple realization problem.

Example 3.6. (Pólya process). Let N be a *Pólya process* on $[0,1]$: N is a mixed Poisson process with directing measure $M(dx) = Y\,dx$, where Y is a random variable with gamma density (3.15). Let $N_1, N_2, \ldots$ be i.i.d. copies of N; the problem is to estimate α and β from complete observation of the N_i. By conditional uniformity the only data from N_i that need be retained are $Z_i = N_i([0,1])$. The Z_i are i.i.d. with mixed Poisson distribution

$$P_{\alpha,\beta}\{Z_i = k\} = \int_0^\infty f_{\alpha,\beta}(x)e^{-x}\frac{x^k}{k!}\,dx$$

and the problem reduces to a standard problem of estimating the parameters of a gamma mixing distribution from mixed Poisson data (see Simar, 1976, and Section 7.1 for details). □

Nonparametric Estimation

We present first a single realization problem; in such cases the quantitative characteristic being estimated is often a real parameter.

Example 3.7. (Stationary point process). Let N be a stationary point process on $\mathbf{R}$ (Definition 1.58) with unknown but finite intensity ν. Thus the mean measure of N is known up to the parameter ν, but the set $\mathcal{P}$ of candidate laws of N is very large indeed. The model is complete observation of a single realization of N. The obvious estimator of ν given the asynchronous data $\mathcal{F}_t^N$ is $\hat{\nu} = N_t/t$. Whether these estimators are even weakly consistent (i.e., satisfy $\hat{\nu} \to \nu$ in probability) depends on $\mathcal{L}_N$. If N is P-ergodic (Definition 1.67), then by Theorem 1.68 there is even strong consistency; if not (which is the case for the Pólya process of Example 3.6 when it is defined on $\mathbf{R}_+$), recovery of ν from single realizations may be impossible. In the ergodic case asymptotic normality does not hold without further hypotheses, typically to the effect that $N(A)$ and $N(B)$ are nearly independent if A and B are widely separated; these are termed mixing conditions. □

Our second example involves multiple realizations.

Example 3.8. (Superposed Bernoulli processes). Let E be a compact space and let $N_1, N_2, \ldots$ be i.i.d. copies of the superposition of Bernoulli processes given by $N = \sum_{j=1}^k U_j \varepsilon_{X_j}$, where k is a known, positive integer, the X_j are i.i.d random elements of E with unknown distribution F, and the U_j are i.i.d. Bernoulli random variables with $p = P\{U_j = 1\}$ also unknown; further, the X_j and U_j are independent (see Section 1.6 for details). The aim is to estimate p and F from data representing complete observation of $N_1, \ldots, N_n$. From the empirical standpoint one chooses the estimators

$$\hat{p} = (kn)^{-1}\sum_{i=1}^n N_i(E) \tag{3.20}$$

and

$$\hat{F}(\cdot) = (nk\hat{p})^{-1} \sum_{i=1}^{n} \frac{N_i(\cdot)}{N_i(E)} 1(N_i(E) > 0)$$

which are maximum likelihood estimators for the statistical model $\mathcal{P} = \{P_{p,F} : p \in (0,1], F \text{ is a probability distribution on } E\}$. It is straightforward to demonstrate strong consistency and joint asymptotic normality.

An alternative is to estimate the Laplace functional

$$L_{p,F}(f) = E_{p,F}[\exp(-N(f))] = (1 - p + p \int e^{-f} \, dF)^k$$

by the empirical Laplace functionals $\hat{L}$ introduced in (3.13), which have strong consistency and asymptotic normality properties described in Section 4.2. Then, since

$$p = 1 - P\{N(E) = 0\}^k \tag{3.21}$$

and with the notation $\ell_F(f) = \int e^{-f} \, dF$, one arrives at the estimators

$$\hat{p} = 1 - \left[n^{-1} \sum_{i=1}^{n} 1(N_i(E) = 0) \right]^{1/k} \tag{3.22}$$

and

$$\hat{\ell}(f) = \frac{1}{\hat{p}} (\hat{L}(f)^{1/k} + \hat{p} - 1) \tag{3.23}$$

by substitution. Note that (3.22) is an *empirical zero-probability functional* corresponding to (3.21). Consistency and asymptotic normality of the estimators (3.22) and (3.23) follow from results in Sections 4.2 and 4.3. □

This example shows a common property of point process inference problems: rarely is there only one way to approach a problem (although different approaches may yield the same final result).

Parametric Testing of Quantitative Hypotheses

We pursue some of the preceding examples.

Example 3.9. (Renewal process). To continue Example 3.4, tests involving the parameters α and β are carried out using established methodology for gamma distributions. For synchronous data the problem is one for i.i.d gamma distributed random variables. If α and β are both unknown, the situation is cumbersome if exact results are required, whereas if either (say α) is known, then there exist (from theory of exponential families) uniformly most powerful tests for hypotheses such as $H_0: \beta \geq \beta_0$. [In the Neyman-Pearson framework, *power*—the probability of rejecting H_0—is required to fall below a specified value whenever the null hypothesis holds and is to be maximized, subject to this constraint, when the alternative hypothesis holds. A test, which is defined by a *test statistic* constructed from the data and a *critical region* (if the test statistic falls in the critical region, H_0 is rejected), is *uniformly most powerful* if its power is maximal for all parameter values associated with the alternative. As a general rule, uniformly most powerful tests exist only for tests of one-sided hypotheses concerning real parameters.] □

Example 3.10. (Stochastic integral). Let N' be the point process of Example 3.5 and assume that the process X is not observable. To test the hypotheses $H_0: c \geq c_0$, $H_1: c < c_0$ given, say, asynchronous data $\mathcal{F}_t^N$, it is evident intuitively that the critical region Ω_c should have the form $\Omega_c = \{N_t' \leq \delta\}$ for some critical value δ depending on c_0, t, and the power bound under H_0. The issues are calculation of δ and of power under the alternative hypothesis. This requires calculation or approximation of the distribution of N_t' under each probability P_c, a problem treated in Exercise 3.16. □

Example 3.11. (p-thinned Poisson process). Let N be a Poisson process on a general space E with known mean measure μ, let $p \in (0,1]$ be unknown, and let the observations be a single realization of the p-thinned process N' (Definition 1.38 and Model 3.3). To test $H_0: p \geq p_0$, $H_1: p < p_0$ given the data $\mathcal{F}^{N'}(A)$, we use conditional uniformity and the fact that the μ is known to conclude that $N'(A)$ is a sufficient statistic. (Heuristically, a *sufficient statistic* for a parameter comprises all information that the data contain concerning that parameter. The technical criterion is that the conditional distribution of the data given a sufficient statistic not depend on the parameter.) Clearly, the critical region should have the form $\Omega_c = \{N'(A) \leq \delta\}$ for appropriately chosen δ. But under P_p, $N'(A)$ has a Poisson distribution with mean $p\mu(A)$ and hence we are really testing $H_0': E[N'(A)] \geq p_0\mu(A)$, $H_1': E[N'(A)] < p_0\mu(A)$ given a sample of size 1 of $N'(A)$, which is a standard problem. □

Nonparametric Tests of Quantitative Hypotheses

Example 3.12. (Stationary point process). In the context of Example 3.7, consider testing just the simple hypotheses $H_0: \nu = \nu_0$, $H_1: \nu = \nu_1 < \nu_0$ given asynchronous data $\mathcal{F}_t^N$. It seems clear that one can consider only critical regions of the form $\Omega_c = \{\sup_{s \leq t}(N_s/s) \leq \delta\}$, but calculation of δ

and of the power of the test, even for simpler critical regions $\Omega_c = \{N_t/t \leqslant \delta\}$, are formidable and perhaps impossible. □

Example 3.13. (Superposed Bernoulli processes). For the processes of Example 3.8, tests involving p are relatively straightforward because both of the estimators of p constructed there have distributions not depending on F. Indeed, such tests in either case are fundamentally tests for binomial distributions, for which there exists substantial theory. Tests for F are more difficult, but can be effected using methods applicable to empirical processes on general spaces (see Section 4.1). □

Tests of Qualitative Hypotheses

Here problems of computation of power and lack of power under the alternative become acute. We mention two examples related to Section 3.5.

Example 3.14. (Renewal process). Let N be a renewal process on $\mathbf{R}_+$ and consider testing the composite null hypothesis that N is a Poisson process against various alternatives, given a single realization of N. Alternatives of potential interest range from rather specific (e.g., that N has a gamma—but not exponential—interarrival distribution), for which the problem is parametric, to broad (e.g., that the interarrival distribution has a hazard function that is increasing) to very broad (i.e., N is simply not Poisson). To the extent that one views the interarrival distribution as an infinite-dimensional quantitative characteristic of N, the problem belongs in the category of nonparametric testing of quantitative hypotheses; however, the tests below are tests of qualitative structure that characterizes Poisson processes among renewal processes.

For synchronous data $\mathcal{F}^N_{T_n}$ the problem is to test the null hypothesis that the interarrival times $U_1, \ldots, U_n$ have an exponential distribution. To test this hypothesis directly, one could use a χ^2 goodness-of-fit test or tests based on the Kolmogorov-Smirnov statistic, the Cramér-von Mises statistic or the Anderson-Darling statistic (see notes for Chapter 4), but for any of these the unknown rate λ of N under the Poisson null hypothesis must be estimated and incorporated into the test statistic, which complicates matters.

The *uniform conditional test* removes λ from consideration at the expense of introducing a conditional test. (In a conditional test all statements, including those about power, are valid only conditionally on the value of some statistic.) The uniform conditional test is based on the qualitative property of conditional uniformity (Exercise 1.1): conditional on the value $T_n = t$, the ratios $V_i = T_i/T_n$, $i = 1, \ldots, n-1$, under the null hypothesis but regardless of the value of λ, are order statistics from the uniform distribution on [0,1]. This property can be tested using previously mentioned statistics; no unknown parameters need be estimated. The power of the test under alternatives is not well understood, however.

A rather better test is the *gap test* based on the spacings of the order statistics $U_{(j)}$ associated with the interarrival times U_i. Under the Poisson null hypothesis the normalized spacings $U_j^n = (n+1-j)(U_{(j)} - U_{(j-1)})$, $j = 1,...,n$, are independent and exponentially distributed (Exercise 3.34), which one can test using the uniform conditional test. The test based on transformed data has been reported on empirical evidence to have power superior to that based on the untransformed data, the null distribution theory is known (Pyke, 1965), and there is some information available concerning power under renewal alternatives.

The uniform conditional test for asynchronous data $\mathcal{F}_t^N$ is constructed as follows. Under the Poisson null hypothesis and regardless of λ, conditional on $N_t = n$ the values $V_i = T_i/t$, $i = 1,...,n$, are order statistics from the uniform distribution on [0,1], which may be tested as described previously. Testing for conditional uniformity using a χ^2 goodness-of-fit test is equivalent to the *dispersion test for homogeneity* (which is based on a form of integral data). The interval $[0, t]$ is partitioned into subintervals $I_1,...,I_k$ of equal length and the test statistic is the dispersion $D = (1/\bar{N}) \sum_{j=1}^{k} [N(I_j) - \bar{N}]^2$, where $\bar{N} = (1/k) \sum_{j=1}^{k} N(I_j)$. □

Things are decidedly worse with less stringent underlying assumptions.

Example 3.15. (Stationary point process). Let N be a stationary point process on **R** with known intensity $\nu = 1$, and consider testing the null hypothesis that N is a Poisson process given a single realization. Here, even though the null hypothesis is simple, it is not clear how to proceed. Any of the tests discussed in Example 3.14 could be used, but except for a few alternatives, nothing is known about their power. Presumably one must either narrow the structural assumptions or consider restricted alternatives other than renewal alternatives, or both, or more. Some specific examples appear in later chapters. □

As we have stated before, this problem "sampler" is in no sense exhaustive. Other specific examples appear later in the chapter, in other chapters, in exercises, and in the literature.

3.3 STATE ESTIMATION FOR POINT PROCESSES

Let N be a point process on E with *known* probability law $\mathcal{L} = \mathcal{L}_N$. Suppose that N is only partially observed, specifically that one of the first two mechanisms in Section 3.1 prevails: the observations are either the restriction of N to a subset A of E:

$$\mathcal{F}^N(A) = \sigma(N(B) : B \subset A) \tag{3.24}$$

or the integral data

$$\mathcal{F}^N(g_1, \ldots, g_k) = \sigma(N(g_1), \ldots, N(g_k)) \tag{3.25}$$

In a state estimation problem the objective is to *reconstruct in optimal fashion* either the unobserved portions of N or an associated but unobservable random process, from the observations (3.24) or (3.25). This is done realization by realization (ω-wise); the term *state estimation* emphasizes that unobserved random variables, rather than statistical parameters, are being estimated. Indeed, the law of N must be assumed to be known completely. (In Section 3.4 we address simultaneous statistical inference and state estimation.)

A state estimator of an unobservable random variable is another random variable, measurable with respect to the observations, that approximates the unobservable state. Having optimal state estimators necessitates an error criterion; the universally accepted criterion is *mean squared error* (MSE). To illustrate, let the observations be $\mathcal{F}^N(A)$ and suppose that we seek to estimate the unobserved portion of N [i.e., $N(\cdot \cap A^c)$]. For each function f on A^c we require a state estimator $\hat{N}(f)$ satisfying

a) $\hat{N}(f)$ is measurable with respect to $\mathcal{F}^N(A)$ [physically, $\hat{N}(f)$ is determined by the observations];
b) For every $Z \in \mathcal{F}^N(A)$,

$$E[(\hat{N}(f) - N(f))^2] \leq E[(Z - N(f))^2] \tag{3.26}$$

Furthermore, $f \to N(f)$ is an integral having linearity and continuity properties that are desirable be retained by the optimal state estimators.

For fixed f the solution $\hat{N}(f)$ to (3.26) is termed the *minimum mean squared error* (MMSE) estimator of $N(f)$. Of course, the MMSE estimator is the conditional expectation $E[N(f)|\mathcal{F}^N(A)]$. Even the linearity and continuity properties hold nearly as desired (it is still necessary to manipulate null sets to ensure that $\hat{N}(\cdot) = E[N(\cdot)|\mathcal{F}^N(A)]$ is almost surely a measure; in specific problems we are always able to do so). If the story ended here there would be no story. In state estimation problems the central issues are *computation* of state estimators and *description* of how state estimators evolve as the point process is observed over larger sets. We illustrate in symbolic form and more explicitly in examples.

For observations $\mathcal{F}^N(A)$ the MMSE state estimator $\hat{N}(f)$, as a conditional expectation, is some function of A, f, and—especially—the observations $N_A(\cdot) = N(\cdot \cap A)$; symbolically,

$$\hat{N}(f) = H(A, f, N_A). \tag{3.27}$$

The first computational issue is to calculate the function H, which also depends on $\mathcal{L}$, *explicitly*. This may or may not be possible. According to Section 2.6, the Palm distributions of N are the key tools in such calculations; unfortunately, Palm distributions themselves often cannot be calculated. Even in those cases that H can be computed "explicitly" the resulting expression may be too complicated to use, especially if the set A changes frequently. Thus one is always led to the issue of *recursive computation*, which we take up momentarily.

Example 3.16. (Empirical process). Let $N = \sum_{i=1}^{k} \varepsilon_{X_i}$ be an empirical process (Definition 1.4) on E, with both k and the distribution ρ of the X_i known. Given a subset of A of E and a function f on A^c,

$$E[N(f)|\mathcal{F}^N(A)] = [k - N(A)] \frac{\int_{A^c} f\,d\rho}{\rho(A^c)}, \tag{3.28}$$

provided that $\rho(A) < 1$. In this case there seems little need to consider recursive computation because the state estimator is so simple; nevertheless, the recursive method may be easier and is discussed in Example 3.17. □

In broad terms, a *recursive* method of computation enables one to revise state estimators as additional observations are obtained, rather than compute *da capo* (from the beginning) the new state estimator that incorporates both previous and newly obtained observations. Recursive computation possesses both a "macro-level" component relevant to point processes on general spaces and a "micro-level" component restricted to point processes on $\mathbf{R}_+$. Again we illustrate, first in stylized terms.

The macro component deals with discrete (noninfinitesimal) increases of the set A. Suppose that observations $\mathcal{F}^N(A)$ have been obtained previously and that the MMSE state estimator $E[N(\cdot)|\mathcal{F}^N(A)]$ has been calculated (by whatever method). At a later time N is observed over an additional set A' disjoint from A, so that the cumulative observations are $\mathcal{F}^N(A' + A)$ (recall that summation for sets denotes disjoint union). To compute the revised state estimator $E[N(\cdot)|\mathcal{F}^N(A + A')]$ one could start from scratch, but unless there is a very simple method for explicit computation in (3.27), starting over is prohibitively difficult and time consuming. At a minimum one should compute only the increment $E[N(\cdot)|\mathcal{F}^N(A + A')] - E[N(\cdot)|\mathcal{F}^N(A)]$, especially if A' is not too large. In the process this increment should be decomposed into two parts, one representing only the effect of having observed the point process over a larger set and not a function of the new observations $N_{A'}$ (but certainly a function of the previous observations), and the second incorporating the new observations but in such a manner that contributions of old and new observations are isolated from one another and such that this term vanishes if $N_{A'}$ is zero. In symbolic form, these ideas imply an expression

$$\begin{aligned} &E[N(f)|\mathcal{F}^N(A+A')] - E[N(f)|\mathcal{F}^N(A)] \\ &\quad = J(A, A', f, N_A) + K(A, A', f, N_A, N_{A'}), \end{aligned} \tag{3.29}$$

with K vanishing if the last argument is zero.

The micro aspect of recursive computation centers around making sense of the idea of the new observations being over an infinitesimal set dx and of an infinitesimal analog of (3.29), namely

$$\begin{aligned} &E[N(f)|\mathcal{F}^N(A+\{dx\})] - E[N(f)|\mathcal{F}^N(A)] \\ &\quad = J(A, dx, f, N_A) + K(A, x, f, N_A, N(dx)), \end{aligned} \tag{3.30}$$

where $x \notin A$. (This expression should not be taken literally; it is intended only to be suggestive.) If the terms on the right-hand side were measures in dx, then it would be possible to construct, by integration, any state estimator from infinitesimal building blocks of the form (3.30), provided that there were an agreed-upon starting point and an agreed-upon linear order in which observations were obtained, "one point at a time." This occurs only for point processes on $\mathbf{R}_+$, and then (3.30) becomes a stochastic differential equation

$$dE[N(f)|\mathcal{F}_t^N] = X_t dt + Y_t dM_t, \tag{3.31}$$

where X and Y are (predictable) random processes depending on f and the past history of N, and where M is the innovation martingale (Theorem 2.14) associated with N. The processes X and Y are the *predictable change* and *filter gain*, respectively. When there is no innovation, the second term on the right-hand side of (3.31) vanishes and there is only predictable change in the state estimator, resulting only from observing the point process over a larger set. (Note that (3.30) and (3.31) are only imperfectly analogous; while each contains a predictable change, the former has a term vanishing if there are no new points of N observed [i.e., $N(d_x) = 0$] but the latter has a term vanishing if there is no innovation ($dM_t = 0$), which is more sensible, but applies only to processes on $\mathbf{R}_+$.)

We illustrate by expanding Example 3.16.

Example 3.17. (Empirical process). Let N be the empirical process of Example 3.16 and suppose that A and A' are disjoint with $\rho(A+A') < 1$.

For notational simplicity let $J(B) = \int_{B^c} f d\rho / \rho(B^c)$. Then for f supported outside A,

$$E[N(f)|\mathcal{F}^N(A+A')] - E[N(f)|\mathcal{F}^N(A)]$$
$$= [k - N(A)][J(A+A') - J(A)] - N(A')J(A+A'),$$

which is evidently of the form (3.29).

Suppose now that $E = \mathbf{R}_+$ and ρ has hazard function h; then N has $\mathcal{F}^N$-stochastic intensity $\lambda_t = h(t)(k - N_{t-})$. Letting $J(t) = [1-\rho(t)]^{-1} \int_t^\infty f d\rho$, we obtain a specific example of (3.31):

$$dE[N(f)|\mathcal{F}_t^N] = h(t)(k - N_{t-})f(t)\,dt - J(t)\,dM_t \quad \square \tag{3.32}$$

Expressions such as (3.31) and (3.32), as well as the discussion of conditioning in Section 2.6, identify the fundamental role of stochastic intensities and Palm distributions in state estimation for point processes. Further confirmation appears in later chapters; a prominent example is state estimation for Cox processes, treated in Section 7.2. Yet another link is Theorem 2.34, the representation theorem for point process martingales. Evidently, a process $(E[N(f)|\mathcal{F}_t^N])$ is a martingale; the representation of it as the integral of a predictable process with respect to the innovation martingale is closely related to the stochastic differential equation (3.31). See also Section 5.4, where we solve a very general state estimation problem using the martingale representation theorem.

The state estimation problem for Cox processes concerns state estimation for processes that are related to N but not functionals of N and may be completely unobservable. For such processes, exact reconstruction as $A \uparrow E$ is generally impossible and the resultant martingales are therefore more interesting. The issues outlined in this section are equally pertinent here and sometimes more difficult. We illustrate with an example treated in Sections 7.2 and 7.4.

Example 3.18. (Mixed Poisson process). Let N be a mixed Poisson process on a compact space E, with directing measure $Y\nu$, where ν is a known, finite, diffuse measure on E and Y is a positive random variable with known distribution F. The observations are $\mathcal{F}^N(E)$; the multiplier Y, which can be interpreted as an underlying random mechanism, is not observable, yet may be of physical importance. General theory developed in Chapter 7 implies that

$$E[Y|\mathcal{F}^N(E)] = \frac{\int F(dx) e^{-x\nu(E)} x^{N(E)+1}}{\int F(dx) e^{-x\nu(E)} x^{N(E)}} \tag{3.33}$$

The MMSE estimator of the directing measure M is therefore the diffuse random measure

$$E[M|\mathcal{F}^N(E)] = E[Y|\mathcal{F}^N(E)]\nu \tag{3.34}$$

Recursive versions of (3.33) and (3.34) appear in Chapter 7. □

3.4 COMBINED STATISTICAL INFERENCE AND STATE ESTIMATION

Without knowledge of the probability law of the point process N, calculation of MMSE state estimators is not possible. Nonetheless, there are many situations in which the law of N is unknown but state estimation is the paramount goal. In such cases statistical inference and state estimation must be effected simultaneously, which we call the problem of *combined inference and state estimation*. In principle this problem is solvable if there exist consistent estimators of those attributes of the law of N needed to compute MMSE state estimators. It is approached using the *principle of separation*: first, derive the relevant state estimation formula under the assumption that $\mathcal{L}_N$ is known, then examine it to determine which characteristics are required to calculate the state estimator. Next, use the observations (together with data from other copies of the process, if available) to estimate these characteristics. Construct a *pseudo-state estimator* by substituting estimated for exact attributes of $\mathcal{L}_N$, at the same time using the observations in the same way they would be used in the "true" state estimator, and finally, analyze properties of the pseudo-state estimator in comparison to the true state estimator, especially asymptotically.

The combined problem therefore incorporates all the questions, classifications, and difficulties already introduced, along with some new wrinkles because the observations of N play a dual role as data for statistical inference and as partial observations on which state estimation is based. Rather than describe fully the resultant classification, we merely mention and illustrate principal cases. The important distinction is whether there is but a single realization of N or there are i.i.d. copies. Other distinctions, for example whether the statistical inference problem is parametric or nonparametric, are secondary.

For several reasons the single realization case is the less interesting and less important. As discussed in Section 3.2, single realization data are useful mainly for estimating real parameters (even though the setting may be nonparametric), but most state estimation formulas involve more about the law of N than just a few real parameters. Furthermore, the roles of the observations as data for statistical inference and for state estimation are difficult to unravel (in the multiple-copy case there is a convenient way to sidestep this issue). Finally, asymptotics are less interesting than in the multiple-copy case in terms of both the kinds of results that can be proved and the assumptions required to prove them. Even so, the single realization case is devoid of neither theoretical substance nor practical applications, so we illustrate it in symbolic form.

Let N be a point process on a noncompact space E with law $\mathcal{L} = \mathcal{L}_N$ that is not completely known, but whose unknown properties can be estimated consistently from $\mathcal{F}^N(A)$ as $A \uparrow E$. Assume that the observations are $\mathcal{F}^N(A)$ with A compact and that we wish to perform state estimation for the unobserved portion of N. According to (3.27), there is a function H such that

$$E_{\mathcal{L}}[N(f)|\mathcal{F}^N(A)] = H(\mathcal{L}^0, f, N_A), \tag{3.35}$$

where $\mathcal{L}^0$ denotes characteristics of $\mathcal{L}$ required to compute the MMSE state estimator (the notation emphasizes dependence of the state estimator itself on $\mathcal{L}$). The functional form of H—and this is a crucial point!—is fully determined by the structural assumptions inherent in the statistical model and can be computed without knowledge of the unknown characteristics $\mathcal{L}^0$. Let $\hat{\mathcal{L}}^0$ be an estimator of $\hat{\mathcal{L}}^0$ based on the observations $\mathcal{F}^N(A)$; then the pseudo-state estimator is

$$\hat{E}[N(f)|\mathcal{F}^N(A)] = H(\hat{\mathcal{L}}^0, f, N_A) \tag{3.36}$$

The notation employs the caret to distinguish an estimator. There is evident dual dependence of the pseudo-state estimator (3.36) on N_A, once as data in construction of $\hat{\mathcal{L}}^0$ and again as observations serving as "input" for state estimation. If the true state estimator can be computed, and if the estimator $\hat{\mathcal{L}}^0$ can, then so can the pseudo-state estimator.

What remains is to analyze the difference between the pseudo-state estimator and the true state estimator. Finite sample results are few and without knowledge of the mean squared error $E_{\mathcal{L}}[(E_{\mathcal{L}}[N(f)|\mathcal{F}^N(A)] - N(f))^2]$, which is often not available, rather uninformative. Hence one is led to seek asymptotic results as $A \uparrow E$, the goals, given that $\hat{\mathcal{L}}^0 \to \mathcal{L}^0$ in an appropriate sense, being to show that $\hat{E}[N(f)|\mathcal{F}^N(A)] - E_{\mathcal{L}}[N(f)|\mathcal{F}^N(A)]$ converges to zero and to relate the rates of convergence. But then f cannot remain fixed, for if it does, then once A is large enough that it contains the support of f (which happens eventually because for finiteness it is necessary to assume that the support of f is compact), then $E_{\mathcal{L}}[N(f)|\mathcal{F}^N(A)]$—and consequently also $\hat{E}[N(f)|\mathcal{F}^N(A)]$—degenerates to $N(f)$, which has become observable. For f to vary with A requires that the variation be systematic, which entails special structure on E (e.g., that of a group). Then H must have suitable invariance properties, while N must presumably be ergodic or more, so the restrictions mount.

The picture is less gloomy if instead of unobserved portions of N one is estimating an underlying but unobservable random mechanism associated with N. Here is an illustration.

Example 3.19. (Yule process). Let N be a Yule process with unknown birth rate $c > 0$ (N_t is the size at t of a population that, when its size is i, reproduces at rate ci). By Exercise 2.12 there is a random variable Y satisfying $e^{-ct}N_t \to Y$ almost surely under P_c. Also, N is a mixed Poisson process directed by the random measure $M = Y\nu$, where $\nu_t = e^{ct} - 1$. Suppose that we wish to perform state estimation for Y having observed one realization of N over $[0,t]$; keep in mind that c is unknown. The property that the $\mathcal{F}^N$-stochastic intensity of N is $\lambda_t = cN_{t-}$, together with the substitution principle embodied in (3.14), leads to the (martingale) estimators

$$\hat{c} = \frac{N_t}{\int_0^t N_s\,ds} \tag{3.37}$$

This is also essentially the maximum likelihood estimator obtained using the property that interarrival times in N are independent and exponentially (but not identically) distributed. Let F_c denote the distribution of Y under P_c; then by (3.33),

$$E_c[Y|\mathcal{F}_t^N] = \frac{\int F_c(dx)e^{-x\nu(t)}x^{N_t+1}}{\int F_c(dx)e^{-x\nu(t)}x^{N_t}}$$

and it is straightforward to combine this expression with (3.37) to obtain a pseudo-state estimator. □

For the case of i.i.d. copies $N_1, N_2, \ldots$ of N (here we take E to be compact) the outlook is brighter still. Suppose that $N_1,\ldots,N_n$ have been observed previously over E and that N_{n+1} is now observed over the subset A. Let $\mathcal{L}^0$ again signify aspects of $\mathcal{L} = \mathcal{L}_N$ not specified by model assumptions. (For example, it may be assumed only that N is a Cox process with diffuse directing measure.) The true state estimator

$$E_{\mathcal{L}}[N_{n+1}(f)|\mathcal{F}^{N_{n+1}}(A)] = H(\mathcal{L}^0, f, (N_{n+1})_A)$$

is now superseded by the pseudo-state estimator

$$\hat{E}[N_{n+1}(f)|\mathcal{F}^{N_{n+1}}(A)] = H(\hat{\mathcal{L}}^0, f, (N_{n+1})_A), \tag{3.38}$$

where $\hat{\mathcal{L}}^0$ is an estimator of $\mathcal{L}$ based on $N_1, \ldots, N_n$: for example, the appropriate functional of the empirical measure $\hat{\mathcal{L}} = n^{-1}\sum_{i=1}^n \varepsilon_{N_i}$. Actually, we have the option whether to include the partial observations $(N_{n+1})_A$ when constructing $\hat{\mathcal{L}}^0$. In practice one never deliberately ignores data, so $(N_{n+1})_A$ should be included, but for proving results, however, basing $\hat{\mathcal{L}}^0$ only on $N_1, \ldots, N_n$ renders the two stochastic arguments in (3.38) independent, and proofs become easier and less complicated. Asymptotically, there is no difference provided that the estimators $\hat{\mathcal{L}}^0$ are consistent, as they are in all specific cases we consider.

The limit theorems are more clearly delineated and valid under fewer structural assumptions than in the single-realization model. Typically, it will hold that $\hat{\mathcal{L}}^0 \to \mathcal{L}^0$ at rate $n^{-1/2}$, in the sense that $n^{1/2}[\hat{\mathcal{L}}^0 - \mathcal{L}^0]$ has a nondegenerate Gaussian limit distribution. Then the main goal is to show that the error process $n^{1/2}(\hat{E}[N_{n+1}(\cdot)|\mathcal{F}^{N_{n+1}}(A)] - E_{\mathcal{L}}[N_{n+1}(\cdot)|\mathcal{F}^{N_{n+1}}(A)])$ also has a nondegenerate limit distribution (often a mixture of Gaussian distributions). Such a rate is evidently the best possible, because the effectiveness of pseudo-state estimators as approximations to the true state estimators surely cannot exceed that of the estimation procedure itself. In some cases, however, results this strong have not been obtained and one must settle for slower rates of convergence. Section 7.4 exhibits both situations for Cox processes.

In "highly nonparametric" contexts the only functional form H that can be employed might involve the Palm distributions of N, giving rise to the problem of estimating Palm distributions from the data N_i; this problem and some implications for combined inference and state estimation are treated in Sections 4.4 and 7.4.

We conclude the section with a simple example.

Example 3.20. (Empirical process). Let $N_1, N_2, \ldots$ be i.i.d. copies of the empirical process N of Example 3.16. Assume that k is known but that the distribution ρ is entirely unknown. With $N_1, \ldots, N_n$ previously observed, the maximum likelihood estimator of ρ is the empirical measure $\hat{\rho} = (nk)^{-1} \Sigma_{i=1}^n N_i$, so that if N_{n+1} is now observed over A, then the pseudo-state estimator associated with the state estimator (3.28) is

$$\hat{E}[N_{n+1}(f)|\mathcal{F}^{N_{n+1}}(A)] = [k - N_{n+1}(A)] \frac{\int_{A^c} f \, d\hat{\rho}}{\hat{\rho}(A^c)}; \tag{3.39}$$

asymptotic properties are readily derivable. □

3.5 EXAMPLE: INFERENCE FOR ORDINARY POISSON PROCESSES

The sample problems in preceding sections have been chosen deliberately to depict a variety of point processes, but in consequence lack the coherence arising from a narrower context. In this section, by contrast, we consider only the most fundamental class of point processes, the ordinary (i.e., stationary) Poisson processes on $\mathbf{R}_+$, and address a number of the problems posed in Sections 3.1 – 3.4. For brevity most proofs are relegated to the exercises. The statistical model is that $N = \Sigma_{n=1}^{\infty} \varepsilon_{T_n}$ is a Poisson process on $\mathbf{R}_+$ with unknown rate $\lambda \in (0,\infty)$. The family of candidate probability laws is $\mathcal{P} = \{P_\lambda : \lambda \in (0,\infty)\}$; both synchronous and asynchronous data are examined. In terms of the classifications in Sections 3.1 and 3.2, the underlying space is noncompact and the statistical setting is parametric; since much of the rest of the book focuses on multiple realizations, we restrict attention here to statistical inference, state estimation, and, combined inference and state estimation based on single realizations of N. Asymptotic properties are as $n \to \infty$ for synchronous data and as $t \to \infty$ for asynchronous data.

Estimation of λ

Estimation of λ from synchronous samples $\mathcal{F}_{T_n}^N = \sigma(T_1, \dots, T_n) = \sigma(U_1, \dots, U_n)$ is a special case of Example 3.4: the interarrival times $U_1, \dots, U_n$ are independent and identically exponentially distributed with mean $\theta = 1/\lambda$. [The conventional statistical parameterization for exponential densities is $f(x,\theta) = (1/\theta)e^{-x/\theta}$, $\theta \in (0,\infty)$.] The following properties are well known.

Proposition 3.21. a) For synchronous data $\mathcal{F}_{T_n}^N$, the maximum likelihood estimator of λ is $\hat{\lambda} = 1/\hat{\theta}$, where $\hat{\theta} = n^{-1}\Sigma_{i=1}^n U_i$.

b) For each λ, $\hat{\lambda} \to \lambda$ almost surely under P_λ.

c) Under P_λ, $n^{1/2}[\hat{\lambda} - \lambda] \xrightarrow{d} N(0,\lambda^2)$, where $N(\mu,\sigma^2)$ denotes the normal distribution with mean μ and variance σ^2. □

For each n, $n\hat{\theta}$ has an Erlang $(n,1/\theta)$ distribution, so one can construct exact confidence intervals for λ. Alternatively, the asymptotic normality in Proposition 3.21c) can be used to construct (approximate) large-sample confidence intervals.

It seems evident that for asynchronous data $\mathcal{F}_t^N$ the appropriate estimator of λ is

$$\hat{\lambda} = \frac{N_t}{t} \tag{3.40}$$

We will discuss shortly four justifications for (3.40)—all principles of estimation lead to the same estimator—but first here is a preliminary result.

Lemma 3.22. Given the data $\mathcal{F}_t^N$, the statistic N_t is sufficient for the parameter λ. □

The empirical method of inference appeals to Lemma 3.22 in order to base inference for λ solely on N_t, which has a Poisson distribution with mean

λt. For a Poisson distribution with unknown mean μ and, as we have, a sample of size 1, the maximum likelihood estimator and method of moments estimator are both simply the single sample value. Thus $(\lambda t)^{\wedge} = N_t$, which gives (3.40).

For the intensity method, under P_λ, N has deterministic, time-independent stochastic intensity $\lambda_t \equiv \lambda$. Theorem 2.31 implies that for each λ we have $P_\lambda \ll P_1$ (and vice versa) with

$$\log \frac{dP_\lambda}{dP_1} = t(1-\lambda) + N_t(\log \lambda) \tag{3.41}$$

(In Theorem 2.31, P_1 is denoted by P_0.) Maximizing the likelihood function (3.41) with respect to λ again yields $\hat{\lambda}$ of (3.40). Finally, the martingale estimation substitution principle (3.14) asserts that for each s, dN_s estimates $\lambda\, ds$ within "innovation" noise, and therefore $N_t = \int_0^t dN_s$ estimates $\lambda t = \int_0^t \lambda\, ds$ (see also Section 5.1).

The following properties are elementary.

Proposition 3.23. Let $\hat{\lambda}$ be given by (3.40). Then under each probability P_λ,

a) $\hat{\lambda} \to \lambda$ almost surely;

b) $t^{1/2}[\hat{\lambda} - \lambda] \xrightarrow{d} N(0,\lambda)$ □

Example 3.7 shows that the estimators $\hat{\lambda}$ of (3.40) have a certain robustness property. ("Robustness" means insensitivity to "minor" perturbations of underlying structural or distributional assumptions. A robust procedure functions nearly as reliably when model assumptions are not grossly violated as it does when they are satisfied exactly.) If the estimators (3.40) were used to analyze data from a point process that is not Poisson but is stationary and ergodic, they would remain strongly consistent as estimators of the intensity. Similarly, $1/\hat{\lambda}$ is a robust estimator of the mean interarrival time if N is a renewal process rather than a Poisson process.

Hypothesis Tests for λ

The estimation procedures just described yield corresponding tests of hypotheses involving λ; we shall not explain them in detail. Instead, we present some possibly less familiar ideas associated with asymptotics of tests based on likelihood ratios. Some sequential tests will also be discussed.

To begin, consider testing the simple hypotheses

$$\begin{aligned} H_0 &: \lambda = \lambda_0 \\ H_1 &: \lambda = \lambda_1 \end{aligned} \tag{3.42}$$

given the data $\mathcal{F}_t^N$. By (2.49) the log-likelihood ratio is

$$\log \frac{dP_{\lambda_1}}{dP_{\lambda_0}} = t(\lambda_0 - \lambda_1) + N_t\left(\log \frac{\lambda_1}{\lambda_0}\right) \tag{3.43}$$

and "finite sample" likelihood ratio tests are easy to devise (except that N_t has a discrete distribution, so that exact choice of power under H_0 may be impossible). The following result shows that a frontal attack on asymptotics is doomed.

Proposition 3.24. (Dichotomy theorem). For $\lambda_0 \neq \lambda_1$, $P_{\lambda_0} \perp P_{\lambda_1}$ on $\mathcal{F}^N_\infty$.

Proof: (Note that singularity occurs on $\mathcal{F}^N_\infty$ even though P_{λ_0} and P_{λ_1} are mutually absolutely continuous on $\mathcal{F}^N_t$ for every $t < \infty$.) It suffices to observe that the probabilities P_λ are concentrated on disjoint sets, which follows from the strong consistency statement in Proposition 3.23a). □

That is, any two of the probabilities P_λ are identical or orthogonal. Given enough data (i.e., for t sufficiently large), any reasonable approach can (asymptotically) decide the test (3.42) without error.

Much more interesting asymptotics arise if one considers instead local behavior of likelihood ratios near the null hypothesis by permitting alternative hypotheses that vary with t and converge to the null hypothesis as $t \to \infty$. Such methods and results fall under the rubrics *local asymptotic normality* and *contiguity*, and serve, among other things, to delineate quite precisely the asymptotic capabilities of likelihood ratio tests. It is not difficult to guess that alternative hypotheses should converge to the null hypothesis at rate $t^{-1/2}$, so we replace (3.42) by

$$\begin{aligned} &H_0 : \lambda = \lambda_0 \\ &H_1 : \lambda = \lambda_0 + h(\lambda_0/t)^{1/2}, \end{aligned} \tag{3.44}$$

where $h \in \mathbf{R}$ is fixed and it is agreed that the data for (3.44) are $\mathcal{F}^N_t$. These alternatives are called *contiguous alternatives*. Let L_t be the logarithm of the corresponding likelihood ratio.

Proposition 3.25. Under the hypotheses of the preceding paragraph, with respect to P_{λ_0}, $L_t \xrightarrow{d} N(-h^2/2, h^2)$.

Proof: From (3.43),

$$L_t = N_t[\log\left(1 + \frac{h}{(\lambda_0 t)^{1/2}}\right)] - (\lambda_0 t)^{1/2}h$$

$$= N_t\left[\frac{h}{(\lambda_0 t)^{1/2}} - \frac{h^2}{2\lambda_0 t}\right] - (\lambda_0 t)^{1/2}h + o_p(1)$$

[the term $o_p(1)$ converges in P_{λ_0}-probability to zero and is omitted henceforth]

$$= \frac{h}{\lambda_0^{1/2}} t^{1/2}[\hat{\lambda} - \lambda_0] - \frac{\hat{\lambda}}{\lambda_0}\frac{h^2}{2}$$

$$\xrightarrow{d} \frac{h}{\lambda_0^{1/2}} N(0,\lambda_0) - \frac{h^2}{2}$$

by Proposition 3.23, which completes the proof. □

The testing implication of Proposition 3.25 is that likelihood ratio tests can discriminate between alternatives that are contiguous at rate slower than $t^{-1/2}$ but not at a faster rate.

Likelihood ratios can also be used to construct sequential tests of hypotheses such as (3.42). Point processes are so often observed in real time that one is obliged to consider sequential methods, especially since an assured amount of information about the unknown rate can be obtained thereby. (By contrast, observation for a fixed length of time, because of the resultant random sample size N_t, yields statements whose precision is not known until the data are analyzed; this is one conditional aspect of the uniform conditional test discussed in Example 3.14.) We now construct a sequential probability ratio test (SPRT) for the simple hypotheses $H_0 : \lambda = \lambda_0$, $H_1 : \lambda = \lambda_0 + \Delta$, where $\Delta < 0$. The probabilities α for rejecting H_0 when it holds and β of rejecting H_1 when it holds are *both* specified in advance. (For fixed-sample-size tests, the Neyman-Pearson theory dictates that α be prescribed and minimizes β; the value that results may or may not be acceptably low.) Description of the SPRT involves specifying boundaries ℓ and u: if ever the likelihood ratio (3.43), viewed as a stochastic process, falls below ℓ, the test is terminated and H_0 is accepted; if it exceeds u, then H_0 is rejected. The termination time is a stopping time T that can be shown to be finite almost surely under H_0 and H_1, but is not bounded above. One observes the data $\mathcal{F}_T^N$. It can further be shown that (approximately) $\ell = \log[\beta/(1 - \alpha)]$ and $u = \log[(1 - \beta)/\alpha]$.

Other Hypothesis Tests

Examples 3.14 and 3.15 discuss testing whether a point process is Poisson; here we mention some related problems and techniques that have been applied to them.

1) *Tests for trend.* It is often of interest to test whether a point process is stationary against the alternative that there is some kind of trend. In full generality, even if it were well posed, the question is intractable, but sufficiently specific cases can be handled.

Example 3.26. (Poisson process with trend). Let N be a possibly nonhomogeneous Poisson process with intensity function $\lambda_t = e^{\alpha+\beta t}$, where α and β are both unknown. We wish to test the null hypothesis $H_0: \beta = 0$ that there is no trend against alternatives under which $\beta \neq 0$. This can be done using a likelihood ratio test. Of course, one must estimate α under H_0 (which is easy: under H_0, N is an ordinary Poisson process with rate $\lambda = e^{\alpha}$) and estimate α and β under H_1, which is hardly more difficult: the log-likelihood function is

$$L = -\frac{e^{\alpha}}{\beta}(e^{\beta t} - 1) + \alpha N_t + \beta \sum_{i=1}^{N_t} T_i; \tag{3.45}$$

one then substitutes into the appropriate log-likelihood ratio. Only large-sample distribution theory is available. □

2) *Comparison of independent Poisson processes.* This two-sample problem can be approached in a variety of ways, some leading to conditional tests. For independent Poisson processes $N(1)$, $N(2)$ with rates $\lambda(1)$, $\lambda(2)$, observed over time intervals $[0,t_1]$, $[0,t_2]$, respectively, conditional on $N_{t_1}(1) + N_{t_2}(2) = n$, $N_{t_1}(1)$ has a binomial distribution with parameters n and $p = t_1\lambda(1)/[t_1\lambda(1) + t_2\lambda(2)]$. The hypothesis $\lambda(1) = \lambda(2)$ is equivalent to $p = t_1/(t_1 + t_2)$, a known value, and can be tested with techniques available for binomial distributions. Similarly, if $N(i)$ is observed until n_i arrivals occur, then the "variance ratio" $R = (n_1/n_2)/[T_{n_1}(1)/T_{n_2}(2)]$ has an F-distribution; note that use of R allows unconditional tests. Nonparametric tests such as the sign test or the Wilcoxon signed rank test have the advantage of robustness if, for example, the $N(i)$ are not Poisson but are renewal processes.

State Estimation

Independence of increments in a Poisson process renders the state estimation problem for observations (3.24) entirely trivial: for observations $\mathcal{F}_t^N$ and f vanishing on $[0,t]$, $E_\lambda[N(f)|\mathcal{F}_t^N] = \lambda \int f(x)\,dx$. By contrast, state estimation given integral data (3.25) is no less difficult than for general Poisson processes, so discussion of it is postponed to Section 6.4.

Combined Inference and State Estimation

There is sufficient structure in a Poisson process to produce nondegenerate asymptotics for the single-realization problem of combined inference and state estimation. Let f be a function vanishing outside $[0,1]$ and for each t, let $f_t(x) = f(x - t)$. The behavior as $t \to \infty$ of the difference between the true state estimator

$$E_\lambda[N(f_t)|\mathcal{F}_t^N] = \lambda \int f \tag{3.46}$$

and the pseudo-state estimator

$$\hat{E}[N(f_t)|\mathcal{F}_t^N] = \hat{\lambda} \int f, \tag{3.47}$$

where $\hat{\lambda}$ is given by (3.40), is described by the final result of the chapter.

Proposition 3.27. Under each P_λ,

$$t^{1/2}\{\hat{E}[N(f_t)|\mathcal{F}_t^N] - E_\lambda[N(f_t)|\mathcal{F}_t^N]\} \xrightarrow{d} N(0,\lambda(\int f)^2) \quad \square$$

A few more Poisson process inference problems appear in the exercises. The theme common to all is "exploitation of special structure."

EXERCISES

3.1. Let $(\Omega,\mathcal{F},P)$ be a probability space, let $\mathcal{G}$ be a sub-σ-algebra of $\mathcal{F}$, and let Z be a random variable with $E[Z^2] < \infty$. Prove that the MMSE state estimator of Z given the observations $\mathcal{G}$ is the conditional expectation $E[Z|\mathcal{G}]$ (defined by the properties that $E[Z|\mathcal{G}]$ is $\mathcal{G}$-measurable and $E[Z;A] = E[E[Z|\mathcal{G}];A]$ for all $A \in \mathcal{G}$).

3.2. Let $X, Y_1, \dots, Y_n$ be random variables with a joint normal distribution. Prove that the conditional expectation $E[X|Y_1, \dots, Y_n]$ is the orthogonal projection of X onto the vector space $V = \text{sp}\{Y_1,\dots,Y_n\}$. Discuss explicit computation of $E[X|Y_1, \dots, Y_n]$.

3.3. Let N be the Pólya process of Example 3.6, defined on $\mathbf{R}$ rather than $\mathbf{R}_+$.

a) Prove that N is stationary.

b) Prove that N is not ergodic and that, in fact, $N([0,t])/t \to Y$ almost surely as $t \to \infty$.

c) Let $\mathcal{I}$ be the invariant σ-algebra for N (Definition 1.67). Prove that $\mathcal{I} = \sigma(Y)$.

3.4. Consider the Bernoulli process $N = U\varepsilon_X$ on $\mathbf{R}_+$, where X has distribution F and U is independent of X with $P\{U = 1\} = 1 - P\{U = 0\} = p$. Assume that F and p are known.

a) For each t calculate the MMSE state estimators $P\{X \in A,\ U = 1|\mathcal{F}_t^N\}$, $A \in \mathcal{B}(\mathbf{R}_+)$.

b) Describe asymptotic behavior of these estimators as $t \to \infty$.

3.5. Establish strong consistency and asymptotic normality of the estimators $\hat{p}$ given by (3.20).

3.6. Let $N_1, N_2, \dots$ be superposed Bernoulli processes as in Example 3.8, with k

there known but F and p unknown. Suppose that $N_1, \ldots, N_n$ have been observed over all of E.

a) Show that $\Sigma_{i=1}^{n} N_i(E)$ is a sufficient statistic for p.

b) Construct and analyze a test of the hypotheses $H_0 : p = p_0, H_1 : |p - p_0| > \Delta$, where $\Delta > 0$ is prescribed.

3.7. Verify (3.28).

3.8. Verify (3.32).

3.9. Let $\tilde{N}$ be a Poisson process on $\mathbf{R}_+$ with unknown rate λ and let T be a positive random variable, independent of $\tilde{N}$, with known distribution G. The observations are the point process $N_t = \tilde{N}_{t \wedge T}$.

a) Devise estimators $\hat{\lambda}$ given the data $\mathcal{F}_t^N$.

b) Show that *no* estimators of λ can be even weakly consistent.

3.10. (Continuation of Exercise 3.9) Suppose that λ is known.

a) Calculate for each t the MMSE state estimator $E[\tilde{N}_t | \mathcal{F}_t^N]$.

b) Obtain a recursive representation for these state estimators.

3.11. For the Yule process of Example 3.19, show that the estimators $\hat{c}$ of (3.37) are strongly consistent and asymptotically normal.

3.12. Consider the counter model of Exercise 2.22 with deterministic deadtime τ. Suppose that the rate λ of the arrival process $\tilde{N}$ is unknown and that only the point process $N = \Sigma_{n=1}^{\infty} \varepsilon_{T_n}$ of recorded arrivals is observable.

a) Show that N is a renewal process and calculate the interarrival distribution.

b) Using a), or otherwise, develop estimators $\hat{\lambda}$ of λ given both synchronous data $\mathcal{F}_{T_n}^N$ and asynchronous data $\mathcal{F}_t^N$, and establish their respective asymptotic properties.

3.13. (Continuation of Exercise 3.12) Suppose now that λ is known but the deadtime τ is unknown.

a) For each n calculate the maximum likelihood estimator $\hat{\tau}$ of τ given the data $\mathcal{F}_{T_n}^N$.

b) Describe asymptotic properties of the estimators in a).

c) Develop a likelihood ratio test, given data $\mathcal{F}_{T_n}^N$, of the hypotheses $H_0 : \tau = \tau_0, H_1 : \tau = \tau_1 < \tau_0$.

3.14. (Continuation of Exercise 3.12) Assume that λ and τ are both known and consider state estimation for the arrival process $\tilde{N}$ given observation of the record process N.

a) For each t calculate the MMSE state estimator $E[\tilde{N}_t | \mathcal{F}_t^N]$.

b) Devise a recursive method (it need not be a stochastic differential equation) for computation of these state estimators.

3.15. Let N' be the partially observed Poisson process of Example 3.5 and Exercise 2.21.

a) Prove that N' is a renewal process; calculate the Laplace transform

$$\ell(\alpha) = \int_0^\infty e^{-\alpha t}[1 - G(t)]\, dt,$$

where G is the interarrival distribution; compute the mean interarrival time.
b) Suppose that the transition rates a, b are known but that the rate c of the underlying Poisson process is not. Use a) to construct estimators $\hat{c}$ given the data $\mathcal{F}_{T_n}^{N'}$.
c) Characterize the asymptotic behavior of the estimators in b).

3.16. (Continuation of Exercise 3.15) Assume that a, b, and c are all known.
a) Prove that N' is a Cox process.
b) Use a) or some other method to calculate the distribution of N'_t for each t.

3.17. (Continuation of Exercise 3.15) Assume once more that c is unknown (but a and b are known).
a) Prove that the estimators $\hat{c}$ of (3.18), based on the data $\mathcal{F}_t^{N'}$, are strongly consistent and asymptotically normal.
b) Prove that the same is true of the estimators $\hat{c}$ of (3.19), which correspond to the data $\mathcal{F}_t^N \vee \mathcal{F}_t^X$.

3.18. Let N be a Poisson process with unknown rate c and let X be a binary Markov process independent of N with known generator

$$A = \begin{bmatrix} -a & a \\ b & -b \end{bmatrix}; \tag{3.48}$$

so far the setup is as in Exercises 2.21 and 3.15 – 3.17. Suppose, however, that one observes the process $Z_t = X_t N_t$, which of course is not a point process. Devise estimators $\hat{c}$ of c given the observations $\mathcal{F}_t^Z$ and discuss their asymptotic behavior as $t \to \infty$.

3.19. (Continuation of Exercise 3.18) Suppose now that the rate c of N is known.
a) For each t calculate the MMSE state estimators $E[N_u|\mathcal{F}_t^Z]$ for all $u \geq 0$.
b) Describe their asymptotic properties as $t \to \infty$.
c) Use a) and results from Exercise 3.18 to construct pseudo-state estimators $\hat{E}[N_u|\mathcal{F}_t^Z]$ for each t and u.
d) Characterize asymptotic behavior of the difference between the true and pseudo-state estimators as $t \to \infty$.

3.20. Let (X_t) be a Markov process with states 0, 1 and generator (3.48); assume that a is known but b is not. Suppose that $X_0 = 1$ and let $T_1, T_2, \ldots$ be the successive times at which X returns to state 1; the observations are the point process $N = \Sigma_{n=1}^{\infty} \varepsilon_{T_n}$.
a) Prove that N is a renewal process and calculate the interarrival distribution.
b) Use a) to devise estimators $\hat{b}$ of b given the synchronous data $\mathcal{F}_{T_n}^N$; discuss their large-sample properties.
c) Suppose now that a is unknown as well. Construct and analyze a test, based on $\mathcal{F}_{T_n}^N$, of the hypothesis $H_0 : a = b$.

3.21. Let N be a stationary point process on **R** with unknown intensity ν (see Example 3.7).
a) Prove that with respect to the Palm measure P^* the sequence (U_i) of interarrival times is strictly stationary.
b) Show that if N is P-ergodic (Definition 1.67), then (U_i) is P^*-ergodic.

c) Construct estimators of ν given the data $\sigma(U_1, \ldots, U_n)$ and describe their behavior as $n \to \infty$ under the assumption that N is ergodic.

3.22. Let N be the point process on $\mathbf{R}$ given by $N = \Sigma_{n=-\infty}^{\infty} \varepsilon_{X+nY\theta}$, where X and Y are independent random variables uniformly distributed on $[0,1]$ and $\theta > 0$ is unknown.
a) Prove that N is stationary.
b) Calculate the intensity $\nu(\theta)$ of N.
c) Discuss the problem of estimation of ν given a single realization of N over $[0,t]$.

3.23. Let N be the point process (see Exercise 3.22) $N = \Sigma_{n=-\infty}^{\infty} \varepsilon_{X+nY}$, where X, Y are independent and uniformly distributed on $[0,1]$ and let $\mathcal{F}_t^N = \sigma(N(0,u] : 0 \leqslant u \leqslant t)$. For each t calculate the MMSE state estimators $E[X \mid \mathcal{F}_t^N]$ and $E[Y \mid \mathcal{F}_t^N]$.

3.24. Consider the following variant of the hypothesis-testing problem of Example 3.12. The observations are a single realization over $[0,t]$ of a stationary point process N with unknown intensity ν, and the hypotheses are $H_0 : \nu = \nu_0$, $H_1 : \nu = \nu_1 > \nu_0$. Consider a test based on the critical region $\Omega_c = \{N_t \geq \delta\}$.
a) Show that given α it is possible to choose δ in such a manner that $P\{\Omega_c\} < \alpha$ for all stationary point processes satisfying the null hypothesis.
b) Show that, however, regardless of the value of δ there are stationary point processes [*Hint*: Consider mixed Poisson processes] satisfying the alternative hypothesis for which $P(Q_c)$ is arbitrarily small. What does this mean concerning the power of the test?
c) Show that b) remains true even for the critical region $\Omega_c = \{N_s/s \geq \delta$ for some $s \leq t\}$.

3.25. Prove the following results concerning estimation of the rate of an ordinary Poisson process.
a) Proposition 3.21.
b) Lemma 3.22.
c) Proposition 3.23.

3.26. Let N be a Poisson process on $\mathbf{R}_+$ with unknown rate λ and suppose that we have the asynchronous data $\mathcal{F}_k^N$ corresponding to observation of N over $[0, k]$. Here we examine alternatives to the estimators $\hat{\lambda}$ of (3.40). Since $P_\lambda\{N(i,i+1] = 0\} = e^{-\lambda}$ for each i, one could estimate λ using $\hat{\lambda}^* = -\log [k^{-1}\Sigma_{i=1}^{k} 1(N(i,i+1]=0)]$.
a) Show that the estimators $\hat{\lambda}^*$ are strongly consistent and asymptotically normal (as $k \to \infty$).
b) Compare the limiting variances of $n^{1/2}[\hat{\lambda} - \lambda]$ (see Proposition 3.23) and $k^{1/2}[\hat{\lambda}^* - \lambda]$ and argue when and why one of the estimators should be preferred to the other.

3.27. From the likelihood function L of (3.45) associated with observation of a nonhomogeneous Poisson process N with intensity function $\lambda_u = \exp[\alpha + \beta u]$, determine the maximum likelihood estimators $\hat{\alpha}$ and $\hat{\beta}$.

3.28. Let N be a Poisson process on $\mathbf{R}_+$ with rate λ and let A be a bounded set. Calculate the MMSE state estimators $E[N(B) | \mathcal{F}^N(A)]$.

3.29. Let $N(1), N(2)$ be independent Poisson processes on $\mathbf{R}_+$ with rates $\lambda(1), \lambda(2)$, respectively, and let $N = N(1) + N(2)$. For each t calculate $E[N_t(1)|\mathscr{F}_t^N]$; develop a recursive method of computation.

3.30. Let $N(1), N(2), \ldots$ be i.i.d. Poisson processes on $\mathbf{R}_+$ with known rate λ and suppose that one can observe only the superposition $N = \Sigma_{i=1}^{m} N_i$, where $m \geq 0$ is an unknown integer.

a) Show that for each t, N_t is a sufficient statistic for m given the data $\mathscr{F}_t^N$.

b) Calculate the maximum likelihood estimators $\hat{m}$; discuss their properties as $t \to \infty$.

3.31. Let $N(1), N(2), \ldots$ be i.i.d. Poisson processes on $\mathbf{R}_+$ with known rate λ and suppose that one observes only the superposition $N = \Sigma_{i=1}^{M} N_i$, where M is a nonnegative, integer-valued random variable independent of the $N(i)$. For each t compute the MMSE state estimator $E[M|\mathscr{F}_t^N]$.

3.32. Let $\tilde{N} = \sum_{n=-\infty}^{\infty} \varepsilon_{T_n}$ be a stationary Poisson process on $\mathbf{R}$ with unknown rate λ and let Z_i, $i \in \mathbf{Z}$, be i.i.d. random variables, independent of N, with unknown distribution G.

a) Discuss estimation of λ from observations of the point process $N = \Sigma_{n=-\infty}^{\infty} \varepsilon_{T_n + Z_n}$. What about estimation of G?

b) Suppose now that λ and G are known. For each f and t compute $E[\tilde{N}(f)|\mathscr{F}_t^N]$.

3.33. Formulate and prove a theorem on local asymptotic normality (as $n \to \infty$) of likelihood ratios associated with Proposition 3.21.

3.34. Let $U_1, \ldots, U_n$ be independent, exponentially distributed random variables with parameter $\lambda = 1$, let $U_{(1)}, \ldots, U_{(n)}$ be their order statistics, and define the normalized *spacings* $U'_j = (n+1-j)[U_{(j)} - U_{(j-1)}]$, $j = 1, \ldots, n$. Prove that $U'_1, \ldots, U'_n$ are independent and identically exponentially distributed.

NOTES

As general references on mathematical statistics we mention Bickel and Doksum (1977), Cox and Hinckley (1974), Ferguson (1967), Lehmann (1959, 1983), Wilks (1962), and Zacks (1971); of course, there are many others. Statistical inference for point processes is treated in some generality in Brillinger (1978) (which draws extensive and not always convincing parallels between inference for point processes and inference for time series), Cox and Lewis (1966), and Krickeberg (1982), but only the latter includes general spaces. The literature on martingale inference for point processes on $\mathbf{R}_+$ is discussed in notes for Chapter 5.

Section 3.1

Inference based on integral data is introduced in Karr (1985e). Model 3.1 provides the setting for all of Chapter 4, with Model 3.3 occupying Section 4.5 there. As examples in this chapter suggest, only in the presence of rather stringent structural assumptions is single-realization inference (Model 3.2) feasible. Cox and Lewis (1966) contains the synchronous/asynchronous differentiation.

Section 3.2

General sources on nonparametric statistics include Hollander and Wolfe (1973) as well as the broader books of Bickel and Doksum (1977) and Wilks (1962); each tries at least implicitly to explain a parametric-nonparametric distinction.

Two facets of classical statistics are absent from this book, one by design, the other by the nature of point processes. The former is Bayesian inference, which we have omitted with the knowledge that rather little exists concerning Bayesian statistical inference for point processes. (Of course, state estimation is the Bayesian method *par excellence,* but not with an arbitrary prior distribution.) Second, there is neither a body of problems nor a set of methods corresponding to the linear models/regression part of classical statistics, because of the discrete, nonnegative character of point processes. In Chapter 5 we do describe one model with a (partial) regression-like structure, but it is analyzed using martingale techniques.

Foundation issues and concepts for estimation and hypothesis testing are treated in the general sources mentioned above; there one can find general principles such as maximum likelihood estimation, the method of moments and likelihood ratio tests, along with more specific notions such as consistency, efficiency, null and alternative hypotheses, power of a test, p-values, ... For estimation problems we concentrate nearly exclusively on point estimation (as opposed to construction of confidence regions) and usually ignore optimality questions such as derivation of minimum variance estimators. For parametric problems these can often be resolved via classical theory (treated very mathematically in Ibragimov and Has'minskii, 1981), while for nonparametric problems they are typically just too difficult. In the latter case we are satisfied with estimators that are consistent and asymptotically normal in the sense that an *error process* converges in distribution to a Gaussian process. Similarly, for tests of hypotheses, computation of power under the null hypothesis and analysis of asymptotics may be all that one can achieve.

Now for the specific examples. Inference procedures for gamma distributions (Examples 3.4 and 3.9) are presented in Barlow and Proschan (1975), Bickel and Doksum (1977), and Wilks (1962). (Here as well as in Chapter 8, the reliability literature is often the best source of details; gamma distributions, for example, often appear in statistics books only as an incompletely presented example of a two-parameter exponential family.) Serfozo (1975) contains a powerful and general method for inferring asymptotic properties of continuous-time stochastic processes from those of discrete parameter processes embedded at random times satisfying minimal stability conditions. The observed stochastic integral of Examples 3.5 and 3.10, treated as a partially Poisson process, is treated by Grandell (1976) and Karr (1982) (see also Karr, 1984a; Exercises 2.21 and 3.15 – 3.17; and Chapters 6, 7, and 10). Smith and Karr (1985) include numerical calculations of maximum likelihood estimators of all three parameters for precipitation data. In Example 3.6 the issue is estimation of an unknown mixing distribution from i.i.d. samples when the distributions being mixed are known; Albrecht (1982b), Lindsay (1983a, b), Simar (1976), and Tucker (1963) address mixed Poisson distributions specifically. Estimation of the intensity of a stationary point process on **R** is treated in Brillinger (1978), Cox (1965), and Cox and Lewis (1966), and in general spaces by Krickeberg (1982) and in Chapter 9.

The problem of testing whether a renewal process is Poisson (Example 3.14) has an extensive literature; see Cox and Lewis (1966) and Lewis (1965, 1972a) for expository treatments and further references. Epstein (1960) is an early source. The uniform conditional test and dispersion test are described in Lewis (1965) and the gap test in Lewis (1972a) and Pyke (1965); these tests seem generally believed to work best against some fairly specific but not very explicitly described alternatives. Tests for uniformity (and for other specific distributions) are discussed in the notes for Chapter 4; to use them to test directly whether interarrival times are exponential entails the complication of estimating the unknown mean. Less has been written about the more difficult problem of testing whether a stationary point process is Poisson (Example 3.15). Davies (1977) and Ripley (1976d) use moment and cumulative measures (see Chapter 9) to analyze narrowly defined alternative hypotheses. Sundt (1982) shows that the easily posed problem of testing whether a mixed Poisson process is Poisson is intractable.

Section 3.3

Brémaud (1981), Liptser and Shiryayev (1978), and Snyder (1975) are the main expository sources on state estimation for point processes, although each treats only point processes on $\mathbf{R}_+$. All of them examine recursive methods in detail; the third is particularly rich in intuition, interpretation, and illustrative examples. In connection with state estimation for mixed Poisson processes (Example 3.18) (see Grandell, 1976; Karr, 1983, 1984b; Krickeberg, 1982; Snyder, 1975; and Section 7.2). Section 5.5 contains the solution of a very general state estimation problem on $\mathbf{R}_+$. Additional insight can be gained from the voluminous literature on state estimation problems involving diffusion processes.

Section 3.4

The problem of combined inference and state estimation is not new, but seems not to have previously been analyzed systematically. Results for specific models appear in Sections 4.4, 6.3, 7.4, 8.3, and 10.3. Inference for Yule processes (Example 3.19) is discussed in Basawa and Prakasa Rao (1980) and Feigin (1976).

Section 3.5

Birnbaum (1954) is one of the earliest substantive references to Poisson processes per se, although obviously the methods of Propositions 3.21 and 3.23 predate it. Good general sources are Cox and Lewis (1966) and Lewis (1965, 1972a). Sufficiency (Lemma 3.22) and minimal sufficient statistics are discussed [e.g., in Bickel and Doksum (1977) and Zacks (1971)]. Proposition 3.24 is a special case of Theorem 6.13, references for which are given in the notes for Section 6.2. Local asymptotic normality (Proposition 3.25) is due to LeCam (1960a); Greenwood and Shiryayev (1985) and Roussas (1972) are oriented to applications to inference for stochastic processes. Sequential estimation and test procedures for Poisson processes are described in Birnbaum (1954), Dvoretsky et al. (1953a, b) and Vardi (1979). Wald

(1947) is the seminal reference on sequential methods in general, with Govindarajulu (1975) a more modern treatment. Bartholomew (1956), Boswell (1966), Cox and Lewis (1966), and Saw (1975) address tests of Poisson processes for trend. Comparison of (independent) Poisson processes is considered by Birnbuam (1954), Durbin (1961), Lewis (1965), and Qureishi (1964); the methods are based on conditional tests involving binomial distributions (for asynchronous data) or techniques for comparing χ^2 distributions (for synchronous data). Section 5.3 presents different methods that may have desirable robustness properties.

Exercises

3.11. Maximum likelihood estimation of c can be based on the property that interarrival times for a Yule process are independent and exponentially (because a Yule process is a Markov process with deterministic embedded Markov chain) (see Basawa and Prakasa Rao, 1980).

3.12–3.14. See Feller (1966) for fundamental properties of this counter model.

3.15. See Karr (1982) and Kingman (1963); the latter characterizes processes that are simultaneously renewal processes and Cox processes (see also Exercise 3.16 and Theorem 8.24).

3.16, 3.17. See Grandell (1976) and Karr (1982); the latter treats the estimators $\hat{c}$ of (3.18) and (3.19).

3.19. The "linear interpolation" property here is yet another manifestation of conditional uniformity.

3.20. Karr (1984a) considers additional aspects, including state estimation.

3.21. See Krickeberg (1982), Neveu (1977), and Chapter 9.

3.24. This is a fairly graphic demonstration that some kinds of structural restrictions are not stringent enough to permit construction of tests with reasonable power.

3.26. Empirical zero-probability functionals are treated in generality in Section 4.3.

3.27. Cox and Lewis (1966) treat this model as a particular alternative hypothesis representing trend; for small t the trend is approximately linear—the most elementary sort of trend—without the intensity function's becoming negative.

3.30. See Cox and Lewis (1966). This exercise shows that seemingly different problems can be identical.

3.31. See also Exercises 1.20 and 6.24. M. Brown (1970) examines estimation of G when both $\tilde{N}$ and N are observable but points cannot be paired between them.

3.34. See Pyke (1965).

4

Empirical Processes Associated with a Sequence of Point Processes

Let $N_1, N_2, \ldots$ be i.i.d copies of a point process N on a *compact* space E and let $\mathcal{L}$ denote their common, unknown probability law. This chapter is devoted to analysis of empirical measures $\hat{\mathcal{L}} = n^{-1} \sum_{i=1}^{n} \varepsilon_{N_i}$ as estimators of $\mathcal{L}$, to their asymptotic properties, and to consequences of the principle that a functional $H(\mathcal{L})$ should be estimated by the corresponding functional $H(\hat{\mathcal{L}})$ of the empirical measure. We are in the case of Model 3.1: the underlying space is compact and the data are completely observed (except in Section 4.5) copies of N. Denote by $\underline{L}$ the class of candidate probability laws for N, which in most of the chapter is not restricted at all; then the family of probability measures possibly governing the observations is $\mathcal{P} = \{P_{\mathcal{L}} : \mathcal{L} \in \underline{L}\}$, where $P_{\mathcal{L}} = \mathcal{L}^{\mathbf{N}}$.

When $\underline{L}$ is the set of all probabilities on $\mathbf{M}_p$, the empirical measure $\hat{\mathcal{L}}$ is the maximum likelihood estimator of $\mathcal{L}$ (Exercise 4.2). If, therefore, the functional H is smooth, then $H(\hat{\mathcal{L}})$ is likewise the maximum likelihood estimator of $H(\mathcal{L})$. However, when $\underline{L}$ is restricted, the maximum likelihood estimator might be difficult or impossible to express as a functional of $\hat{\mathcal{L}}$. For example, if $\underline{L} = \{\mathcal{L}_\theta\}$ is a parametric family, then provided that there exists a maximum likelihood estimator $\hat{\theta}$ of θ and that the mapping $\theta \to \mathcal{L}_\theta$ is smooth, the maximum likelihood estimator of $\mathcal{L}$ is $\mathcal{L}_{\hat{\theta}}$, but there may be no functional H satisfying $\theta = H(\mathcal{L}_\theta)$ for which $H(\hat{\mathcal{L}})$ is well defined. (Similar phenomena arise in density estimation.) Nevertheless, even in such cases empirical measures—because of robustness properties—remain in contention as estimators of $\mathcal{L}$. Indeed, in some cases they have properties superior to those of parametric estimators (see Example 4.15).

The method of moments basis for the empirical measure is that $E_{\mathcal{L}}[\hat{\mathcal{L}}(\Gamma)] = P_{\mathcal{L}}\{N \in \Gamma\} = \mathcal{L}(\Gamma)$ for $\Gamma \in \mathcal{M}_p$ and each $\mathcal{L}$. Although it does not follow in general that $H(\hat{\mathcal{L}})$ is also the method of moments estimator of $H(\mathcal{L})$, this is the case if H is an integral functional [i.e.,

$H(\mathcal{L}) = \int_{\mathbf{M}_p} \widetilde{H}(\mu)\mathcal{L}(d\mu)$, where $\widetilde{H}: \mathbf{M}_p \to \mathbf{R}$]. Empirical Laplace functionals and empirical zero-probability functionals, perhaps the most important functionals, are of this kind.

Each empirical measure $\hat{\mathcal{L}}$ is a random, purely atomic probability on $\mathbf{M}_p$—and $n\hat{\mathcal{L}}$ is a point process on $\mathbf{M}_p$; therefore, the general theory developed in Chapter 1 applies. That the N_i are themselves point processes, although obviously central to our development, is less crucial to some tools and ideas used in this chapter. There is an extensive literature dealing with empirical processes engendered by i.i.d. random elements of general measurable spaces, which can be utilized in conjunction with the special structure of $\mathbf{M}_p$ to derive characteristics of the $\hat{\mathcal{L}}$. Moreover, the line of deduction can be reversed: a random element X_i of E is the "mean" of the point process $N_i = \varepsilon_{X_i}$, so point process techniques can be used to establish properties of ordinary empirical processes. Both points of view are pursued.

Using classical methods, one easily obtains the following results, which are prototypes for those in Section 4.1. We begin with strong consistency.

Proposition 4.1. For each $\mathcal{L}$ and each $\Gamma \in \mathcal{M}_p$, $\hat{\mathcal{L}}(\Gamma) \to \mathcal{L}(\Gamma)$ almost surely with respect to $P_{\mathcal{L}}$. Furthermore,

$$\mathcal{L}_n \xrightarrow{w} \mathcal{L} \tag{4.1}$$

where $\xrightarrow{w}$ denotes weak convergence of probability measures on $\mathbf{M}_p$.

Proof: The first assertion is immediate from the strong law of large numbers. Concerning the second, by separability of $\mathbf{M}_p$ there exists a sequence of functions h_j in $C(\mathbf{M}_p)$ such that for probability measures η_n, η on $\mathbf{M}_p$, $\eta_n \xrightarrow{w} \eta$ if and only if $\eta_n(h_j) \to \eta(h_j)$ for each j; hence (4.1) holds. □

Next is asymptotic normality.

Proposition 4.2. For each $\mathcal{L}$ and for $\Gamma_1, \ldots, \Gamma_k \in \mathcal{M}_p$,

$$n^{1/2}[(\hat{\mathcal{L}}(\Gamma_1), \ldots, \hat{\mathcal{L}}(\Gamma_k)) - (\mathcal{L}(\Gamma_1), \ldots, \mathcal{L}(\Gamma_k))] \xrightarrow{d} N(0,R) \tag{4.2}$$

where $R(i,j) = \mathcal{L}(\Gamma_i \cap \Gamma_j) - \mathcal{L}(\Gamma_i)\mathcal{L}(\Gamma_j)$. □

Finally, there is a law of the iterated logarithm (see also Proposition 4.31).

Proposition 4.3. For each $\mathcal{L}$ and each $\Gamma \in \mathcal{M}_p$,

$$\limsup_{n\to\infty} \frac{n^{1/2}[\hat{\mathcal{L}}(\Gamma) - \mathcal{L}(\Gamma)]}{(2 \log \log n)^{1/2}} = \mathcal{L}(\Gamma)[1 - \mathcal{L}(\Gamma)] \tag{4.3}$$

almost surely with respect to $P_{\mathcal{L}}$. □

The main objective of Section 4.1 is to strengthen Propositions 4.1 and 4.2 along two lines. First, we examine not only set-indexed empirical measures $\{\hat{\mathcal{L}}(\Gamma) : \Gamma \in \mathcal{M}_p\}$ but also function-indexed analogs $\{\hat{\mathcal{L}}(h) : h \in C_K(\mathbf{M}_p)\}$, which in many ways are the more tractable. Second, we develop "process" versions of Propositions 4.1 and 4.2. In the former case this leads to strong uniform consistency results such as Theorems 4.5 and 4.9, while for the latter we derive "infinite-dimensional" central limit theorems, of which Theorems 4.10 and 4.11 are examples. Sections 4.2 – 4.4 concern estimation of functionals of $\mathcal{L}$: the Laplace functional, zero-probability functional, and (reduced) Palm distributions, respectively. As noted already, the first two are integral functionals of $\mathcal{L}$; consequently, estimators are easy to formulate and their properties are reasonably straightforward to obtain. Palm distributions, as derivatives, present a more difficult problem. Our approach is to view them as limits of ratios of integral functionals of $\mathcal{L}$: $Q(\mu, \cdot) = \lim H_n(\mathcal{L})/\widetilde{H}_n(\mathcal{L})$, which are estimated by corresponding limits $\hat{Q}(\mu, \cdot) = \lim H_n(\hat{\mathcal{L}})/\widetilde{H}_n(\hat{\mathcal{L}})$. Section 4.5 applies material from Section 4.2 to the case that the observed point processes are i.i.d. p-thinnings of the N_i, where both $\mathcal{L}_N$ and the thinning function p are unknown. In Section 4.6 we explore some implications of point process results concerning ordinary empirical processes and the application of conditional uniformity to produce strong approximations of Poisson processes by Wiener processes.

4.1 EMPIRICAL MEASURES

This section surveys fundamental asymptotic properties of empirical measures. We merge methodology exploiting special structure of the space $\mathbf{M}_p$ in which the point processes N_i take their values with the theory of empirical processes associated with i.i.d. random elements of general measurable spaces. Interpretation may be eased if it is borne in mind that empirical measures are analogous to empirical distribution functions, that strong uniform consistency results are versions of the Glivenko-Cantelli theorem, and that for statistical purposes the central limit theorems are directed to obtaining, among other things, analogs of the Kolmogorov-Smirnov limit distribution.

The setting is this: Let E be a compact metric space. Under the vague topology $\mathbf{M}_p$ is a complete, separable metric space and is even locally compact (Exercise 4.1). Let $N_1, N_2, \ldots$ be i.i.d. copies of a point process N on E with law $\mathcal{L}$. Our model is fully nonparametric: $\mathcal{L}$ is presumed to have no particular structure or form.

Definition 4.4. The *empirical measures* engendered by (N_i) are the random probability measures $\hat{\mathcal{L}} = n^{-1} \sum_{i=1}^{n} \varepsilon_{N_i}$, $n = 1, 2, \ldots$, on $\mathbf{M}_p$.

The notation accentuates the role of empirical measures as estimators of $\mathcal{L}$; as throughout the book, the sample size is suppressed. In our unrestricted model, for each n the n-sample empirical measure $\hat{\mathcal{L}}$ is the maximum likelihood estimator of $\mathcal{L}$ given the data $N_1, \ldots, N_n$. One approach to properties of the $\hat{\mathcal{L}}$ is to view them as random elements of $\mathbf{M}_p$ and error processes $n^{1/2}[\hat{\mathcal{L}} - \mathcal{L}]$ as random elements of the space V of finite signed measures on E, and to exploit directly special structure of $\mathbf{M}_p$ and V using results from Chapter 1. Alternatively, one can regard the $\hat{\mathcal{L}}$ as empirical measures arising from the i.i.d. random elements N_i of $\mathbf{M}_p$, and rely instead on the substantial theory of empirical processes on general measurable spaces (see, e.g., Gaenssler, 1984) and only later use special structure of $\mathbf{M}_p$.

Consistency and asymptotic normality as proved in Propositions 4.1 and 4.2 are weak and uninformative from an esthetic point of view and for purposes of statistical inference. For estimation of $\mathcal{L}$ much more information is contained in uniform convergence statements $\sup\{|\hat{\mathcal{L}}(h) - \mathcal{L}(h)| : h \in \mathcal{H}\} \to 0$, where $\mathcal{H}$ is a reasonably large subset of $\mathbf{M}_p$. Similarly, for testing the hypothesis $H_0 : \mathcal{L} = \mathcal{L}_0$ the test statistic $D_n = \sup\{|\hat{\mathcal{L}}(h) - \mathcal{L}_0(h)| : h \in \mathcal{H}\}$ is more sensitive than a statistic $|\hat{\mathcal{L}}(h) - \mathcal{L}_0(h)|$ with h fixed. Use of D_n as a test statistic necessitates knowledge at least of its distribution under H_0, which is one motivation for the central limit theorems presented in this section.

Concerning behavior of empirical measures on sets, we seek classes $\mathcal{C}$ of subsets of $\mathbf{M}_p$ such that the *discrepancy*

$$D_n(\mathcal{C}, \mathcal{L}) = \sup\{|\hat{\mathcal{L}}(\Gamma) - \mathcal{L}(\Gamma)| : \Gamma \in \mathcal{C}\} \tag{4.4}$$

satisfies $P_{\mathcal{L}}\{D_n(\mathcal{C}, \mathcal{L}) \to 0\} = 1$. Similar issues arise for function-indexed empirical measures $\hat{\mathcal{L}}(h)$, $h \in C_+(\mathbf{M}_p)$.

Uniformity for functions will be applied in Section 4.2 to convergence of empirical Laplace functionals; uniformity for sets is applied in Section 4.3 to empirical zero-probability functionals. To obtain uniform convergence we pursue a direct attack that yields a strong uniform consistency theorem for function-indexed processes; perform more detailed analysis of the weak convergence $\hat{\mathcal{L}}(\omega) \xrightarrow{w} \mathcal{L}$ with ω fixed (Proposition 4.1), which provides information concerning setwise convergence and uniform convergence for both sets and functions; and apply the general theory of empirical processes, mainly to derive uniformity results for sets. Rather than present the most general results known, we strike a balance between difficulty and generality and illustrate a range of techniques.

We begin with the direct approach.

Theorem 4.5. Let $\mathcal{H}$ be a subset of $C(\mathbf{M}_p)$ that is uniformly bounded in the topology of uniform convergence on compact subsets. Then for each

$\mathcal{L}$, almost surely under $P_{\mathcal{L}}$,

$$\sup\{|\hat{\mathcal{L}}(h) - \mathcal{L}(h)| : h \in \mathcal{H}\} \to 0 \tag{4.5}$$

Proof: For each k let $\mathbf{M}_p(k) = \{\mu \in \mathbf{M}_p : \mu(E) \le k\}$, a compact set. Given $\varepsilon > 0$ choose and fix k such that $\mathcal{L}(\mathbf{M}_p \backslash \mathbf{M}_p(k)) < \varepsilon$; by (4.1) almost surely $\hat{\mathcal{L}}(\mathbf{M}_p \backslash \mathbf{M}_p(k)) < 2\varepsilon$ for all n sufficiently large. Letting $\tilde{h} = h 1_{\mathbf{M}_p(k)}$, we have that for large n,

$$\begin{aligned}\sup_{\mathcal{H}} |\hat{\mathcal{L}}(h) - \mathcal{L}(h)| &\le \sup_{\mathcal{H}} |\hat{\mathcal{L}}(\tilde{h}) - \mathcal{L}(\tilde{h})| + \sup_{\mathcal{H}} \hat{\mathcal{L}}(|h - \tilde{h}|) + \sup_{\mathcal{H}} \mathcal{L}(|h - \tilde{h}|) \\ &\le \sup_{\mathcal{H}} |\hat{\mathcal{L}}(\tilde{h}) - \mathcal{L}(\tilde{h})| + 6\,\varepsilon\, C,\end{aligned}$$

where $C = \sup\{\|h\|_\infty : h \in \mathcal{H}\}$ and is finite. To show that the remaining supremum converges to zero almost surely, again let $\varepsilon > 0$ be fixed. Then by the Arzéla-Ascoli theorem there is $\delta > 0$ such that whenever $d(\mu, \nu) < \delta$ in $\mathbf{M}_p(k)$ [d denotes the Prohorov metric (see Kallenberg, 1983)], then

$$\sup\{|\tilde{h}(\mu) - \tilde{h}(\nu)| : h \in \mathcal{H}\} < \varepsilon. \tag{4.6}$$

By compactness, $\mathbf{M}_p(k)$ can be partitioned as $\mathbf{M}_p(k) = \Sigma_{j=1}^M \Gamma_j$, where diam $\Gamma_j < \varepsilon$ for each j; also, let $\mu_j \in \Gamma_j$ be fixed. The multidimensional Glivenko-Cantelli theorem implies that almost surely

$$\sup\{|\hat{\mathcal{L}}\left(\sum_{j=1}^M \alpha_j 1_{\Gamma_j}\right) - \mathcal{L}\left(\sum_{j=1}^M \alpha_j 1_{\Gamma_j}\right)| : |\alpha_i| \le C,\ i = 1, \ldots, M\} \to 0; \tag{4.7}$$

(this could also be argued using the Vapnik/Chervonenkis criterion to be introduced presently). For $h \in \mathcal{H}$,

$$\begin{aligned}|\hat{\mathcal{L}}(\tilde{h}) - \mathcal{L}(\tilde{h})| \le &\left|\hat{\mathcal{L}}(\tilde{h}) - \hat{\mathcal{L}}\left(\sum_{j=1}^M \tilde{h}(\mu_j) 1_{\Gamma_j}\right)\right| \\ &+ \left|\hat{\mathcal{L}}\left(\sum_{j=1}^M \tilde{h}(\mu_j) 1_{\Gamma_j}\right) - \mathcal{L}\left(\sum_{j=1}^M \tilde{h}(\mu_j) 1_{\Gamma_j}\right)\right| \\ &+ \left|\mathcal{L}\left(\sum_{j=1}^M \tilde{h}(\mu_j) 1_{\Gamma_j}\right) - \mathcal{L}(\tilde{h})\right| \\ \le 2\varepsilon + &\left|\hat{\mathcal{L}}\left(\sum_{j=1}^M \tilde{h}(\mu_j) 1_{\Gamma_j}\right) - \mathcal{L}\left(\sum_{j=1}^M \tilde{h}(\mu_j) 1_{\Gamma_j}\right)\right|\end{aligned}$$

and now only application of (4.7) is needed to complete the proof. □

We next consider what can be extracted from more detailed study of the conclusion $\hat{\mathcal{L}} \xrightarrow{w} \mathcal{L}$ in (4.1). As is well known, (4.1) implies that

$$\hat{\mathcal{L}}(\Gamma) \to \mathcal{L}(\Gamma) \tag{4.8}$$

for every $\mathscr{L}$*-continuity set* Γ [satisfying $\mathscr{L}(\partial\Gamma)=0$, where $\partial\Gamma$ is the boundary of Γ]. There are uniform versions of (4.8), as well as corresponding statements for functions, that derive solely from (4.1). We list one such result, due to Billingsley and Tøpsøe (1967).

Theorem 4.6. a) Let $\mathscr{H}$ be a subset of $C(\mathbf{M}_p)$ such that $\sup\{\|h\|_\infty : h \in \mathscr{H}\} < \infty$ and such that for every $\varepsilon > 0$,

$$\lim_{\delta \downarrow 0} \sup_{h \in \mathscr{H}} \mathscr{L}(\{\mu : \sup\{|h(\mu) - h(\nu)| : d(\mu,\nu) < \delta\} > \varepsilon\}) = 0 \qquad (4.9)$$

Then almost surely $\sup\{|\hat{\mathscr{L}}(h) - \mathscr{L}(h)| : h \in \mathscr{H}\} \to 0$.

b) Let $\mathscr{C}$ be a class of bounded Borel subsets of $\mathbf{M}_p$ such that $\mathscr{L}(\partial\Gamma)=0$ for every $\Gamma \in \mathscr{C}$ and suppose that $\{\partial\Gamma : \Gamma \in \mathscr{C}\}$ is compact in the Hausdorff topology (on the set of bounded, closed subsets of $\mathbf{M}_p$). Then almost surely $\sup\{|\hat{\mathscr{L}}(\Gamma) - \mathscr{L}(\Gamma)| : \Gamma \in \mathscr{C}\} \to 0$. □

The hypothesis (4.9) is less awkward than it appears (it is better stated in terms of oscillation functions), but nonetheless not easy to verify; moreover, it involves intricate properties of $\mathscr{L}$ that one might be unwilling to stipulate. Theorem 4.5 does not have this defect, nor does Theorem 4.9. Theorem 4.6b) is even more cumbersome than Theorem 4.6a). Yet Proposition 4.1 implies that for any $\Gamma \in \mathscr{M}_p$, not even an $\mathscr{L}$-continuity set, $\hat{\mathscr{L}}(\Gamma) \to \mathscr{L}(\Gamma)$ almost surely, so there *must* be uniform convergence results that are both usable and general.

There are, and they come from the theory of empirical processes. We will describe the setting and one result in some detail; application to empirical zero-probability functionals is given in Section 4.3. We seek classes $\mathscr{C} \subset \mathscr{M}_p$ such that the discrepancy $D_n(\mathscr{C},\mathscr{L})$ defined by (4.4) converges to zero almost surely under $P_{\mathscr{L}}$ for each $\mathscr{L}$. Obviously, $\mathscr{C}$ must be restricted, for if $\mathscr{C}$ were $\mathscr{M}_p$, then $D_n(\mathscr{C},\mathscr{L})$ would be the variation distance between $\hat{\mathscr{L}}$ and $\mathscr{L}$ (see Section 1.6) and if $\mathscr{L}$ were diffuse (the usual case in applications), no convergence would obtain. As it turns out (not surprisingly; the most concrete way to visualize the information contained in a σ-algebra is as the capability to separate points), one way to limit the size of $\mathscr{C}$ is to restrict its power to "shatter" finite subsets of $\mathbf{M}_p$. The associated theory, due originally to Vapnik and Chervonenkis (1971), is very general; we merely outline its main features, retaining the point process context even though the theory applies much more generally.

The keystone is a combinatorial way of limiting the size of $\mathscr{C}$. Given a finite subset Λ of $\mathbf{M}_p$, let $\Delta(\mathscr{C},\Lambda) = |\{\Lambda \cap \Gamma : \Gamma \in \mathscr{C}\}|$ be the number of subsets of Λ that can be realized by intersecting Λ with some element of $\mathscr{C}$.

(Here $|A|$ is the cardinality of A.) Let

$$v_r(\mathscr{C}) = \sup\{\Delta(\mathscr{C},\Lambda) : |\Lambda| = r\}; \tag{4.10}$$

than evidently $v_r(\mathscr{C}) \leq 2^r$ for all r. A set Λ with $|\Lambda| = r$ and $\Delta(\mathscr{C},\Lambda) = 2^r$ is *shattered* by $\mathscr{C}$. The larger $\mathscr{C}$, the more sets it can shatter. Vapnik and Chervonenkis (1971) established the following dichotomy.

Proposition 4.7. For a given class $\mathscr{C}$, either $v_r(\mathscr{C}) = 2^r$ for all r or there is an integer s such that

$$v_r(\mathscr{C}) \leq r^s + 1 \tag{4.11}$$

for all r. □

This leads to the following terminology.

Definition 4.8. The class $\mathscr{C}$ is a *Vapnik-Chervonenkis class* (V/C class) if (4.11) holds for some s. The smallest such s is the *V/C exponent* of $\mathscr{C}$. □

Clearly, a V/C class is restricted in size; that this is the "right" condition in order to derive strong uniform consistency is not obvious, but true notwithstanding, modulo one measurability detail: there is no way other than assumption to ensure that the discrepancy $D_n(\mathscr{C},\mathscr{L})$ is a random variable.

Here is the main result.

Theorem 4.9. If $\mathscr{C}$ is a V/C subclass of $\mathscr{M}_p$ such that $D_n(\mathscr{C},\mathscr{L})$ is a random variable, then for each $\mathscr{L}$, $D_n(\mathscr{C},\mathscr{L}) \to 0$ almost surely with respect to $P_{\mathscr{L}}$. □

For the proof, we refer to Gaenssler (1984).

The great virtue of Theorem 4.9 is that the law of the N_i does not enter the hypotheses: if $\mathscr{C}$ is a V/C class, then strong uniform consistency holds. Consequently, $D_n(\mathscr{C},\mathscr{L}_0)$ is a suitable statistic for testing the hypothesis $H_0 : \mathscr{L} = \mathscr{L}_0$. However, as noted before, without at least a large-sample approximation to its distribution under $P_{\mathscr{L}_0}$ its usefulness is severely curtailed, which leads one to seek central limit theorems complementing the consistency properties.

There is no hope for a central limit theorem entirely independent of $\mathscr{L}$ because as Proposition 4.2 demonstrates, limiting covariances depend on $\mathscr{L}$. For set-indexed processes additional issues arise that we address momentarily, but first here is one central limit theorem for function-indexed empirical measures. Discussion of the metric entropy condition imposed in it appears in Section 4.2.

Theorem 4.10. Let k be fixed and let $\mathbf{M}_p(k) = \{\mu : \mu(E) \leq k\}$. Let $\mathcal{H}$ be a compact set of continuous functions on $\mathbf{M}_p(k)$ that has finite metric entropy with respect to the uniform metric on $C(\mathbf{M}_p(k))$. Then

$$\{n^{1/2}[\hat{\mathcal{L}}(h) - \mathcal{L}(h)] : h \in \mathcal{H}\} \xrightarrow{d} \{G(h) : h \in \mathcal{H}\}, \tag{4.12}$$

as random elements of $C(\mathcal{H})$ (endowed with the uniform topology), where G is a continuous Gaussian process with mean zero and covariance function

$$R(h_1,h_2) = \mathcal{L}(h_1h_2) - \mathcal{L}(h_1)\mathcal{L}(h_2) \quad \square$$

The proof is substantially that of Theorem 4.13, which is proved in Section 4.2, so we omit it.

For set-indexed processes new complications arise. Given a class of $\mathcal{C}$ of subsets of $\mathbf{M}_p$, define *empirical error processes*

$$E_n(\Gamma) = n^{1/2}[\hat{\mathcal{L}}(\Gamma) - \mathcal{L}(\Gamma)] \tag{4.13}$$

parameterized by $\mathcal{C}$. The goal is to identify conditions on $\mathcal{C}$ and $\mathcal{L}$ under which the E_n converge in distribution to a continuous Gaussian process $G = \{G(\Gamma) : \Gamma \in \mathcal{C}\}$, which by Proposition 4.2 has covariance function

$$R(\Gamma_1,\Gamma_2) = \mathcal{L}(\Gamma_1 \cap \Gamma_2) - \mathcal{L}(\Gamma_1)\mathcal{L}(\Gamma_2) \tag{4.14}$$

In order that this convergence make sense, there must be a metric space V in which the E_n and G take their values; in practice V is not separable and G must then for technical reasons be confined to a separable subspace V_0. Continuity of G entails a metric on $\mathcal{C}$ itself. In the process, because V is "large," measurability of the E_n as random elements of V must also be confronted.

We cannot describe fully here how these problems are resolved within the milieu of empirical processes, but we do sketch how the metric on $\mathcal{C}$ and the spaces V_0 and V, all depending on $\mathcal{L}$—a major shortcoming in terms of inference—are constructed. We then present one central limit theorem that has some potential for inference.

For the (pseudo-) metric on $\mathcal{C}$, one takes $\tilde{d}(\Gamma_1,\Gamma_2) = \mathcal{L}(\Gamma_1 \Delta \Gamma_2)$; the covariance function R of (4.14) is then uniformly continuous. The space V_0 is the linear space of functions $\varphi : \mathcal{C} \to \mathbf{R}$ that are bounded and uniformly continuous with respect to $\tilde{d}$. Then, if $\mathcal{C}$ satisfies a strengthened form of compactness, existence of a Gaussian process G taking values in V_0 (hence continuous) and admitting covariance function R can be shown. Next, V is the linear space of functions $\psi = \varphi + \Sigma_{i=1}^k a_i \varepsilon_{\mu_i}$, where $\varphi \in V_0$, $k \geq 0$, the a_i

are real (not necessarily positive), and the μ_i are elements of $\mathbf{M}_p$. One endows V with the metric $d(\psi_1, \psi_2) = \sup\{|\psi_1(\Gamma) - \psi_2(\Gamma)| : \Gamma \in \mathscr{C}\}$. While the E_n take values in V, it is not true without further restrictions on $\mathscr{C}$ that they are measurable with respect to the Borel σ-algebra $\mathscr{V}$ on V, which is not countably generated. Nor is separability of (V_0,d) immediate; it must be assumed directly or indirectly.

Finally, it is necessary to deal with convergence in distribution of random elements of nonseparable metric spaces. Given a nonseparable metric space (V,d), let $\mathscr{V}_b$ denote the σ-algebra generated by the bounded d-balls; then $\mathscr{V}_b$ is strictly contained in the Borel σ-algebra $\mathscr{V}$. For random elements $X, X_1, X_2, \ldots$ measurable with respect to a σ-algebra $\mathscr{G}$ satisfying $\mathscr{V}_b \subset \mathscr{G} \subset \mathscr{V}$, we say that (X_n) converges in distribution to X (still written $X_n \xrightarrow{d} X$) if there is a separable subspace V_0 of V such that $P\{X \in V_0\} = 1$ and if $E[g(X_n)] \to E[g(X)]$ for every bounded, continuous, and $\mathscr{V}_b$-measurable function g on V. See Gaenssler (1984) for details; in general, properties mimic those of convergence in distribution for random elements of separable metric spaces.

If the metric space $(\mathscr{C},\tilde{d})$ is totally bounded (this is an important limitation on the size of $\mathscr{C}$), then (V_0,d) is separable and the error processes E_n are measurable random elements of $(V,\mathscr{V}_b)$.

The following result imposes restrictions stringent enough to overcome all the difficulties.

Theorem 4.11. Let $\ell \geq 1$ be fixed, let K be a compact subset of $\mathbf{R}^\ell$, and let $F : \mathbf{M}_p \times K \to \mathbf{R}$ be a function fulfilling the conditions:

a) For each $z \in K$, $F(\cdot, z)$ is continuous on $\mathbf{M}_p$;

b) F is Lipschitz in the sense that

$$\sup_{\mu \in \mathbf{M}_p} \sup_{z \neq z' \in K} \frac{|F(\mu,z) - F(\mu,z')|}{|z - z'|} < \infty;$$

c) Uniformly in $z \in K$, as $\varepsilon \downarrow 0$

$$P_{\mathscr{L}}\{-\varepsilon \leq F(N,z) < \varepsilon\} = O(\varepsilon) \tag{4.15}$$

Let $\mathscr{C} = \{\{\mu : F(\mu,z) \geq 0\} : z \in K\}$. Then the metric space (V_0,d) is separable, the error processes E_n are measurable random elements of $(V,\mathscr{V}_b)$, there exists a $\tilde{d}$-continuous Gaussian process G with covariance function R given by (4.14), and finally, $E_n \xrightarrow{d} G$ under $P_{\mathscr{L}}$. □

4.2 EMPIRICAL LAPLACE FUNCTIONALS

Given the i.i.d. copies $N_1, N_2, \ldots$ of N, the Laplace functional $L(f) = E_{\mathscr{L}}[e^{-N(f)}]$ is an integral functional of $\mathscr{L}$:

$$L(f) = \int_{\mathbf{M}_p} e^{-\mu(f)} \mathcal{L}(d\mu)$$

Since the empirical measure $\hat{\mathcal{L}}$ is both maximum likelihood estimator and a method of moments estimator of $\mathcal{L}$, the *empirical Laplace functional*

$$\hat{L}(f) = \int_{\mathbf{M}_p} e^{-\mu(f)} \hat{\mathcal{L}}(d\mu) = n^{-1} \sum_{i=1}^{n} e^{-N_i(f)} \tag{4.16}$$

is similarly maximum likelihood and method of moments estimator of L. In this section we develop asymptotic properties of empirical Laplace functionals, some of them consequences of results in Section 4.1, others derived directly. Our interpretation throughout is that empirical Laplace functionals are stochastic processes parametrized by $C_+(E)$; the goal is to derive limit theorems in this setting. As it turns out, $C_+(E)$ is too large to yield sufficiently precise results and one must work instead with (for consistency) compact subsets of $C_+(E)$ or (for asymptotic normality) compact subsets with finite metric entropy.

We envision empirical Laplace functionals as tools for statistical estimation and hypothesis testing, in particular since—as Theorems 4.12 and 4.13 confirm—there is a rather satisfactory asymptotic theory for them, especially as compared to empirical measures themselves. The reasons for this are that the Laplace functional and the empirical Laplace functionals depend continuously on $f \in C_+(E)$, as do function-indexed empirical processes examined in Section 4.1, and that $C_+(E)$ is smaller than $C_+(\mathbf{M}_p)$. For statistical inference (and, on esthetic grounds, in general) hypotheses of theorems should exhibit only minimal dependence on the unknown Laplace functional; in this respect as well, the results for empirical Laplace functionals are very acceptable: Theorem 4.12 entails no assumptions on L, while Theorem 4.13 imposes only a first moment assumption.

We now formulate and prove the main results of the section. By Theorem 1.12 we choose L)$=L_{N_i}$ for each i) as the index for the statistical model, which remains unrestricted, and denote by P_L the corresponding probability measure for the sequence (N_i).

Here is the main strong uniform consistency property.

Theorem 4.12. Let $\mathcal{K}$ be a compact subset of $C_+(E)$; then for each L, $\sup\{|\hat{L}(f) - L(f)| : f \in \mathcal{K}\} \to 0$ almost surely with respect to P_L.

Proof: Define a mapping $f \to h_f$ of $\mathcal{K}$ into $C(\mathbf{M}_p)$ by $h_f(\mu) = e^{-\mu(f)}$ and observe that $L(f) = \mathcal{L}(h_f) = \int h_f d\mathcal{L}$, and, similarly, $\hat{L}(f) = \hat{\mathcal{L}}(h_f)$. By Theorem 4.5 it suffices to show that $\mathcal{K}' = \{h_f : f \in \mathcal{K}\}$ is a uniformly bounded and compact subset of $C(\mathbf{M}_p)$. The former is evident since $0 \le h_f \le 1$ for each f, while since the mapping $f \to h_f$ is continuous, $\mathcal{K}'$ is the continuous image of a compact set and hence is compact. □

Theorem 4.12 is useful for statistical estimation of L as well as for hypothesis testing. In the latter context the hypothesis $H_0: L = L_0$ (where L_0 is a specific Laplace functional) can be tested using the discrepancy

$$D_n(\mathcal{H},L_0) = \sup\{|\hat{L}(f) - L_0(f)| : f \in \mathcal{H}\}, \tag{4.17}$$

provided that some knowledge is available concerning its H_0-distribution. Among other things Theorem 4.13 is a step in this direction.

Theorem 4.13. Let $\mathcal{H}$ be a compact subset of $C_+(E)$ having finite metric entropy with respect to the uniform metric (see discussion following the theorem) and assume that $E_L[N(E)] < \infty$. Then under P_L,

$$\{n^{1/2}[\hat{L}(f) - L(f)] : f \in \mathcal{H}\} \xrightarrow{d} G = \{G(f) : f \in \mathcal{H}\} \tag{4.18}$$

as random elements of the metric space $C(\mathcal{H})$, where G is a continuous Gaussian process with mean zero and covariance function

$$R(f,g) = L(f+g) - L(f)L(g) \tag{4.19}$$

Proof: The convergence (4.18) is a direct consequence of Araujo and Giné (1980, Theorem 7.16 and Corollary 7.17), for which one views the functionals $e^{-N_i(\cdot)} - L(\cdot)$ as i.i.d. centered random elements of $C(\mathcal{H})$; the hypothesis $E_L[N(E)] < \infty$ is required to fulfill the assumptions there. These results also imply the existence of a continuous Gaussian process G with covariance function R. □

Exactly the same argument proves Theorem 4.10 as well; there the compact space $\mathbf{M}_p(k)$ plays the role of E in Theorem 4.13. In both cases we are dealing with i.i.d. random elements of the set of continuous functions on a compact metric space.

Metric entropy of a compact subset A of a metric space (V,d) is a measure of the "massiveness" of A; a set with finite metric entropy is "nonmassive" in a particular manner. For each $\varepsilon > 0$, let $n(A,\varepsilon)$ be the smallest integer m for which there exist open sets $O_1, \ldots, O_m$ with diam $O_j < 2\varepsilon$ for each j, such that $A \subset \cup_{j=1}^m O_j$; roughly speaking, $n(A,\varepsilon)$ is the number of balls of radius ε needed to cover A. If

$$\int_0^1 [\log n(A,\varepsilon^2)]^{1/2}\, d\varepsilon < \infty, \tag{4.20}$$

then A has *finite metric entropy* with respect to d. Compactness of A implies that $n(A,\varepsilon^2) < \infty$ for each ε and evidently $n(A,\varepsilon^2)$ increases as ε decreases, and so only the lower limit of integration matters in (4.20). The role of (4.20) in establishing central limit theorems is to ensure continuity of the limit process G and then, by approximation, to provide tightness of the processes $n^{1/2}[\hat{L} - L]$; convergence of finite-dimensional distributions is elementary, and convergence in distribution ensues.

Metric entropy of sets of continuous functions is treated in Kolmogorov and Tihomirov (1961). For $E = [0,1]^d$ the set of functions with order of smoothness $p + \alpha$ (derivatives of orders $1,\ldots,p$ exist and are uniformly bounded, and those of order p are Lipschitz with exponent α) has finite metric entropy provided that $p + \alpha > d/2$; in particular, given $\alpha > 1/2$, the set $\mathcal{H}$ of functions f on $[0,1]$ with $\|f\|_\infty \leq 1$ and $|f(x) - f(y)| \leq |x - y|^\alpha$ for all x,y has finite metric entropy. Less is known for a general compact space E, but it is always possible (Exercise 4.14) to construct a compact set $\mathcal{H} \subset C_+(E)$ such that

a) $\mathcal{H}$ has finite metric entropy;
b) point processes N_1, N_2 on E for which $L_{N_1}(f) = L_{N_2}(f)$ for all $f \in \mathcal{H}$ are identically distributed.

Concerning the discrepancy $D_n(\mathcal{H},L_0)$ of (4.17), Theorem 4.13 and the continuous mapping theorem have the following consequence.

Corollary 4.14. Under the assumptions of Theorem 4.13,

$$n^{1/2} D_n(\mathcal{H},L_0) \xrightarrow{d} \sup\{|G(f)| : f \in \mathcal{H}\} \quad \square \tag{4.21}$$

One immense stumbling block remains. To use $D_n(\mathcal{H},L_0)$ as a test statistic for the hypothesis $H_0 : L = L_0$, one must calculate the distribution of the supremum $M = \sup\{|G(f)| : f \in \mathcal{H}\}$. In general, this is impossible. Even for Gaussian processes parametrized by **R**, the exact distribution of the supremum is known only for a small number of cases (stated in Adler, 1981, to be five), such as the Wiener process and the Brownian bridge (see Section 4.6). For Gaussian processes with a multidimensional Euclidean parameter, only properties of the tail of **M**, which may, though, be adequate approximations for application to testing, are available.

Whether analogs of Theorems 4.12 and 4.13 remain valid for composite hypotheses is unknown.

On the bright side, empirical Laplace functionals sometimes exhibit good behavior even in parametric situations when specially tailored alternatives exist.

Example 4.15. Let $E = [0,1]$ and suppose that $N_1, N_2, \ldots$ are i.i.d. homogeneous Poisson processes with unknown rate λ. In the absence of this information one would use the empirical Laplace functional $\hat{L}$ given by (4.14). Under the Poisson model the maximum likelihood estimator of λ is $\hat{\lambda} = n^{-1} \Sigma_{i=1}^n N_i(E)$, which by substitution yields as restricted estimator of the

Laplace functional

$$\hat{L}_0(f) = \exp[-\hat{\lambda} \int_0^1 (1 - e^{-f})] \tag{4.22}$$

Concerning asymptotic behavior of $\hat{L}_0$, one can show that for all λ,

$$\sup_{f \in C_+} |\hat{L}_0(f) - L_\lambda(f)| \leq |\hat{\lambda} - \lambda| \sup_{f \in C_+} \int_0^1 (1 - e^{-f}) \to 0 \tag{4.23}$$

almost surely, which is stronger than can be concluded using Theorem 4.12, and that if Y [with distribution $N(0,\lambda)$] is such that $n^{1/2}[\hat{\lambda} - \lambda] \xrightarrow{d} Y$ under P_λ, then the error processes $\{n^{1/2}[\hat{L}_0(f) - L_\lambda(f)] : f \in C_+\}$ converge in distribution to the process

$$G(f) = Y[\int_0^1 (1 - e^{-f})] \exp[-\lambda \int_0^1 (1 - e^{-f})]$$

On qualitative grounds, $\hat{L}_0$ is preferred to $\hat{L}$ because it is tailored to the Poisson model, but this estimator is risky because it may be useless if the Poisson assumption fails. Correspondingly, the empirical Laplace functional has the advantage of robustness, but perhaps at the price of sacrificing its behavior when the Poisson hypothesis holds. This happens sometimes but, perhaps surprisingly, not always. The proper tool for quantitative comparison is *asymptotic relative efficiency*, in other words, comparison of asymptotic variances. For fixed f, the asymptotic variance of $\hat{L}(f)$ is [by (4.19)]

$$L(2f) - L(f)^2 = \exp[-\lambda \int_0^1 (1 - e^{-2f})] - \exp[-2\lambda \int_0^1 (1 - e^{-f})],$$

while that of $\hat{L}_0(f)$ is $\lambda[\int_0^1 (1 - e^{-f})]^2 \exp[-2\lambda \int_0^1 (1 - e^{-f})]$. Hence the asymptotic efficiency of $\hat{L}(f)$ relative to $\hat{L}_0(f)$ is

$$e(\hat{L}(f), \hat{L}_0(f), \lambda) = \frac{\lambda[\int_0^1 (1 - e^{-f})]^2 \exp[-\lambda \int_0^1 (1 - e^{-f})]}{1 - \exp[-\lambda \int_0^1 (1 - e^{-f})]}$$

For small values of λ the asymptotic relative efficiency is less than 1, indicating that the estimator $\hat{L}_0(f)$ is superior; however, as $\lambda \to \infty$ the asymptotic relative efficiency converges to zero and hence for large λ the empirical Laplace functional $\hat{L}(f)$ is the better estimator. □

4.3 EMPIRICAL ZERO-PROBABILITY FUNCTIONALS

From the integral representation

$$z_{\mathscr{L}}(A) = P_{\mathscr{L}}\{N(A) = 0\} = \int_{\mathbf{M}_p} 1(\mu(A) = 0) \mathscr{L}(d\mu)$$

of the zero-probability functional, we obtain by substitution the *empirical zero-probability functionals*

$$\hat{z}(A) = \int_{\mathbf{M}_p} 1(\mu(A)=0)\hat{\mathcal{L}}(d\mu) = n^{-1}\sum_{i=1}^{n} 1(N_i(A)=0), \qquad (4.24)$$

where $A \in \mathscr{C}$. By contrast with the continuous integrand $\mu \to \exp[-\mu(f)]$ appearing in empirical Laplace functionals, here there is the discontinuous integrand $\mu \to 1(\mu(A)=0)$; the difficulty, discontinuity of $\mu \to \mu(A)$—rather than that of the indicator function—is so substantial that beyond finite-dimensional results we have only one fairly general strong uniform consistency theorem (Theorem 4.17) and a much less general central limit theorem (Theorem 4.18). The latter, however, is more general than may first seem apparent.

We record the elementary results.

Proposition 4.16. Let $A_1, \ldots, A_k$ be sets in $\mathscr{C}$. Then for each $\mathcal{L}$,

a) $(\hat{z}(A_1), \ldots, \hat{z}(A_k)) \to (z_{\mathcal{L}}(A_1), \ldots, z_{\mathcal{L}}(A_k))$ in $[0,1]^k$, almost surely with respect to $P_{\mathcal{L}}$;

b) As random k-vectors,

$$n^{1/2}[(\hat{z}(A_1), \ldots, \hat{z}(A_k)) - (z_{\mathcal{L}}(A_1), \ldots, z_{\mathcal{L}}(A_k))] \xrightarrow{d} Y$$

where Y has a normal distribution with mean zero and covariance matrix $R(i,j) = z_{\mathcal{L}}(A_i \cup A_j) - z_{\mathcal{L}}(A_i)z_{\mathcal{L}}(A_j)$. □

Inspection of R reveals the problems that must be contended with in extending Proposition 4.16 to infinite subclasses $\mathscr{D}$ of $\mathscr{C}$. To begin, one requires a metric on $\mathscr{D}$ with respect to which the covariance function

$$R(A_1,A_2) = z_{\mathcal{L}}(A_1 \cup A_2) - z_{\mathcal{L}}(A_1)z_{\mathcal{L}}(A_2)$$

is continuous. (Without continuity of R there is no hope for continuity of a Gaussian process G with covariance function R.) Then the empirical zero-probability functionals and the Gaussian limit G must take values in a suitable metric space, ... (see discussion in Section 4.1). All this can be overcome, but only (it seems) under severe restrictions on $\mathscr{D}$. Before doing so, however, we show how the Vapnik-Chervonenkis approach (Theorem 4.9), implemented properly, yields strong uniform consistency for empirical zero-probability functionals.

Theorem 4.17. Let $\mathscr{D}$ be a V/C subclass of $\mathscr{C}$ (Definition 4.8); then for each $\mathcal{L}$, almost surely with respect to $P_{\mathcal{L}}$,

$$\sup\{|\hat{z}(A) - z_{\mathcal{L}}(A)| : A \in \mathscr{D}\} \to 0 \qquad (4.25)$$

Proof: The key point is to show that for each positive integer

k, $\mathscr{C}_k = \{\{\mu : \mu(E) = k, \mu(A) = 0\} : A \in \mathscr{D}\}$ is a V/C class of subsets of $\mathbf{M}_p(k) = \{\mu \in \mathbf{M}_p : \mu(E) = k\}$. Given a finite subset $\{\mu_1, \dots, \mu_\ell\}$of $\mathbf{M}_p(k)$, with $\mu_i = \Sigma_{j=1}^k \varepsilon_{x_{ij}}$, then for each i and A, $\mu_i(A) = 0$ if and only if $(x_{i1}, \dots, x_{ik})$ belongs to $(E\setminus A)^k$, and hence it suffices to show that $\{(E\setminus A)^k : A \in \mathscr{D}\}$ is a V/C class in E^k. But $\{E\setminus A : A \in \mathscr{D}\}$ is a V/C class in E, so what needs to be shown is that if $\mathscr{H}$ is a V/C class in E, then $\mathscr{H}^k = \{B^k : B \in \mathscr{H}\}$ is a V/C class in E^k. Indeed, if r is such that $\mathscr{H}$ can shatter no subset of E with cardinality r (or greater), then $\mathscr{H}^k$ cannot shatter any subsets of E^k of cardinality $(r-1)^k + 1$, for given $G \subset E^k$ with $|G| = (r-1)^k + 1$, there is at least one value of i such that $G_i = \{x_i{:}x{\in}G\}$ satisfies $|G_i| \geqslant r$. Choose $X(1),\dots,x(r) \in G$ such that $x_i(1),\dots,x_i(r)$ are distinct. If G were shattered by $\mathscr{H}^k$, then $\{x(1), \dots, x(r)\}$ would be shattered by sets in $\mathscr{H}^k$, which would engender a shattering of $\{x_i(1), \dots, x_i(r)\}$ by sets in $\mathscr{H}$. Hence $\mathscr{C}_k$ is a V/C class in $\mathbf{M}_p(k)$.

By Theorem 4.9 applied to the subprobability measures $\mathscr{L}_k(\Gamma) = \mathscr{L}(\Gamma \cap \mathbf{M}_p(k)) = P_{\mathscr{L}}\{N \in \Gamma, N(E) = k\}$, with associated empirical measures

$$\hat{\mathscr{L}}_k(\Gamma) = n^{-1} \sum_{i=1}^{n} 1(N_i \in \Gamma, N_i(E) = k),$$

we infer that

$$\sup_{A\in\mathscr{D}} |\hat{\mathscr{L}}_k(\{\mu : \mu(A) = 0\}) - \mathscr{L}_k(\{\mu : \mu(A) = 0\})| = \sup_{\Gamma\in\mathscr{C}_k} |\hat{\mathscr{L}}_k(\Gamma) - \mathscr{L}_k(\Gamma)| \to 0 \tag{4.26}$$

almost surely for each k. Since $\mathscr{L}_k(\mathbf{M}_p) = P_{\mathscr{L}}\{N(E) = k\}$ and analogously for $\hat{\mathscr{L}}_k(\mathbf{M}_p)$, standard arguments imply that

$$\sum_{k=0}^{\infty} |\hat{\mathscr{L}}_k(\mathbf{M}_p) - \mathscr{L}_k(\mathbf{M}_p)| \to 0$$

Thus given $\varepsilon > 0$ and k_0 large enough that $\Sigma_{k_0+1}^{\infty} \mathscr{L}_k(\mathbf{M}_p) < \varepsilon$, then almost surely $\Sigma_{k_0+1}^{\infty} \hat{\mathscr{L}}_k(\mathbf{M}_p) < 2\varepsilon$ for all n sufficiently large; for such n,

$$\sup_{A\in\mathscr{D}} |\hat{z}(A) - z_{\mathscr{L}}(A)| \leq 3\varepsilon + \sum_{k=0}^{k_0} \sup_{\Gamma\in\mathscr{C}_k} |\hat{\mathscr{L}}_k(\Gamma) - \mathscr{L}_k(\Gamma)| \tag{4.27}$$

and the proof is concluded by combining (4.26) and (4.27). □

We present a central limit analog of Theorem 4.17, in what seems at first to be a very, very special case; thereafter, we argue that the result is less restrictive than it appears initially.

Theorem 4.18. Let $E=[0,1]$ and let $\mathcal{D}=\{D_t:0\leq t\leq 1\}$, where $D_t=(t,1]$, and assume that the mean measure $\mu(\cdot)=E_{\mathcal{L}}[N(\cdot)]$ is diffuse. Then with $z=z_{\mathcal{L}}$,

$$\{n^{1/2}[\hat{z}(D_t)-z(D_t)]\}\xrightarrow{d}\left\{Y+[1-z(D_0)]B\left(\frac{z(D_t)-z(D_0)}{1-z(D_0)}\right)\right\}, \quad (4.28)$$

where Y is a random variable with distribution $N(0,z(D_0)[1-z(D_0)])$ and B is a Brownian bridge on $[0,1]$ independent of Y.

Proof: For each i, let $X_i=\inf\{u:N_i([0,u])=N_i(E)\}$ and note that $X_i=0$ if $N_i(E)=0$. The X_i are i.i.d. random variables taking values in $[0,1]$, with distribution $F(t)=P_{\mathcal{L}}\{X_i\leqslant t\}=z(D_t)$; hence the $\hat{z}(D_t)$ are the empirical distribution functions $\hat{F}(t)$ associated with the sequence (X_i) and, moreover, $n^{1/2}[\hat{z}(D_t)-z(D_t)]=n^{1/2}[\hat{F}(t)-F(t)]$. By Proposition 4.2,

$$\begin{aligned} n^{1/2}[\hat{z}(D_0)-z(D_0)] &= n^{1/2}[\hat{F}(0)-F(0)] \\ &\xrightarrow{d} N(0,F(0)[1-F(0)]) \end{aligned} \quad (4.29)$$

On $(0,1]$ the distribution function F is continuous (Exercise 4.9) and hence by standard theory of empirical processes on **R** (see Csörgö and Révész, 1981, or Gaenssler, 1984), if $G_n(t)=[\hat{F}_n(t)-\hat{F}_n(0)]/[1-\hat{F}_n(0)]$ and $G(t)=[F(t)-F(0)]/[1-F(0)]$, then

$$\{n^{1/2}[G_n(t)-G(t)]\}\xrightarrow{d}\{B(G(t))\}, \quad (4.30)$$

where B is a Brownian bridge independent of the limit random variable in (4.29). Now, (4.28) follows from (4.29) and (4.30) by straightforward calculations. □

One can generalize Theorem 4.18 as follows. Let E be arbitrary and let $\mathcal{D}=\{D_t:0\leq t\leq 1\}$ be a linearly ordered, decreasing family of subsets of E with $E_{\mathcal{L}}[N(D_1)]=0$. Then one can interpret $1(N_i(D_t)=0)$ as $1(X_i\leq t)$ for some random variable X_i with distribution function $F(t)=z_{\mathcal{L}}(D_t)$ and (4.28) holds even without notational changes.

4.4 ESTIMATION OF REDUCED PALM DISTRIBUTIONS

As described in Section 2.6, reduced Palm distributions play a fundamental role in calculation of conditional distributions for a point process N given observations $\mathcal{F}^N(B^c)$ and are, furthermore, central to state estimation for Cox processes on general spaces (Section 7.2, especially Theorem 7.6) and to combined inference and state estimation for Cox processes (Section 7.4). Thus it is important to estimate the reduced Palm distributions given observation of i.i.d. copies N_i. We maintain the approach of the preceding sections, expressing Palm distributions as functionals of the probability law of the N_i and taking as estimators of them the same functionals of the empirical measures. However, now the situation is more complicated because the reduced Palm distributions $Q_N(\mu, \cdot)$ are derivatives in μ. We represent them as limits of integral functionals of $\mathcal{L}$ that are estimated, once again, with corresponding functionals of $\hat{\mathcal{L}}$. As in the remainder of the chapter the setting is fully nonparametric; however, we do suppose throughout that N is simple.

Rather than estimate reduced Palm distributions directly we estimate their Laplace functionals

$$L(\mu,g) = L_{N_\mu}(g) = E[e^{-N_\mu(g)}], \tag{4.31}$$

where the N_μ are the *reduced* Palm processes of N. Our estimators not only are simple conceptually—they are essentially histogram estimators—and computable but also are strongly uniformly consistent and asymptotically normal.

With $L(\mu,g)$ written (see Section 1.7) in the heuristic form

$$L(\mu,g) = E[e^{-\{N(g)-\mu(g)\}}|N-\mu \geq 0] \tag{4.31a}$$

we identify the main issue: approximation of the null event $\{N-\mu \geq 0\}$ by events of positive probability. For each n, partition the space E (recall that it is compact) as

$$E = \sum_{j=1}^{\ell_n} A_{nj} \tag{4.32}$$

We employ the estimators

$$\hat{L}(\mu,g) = \frac{e^{\mu(g)} \sum_{i=1}^{n} e^{-N_i(g)} \prod_{j=1}^{\ell_n} 1(N_i(A_{nj}) \geq \mu(A_{nj}))}{\sum_{i=1}^{n} \prod_{j=1}^{\ell_n} 1(N_i(A_{nj}) \geq \mu(A_{nj}))} \tag{4.33}$$

If the partitions (4.32) become successively finer as $n \to \infty$, then for given μ eventually all its points lie in distinct partition sets, so that each $N_i(A_{nj})$ is either zero or 1, and the product of indicator functions is indeed an approximation to $1(N_i - \mu \geq 0)$.

We now establish large-sample properties of these estimators.

Theorem 4.19. Assume that

a) For each $g \in C_+$, $\mu \to L(\mu,g)$ is continuous on $\mathbf{M}_p$;
b) As $n \to \infty$, $\max_j \operatorname{diam} A_{nj} \to 0$;
c) There is $t > 0$ such that $E[e^{tN(E)}] < \infty$;
d) For each k,

$$\sum_{n=1}^{\infty} \frac{(\ell_n)^k}{n^2} < \infty \tag{4.34}$$

Then for compact subsets $\mathcal{K}$ of C_+ and $\mathcal{K}'$ of $\mathbf{M}_p$, almost surely

$$\sup\{|\hat{L}(\mu,g) - L(\mu,g)| : \mu \in \mathcal{K}', g \in \mathcal{K}\} \to 0 \tag{4.35}$$

Proof: For each k let $\mathbf{M}_p(k) = \{\mu \in \mathbf{M}_p : \mu(E) = k\}$; in proving (4.35) we may and do assume that $\mathcal{K}'$ is a subset of $\mathbf{M}_p(k)$ for some fixed k. Following more or less standard arguments we form the decomposition

$$\begin{aligned}\hat{L}(\mu,g) &= \frac{e^{\mu(g)}E[e^{-N(g)}\Pi_{j=1}^{\ell_n} 1(N(A_{nj}) \geq \mu(A_{nj}))]}{E[\Pi_{j=1}^{\ell_n} 1(N(A_{nj}) \geq \mu(A_{nj}))]} \\ &\times \frac{n^{-1}\Sigma_{i=1}^{n} e^{-N_i(g)}\Pi_{j=1}^{\ell_n} 1(N_i(A_{nj}) \geq \mu(A_{nj}))}{E[e^{-N(g)}\Pi_{j=1}^{\ell_n} 1(N(A_{nj}) \geq \mu(A_{nj}))]} \\ &\times \frac{E[\Pi_{j=1}^{\ell_n} 1(N(A_{nj}) \geq \mu(A_{nj}))]}{n^{-1}\Sigma_{i=1}^{n} \Pi_{j=1}^{\ell_n} 1(N_i(A_{nj}) \geq \mu(A_{nj}))} \\ &= I \times \frac{II}{III}\end{aligned}$$

We shall show that $I \to L(\mu,g)$, while $II \to 1$ and $III \to 1$, the latter two in the almost sure sense.

Suppose now that μ lies in $\mathbf{M}_p(k)$. Then in view of b), almost surely for all sufficiently large n,

$$\prod_{j=1}^{\ell_n} 1(N(A_{nj}) \geq \mu(A_{nj})) = \frac{1}{k!}\int_{E^k} \prod_{j=1}^{\ell_n} 1\left(\sum_{i=1}^{k} \varepsilon_{x_i}(A_{nj}) \geq \mu(A_{nj})\right) N^{(k)}(dx),$$

so that with $\Gamma_n(\mu) = \{\nu : \Pi_{j=1}^{\ell_n} 1(\nu(A_{nj}) \geq \mu(A_{nj})) = 1\}$,

$$\begin{aligned} E\left[\prod_{j=1}^{\ell_n} 1(N(A_{nj}) \geq \mu(A_{nj}))\right] &\sim \left(\frac{1}{k!}\right) E\left[\int_{E^k} 1_{\Gamma_n(\mu)}\left(\sum_{i=1}^{k} \varepsilon_{x_i}\right) N^{(k)}(dx)\right] \\ &= K_N(\Gamma_n(\mu) \cap \mathbf{M}_p(k))\end{aligned}$$

(See Section 1.7 for notation, definitions, and key computational formulas.) Similarly,

$$e^{\mu(g)}E[e^{-N(g)}\prod_{j=1}^{\ell_n} 1(N(A_{nj}) \geq \mu(A_{nj}))] \sim \int_{\Gamma_n(\mu) \cap \mathbf{M}_p(k)} K_N(d\nu)L(\nu,g) \tag{4.36}$$

and therefore

$$|I - L(\mu,g)| \leq K_N(\Gamma_n(\mu))^{-1}\int_{\Gamma_n(\mu)} |L(\eta,g) - L(\mu,g)| K_N(d\eta) \to 0$$

by a) and b), since $\Gamma_n(\mu) \downarrow \{N - \mu \geq 0\}$.

To deal with the term III we note that the sets $\Gamma_n(\mu)$ may be chosen so that there are at most $(\ell_n)^k$ of them, denoted by $\Gamma_{n\ell}$, so that given $\varepsilon > 0$

$$P\{\sup_{\mu} |III - 1| > \varepsilon\} = P\left\{ \max_{\ell} \left| \frac{n^{-1}\Sigma_{i=1}^{n} 1(N_i \in \Gamma_{n\ell})}{P\{N \in \Gamma_{n\ell}\}} - 1 \right| > \varepsilon \right\}$$

$$\leq \sum_{\ell} P\left\{ \left| \frac{n^{-1}\Sigma_{i=1}^{n} 1(N_i \in \Gamma_{n\ell})}{P\{N \in \Gamma_{n\ell}\}} - 1 \right| > \varepsilon \right\}$$

$$\leq \frac{\text{constant} \times (\ell_n)^k}{n^2}$$

by Chebyshev's inequality; hence $III \to 0$ almost surely by d) and the Borel-Cantelli lemma.

The argument that $I \to 0$ almost surely is deduced from that for III by an absolutely continuous change of probability measure (see Karr, 1985a, for details).

Uniformity in $\mu \in \mathcal{K}'$ is established by standard approximation/equicontinuity reasoning; hence (4.35) holds. □

Broader central limit theorems than the one we are about to present can be derived; the version here is a compromise between generality and usefulness.

Theorem 4.20. Let g be a fixed element of C_+ and let the sets $\Gamma_n(\mu)$ be as in the proof of Theorem 4.19. Assume that

a) $\mu \to L(\mu, g)$ is continuous on $\mathbf{M}_p$;

b) Hypotheses b) and c) of Theorem 4.19 are fulfilled;

c) As $n \to \infty$,

$$n^{1/2} \int_{\mathbf{M}_p} \left| \frac{E[e^{-\{N(g)-\mu(g)\}} 1(N \in \Gamma_n(\mu))]}{P\{N \in \Gamma_n(\mu)\}} - L(\mu, g) \right| K_N(d\mu) \to 0 \quad (4.37)$$

Then as signed random measures on $\mathbf{M}_p$,

$$n^{1/2}[\hat{L}(\mu,g) - L(\mu,g)]K_N(d\mu) \xrightarrow{d} G_1(d\mu) - L(\mu,g)G_2(d\mu), \quad (4.38)$$

where $G = (G_1, G_2)$ is a bivariate Gaussian random measure on $\mathbf{M}_p$ with covariance function R given in (4.39).

Proof: Recall that g is fixed. Introduce the centered random measures $M_i(\Gamma) = e^{-N_i(g)} 1(N_i \in \Gamma) - E[e^{-N(g)} 1(N \in \Gamma)]$ and $\tilde{M}_i(\Gamma) = 1(N_i \in \Gamma) - P\{N \in \Gamma\}$ on $\mathbf{M}_p$. The pairs $(M_i, \tilde{M}_i)$ are i.i.d. and hence by Karr (1979, Theorem 2.2) (see also Proposition 4.30) there is a mean-zero Gaussian process $G = (G_1, G_2)$, with each component a random measure, such that $n^{-1/2} \Sigma_{i=1}^{n} (M_i, \tilde{M}_i) \xrightarrow{d} G$. Derivation of the covariance function R of G is elementary, with the result that if $r(f,\Gamma) = E[e^{-N(f)} 1(N \in \Gamma)]$, then the 4-vector $(G(\Lambda_1,\Gamma_1), G(\Lambda_2,\Gamma_2))$ has covariance matrix

$$R((\Lambda_1,\Gamma_1),(\Lambda_2,\Gamma_2)) = \begin{bmatrix} R_{11} & R_{12} \\ R_{21} & R_{22} \end{bmatrix}, \tag{4.39a}$$

where

$$\mathrm{R}_{11} = \begin{bmatrix} r(2g,\Gamma_1) - r(g,\Gamma_1)^2 & r(g, \Gamma_1 \cap \Lambda_1) - r(g,\Gamma_1)r(0,\Lambda_1) \\ r(g, \Gamma_1 \cap \Lambda_1) - r(g,\Gamma_1)r(0,\Lambda_1) & r(0,\Lambda_1) - r(0,\Lambda_1)^2 \end{bmatrix}, \tag{4.39b}$$

where R_{22} is defined analogously with Γ_1, Λ_1 replaced by Γ_2, Λ_2, respectively, and where

$$\mathrm{R}_{12} = \begin{bmatrix} r(2g, \Gamma_1 \cap \Gamma_2) - r(g,\Gamma_1)r(g,\Gamma_2) & r(g, \Gamma_1 \cap \Lambda_2) - r(g,\Gamma_1)r(0,\Lambda_2) \\ r(g, \Gamma_1 \cap \Gamma_2) - r(g,\Lambda_1)r(0,\Gamma_2) & r(0,\Lambda_1 \cap \Lambda_2) - r(0,\Lambda_1)r(0,\Lambda_2) \end{bmatrix} \tag{4.39c}$$

with R_{21} defined by obvious analogy. Note also that for fixed f, $r(f,\cdot)$ is a measure on $\mathbf{M}_p$ with $r(f,\cdot) \ll r(0,\cdot)$.

Suppose that H is a bounded, continuous function on $\mathbf{M}_p$; the argument that follows is a function space version of the "delta method." For each n,

$$n^{1/2} \int H(\mu)[\hat{L}(\mu,g) - L(\mu,g)]K_N(d\mu)$$
$$= n^{1/2} \int H(\mu)e^{\mu(g)} \left[\frac{\Sigma_{i=1}^n e^{-N_i(g)} 1(N_i \in \Gamma_n(\mu))}{\Sigma_{i=1}^n 1(N_i \in \Gamma_n(\mu))} - \frac{r(g,\Gamma_n(\mu))}{r(0,\Gamma_n(\mu))} \right] K_N(d\mu)$$
$$+ n^{1/2} \int H(\mu) \left[\frac{E[e^{-\{N(g)-\mu(g)\}} 1(N \in \Gamma_n(\mu))]}{P\{N \in \Gamma_n(\mu)\}} - L(\mu,g) \right] K_N(d\mu), \tag{4.40}$$

and the second term, which represents the asymptotic bias of the estimators $\hat{L}(\mu,g)$, converges to zero by (4.37). Given measures τ,σ on $\mathbf{M}_p$, let

$$h_n(\tau,\sigma) = \int H(\mu)e^{\mu(g)} \frac{\tau(\Gamma_n(\mu))}{\sigma(\Gamma_n(\mu))} K_N(d\mu)$$

Then concerning the remaining term in (4.40) we have

$$n^{1/2} \int H(\mu)e^{\mu(g)} \left[\frac{\Sigma_{i=1}^n e^{-N_i(g)} 1(N_i \in \Gamma_n(\mu))}{\Sigma_{i=1}^n 1(N_i \in \Gamma_n(\mu))} - \frac{r(g,\Gamma_n(\mu))}{r(0,\Gamma_n(\mu))} \right] K_N(d\mu)$$
$$= n^{1/2} \left[h_n\left(n^{-1} \sum_{i=1}^n e^{-N_i(g)} 1(N_i \in (\cdot)), n^{-1} \sum_{i=1}^n 1(N_i \in (\cdot)) \right) - h_n(r(g,\cdot), r(0,\cdot)) \right]$$
$$= n^{1/2} h_n'(r(g,\cdot), r(0,\cdot)) \left[n^{-1} \sum_{i=1}^n (M_i, \tilde{M}_i) \right] + o_p(1), \tag{4.41}$$

where h_n' is the Fréchet derivative of h_n and the term $o_p(1)$ converges to zero in probability. It is straightforward to verify that for measures η, λ, σ, and τ on $\mathbf{M}_p$, the latter three absolutely continuous with respect to the first,

$$\lim h_n'(\lambda,\eta)[\tau,\sigma] = \int H(\mu)e^{\mu(g)}\frac{d\tau}{d\eta}(\mu)K_N(d\mu)$$
$$- \int H(\mu)e^{\mu(g)}\frac{d\lambda}{d\eta}(\mu)\frac{d\sigma}{d\eta}(\mu)K_N(d\mu). \quad (4.42)$$

Combination of (4.41) and (4.42) produces

$$n^{1/2}\int H(\mu)[\hat{L}(\mu,g) - L(\mu,g)]K_N(d\mu)$$
$$\xrightarrow{d} \int H(\mu)G_1(d\mu) - \int H(\mu)L(\mu,g)G_2(d\mu),$$

where G is the bivariate Gaussian random measure introduced above. Since H was arbitrary, the proof is complete. □

4.5 EMPIRICAL INFERENCE FOR THINNED POINT PROCESSES

Our setting is now Model 3.3: there are underlying point processes N_i constituting i.i.d. copies of a point process N with unknown law $\mathcal{L}$ (and Laplace functional L), but these processes are not observable. Instead, the observations are p-thinned processes (Definition 1.38) N_1', N_2', ... constructed from the N_i, where $p: E \to [0,1]$ is an unknown function. Hence the N_i' are i.i.d. copies of a p-thinning N' of N, which has law $\mathcal{L}'$, Laplace functional L', mean measure μ', ... Principal problems—once more in a full nonparametric model—are to estimate from observation of the N_i' the Laplace functional L of the underlying processes N_i *and* the thinning function p. For reference we recall that [see (1.34) and (1.35)]

$$L'(f) = L[-\log(1 - p + pe^{-f})], \quad (4.43)$$

which provided that $p(x) > 0$ for all x can be inverted as

$$L(g) = L'\left(-\log\frac{p - 1 + e^{-g}}{p}\right), \quad (4.44)$$

and that

$$\mu'(dx) = p(x)\mu(dx), \quad (4.45)$$

so that estimation of p is a density estimation problem. Indeed, in light of (4.44) and results in Section 4.2 concerning empirical Laplace functionals, estimation of p is the novel aspect of the section.

Not only at the level of generality of the preceding paragraph, but even in specific cases the model formulated there is not identifiable. (A statistical model with probability measures $\{P_\alpha : \alpha \in I\}$ and data $\mathcal{F}$ is *identifiable* if $P_{\alpha_1} \neq P_{\alpha_2}$ on $\mathcal{F}$ whenever $\alpha_1 \neq \alpha_2$ in I. Identifiability is nearly indispensable for meaningful inference.) For the problem at hand the set of candidate laws for the N_i' is $\{\mathcal{L}_{L,p}\}$, with L varying over the set of Laplace functionals of point processes on E and p over the set of measurable functions from E to $[0,1]$. For the data $\mathcal{F}$ representing the entire sequence (N_i') it is possible that $P_{L_1,p_1} = P_{L_2,p_2}$ on $\mathcal{F}$ even though $L_1 \neq L_2$ and $p_1 \neq p_2$ (Exercise 4.15; in the extreme case that $p \equiv 0$ nothing about L can be discerned from the N_i'). One can bring about identifiability in a number of ways; we choose the following conditions, which are stipulated for the remainder of the section.

Assumptions 4.21. a) The mean measure μ of the N_i is known; b) $\mu(\{x : p(x) = 0\}) = 0$.

Assumption 4.21b) is essentially unavoidable (we actually assume that p is bounded away from zero), but there are alternatives to Assumption 4.21a). We impose this particular choice because it entails only first moment knowledge and because together with (4.45) it reduces estimation of p to a "true" density estimation problem; our approach, based on histogram estimators, remains "primitive." Finally, if the N_i were observable, then μ could be estimated easily, using techniques in Propositions 4.28 – 4.31.

We now construct estimators of p and L. For the former, as in Section 4.4, for each n let E (recall that it is a compact metric space) be partitioned as $E = \Sigma_{j=1}^{\ell_n} A_{nj}$: since μ is known we may and do assume that $\mu(A_{nj}) > 0$ for each n and j. Then for $x \in A_{nj}$ we take

$$\hat{p}(x) = \frac{1}{n\mu(A_{nj})} \sum_{i=1}^{n} N_i'(A_{nj}): \tag{4.46}$$

the motivation is (4.45). Let $(\hat{L}')$ be the empirical Laplace functionals associated with the N_i':

$$\hat{L}'(f) = n^{-1} \sum_{i=1}^{n} \exp[-N_i'(f)]$$

On the basis of (4.44) we take as estimator of the Laplace functional L of the N_i,

$$\hat{L}(g) = \hat{L}'\left(-\log \frac{\hat{p} - 1 + e^{-g}}{\hat{p}}\right), \tag{4.47}$$

where $g \in C_+(E)$ and $\hat{p}$ is given by (4.46).

Results proved in Section 4.2 for empirical Laplace functionals allow us to concentrate on the estimators $\hat{p}$; both they and the $\hat{L}$ are strongly uniformly consistent under rather broad assumptions.

Theorem 4.22. Suppose that Assumptions 4.21 hold and that in addition

i) μ is diffuse.
ii) p is continuous and $p(x) > 0$ for all $x \in E$.
iii) $\max_j \operatorname{diam} A_{nj} \to 0$.
iv) $E[N(E)^4] < \infty$.
v) There is $\delta < 1$ such that as $n \to \infty$,

$$\ell_n \max_j \left\{ \frac{E[N(A_{nj})^3]}{\mu(A_{nj})^3} \right\} = O(n^\delta) \tag{4.48}$$

Then almost surely

a) $\sup\{|\hat{p}(x) - p(x)| : x \in E\} \to 0$;
b) For each compact subset $\mathcal{K}$ of $C_+(E)$, $\sup\{|\hat{L}(g) - L(g)| : g \in \mathcal{K}\} \to 0$.

Proof: We prove a) first: it is then quite easy to deduce b) from a) and Theorem 4.12 (although one does need the full strength of the uniform convergence in Theorem 4.12). Let $p_n(x) = \mu'(A_{nj})/\mu(A_{nj})$, $x \in A_{nj}$; then i) and the martingale convergence theorem imply that $\|p_n - p\|_\infty \to 0$. Thus it suffices to show that for

$$R_n(x) = \frac{1}{n\mu'(A_{nj})} \sum_{i=1}^n N_i'(A_{nj}), \qquad x \in A_{nj},$$

we have $\|R_n - 1\|_\infty \to 0$ almost surely.

Let $\varepsilon > 0$ be fixed; then for each n

$$\begin{aligned}
P\{\sup_x |R_n(x) - 1| > \varepsilon\} &= P\{\max_j \mu'(A_{nj})^{-1} |\sum_{i=1}^n [N_i'(A_{nj}) - \mu'(A_{nj})]| > n\varepsilon\} \\
&\le \frac{\ell_n}{\varepsilon^4 n^4} \max \left\{ \frac{E[(\Sigma_{i=1}^n [N'_i(A_{nj}) - \mu'(A_{nj})])^4]}{\mu'(A_{nj})^4} \right\} \\
&= O\left(n^{-2} \ell_n \max \left\{ \frac{\operatorname{Var}(N'(A_{nj}))^2}{\mu'(A_{nj})^4} \right\} \right) \\
&= O\left(n^{-2} \ell_n \max \left\{ \frac{E[N(A_{nj})^3]}{\mu(A_{nj})^3} \right\} \right)
\end{aligned}$$

by the Cauchy-Schwarz inequality and since p is bounded away from zero. From (4.48) and the standard Borel-Cantelli argument we conclude that $\|R_n - 1\|_\infty \to 0$ almost surely.

To prove b), observe that for each g,

$$|\hat{L}(g) - L(g)| \le \left|\hat{L}'\left(-\log\frac{\hat{p}-1+e^{-g}}{\hat{p}}\right) - \hat{L}'\left(-\log\frac{p-1+e^{-g}}{p}\right)\right| + \left|\hat{L}'\left(-\log\frac{p-1+e^{-g}}{p}\right) - L'\left(-\log\frac{p-1+e^{-g}}{p}\right)\right|$$

By a), $\hat{p}(x) > 0$ for all x once n is sufficiently large [hence $\hat{L}$ is well defined, which is not obvious from (4.47)]. The set $\mathcal{K}' = \{-\log[(p-1+e^{-g})/\mathrm{p}]: g \in \mathcal{K}\}$ is a compact subset of $C_+(E)$ by compactness of $\mathcal{K}$, and hence the second term above converges to zero almost surely, uniformly in g, by Theorem 4.12. For each g

$$\left|\hat{L}'\left(-\log\frac{\hat{p}-1+e^{-g}}{\hat{p}}\right) - \hat{L}'\left(-\log\frac{p-1+e^{-g}}{p}\right)\right| \le \left[n^{-1}\sum_{i=1}^{n} N_i'(E)\right]\left[\left\|\frac{\hat{p}}{p}\right\|_\infty + \left\|\log\frac{p-1+e^{-g}}{\hat{p}-1+e^{-g}}\right\|_\infty\right];$$

in this expression the first factor converges to the finite constant $\mu'(E)$ and the second converges to zero uniformly in $g \in \mathcal{K}$ by a) and the Arzéla-Ascoli theorem. □

The estimators $\hat{p}$ may exceed one for small values of n; evidently, Theorem 4.22 remains valid if $\hat{p}$ is replaced by $\min\{\hat{p},1\}$. Of the conditions of the theorem, only (4.48) requires comment; it adjudicates the rate at which $\ell_n \to \infty$, a faster rate being better for the convergence $\| p_n - p \|_\infty \to 0$, but a slow rate necessary in order that the numerator of $\hat{p}$ be effective as estimator of $\mu'(A_{nj})$. In Example 4.26 the structure of the increments of the N_i forces $\delta < 1/3$. Obviously, (4.48) can be relaxed if one desires only weak uniform consistency.

We now consider central limit aspects.

Theorem 4.23. Assume that

i) $|p(x) - p(y)| \le Cd(x,y)$ for some constant C, where d is the metric on E.

ii) As $n \to \infty$

$$n^{1/2} \max_j \{\mu(A_{nj})^{1/2} \operatorname{diam} A_{nj}\} \to 0 \tag{4.49}$$

while

$$n^{1/2} \min_j \mu(A_{nj}) \to \infty \tag{4.50}$$

iii) As $A \downarrow \emptyset$,

$$\frac{\mathrm{Var}(N(A))}{\mu(A)} \to 1 \tag{4.51}$$

Let $x \in E$ be fixed and let $j = j(n,x)$ satisfy $x \in A_{nj}$. Then

$$[n\mu(A_{nj})]^{1/2}[\hat{p}(x) - p(x)] \xrightarrow{d} N(0,p(x)) \tag{4.52}$$

Proof: In the decomposition

$$[n\mu(A_{nj})]^{1/2}[\hat{p}(x) - p(x)] = [n\mu(A_{nj})]^{-1/2}\sum_{i=1}^{n}[N_i'(A_{nj}) - \mu'(A_{nj})] + [n\mu(A_{nj})]^{1/2}\left[\frac{\mu'(A_{nj})}{\mu(A_{nj})} - p(x)\right],$$

the second term converges to zero by i) and (4.49); see the comments following the proof. Since (4.51) implies that $\mathrm{Var}(N'(A_{nj}))/\mu(A_{nj}) \to p(x)$, it suffices to show that the triangular array

$$X_{ni} = \frac{N_i'(A_{nj}) - \mu'(A_{nj})}{[n\_\mathrm{Var}(N'(A_{nj}))]^{1/2}}, \qquad i \leq n,$$

satisfies the central limit theorem. For each n, $E[X_{ni}] = 0$ for all i, $\Sigma_{i=1}^{n}\mathrm{Var}(X_{ni}) = 1$ and $P\{|X_{ni}| > \varepsilon\} \leq 1/(n\varepsilon^2)$ uniformly in i, so it remains only to verify the Lindeberg condition. For each $\eta > 0$,

$$\sum_{i=1}^{n} E[X_{ni}^2 1(|X_{ni}| > \eta)] = nE[X_{n1}^2 1(|X_{n1}| > \eta)] = E[Y_{n1}^2 1(|Y_{n1}| > \eta^{1/2})],$$

where $Y_{n1} = \mathrm{Var}(N_1'(A_{nj}))^{1/2}[N_1'(A_{nj}) - \mu'(A_{nj})]$, and this last quantity evidently converges to zero. □

Under a condition analogous to (4.51), the limits (4.52) are independent for distinct values of x.

Proposition 4.24. Suppose that in addition to the hypotheses of Theorem 4.23, for disjoint sets $A \downarrow \emptyset$, $B \downarrow \emptyset$,

$$\mathrm{Cov}(N(A),N(B)) = o(\{\mu(A)\mu(B)\}^{1/2}) \tag{4.53}$$

Then for $x \neq y$ the limit random variables in (4.52) associated with x and y are independent. □

Nevertheless, a limit distribution for $\|\hat{p} - p\|_\infty$, suitably normalized, exists.

Theorem 4.25. Assume that the hypotheses of Theorem 4.23 and Proposition 4.24 are fulfilled and let

$$b(n) = (2 \log n - \log \log n - \log 2\pi)^{1/2}$$

Then for each t,

$$\lim_n P\{b(\ell_n)[\sup_x [n\mu(A_{nj_x})]^{1/2}|\hat{p}(x) - p(x)| - b(\ell_n)] \leq t\}$$

$$= \exp[-\int_E p(x)^{1/2} e^{-t/p(x)} \mu(dx)] \quad \square$$

The proof is given in Karr (1985b).

Our discussion of statistical estimation concludes with an example.

Example 4.26. (Mixed Poisson process). Let N be a mixed Poisson process on E with directing measure $m = Y\mu$, where we assume that $E[Y] = 1$, so that μ is the mean measure of N. The thinned processes N_i' are mixed Poisson processes directed by the random measures $N_i'(dx) = Y_i p(x)\mu(dx)$. Although there are nonparametric techniques available for estimation of $p(x)\mu(dx)$ and the distribution of the Y_i that are tailored to mixed Poisson processes (see Section 7.1), if the mixed Poisson assumption were suspect it would be desirable to use the more robust estimators from this section. Concerning the various conditions, we note first that

$$E[N(A)^3] = E[Y^3]\mu(A)^3 + 3E[Y^2]\mu(A)^2 + \mu(A) \tag{4.54}$$

and that for any A and B,

$$\mathrm{Cov}(N(A),N(B)) = \mathrm{Var}(Y)\mu(A)\mu(B) + \mu(A \cap B) \tag{4.55}$$

Hence (4.48) is satisfied by the choices $\ell_n = n^{1/4}$, $\mu(A_{nj}) = \mu(E)n^{-1/4}$, in which case strong uniform consistency obtains. From (4.55), $\mathrm{Var}(N(A)) = \mathrm{Var}(Y)\mu(A)^2 + \mu(A)$, so that (4.51) is satisfied; similarly, for $A \cap B = \emptyset$, $\mathrm{Cov}(N(A),N(B)) = \mathrm{Var}(Y)\mu(A)\mu(B)$ and (4.53) holds. We may satisfy (4.49) with $\ell_n = n^{2/5}$, $\mu(A_{nj}) = \mu(E)n^{-2/5}$, and $\mathrm{diam}\, A_{nj} \sim n^{-2/5}$. Unfortunately, these choices fail to fulfill (4.48), although they do satisfy an analogous condition for weak uniform consistency. $\square$

The obvious state estimation problem for thinned point processes is this: given observation of a p-thinning N' of an unobservable point process N, with both the law $\mathcal{L}$ of N and the thinning function p known, reconstruct N. Of course, since N' is observable we only need reconstruct $N - N'$. A general solution is expressible in terms of the reduced Palm distributions (Definition 1.52) of N.

Theorem 4.27. Let N be a point process with reduced Palm distributions $Q(\mu,d\nu)$ and let N' be a p-thinning of N. Then

$$P\{N - N' \in d\nu | \mathcal{F}^{N'}\} = \frac{\exp[-\nu(-\log(1-p))]Q(N',d\nu)}{\int \exp[-\eta(-\log(1-p))]Q(N',d\eta)} \tag{4.56}$$

Proof: In fact, (4.56) is merely Bayes' theorem. Knowledge of N' implies that N has a point at each point of N', on which we may condition; then $Q(N',d\nu)$ is the probability that the "thinned" part $N - N'$ of N is ν, and in order that this result in the given observation N', all points of ν must have been deleted, which for $\nu = \Sigma_{i=1}^{r} \varepsilon_{x_i}$ occurs with probability $\Pi_{i=1}^{r} [1-p(x_i)] = \exp[-\nu(-\log(1-p))]$. For more detailed justification of the Bayes' theorem computation, see Kallianpur and Striebel (1968) or Liptser and Shiryayev (1978). □

Note that the denominator of (4.56) is the Laplace functional of the reduced Palm distribution $Q(N',\cdot)$ and could be estimated using the techniques presented in Section 4.4.

4.6 EMPIRICAL PROCESSES: SOME ADDITIONAL ASPECTS

In Section 4.1 we saw how the theory of empirical processes associated with sequences of i.i.d. random elements of general measurable spaces yields results for empirical measures engendered by a sequence of i.i.d. point processes. Here we discuss how the direction of deduction can be reversed, with properties of estimators of the mean measure of a point process used to describe those of empirical measures associated with random elements of a compact metric space. In addition, we mention the powerful and recently developed concept of strong approximation for uniform empirical processes and show how—together with conditional uniformity—it can be used to produce a strong approximation of Poisson processes by Wiener processes.

To begin, let E be a compact metric space, let ρ be a probability on the Borel σ-algebra $\mathcal{E}$, and let $X_1, X_2, \ldots$ be i.i.d. random elements of $(E,\mathcal{E})$. The *empirical process* engendered by (X_i) is the sequence of random probability measures $\hat{\rho} = n^{-1} \Sigma_{i=1}^{n} \varepsilon_{X_i}$ on E. Of course, the theory of empirical processes described in Section 4.1 applies to the $\hat{\rho}$, but the identification of X_i with a point process $N_i = \varepsilon_{X_i}$ on E reduces study of the empirical processes $\hat{\rho}$ to a special case of partial sum processes $\bar{N}_n = n^{-1} \Sigma_{i=1}^{n} N_i$, where now (N_i) is any sequence of i.i.d. point processes on E. If $\mathcal{L}$ denotes the probability law of the N_i and $\hat{\mathcal{L}}$ the empirical measure from Section 4.1, then for $f \in C_+(E)$,

$$\bar{N}_n(f) = \int_{\mathbf{M}_p} \mu(f)\hat{\mathcal{L}}(d\mu),$$

and hence $\bar{N}_n$ is the maximum likelihood and method of moments estimator of the mean measure $\mu_{\mathcal{L}}$ of the N_i given the data $N_1, \ldots, N_n$. Asymptotic properties of $\bar{N}_n$ hence imply those of the empirical processes $\hat{\rho}$.

The following results pertain to partial sum processes.

Proposition 4.28. If $\mu_{\mathcal{L}} \in \mathbf{M}_p$, then $\bar{N}_n \xrightarrow{w} \mu_{\mathcal{L}}$ almost surely, where $\xrightarrow{w}$ denotes weak convergence in $\mathbf{M}_p$. □

Proposition 4.29. If $\mu_{\mathcal{L}} \in \mathbf{M}_p$, then for each compact subset $\mathcal{K}$ of $C_+(E)$, $\sup\{|\bar{N}_n(f) - \mu_{\mathcal{L}}(f)| : f \in \mathcal{K}\} \to 0$ almost surely. □

Proposition 4.30. If $E_{\mathcal{L}}[\mathrm{N}(E)^2] < \infty$, then as random measures on E the processes $n^{1/2}[\bar{N}_n - \mu_{\mathcal{L}}]$ converge in distribution to a Gaussian random measure G with covariance function $R(f,g) = \rho_{\mathcal{L}}(f \otimes g)$. □

Each can be proved as a consequence of a corresponding result in Section 4.1 or by a nearly verbatim repetition of a proof from there or Section 4.2. One can complement them with a functional law of the iterated logarithm, whose proof is given in Karr (1979).

Proposition 4.31. Assume that there is $\delta > 0$ such that $E_{\mathcal{L}}[N(E)^{2+\delta}] < \infty$. Then almost surely the set

$$\left\{ \frac{n^{1/2}[\bar{N}_n - n\mu_{\mathcal{L}}]}{(2 \log \log n)^{1/2}} : n \geqslant 3 \right\}$$

is relatively compact in the space of signed Radon measures on E (see Appendix A), with limit set $\{\eta : |\eta(f)| \leq R(f,f)^{1/2}$ for all $f \in C(E)\}$. □

For the special case that $N_i = \varepsilon_{X_i}$ and the X_i have law ρ, $\mu_{\mathcal{L}} = \rho$ and each proposition becomes a statement concerning asymptotic behavior of the empirical processes $\hat{\rho}$.

Suppose now that $E = [0,1]$ and that the X_i are uniformly distributed. If we put $\hat{F}_n(x) = \hat{\rho}([0,x]) = n^{-1} \Sigma_{i=1}^n 1(X_i \leq x)$—the $\hat{F}_n$ are (uniform) *empirical distribution functions*—and $B(x) = G([0,x])$, then

$$\{n^{1/2}[\hat{F}_n(x) - x] : 0 \leq x \leq 1\} \xrightarrow{d} \{B(x) : 0 \leq x \leq 1\} \tag{4.57}$$

The limit process B, already mentioned in Theorem 4.18, has covariance function

$$R(x,y) = x \wedge y - xy \tag{4.58}$$

and is called the *Brownian bridge*, a name derived from the representation

$$B(x) = W(x) - xW(1) \tag{4.59}$$

of B in terms of the Wiener process (Brownian motion) W. An increasingly important tool for analysis of limit properties such as (4.57) is *strong approximation*: a statement of convergence in distribution is replaced by a statement of almost sure convergence accompanied by a rate of convergence (this necessitates a sequence of Brownian bridges B_n). Provided that the rate of convergence is sufficiently rapid, a multitude of results on path properties of the Brownian bridge (e.g. modulus of continuity and oscillation behavior) carries over to the *uniform error processes*

$$\alpha_n(x) = n^{1/2}[\hat{F}_n(x) - x] \tag{4.60}$$

To pursue these ramifications (among them the variance principle) would lead us too far afield at this point; rather, we refer the reader to Csörgö and Révész (1981), especially Chapter 5, and the references there. The following result of Komlós et al. (1975, 1976) contains the best possible rate of convergence.

Theorem 4.32. There exists a probability space on which are defined a sequence (X_i) of i.i.d. uniform random variables and a sequence (B_n) of Brownian bridges such that

$$\sup_{0\leq x\leq 1} |\alpha_n(x) - B_n(x)| = O(n^{-1/2}\log n) \tag{4.61}$$

as $n \to \infty$, almost surely. □

We omit the proof; this is a deep and difficult result.

If Brownian bridges can be used to approximate empirical processes, it seems plausible that partial sum processes for i.i.d. random variables can be approximated by Wiener processes, and indeed this is so (see Komlós et al., 1975, 1976). But then conditional uniformity of Poisson processes should allow one to construct strong approximations of them. In the final result of the chapter we do so (see also Theorem 6.5).

Theorem 4.33. There exists a probability space on which are defined a Poisson process $(N_t)_{t\geq 0}$ and Wiener process W such that

$$\sup_{0\leq x\leq 1} \left| t^{1/2}\left(\frac{N_{xt}}{N_t} - x\right) - t^{-1/2}[W(tx) - xW(t)]\right| = O(t^{-1/2}\log t) \tag{4.62}$$

almost surely.

Note that each process $B_t(x) = t^{-1/2}[W(tx) - xW(t)]$ is a Brownian bridge; thus for large t the process $t^{1/2}[N_{xt}/N_t - x]$ is approximately a Brownian bridge; (4.62) provides a bound on the approximation error.

Proof of Theorem 4.33: Let $Z_1, Z_2, \ldots$ be i.i.d. random variables having a Poisson distribution with mean 1; by the strong approximation theorem of Komlós et al. (1975, 1976) for partial sums of i.i.d. random variables (see also Csörgö and Révész, 1981, Theorem 2.6.1) these are definable on a probability space that also supports a Wiener process W in such a way that for $N_n = \Sigma_{i=1}^n Z_i$,

$$|N_n - n - W(n)| = O(\log n) \tag{4.63}$$

almost surely. "Extend" the sequence (N_n) to become a Poisson process (N_t) by conditional uniformity: in each interval $(n-1, n]$ distribute $Z_n = N_n - N_{n-1}$ points independently and uniformly; these are the points of N in $(n-1, n]$. With n and $k \le n$ fixed and $u \in (k-1, k]$,

$$|N_u - n - W(n)| \le |N_u - N_k| + |u-k| + |W(u) - W(k)| + |N_k - k - W(k)|;$$

consequently,

$$\begin{aligned}\sup_{u\le n} |N_u - u - W(u)| &\le \max_{k\le n} |N_k - k - W(k)| + \max_{k\le n} Z_k + \max_{k\le n}\ \sup_{x\le 1} |W_k - W_{k-x}| + 1 \\ &= O(\log n) + O(\log n) + O((\log n)^{1/2}) + O(1)\end{aligned}$$

by (4.63), an elementary calculation and (Csörgö and Révész, 1981, Corollary 1.2.3). Therefore, $\sup\{|N_u - u - W(u)| : u < t\} = O(\log t)$ almost surely. Thus within error $t^{-1/2}N_t(\log t)$, uniformly in x.

$$\begin{aligned} t^{1/2}\left(\frac{N_{xt}}{N_t} - x\right) &= \frac{t^{1/2}}{N_t}[xt + W(xt) - xN_t] \\ &= t^{-1/2}[W(xt) - xW(t)] + \{t^{-1/2}[N_t - t]\}\{t^{-1}[W(xt) - xW(t)]\}\frac{t}{N_t} \\ &\quad + x\left(\frac{t^{1/t}}{N_t}\right)[t - N_t + W(t)] \\ &= t^{-1/2}[W(xt) - xW(t)] + \{t^{-1/2}[N_t - t]\}\{t^{-1}[W(xt) - xW(t)]\}\frac{t}{N_t} \\ &\quad + O\left(\frac{t^{1/2}\log t}{N_t}\right) \\ &= t^{-1/2}[W(xt) - xW(t)] + O(t^{-1/2}\log t)\end{aligned}$$

since $N_t/t \to 1$, $t^{-1/2}(N_t - t) = O((\log\log t)^{1/2})$ and $t^{-1}[W(xt) - xW(t)] = O(t^{-1/2}(\log\log t)^{1/2})$ by the functional law of the iterated logarithm of Strassen (1964). □

There is further connection between Poisson processes and strong approximation. The earliest strong approximation theorem for uniform empirical processes (Brillinger, 1969) constructed the uniform random variables (a triangular array rather than a single sequence) from a Poisson process with arrival times T_k by having the order statistics for X_{ni} be T_i/T_{n+1}, $i = 1, \dots, n$.

EXERCISES

4.1. Prove that if E is a compact metric space, then the space $\mathbf{M}_p(E)$ is locally compact.

4.2. Let $N_1, N_2, \dots$ be i.i.d. point processes with unknown law $\mathcal{L}$.
a) Show that the empirical measures $\hat{\mathcal{L}}$ are the maximum likelihood estimators of $\mathcal{L}$.
b) Show that if H is a bijection of the set of probability measures on $\mathbf{M}_p$ into a topological space V, then $H(\hat{\mathcal{L}})$ is the maximum likelihood estimator of $H(\mathcal{L})$.

4.3. Prove Propositions 4.2 and 4.3.

4.4. Verify that (4.19) gives the proper covariance function for the limit process of Theorem 4.13.

4.5. In connection with Example 4.15,
a) Prove (4.23).
b) Prove the central limit theorem stated there for the empirical Laplace functional error processes $\{n^{1/2}[\hat{L}_0(f) - L_\lambda(f)] : f \in C_+\}$.
c) Verify the computation of the asymptotic relative efficiency.

4.6. Let $N_1, N_2, \dots$ be i.i.d. copies of a homogeneous Poisson process N on $[0,1]$ with unknown rate λ. With $\hat{\lambda} = n^{-1} \sum_{i=1}^{n} N_i([0, 1])$, *the maximum likelihood estimator of* λ, prove that the specialized empirical zero-probability functionals $\hat{z}(A) = \exp(-\hat{\lambda}|A|)$, where $|A|$ is the Lebesgue measure of A, are strongly uniformly consistent and asymptotically normal. (Establish the latter for as large a class of sets as possible.)

4.7 Let N be a Poisson process on a compact metric space E with diffuse mean measure μ_N. Prove that if $f \geq 0$ is a function such that $\mu_N(\{x : f(x) \notin \mathbf{N}\}) = 0$, then $N(f)$ has a discrete distribution.

4.8 Let $(D, \mathcal{D}, \rho)$ be a finite measure space. Prove that the function $d(A,B) = \rho(A \,\Delta\, B)$ is a pseudometric on $\mathcal{D}$.

4.9. Let N be a point process with mean measure μ and zero-probability functional z. Prove that z is μ-continuous: if $A, A_1, A_2, \dots$ are sets with $\mu(A_k \Delta A) \to 0$, then $z(A_k) \to z(A)$.

4.10. Let $\hat{z}$ be the empirical zero-probability functional engendered by i.i.d. point processes $N_1, \dots, N_n$.
a) Show that if the N_i are Poisson, then for any sets A and B, the correlation between $\hat{z}(A)$ and $\hat{z}(B)$ is nonnegative.
b) Construct an example in which $\hat{z}(A)$ and $\hat{z}(B)$ are negatively correlated whenever $A \cap B = \emptyset$.

4.11. Let $\mathscr{C}$ be the class of subintervals of $[0,1]$.
a) Prove that $\mathscr{C}$ is a Vapnik-chervonenkis class and calculate its V/C exponent.
b) Prove that the collection of sets $\{\mu : \mu(A) = 0\}$, $A \in \mathscr{C}$, is *not* a V/C class in $\mathbf{M}_p([0,1]$. Why does this not violate Theorem 4.17?

4.12 Prove that [0,1] has finite metric entropy with respect to the Euclidean metric.

4.13. Let $\mathscr{K}$ be the set of Lipschitz functions f on [0,1] with $|f| \leqslant 1$ [i.e., $|f(x) - f(y)| \leqslant |x - y|$ for each x, y]. Prove that $\mathscr{K}$ has finite metric entropy with respect to the uniform metric.

4.14. Let E be a compact metric space. Prove that there exists a subset $\mathscr{K}$ of $C_+(E)$ such that
i) $\mathscr{K}$ has finite metric entropy with respect to the uniform metric.
ii) If N_1, N_2 are point processes on E with $N_1(f) \stackrel{d}{=} N_2(f)$ for all $f \in \mathscr{K}$, then $N_1 \stackrel{d}{=} N_2$.

4.15. Show that there exist Poisson processes N_1, N_2 with mean measures $\mu_1 \neq \mu_2$ and functions $p_1 \neq p_2$ such that the p_1-thinning N'_1 of N_1 has the same distribution as the p_2-thinning N'_2 of N_2.

4.16. For the mixed Poisson process model of Example 4.26,
a) Establish (4.54).
b) Prove (4.55) and from it derive the formula given for $\mathrm{Var}(N(A))$.

4.17. Let (N_1,N'_1), (N_2,N'_2), ... be i.i.d. copies of a pair (N,N') consisting of a Poisson process N on a compact metric space E with unknown, diffuse mean measure μ and a p-thinning N' of N, where $p: E \to [0,1]$ is an unknown function. Suppose that both N_i and N'_i are observable for each i and that it is desired to estimate the thinning function p with μ treated as a nuisance parameter. Given partitions $E = \Sigma A_{nj}$ as in Section 4.5, reasonable estimators are given by

$$\hat{p} = \frac{\sum_{i=1}^{n} N'_1(A_{nj})}{\sum_{i=1}^{n} N_i(A_{nj})}, \qquad x \in A_{nj}$$

Using Theorem 4.22 as a model, develop and verify conditions for strong uniform consistency of these estimators.

4.18. Let X be a random variable with distribution function F, and let F^{-1} be the left-continuous inverse of F: $F^{-1}(x) = \inf\{t: F(t) \geqslant x\}$, $0 < x < 1$.
a) Prove that $X \stackrel{d}{=} F^{-1}(U)$, where U is uniformly distributed on $[0,1]$.
b) Prove that if F is continuous, then $F(X) \stackrel{d}{=} U$.

4.19. Take as the definition of the Wiener process W on [0,1] the properties $N(0,x)$ for each x.
i) $x \to W_x$ is continuous almost surely;
ii) W has independent and stationary increments;
iii) W_x has distribution $N(0,x)$ for each x.
Also take as known that there exists such a process.
a) Prove that $E[W_x W_y] = x \wedge y$ for each x and y.

b) Prove that the process $B_x = W_x - xW_1$ has covariance function R given by (4.58).

4.20. Let $X_1, \ldots, X_n$ be i.i.d. random variables uniformly distributed on [0,1], let $\hat{F}(x) = n^{-1} \Sigma_{i=1}^{n} 1(X_i \leq x)$ be the empirical distribution function, and let $\alpha(x) = n^{1/2}[\hat{F}(x) - x]$ be the associated uniform empirical error process. Prove that

a) $\hat{F}(\cdot)$ is a strong Markov process.

b) The process $\{\alpha(x)/(1-x): 0 \leq x \leq 1\}$ is a martingale with respect to the history generated by $\hat{F}$.

4.21. Let (α_n) be as in Exercise 4.20. Prove directly that the finite-dimensional distributions of (α_n) converge to those of a Brownian bridge.

4.22. Let N be a Poisson process on $\mathbf{R}_+$ with rate 1. Prove that the sequence (N_n) defined by $N_n(t) = n^{-1/2}[N_{nt} - nt]$, $0 \leq t \leq 1$, converges in distribution to a Wiener process on [0,1].

4.23. Let W be a Wiener process on [0,1].

a) Show that for each a, $P\{\sup_{t \leq 1} |W_t| \geq a\} = 2P\{W_1 > a\}$.

b) Discuss how to use a) and Exercise 4.22 to approximate the distribution of $\sup\{|N_t - t| : t \leq n\}$, where N is a Poisson process with rate 1.

NOTES

Section 4.1

None of the material on empirical processes is new, but its systematic application to empirical measures engendered by point processes does seem to be novel. Good expository treatments of empirical processes are Csörgö and Révész (1981), Gaenssler (1984), Gaenssler and Stute (1979), and Serfling (1980). Theorem 4.5 is based on ideas applied in Karr (1985b) to empirical Laplace functionals. Uniformity in weak convergence is examined in Billingsley and Tøpsøe (1967), from which Theorem 4.6 is taken, as well as in Gaenssler and Stute (1976) and Tøpsøe (1977). Proposition 4.7, Theorem 4.9, and the idea that combinatorial conditions could substitute for topological hypotheses are all due to Vapnik and Chervonenkis (1971). Theorem 4.10 and the highly similar Theorem 4.13 rest ultimately on Dudley (1974) and Jain and Marcus (1975); Araujo and Giné (1980) is a general presentation of central limit theorems in $C(S)$, where S is a compact metric space. Theorem 4.11 is taken from Gaenssler (1984).

Section 4.2

The principal results, Theorems 4.12 and 4.13, are due to Karr (1985b), where they were applied in the context of thinned point processes. Dudley (1973, 1974, 1978) and Gaenssler (1984) analyze the role of metric entropy conditions in proving central limit theorems for i.i.d. random elements of "large" metric spaces such as function spaces. Metric entropy of sets of continuous functions is discussed in Kolmogorov and Tihomirov (1961) and Lorentz (1966).

Section 4.3

Proposition 4.16 is elementary, while Theorems 4.17 and 4.18 appear here for the first time. General discussion of the Brownian bridge, which also appears in Section 4.6, is given in Karlin and Taylor (1975) and at a more advanced level in Csörgö and Révész (1981), where there is extensive development of sample path properties not only of the Brownian bridge and the Wiener process but also of the two-dimensional Wiener process and the Kiefer process (see Section 6.1).

Section 4.4

The material here comes from Karr (1985a). See Kallenberg (1983), Section 2.6, and Section 7.2 for discussion of reduced Palm distributions in conditioning for point processes, and Grenander (1981) and Tapia and Thompson (1978) concerning histogram density estimators, to which our estimators (4.33) are analogous.

Section 4.5

The principal source is Karr (1985b), which contains Theorems 4.22, 4.23, and 4.25. New here is the general state estimation result for thinned point processes, Theorem 4.27; it illustrates once more the importance of reduced Palm distributions in conditioning. Underlying it is a very general Bayes' theorem of Kallianpur and Striebel (1968) (see also Liptser and Shiryayev, 1978). Brillinger (1979) analyzes related problems, while in Section 6.3 inference for thinned Poisson processes is investigated by specialized methods.

Section 4.6

Propositions 4.28 – 4.31 were given for measure-valued Markov chains in Karr (1979). As part of its systematic presentation of strong approximation — especially for partial sum processes and unform empirical error processes — Csörgö and Révész (1981) contains Theorem 4.32 as well as the associated approximation for partial sums by a Wiener process; the definitive forms of these results are due to Komlós et al. (1975, 1976). It is interesting in terms of point processes to note that in an early version of Theorem 4.32 (Brillinger, 1969) uniformly distributed random variables are constructed from a Poisson process by means of conditional uniformity. Theorem 4.33 is new; a different strong approximation for Poisson processes is given in Theorem 6.5.

This chapter contains almost no mention of hypothesis tests, for which empirical distribution functions are also significant. Suppose that X_1, X_2,... are i.i.d. random variables with unknown distribution function F and let $\hat{F}(x) = n^{-1}\sum_{i=1}^{n} 1(X_i \leq x)$ be the sequence of empirical distribution functions. Tests of the null hypothesis $H_0: F = F_0$ can be based on the *Kolmogorov-Smirnov statistic*

$$n^{1/2}\|\hat{F} - F_0\|_\infty = n^{1/2}\sup\{|\hat{F}(x) - F_0(x)| : x \in \mathbf{R}\},$$

for which universal tail bounds and an exact asymptotic distribution theory under H_0 are known, on the *Cramér-von Mises statistic*

$$W_n^2 = n\int_{\mathbf{R}} [\hat{F}(x) - F_0(x)]^2 F_0(dx),$$

and on the *Anderson-Darling statistic*

$$A_n^2 = n\int_{\mathbf{R}} \left\{\frac{[\hat{F}(x) - F_0(x)]^2}{F_0(x)(1 - F_0(x))}\right\} F_0(dx),$$

which can be extended to permit other weight functions. For details including extension to composite hypotheses (see Durbin, 1973a, b, or Serfling, 1980).

Exercises

4.5; 4.6. Specialized empirical Laplace and zero-probability functionals for Poisson processes on general spaces are developed in Section 6.1.

4.11. See Gaenssler (1984) for this and related examples.

4.16. The mixed Poisson model is treated in detail in Karr (1984b).

4.17. See Karr (1985b).

4.18. The quantile transformation is a standard and important tool for reducing assertions concerning general empirical processes to corresponding properties of uniform empirical processes (see, e.g., Csörgö and Révész, 1981).

4.20. See Durbin (1973a) for these and related properties.

4.21. Convergence of finite-dimensional distributions was shown in Doob (1949), with weak convergence in $C[0,1]$ established in Donsker (1952).

4.22. A "martingale method" proof is contained in Exercise 5.9.

4.23. See Pyke (1959) for an early analysis.

5

Intensity-Based Inference: General Theory

In the same way that Chapter 4 is associated with Chapter 1, this chapter is linked with Chapter 2. Our topic is the "martingale method" of inference for point processes on $\mathbf{R}_+$ that admit a stochastic intensity; except for the nature of the dependence of the stochastic intensity on unknown "parameters" we impose few structural restrictions, so that the level of generality is more comparable to Chapter 4 than to Chapters 6 – 9. As intimated in Chapters 2 and 3, the martingale method can be interpreted as a conditional form of the method of moments, in which martingales are construed as "noise" processes. Thus given an unobservable process, an observable process differing from it by a martingale is a martingale estimator. Although this may seem more a problem of state estimation than statistical inference, in models for which good results are available the unobservable process being estimated is comprised mainly or entirely of an unknown function and is only minimally random. Moreover, in the state estimation context it is not the noise interpretation of a martingale that prevails but rather the innovation interpretation.

The principal advantage of the martingale method is generality: it provides coherent rules for construction of estimators that apply when other methods fail, effective methods for systematic calculation of means, variances, and covariances (perhaps its main contribution to applications), and potent techniques for establishing asymptotic normality of function-valued estimators.

One consequence of our assumptions concerning stochastic intensities is that the family $\mathcal{P} = \{P_\alpha : \alpha \in I\}$ of candidate probability laws is dominated (e.g., by the law of a Poisson process), so that legitimate likelihood functions are available. Typically, as will be seen, the index set is so large that the likelihood function is unbounded, rendering conventional maximum likelihood estimation impossible; in such situations the martingale method is appealing because it remains effective notwithstanding. Alternatively, a problem that is of interest because the martingale method is not based on an optimality principle, one can construct consistent sequences of restricted maximum likelihood estimators, as we do in Theorem 5.18. When the index set is a finite-dimensional set Θ, conventional likelihood techniques apply with only minor modification, but the martingale method remains germane because it is used to establish asymptotic normality of maximum likelihood estimators.

Similarly, in testing hypotheses, likelihood ratio tests can be constructed, but small-sample distribution theory, even under null hypotheses, is difficult. By contrast, the martingale method yields whole families of test statistics as easily as it yields martingale estimators, and asymptotic properties can be derived readily, with martingale central limit theorems used to produce large-sample approximations. Indeed, perhaps the most compelling argument in support of the martingale method is the ease with which estimators and test statistics can be calculated and analyzed. It is practical in a way that likelihood methods are not.

Organization of the sections is as follows. Section 5.1 formulates the fundamental statistical model for point processes with stochastic intensities, outlines the basis of the martingale method of estimation, and introduces two important examples. Large-sample properties of martingale estimators—quadratic mean consistency and asymptotic normality—are described in Section 5.2, which begins with two very general limit theorems for sequences of martingales. Thereafter, asymptotic behavior of martingale estimators is examined in settings of decreasing generality, from intensities that merely converge to infinity to superposed i.i.d. copies of a single process. Finally, consistency and asymptotic normality of maximum likelihood estimators are analyzed; even here martingales play an important role in proofs. Section 5.3 is a parallel treatment of hypothesis testing, with attention restricted to the simplest one- and two-sample problems. In Section 5.4 we consider the problem of state estimation given observations constituting a point process on $\mathbf{R}_+$ admitting a stochastic intensity. An unobservable or partially observable state process is to be reconstructed with minimal error from the point process observations; the central issue is recursive computation of MMSE state estimators. Our principal tool is the martingale representation from Theorem 2.34. One model appears by itself in Section 5.5: the Cox regression model and its point process extension. This model, which has important applications in medical statistics, has been a focus and inspiration of recent research.

5.1 STATISTICAL MODELS INVOLVING STOCHASTIC INTENSITIES

We consider marked point processes on [0,1] with finite mark space $E = \{1, \dots, K\}$. Let $(\Omega, \mathcal{G}, P)$ be a probability space on which are defined

1) A marked point process $N = \Sigma_n \varepsilon_{(T_n, Z_n)}$ with mark space E;
2) A history $\mathcal{H} = (\mathcal{H}_t)$ satisfying Assumptions 2.10a) and b), with $\mathcal{H}_1 \subset \mathcal{G}$;
3) A *bounded*, K-variate, nonnegative, $\mathcal{H}$-predictable process $\lambda = (\lambda_t(1), \dots, \lambda_t(K))$.

As in Chapter 2, put $N_t(k) = \Sigma_n 1(T_n \le t, Z_n = k)$, $k = 1, \dots, K$, and let $N_t = \Sigma_k N_t(k)$. We further stipulate that for each k the $(P,\mathcal{H})$-stochastic intensity of $N(k)$ is $\lambda_t(k)$. Throughout Sections 5.1 – 5.3 the boundedness assumption on λ remains in force.

Since in the statistical model to be formulated momentarily different mechanisms governing the random experiment correspond to different probabilities P_α on $(\Omega,\mathcal{G})$ it is necessary to indicate dependence of the stochastic intensity on the probability as well as the history; whenever the history is clearly understood dependence on it is suppressed. There follow, we recall, the properties:

1) For each k,

$$M_t(k) = N_t(k) - \int_0^t \lambda_s(k)\, ds \tag{5.1}$$

is a square integrable $(P,\mathcal{H})$-martingale.

2) For each k, the predictable variation of $M(k)$ is $\langle M(k) \rangle_t = \int_0^t \lambda_s(k)\, ds$.
3) For $k \ne j$ the martingales $M(k)$ and $M(j)$ are orthogonal.

The statistical model we now introduce incorporates an arbitrary index set I; with respect to the probability P_α $(\alpha \in I)$, N has $\mathcal{H}$-stochastic intensity constructed by multiplying (componentwise) the "baseline" stochastic intensity (λ_t) by another predictable process $(H_t(\alpha))$, which in important applications is deterministic.

Model 5.1. (Stochastic intensity model). Let $(\Omega,\mathcal{G},P)$, N, $\mathcal{H}$, and (λ_t) be given satisfying the conditions above. Let I be an arbitrary index set and for each $\alpha \in I$ let (H_t) be a K-variate, nonnegative, $\mathcal{H}$-predictable process. The *stochastic intensity model* with *baseline intensity* λ is given as follows.

a) The sample space is $(\Omega,\mathcal{G})$.

b) For each $\alpha \in I$ let P_α be a probability on $(\Omega,\mathcal{G})$ such that for each k the point process $N(k)$ has $(P_\alpha,\mathcal{H})$-stochastic intensity

$$\lambda_t(\alpha,k) = H_t(\alpha,k)\lambda_t(k) \tag{5.2}$$

Then the set $\mathcal{P}$ of candidate laws consists of those P_α for which

$$E_\alpha\left[\sum_{k=1}^{K}\int_0^1 H_t(\alpha,k)\lambda_t(k)\,dt\right]<\infty \tag{5.3}$$

c) The observations are the σ-algebras

$$\mathcal{F}_t=\mathcal{F}_t^N\vee\sigma(\lambda_u:0\le u\le t) \tag{5.4}$$

corresponding to complete observation of the marked point process N *and* the baseline stochastic intensity λ over $[0,t]\times E$. □

The general goal, obviously, is to make inferences concerning the unknown index α, or the unknown probability P_α, from the given observations. Using (5.2), we see that α influences N only by multiplying the baseline intensity by the α-dependent factor $H_t(\alpha)$. As the examples below illustrate, the model is very general, but reasonably precise results obtain only when $H(\alpha)$ is not very — or not at all — random. However, before presenting examples we demonstrate that Model 5.1 can always be realized in such a manner that $\mathcal{P}=\{P_\alpha\}$ is a dominated family.

Theorem 5.2. Let $(\Omega,\mathcal{G})$, $\mathcal{H}$, N, λ, P be as above and suppose that for each $\alpha\in I$, $H(\alpha)$ is a K-variate, nonnegative, $\mathcal{H}$-predictable process satisfying

$$\sum_{k=1}^{K}\int_0^1 H_t(\alpha,k)\lambda_t(k)\,dt<\infty \tag{5.5}$$

almost surely with respect to P. Then for each α, with respect to the probability P_α defined by

$$\frac{dP_\alpha}{dP}=\exp\left[\sum_{k=1}^{K}\int_0^1 \lambda(k)[1-H(\alpha,k)]\;+\int_E\int_0^1 \log H(\alpha)\,dN\right], \tag{5.6}$$

N has $\mathcal{H}$-stochastic intensity $\lambda_t(\alpha)=H_t(\alpha)\lambda_t$.

Proof: Commencing with (5.6) we omit variables of integration whenever possible; an omitted measure is Lebesgue measure. Moreover, dN-integrals always include the right-hand endpoint of the interval of integration.

If the process

$$\begin{aligned}Y_t(\alpha)&=\exp\left[\sum_{k=1}^{K}\int_0^t \lambda(k)[1-H(\alpha,k)]\;+\int_0^t \log H(\alpha)\,dN\right]\\&=\exp\left[\sum_{k=1}^{K}\int_0^t \lambda(k)[1-H(\alpha,k)]\;+\sum_{k=1}^{K}\int_0^t \log H(\alpha,k)\,dN(k)\right]\end{aligned}$$

is a martingale over $(\Omega,(\mathcal{H}_t),P)$, then by imitating the proof of Theorem 2.31 we can conclude that under P_α, N has a stochastic intensity $\lambda(\alpha)$. Hence we need to verify that for $s<t$,

$$E\left[\exp\left\{\sum_{k=1}^{K}\int_s^t \lambda(k)[1-H(\alpha,k)] + \int_E\int_s^t \log H(\alpha)\,dN\right\}|\mathcal{H}_s\right] = 1, \quad (5.7)$$

which we do in heuristic but illuminating fashion. By iteration properties of conditional expectations it suffices to establish (5.7) for value of $t-s$ small enough that $N_t(k)-N_s(k)$ is with high probability zero or 1 for each k. In this case

$$E\left[\exp\left\{\sum_{k=1}^{K}\int_s^t \lambda(k)[1-H(\alpha,k)] + \int_E\int_s^t \log H(\alpha)\,dN\right\}|\mathcal{H}_s\right]$$

$$\cong E\left[\exp\left\{\sum_{k=1}^{K}\lambda_s(k)[1-H_s(\alpha,k)](t-s) + \sum_{k=1}^{K}\log H_s(\alpha,k)\Delta N_s(k)\right\}|\mathcal{H}_s\right]$$

$$\cong E\left[\left\{1-(t-s)\sum_{k=1}^{K}\lambda_s(k)[1-H_s(\alpha,k)]\right\}\prod_{k=1}^{K}H_s(\alpha,k)^{\Delta N_s(k)}|\mathcal{H}_s\right]$$

$$= E\left[\left\{1-(t-s)\sum_{k=1}^{K}\lambda_s(k)[1-H_s(\alpha,k)]\right\}\prod_{k=1}^{K}[1-\Delta N_s(k)+H_s(\alpha,k)\,\Delta N_s(k)]|\mathcal{H}_s\right]$$

$$\cong E\left[\left\{1-(t-s)\sum_{k=1}^{K}\lambda_s(k)[1-H_s(\alpha,k)]\right\} - \sum_{k=1}^{K}[1-H_s(\alpha,k)]\,\Delta N_s(k)|\mathcal{H}_s\right]$$

(ignoring terms with two or more "differentials," one of which is $t-s$)

$$= 1 + \sum_{k=1}^{K}[1-H_s(\alpha,k)]E[(t-s)\lambda_s(k)-\Delta N_s(k)|\mathcal{H}_s] = 1$$

by predictability of H and since for each k, $N_t(k)-\int_0^t\lambda(k)$ is a $(P,\mathcal{H})$-martingale. □

For likelihood-based inference (5.5) will be the starting point. Since $\mathcal{P}=\{P_\alpha\}$ is a dominated family we may safely employ the right-hand side of (5.5) as a *likelihood function*, either to be maximized for purposes of estimation of α or to form likelihood ratios for hypothesis tests. However, at least two difficulties must be confronted. First, the log-likelihood function

$$L(\alpha) = \sum_{k=1}^{K}\int_0^1 \lambda(k)[1-H(\alpha,k)] + \int_E\int_0^1 \log H(\alpha)\,dN, \quad (5.8)$$

without further restrictions, is not expressed solely in terms of the observed data (N,λ) and the unknown index α, because of the presence of the *unobservable* processes $H(\alpha,k)$. This difficulty vanishes if $\mathcal{H} = \mathcal{F}^N$ [at least in principle, although explicit computation of $H(\alpha,k)$ from the observations may not be feasible] but also in the more important case that either $H(\alpha)$ is deterministic, as in the multiplicative intensity model introduced shortly or that the random past is decreed to be observable, as in the model treated in Section 5.5. A more ingrained problem is that except in rather degenerate cases the likelihood function is unbounded above; even so, it is possible, using, for example, "sieve" methods to calculate consistent estimators by maximum likelihood techniques (see Theorem 5.18).

Before examples are presented we explain the rationale behind the martingale method of estimation. There are two interrelated aspects: *what* to estimate and *how*. The "how" is the more fundamental in that it comprises the general principle of estimation via the martingale method: *choose an estimator for which the estimation error process is a martingale*. This in turn dictates what can be estimated. Theorem 2.34 identifies as fundamental $(P_\alpha,\mathcal{F}^N)$-martingales the innovations

$$M_t(\alpha,k) = N_t(k) - \int_0^t \lambda_s(\alpha,k)\,ds,$$

in terms of which every $(P_\alpha,\mathcal{F}^N)$-martingale M^* is expressible as the integral of a predictable process:

$$M_t^* = M_0^* + \sum_{k=1}^K \int_0^t Y_s(k)\,dM_s(\alpha,k), \tag{5.9}$$

where the $Y(k)$ are predictable. Therefore, (5.9) defines the class of allowable *error processes*. The integral there is the error associated with estimation of the K-variate process $\int_0^t Y(k)H(\alpha,k)\lambda(k)$ by the stochastic integrals $\int_0^t Y(k)dN(k)$, which nearly brings us to the class of processes to be estimated. Since the baseline intensity is observable there is no sense in its being part of what is estimated; however, one must account for its possibly being equal to zero.

All of this leads to the following formalization.

Definition 5.3. Let Y be a predictable process with

$$\sum_{k=1}^K \int_0^1 |Y(k)|H(\alpha,k)\lambda(k) < \infty \tag{5.10}$$

almost surely with respect to each P_α. The *martingale estimator* of the K-variate process

$$B_t(\alpha,k) = \int_0^t Y_s(k)H_s(\alpha,k)1(\lambda_s(k) > 0)\,ds \tag{5.11}$$

is the process

$$\hat{B}_t(k) = \int_0^t Y_s(k)1(\lambda_s(k) > 0)\lambda_s(k)^{-1}\,dN_s(k) \qquad (5.12)$$

Note that whereas the estimated B is continuous in t, the martingale estimator $\hat{B}$ is purely discontinuous; in some circumstances this may be undesirable.

Confirmation that the error processes $\hat{B}_t(k) - B_t(\alpha,k)$ are martingales is provided, along with additional properties, by the following result.

Proposition 5.4. Let Y be a bounded predictable process and let B, $\hat{B}$ be given by (5.11) and (5.12), respectively. Assume that there is $\delta > 0$ such that for each k and t, $\lambda_t(k) > \delta$ if $\lambda_t(k) > 0$. Then under each probability P_α the processes $\hat{B}(k) - B(\alpha,k)$ are orthogonal mean zero square integrable martingales with

$$<\hat{B}(k) - B(\alpha,k)>_t = \int_0^t Y(k)^2 1(\lambda(k) > 0)\lambda(k)^{-1} H(\alpha,k) \qquad (5.13)$$

Proof: From (5.9),

$$\hat{B}_t(k) - B_t(\alpha,k) = \int_0^t Y(k)1(\lambda(k) > 0)\lambda(k)^{-1}\,dM(\alpha,k)$$

The integrand is bounded and predictable, and $M(\alpha,k)$ is a square integrable martingale; consequently, the stochastic integral is a martingale (Theorem 2.9) and the boundedness assumptions ensure that it is square integrable. By Theorem B.12 and the property that $d<M(\alpha,k)>_t = H_t(\alpha,k)\lambda_t(k)\,dt$ (Proposition 2.32),

$$\begin{aligned} <\hat{B}(k) - B(\alpha,k)>_t &= \int_0^t Y(k)^2 1(\lambda(k) > 0)\lambda(k)^{-2}\,d<M(\alpha,k)> \\ &= \int_0^t Y(k)^2 1(\lambda(k) > 0)\lambda(k)^{-1} H(\alpha,k) \quad \square \end{aligned}$$

We note a simple but extremely useful by-product: means and variances of martingale estimators are straightforward to calculate—and hence also to estimate; this in itself is the most important strength of the martingale method in terms of applications.

Proposition 5.5. For each α, t, and k,

$$E_\alpha[\hat{B}_t(k)] = E_\alpha[B_t(\alpha,k)]$$

and

$$E_\alpha[(\hat{B}_t(k) - B_t(\alpha,k))^2] = \int_0^t E_\alpha[Y(k)^2 1(\lambda(k) > 0)\lambda(k)^{-1}H(\alpha,k)] \quad (5.14)$$

For $k \neq j$ and all t,

$$E_\alpha[(\hat{B}_t(k) - B_t(\alpha,k))(\hat{B}_t(j) - B_t(\alpha,j))] = 0 \quad (5.15)$$

Proof: These follow, respectively from the martingale nature of $\hat{B}(k) - B(\alpha,k)$, of $(\hat{B}(k) - B(\alpha,k))^2 - \langle\hat{B}(k) - B(\alpha,k)\rangle$, and of $([B(k) - B(\alpha,k)]\,[\hat{B}(j) - B(\alpha,j)])$. □

The mean squared error (effectively a variance) in (5.14) is needed in order to apply central limit theorems for martingale estimators (which are discussed in Section 5.2), and therefore must be estimated itself. A beauty of the martingale method is that this is entirely straightforward: the process

$$C_t(\alpha,k) = \int_0^t Y_s(k)^2 1(\lambda_s(k) > 0)\lambda_s(k)^{-1}H_s(\alpha,k)$$

is of the form (5.10) for $\widetilde{Y}_s(k) = Y_s(k)^2/\lambda_s(k)$, and hence has martingale estimator

$$\hat{C}_t(k) = \int_0^t \left[\frac{Y_s(k)}{\lambda_s(k)}\right]^2 1(\lambda_s(k) > 0)\, dN_s(k),$$

which becomes the estimator of the mean squared error.

In a strict sense, Model 5.1 is completely general: by taking $\lambda \equiv 1$ and I the set of all predictable processes H satisfying (5.3) the point process can have an arbitrary stochastic intensity, but of course at this level of generality one can deduce only comparably uninformative conclusions. The interesting and useful cases, as hinted above, are those that the α-dependent factor H is primarily deterministic. We now introduce this case in detail, then discuss two specific instances.

Model 5.6. (Multiplicative intensity model). Let I be a subset of the set of functions α: $[0, 1] \to \mathbf{R}_+^K$, each component of which is left-continuous with right-hand limits and satisfies

$$\int_0^1 \alpha_s(k)\, ds < \infty \quad (5.16)$$

The *multiplicative intensity model* is Model 5.1 with $H_t(\alpha,k) = \alpha_t(k)$ for each k, identically in ω, so that under P_α, $N(k)$ has stochastic intensity

$$\lambda_t(\alpha,k) = \alpha_t(k)\lambda_t(k) \quad (5.17)$$

The goal is to estimate the unknown function $\alpha = (\alpha(1), \ldots, \alpha(K))$; the martingale method—and this is a shortcoming—estimates integrals of the $\alpha(k)$ rather than the functions themselves. Since N and λ are observable, the process

$$B_t(\alpha,k) = \int_0^t \alpha(k)1(\lambda(k)>0)$$

has martingale estimator

$$\hat{B}_t(k) = \int_0^t 1(\lambda(k)>0)\lambda(k)^{-1}\,dN(k) \tag{5.18}$$

If $\lambda(k) > 0$ for all t almost surely with respect to P (and hence with respect to each P_α), the martingale estimator of $B_t(\alpha,k) = \int_0^t \alpha(k)$ is simply $\hat{B}_t(k) = \int_0^t \lambda(k)^{-1}\,dN(k)$. Note that these estimators are *very* easy to calculate, a not insignificant advantage of the martingale method.

The multiplicative intensity model is the broadest setting in which there exists a good asymptotic theory given i.i.d. copies of an underlying process; this theory is described in Section 5.2. Perhaps it is not evident even how to formulate limit theorems; roughly, they arise as stochastic intensities converge to infinity, with scaling effected by the Y-processes in (5.11) and (5.12). For i.i.d. processes one achieves this by a sufficiency reduction in Proposition 5.14 that permits attention to be restricted to superpositions. The i.i.d. assumption is used to verify the hypotheses of general limit theorems that, as explained in Section 5.2, hold rather more broadly.

For the multiplicative intensity model, difficulties with unbounded likelihood functions are apparent, for in this case the log-likelihood function (5.8) becomes

$$L(\alpha) = \sum_{k=1}^{K} \int_0^1 \lambda(k)[1 - \alpha(k)] + \sum_{k=1}^{K} \int_0^1 (\log \alpha(k))\,dN(k)$$

which can be made arbitrarily large by taking each $\alpha(k)$ to be large near the points of N and nearly zero everywhere else. In Section 5.2 we describe one way to circumvent this difficulty.

Two specific instances of the multiplicative intensity model are pursued at some length in the text and exercises; the first relates to empirical distribution functions and yields in a simple case an estimator identical to the empirical distribution function.

Example 5.7. (i.i.d. random variables). Let n be fixed and let $X_1, \ldots, X_n$ be i.i.d. random variables with values in $[0,1]$ and distribution function F admitting *hazard function* $h = F'/(1 - F)$. Let $\alpha = h$, take $K = 1$, and let P be such that the X_i are uniformly distributed on $[0,1]$. With respect to P_h the point process $N=\sum_{i=1}^{n}\varepsilon_X$ has $\mathcal{F}^N$-stochastic intensity $\lambda_t(h)=h(t-)(n-N_{t-})$ (see Example 3.16). As estimator of the integrated hazard function $B_t(h)=\int_0^t h(s)1(N_s<n)\,ds$, The martingale method gives

$$\hat{B}_t=\int_0^t (n-N_{s-})^{-1}\,dN_s=\sum_{j=1}^{N_t}(n-j+1)^{-1}$$

Typically, though, it is more germane to estimate the distribution function $F(t)=1-\exp(-\int_0^t h)$ itself, rather than h. One way to accomplish this is the following. Let $\lambda_t=n-N_{t-}$ be the baseline stochastic intensity. Then if one treats $\hat{B}_t$ as estimator of $\int_0^t h$ (a good approximation for large n) and substitutes into the preceding formula there results

$$\hat{F}(t)=1-\exp(-\int_0^t \lambda^{-1}\,dN),$$

which formally satisfies the differential equation

$$d\hat{F}(t)=[1-\hat{F}(t)]\lambda_t^{-1}\,dN_t \tag{5.19}$$

The exact solution of (5.19), the Nelson-Aalen or *product limit estimator*

$$\hat{F}(t)=1-\prod_{s\le t}\left(1-\frac{\Delta N_s}{n-N_{s-}}\right), \tag{5.20}$$

in this case reduces to the empirical distribution function.

One can also give a direct argument in favour of (5.20). In survival analysis, the X_i represent lifetimes, for example of individuals receiving medical treatment. Just before time s, $n-N_{s-}$ individuals remain alive (they are said to be still *at risk*), so $1-\Delta N_s/(n-N_{s-})$ is an estimator of the probability of an individual's surviving past s given that it has survived until s. Multiplication of these estimated conditional probabilities over $s\le t$ and subtraction from 1 yields (5.20).

The product limit estimator, unlike the empirical distribution function, can easily accommodate *censored survival times* (see Exercise 2.23). Suppose that in addition to the lifetimes $X_1,\dots,X_n$ there are i.i.d. random variables $Y_1,\dots,Y_n$ with (possibly unknown) distribution function G, such that $\{X_1,\dots,X_n\}$ and $\{Y_1,\dots,Y_n\}$ are independent, and that for each j one observes not necessarily X_j but rather $Z_j=\min\{X_j,Y_j\}$ along with $\delta_j=1(X_j\le Y_j)$, so that it is known whether Z_j is a survival time ($\delta_j=1$) or a censoring time ($\delta_j=0$). The point process $N_t=\sum_{j=1}^{n}1(Z_j\le t,\delta_j=1)$ of recorded deaths has stochastic intensity $\lambda_t(h)=h(t-)\lambda_t$, where now $\lambda_t=\sum 1(Z_j\ge t)$ is the number of individuals still at risk (neither dead nor censored) at time t. One may then repeat the steps above to obtain the product limit estimator

$$\hat{F}(t) = 1 - \prod_{s \le t} \left(1 - \frac{\Delta N_s}{\lambda_s}\right), \tag{5.20a}$$

which is no longer an empirical distribution function.

Thus not only can the martingale method accommodate censoring but also its estimator reduces to the classical estimator in the absence of censoring. For this model asymptotics arise as $n \to \infty$ and are discussed in Example 5.17. □

The second special case is the Markov process model of Example 2.33.

Example 5.8. (Independent Markov processes). Let $(X_t(1)), \ldots, (X_t(n))$ be independent Markov processes with finite state space S and unknown generator A, let the mark space be $E = \{(i,j) : i, j \in S; i \neq j\}$, and let $\mathcal{H}_t = \bigvee_{\ell=1}^{n} \sigma(X_s(\ell) : 0 \le s \le t)$. For each $(i,j) \in E$ let $N_t(i,j)$ be the number of observed $i \to j$ transitions among the n processes during the interval $[0,t]$, and let $\lambda_t(i) = \sum_{\ell=1}^{n} 1(X_{t-}(\ell) = i)$ be the number of processes in state i at time $t-$. Then (Example 2.33) the $(P_A, \mathcal{H})$-stochastic intensity of N is

$$\lambda_t(A,i,j) = A(i,j)\lambda_t(i) \tag{5.21}$$

[For P we take the probability corresponding to $A(i,j) = 1$ for all $i \neq j$.] The martingale estimator of $B_t(A,i,j) = A(i,j) \int_0^t 1(\lambda(i) > 0)$ is $\hat{B}_t(i,j) = \int_0^t \lambda(i)^{-1}\, dN(i,j)$; thus one martingale estimator of A is

$$\hat{A}(i,j) = \frac{\int_0^1 \lambda(i)^{-1}\, dN(i,j)}{\int_0^1 1(\lambda(i) > 0)} \tag{5.22}$$

Alternatively, the process $B_t(A,i,j) = A(i,j) \int_0^t \lambda(i)$ has martingale estimator $\hat{B}_t(i,j) = N_t(i,j)$, which engenders the more intuitive and classical estimator

$$\hat{A}^*(i,j) = \frac{N_1(i,j)}{\int_0^1 \lambda(i)}; \tag{5.23}$$

$\hat{A}^*(i,j)$ is the number of $i \to j$ transitions divided by the total occupation time of i, so plausibly estimates the rate of $i \to j$ transitions, namely $A(i,j)$. However, the martingale method accommodates time-dependent generators (Jacobsen, 1982), whereas "classical" methods leading directly to (5.23), which is the nonparametric maximum likelihood estimator (see, e.g., Billingsley, 1961a), cannot.

Asymptotic properties as $n \to \infty$ are discussed in the exercises. □

5.2 CONSISTENCY AND ASYMPTOTIC NORMALITY OF MARTINGALE ESTIMATORS

Initially, our setting is that of Model 5.1, and our concern with limit behavior of martingale estimators of Definition 5.3. We begin with two fundamental martingale limit theorems. Next we develop a general asymptotic theory for martingale estimators associated with a sequence of versions of Model 5.1, with broad rather than specific stability conditions; thereafter we specialize to i.i.d. copies of a multiplicative intensity process and finally to the model of Example 5.7. After brief discussion of Example 5.8 we conclude with examination of asymptotic properties of maximum likelihood estimators in the context of i.i.d. copies of a multiplicative intensity model.

To embark on this program, let $(\Omega,\mathcal{G},P)$ be a probability space on which is defined a history $\mathcal{H}$ satisfying the "conditions habituelles." Several forms of consistency for martingale estimators can deduced from the following inequality.

Proposition 5.9. Let X, Y be adapted, right-continuous, nonnegative processes with Y nondecreasing and predictable and $Y_0 = 0$. Suppose that Y dominates X in the sense that for every finite stopping time T,

$$E[X_T] \leq E[Y_T] \tag{5.24}$$

Then for each ε, $\eta > 0$ and each finite stopping time T,

$$P\{\sup_{s\leq T} X_s \geq \varepsilon\} \leq \frac{\eta}{\varepsilon} + P\{Y_T \geq \eta\} \tag{5.25}$$

Proof: Let $X_t^* = \sup\{X_s : s \leq t\}$; then, evidently,

$$P\{X_T^* \geq \varepsilon\} \leq P\{X_T^* \geq \varepsilon,\ Y_T < \eta\} + P\{Y_T \geq \eta\}$$

With $S = \inf\{t: Y_t \geq \eta\}$, predictability of Y implies that $1(Y_T < \eta)X_T^* \leq X_{T\wedge S}^*$, so the proof is complete if we show that for every finite stopping time T, $\varepsilon P\{X_T^* > \varepsilon\} \leq E[Y_T]$. With T fixed, let $R_n = T \wedge \inf\{s \leq n : X_s \geq \varepsilon\} \wedge n$; then successively using monotonicity of Y, (5.24), and the definition of R_n,

$$E[Y_T] \geq E[Y(R_n)] \geq E[X(R_n)] \geq \varepsilon P\{X(R_n) \geq \varepsilon\} = \varepsilon P\{X_{T\wedge n}^* \geq \varepsilon\}$$

and the proof follows by letting $n \to \infty$. □

Despite its simplicity this inequality is central, because the square of a square integrable martingale is dominated by the predictable variation process.

Before presenting the general central limit theorem for sequences of martingales, let us consider what form it should have. Suppose that M^1,

$M^2, \ldots$ are mean zero square integrable martingales. Under what conditions does there exist a mean zero continuous Gaussian martingale M such that $M^n \xrightarrow{d} M$? Since a martingale has orthogonal increments, the Gaussian limit has independent increments, so there is a continuous, nondecreasing function V, the variance function, which characterizes the law of M, satisfying $E[(M_t)^2] = E[<M>_t] = V_t$ for each t. At a minimum we must have $E[(M_t^n)^2] = E[<M^n>_t] \to V_t$ for every t, but a stronger variance stabilization condition, namely that $<M^n>_t \to V_t$ in probability (equivalently, in distribution) for each t, is needed for "central limit" behavior. One can think of this condition as simultaneously matching second moments and imposing the asymptotic determinism engendered by independence assumptions in the classical central limit theorem. Finally, since M is continuous, jumps in the martingales M^n must be asymptotically negligible. The hypothesis imposed below is stronger than needed to prove the theorem; variants are noted following the proof.

Theorem 5.10. Let $M^1, M^2, \ldots$ be mean zero square integrable martingales (possibly with respect to different histories or even defined on different probability spaces) and let V be a continuous, nondecreasing function on $[0,1]$ with $V_0 = 0$. Suppose that

a) For each t,

$$<M^n>_t \xrightarrow{d} V_t; \tag{5.26}$$

b) There are constants $c_n \downarrow 0$ such that as $n \to \infty$,

$$P\{\sup_{t \le 1} |\Delta M_t^n| \le c_n\} \to 1 \tag{5.27}$$

Then there exists a continuous Gaussian martingale M with

$$<M>_t = V_t \tag{5.28}$$

for all t, such that $M^n \xrightarrow{d} M$ on the function space $D[0,1]$.

Proof: (Sketch). The proof divides into two main steps.

1) We first show that (M^n) is tight in $D[0,1]$; then (5.27) implies that every limit in distribution of a subsequence corresponds to a continuous process. For each stopping time T and each ε, η, $\delta > 0$ the processes $X_t = (M_{T+t}^n - M_T^n)^2$ and $Y_t = <M^n>_{T+t} - <M^n>_T$ satisfy the hypotheses of Proposition 5.9; consequently,

$$P\{\{\sup|M_s^n - M_T^n| : T \le s \le T + \delta\} > \varepsilon\} \le \frac{\eta}{\varepsilon^2} + P\{<M^n>_{T+\delta} - <M^n>_T > \eta\} \tag{5.29}$$

From (5.29) we conclude (see Aldous, 1978; Billingsley, 1968) that tightness of (M^n) ensues from that of $(<M^n>)$, which in turn follows from (5.26).

2) It suffices now to show that for every subsequence $(M^{n'})$ converging in distribution to some $\widetilde{M}$, $\widetilde{M} \stackrel{d}{=} M$, for which it is enough to establish that $\widetilde{M}$ and $\widetilde{M}^2 - V$ are martingales. Except for integrability details that we gloss over, by the continuous mapping theorem we only need demonstrate that each of these processes is the limit in distribution of a sequence of martingales, which for $\widetilde{M}$ is true by definition. By (5.26) the sequences $((M^{n'})^2 - <M^{n'}>)$ and $((M^{n'})^2 - V)$ have the same limit in distribution, and the functional $J: C[0,1] \to D[0,1]$ defined by $J(f) = f^2 - V$ is continuous; hence again by the continuous mapping theorem

$$(M^{n'})^2 - <M^{n'}> = J(M^{n'}) + V - <M^{n'}> \stackrel{d}{\to} J(\widetilde{M}) = \widetilde{M}^2 - V,$$

so $\widetilde{M}^2 - V$ is indeed the limit in distribution of a sequence of martingales. □

Conditions weaker than (5.27) suffice to yield the same conclusion; these include the Lindeberg condition

$$\lim E[\sum_{t \le 1} (\Delta M_t^n)^2 1(|\Delta M_t^n| \ge \varepsilon)] = 0 \tag{5.30}$$

for every $\varepsilon > 0$ and the still weaker asymptotic rarefaction of jumps condition of Rebolledo (1980).

For completeness and later reference, we record the multivariate version of Theorem 5.10.

Theorem 5.11. Let K be fixed and for each n let $M^n(1), \ldots, M^n(K)$ be orthogonal mean zero square integrable martingales, and let $V: [0,1] \to \mathbf{R}_+^K$ be continuous and nondecreasing in each component with $V_0 = 0$. Suppose that for each t and k,

$$<M^n(k)>_t \stackrel{d}{\to} V_t(k) \tag{5.31}$$

and that each sequence $M^n(k)$ satisfies (5.27) [or (5.30)]. Then there exists a K-variate continuous Gaussian process M, with each component a martingale and with

$$<M(k), M(j)> = 1(k = j)V(k), \tag{5.32}$$

such that $M^n \stackrel{d}{\to} M$ on $D[0,1]^K$. □

That is, the components of M are independent Gaussian martingales and $M(k)$ has variance function $V(k)$. Theorem 5.11 is proved by reduction to the one-dimensional case using the Cramér-Wold device (see Rebolledo, 1978).

We now study implications of Proposition 5.9 and Theorem 5.11

concerning asymptotic behavior of martingale estimators. First we present general results for a sequence of versions of Model 5.1 in which intensities are required merely to converge to infinity. Later we show that they apply to data constituting i.i.d. copies of a multiplicative intensity process.

The general formulation for limit theorems is a bit involved (but not difficult once one grasps it), and proceeds as follows. Fixed *independently of the sequence index n* are the underlying probability space $(\Omega,\mathcal{G},P)$, the index set I, the family $\mathcal{P} = \{P_\alpha : \alpha \in I\}$ of candidate probability laws and the mark space $E = \{1, \dots, K\}$. *For each n* there are

1) A marked point process N^n with mark space E;
2) A history $\mathcal{H}^n$ satisfying Assumptions 2.10a) and b);
3) A bounded, nonnegative, K-variate, $\mathcal{H}^n$-predictable process (λ_t^n) such that $N^n(k)$ has $(P,\mathcal{H}^n)$-stochastic intensity $\lambda^n(k)$;
4) For each $\alpha \in I$, a nonnegative, K-variate, $\mathcal{H}^n$-predictable process $(H_t^n(\alpha))$ such that each $N^n(k)$ has $(P_\alpha,\mathcal{H}^n)$-stochastic intensity

$$\lambda_t^n(\alpha,k) = H_t^n(\alpha,k)\lambda_t^n(k) \tag{5.33}$$

Thus the point processes, histories, and stochastic intensities vary with n but the probabilities do not. We assume throughout that

$$E_\alpha\left[\sum_{k=1}^{K}\int_0^1 H_t^n(\alpha,k)\lambda_t^n(k)\,dt\right] < \infty$$

for each n and α. For each n the observations are the point process N^n and the baseline stochastic intensity λ^n.

In addition to these objects we have for each n and k an $\mathcal{H}^n$-predictable process Y^n satisfying (5.10). For the processes

$$B_t^n(\alpha,k) = \int_0^t Y^n(k)H^n\,(\alpha,k)1(\lambda^n(k) > 0) \tag{5.34}$$

we form martingale estimators

$$\hat{B}_t^n(k) = \int_0^t Y^n(k)1(\lambda^n(k) > 0)\lambda^n(k)^{-1}\,dN^n(k) \tag{5.35}$$

Finally, we can pose the questions of interest.

1) Are the martingale estimators $\hat{B}^n$ *consistent* in the sense that $\hat{B}^n - B^n(\alpha) \to 0$ under P_α for each α?
2) Are the estimation error processes $\hat{B}^n - B^n(\alpha)$, suitably scaled, *asymptotically normal* in the sense that under each P_α they satisfy a martingale central limit theorem?

We provide sufficient conditions for positive response to each.

The martingale method leads naturally to quadratic mean consistency, uniformly in t.

Theorem 5.12. Suppose that

$$E_\alpha\left[\int_0^1\left[\frac{Y^n(k)}{\lambda^n(k)}\right]^2 1(\lambda^n(k)>0)\,dN^n(k)\right]\to 0 \tag{5.36}$$

for each α and k. Then the estimators $\hat{B}^n$ are uniformly consistent in quadratic mean: for each α and k, as $n\to\infty$,

$$E_\alpha\left[\sup_{t\le 1}(\hat{B}^n_t - B^n_t(\alpha,k))^2\right]\to 0 \tag{5.37}$$

Proof: By Proposition 5.5, $\hat{B}^n(k)-B^n(\alpha,k)$ is a square integrable P_α-martingale, so that Theorem B.15 applied to the submartingale $\hat{B}^n(k)-B(\alpha,k))^2$ gives

$$E_\alpha[\sup(\hat{B}^n_t(k)-B^n_t(\alpha,k))^2]\le 4E_\alpha[(\hat{B}^n_1(k)-B^n_1(\alpha,k))^2]$$

$$= 4E_\alpha[<\hat{B}^n(k)-B^n(\alpha,k)>_1]$$

$$= 4E_\alpha[\int_0^1 Y^n(k)^2 1(\lambda^n(k)>0)\lambda^n(k)^{-1}H^n(\alpha,k)]$$

[by (5.13)]

$$= 4E_\alpha\left[\int_0^1\left[\frac{Y^n(k)}{\lambda^n(k)}\right]^2 1(\lambda^n(k)>0)\,dN^n(k)\right],$$

the last equality holding because (5.33) gives the P_α-stochastic intensity of N^n. □

Various alternatives exist. Using the Burkholder-Davis-Gundy martingale inequalities one can extend Theorem 5.12 to L^p-uniform consistency for $p\in[1,\infty)$ and even obtain necessary and sufficient conditions. In another direction, weak uniform consistency, i.e.,

$$\sup_{t\le 1}|\hat{B}^n_t(k)-B^n_t(\alpha,k)|\to 0 \tag{5.38}$$

in P_α-probability, can be deduced directly—in sharper form—using (5.13) and Proposition 5.9. The process $(\hat{B}_n(k)-B_n(\alpha,k))^2$ is dominated in the sense of (5.24) by its predictable variation $<\hat{B}_n(k)-B_n(\alpha,k)>$, and (5.25) shows that (5.38) obtains provided only that

$$\int_0^1 Y^n(k)^2 1(\lambda^n(k)>0)\lambda^n(k)^{-1}H^n(\alpha,k)\to 0 \tag{5.39}$$

in P_α-probability for each k.

The first central limit theorem is so general that there is no need for a proof; it amounts only to a restatement of Theorem 5.11.

Theorem 5.13. Suppose that for each α,

$$\sum_{k=1}^{K}\int_0^1 Y^n(k)^2\lambda^n(k)H^n(\alpha,k)<\infty \tag{5.40}$$

almost surely with respect to P_α and let (b_n) be a sequence of positive constants increasing to ∞. Assume that

a) For each α and k the P_α-martingales $b_n^{1/2}[\hat{B}^n(k)-B^n(\alpha,k)]$ satisfy (5.27);

b) For each α there is a function $V(\alpha)$ fulfilling the properties in Theorem 5.11 such that

$$b_n\int_0^t\left[\frac{Y^n(k)}{\lambda^n(k)}\right]^2 H^n(\alpha,k)\lambda^n(k)\xrightarrow{d} V_t(\alpha,k), \tag{5.41}$$

for each t and k.

Then for each α these exists a continuous Gaussian martingale $M(\alpha)$ with

$$<M(\alpha,k),M(\alpha,j)>=1(k=j)V(\alpha,k), \tag{5.42}$$

such that on $D[0,1]^K$,

$$b_n^{1/2}[\hat{B}^n-B^n(\alpha)]\xrightarrow{d} M(\alpha).\quad\square \tag{5.43}$$

An intuitive basis (with $Y^n\equiv 1$) is that a point process with a stochastic intensity is locally conditionally Poisson (see discussion following Theorem 2.22), while a Poisson process, normalized, is approximately a Wiener process (Theorem 4.33); hence the more general point process, when scaled, is locally a Wiener process (i.e., approximately a Gaussian martingale).

Finally, we come to a case sufficiently specific that truly useful results can be derived. The setting is the multiplicative intensity model: we assume a sequence of pairs $(N^{(i)},\lambda^{(i)})$ that, together with histories $\mathcal{H}^{(i)}$, under P_α are i.i.d. copies of a process (N,λ) for which each $N(k)$ has stochastic intensity $\lambda_t(\alpha,k)=\alpha_t(k)\lambda_t(k)$. Recall that α does not vary with i. To analyze limit behavior of martingale estimators we apply the preceding results to the sequence defined by $N^n=\Sigma_{i=1}^n N^{(i)}$, $\lambda^n=\Sigma_{i=1}^n\lambda^{(i)}$, and $\mathcal{H}_t^n=\vee_{i=1}^n\mathcal{H}_t^{(i)}$. That this entails no loss of information is implied by the following result.

Proposition 5.14. a) For each n, (N^n,λ^n) is a sufficient statistic for α given the data $(N^{(1)},\lambda^{(1)}),\ldots,(N^{(n)},\lambda^{(n)})$.

b) For each k, $N^n(k)$ has $(P_\alpha,\mathcal{H}^n)$-stochastic intensity $\alpha(k)\lambda^n(k)$. $\square$

The stochastic intensities λ^n do converge to infinity and we are now in a position to utilize Theorems 5.12 and 5.13 to estimate the deterministic integrals

$$B_t(\alpha,k) = \int_0^t \alpha^s(k)1(E_\alpha[\lambda_s(k)] > 0)\, ds$$

The martingale estimators

$$\hat{B}_t^n = \int_0^t 1(\lambda^n(k) > 0)\lambda^n(k)^{-1}\, dN^n(k), \tag{5.44}$$

are consistent and asymptotically normal.

Theorem 5.15. For each α and k, as $n \to \infty$,

$$E_\alpha\left[\sup_{t\leq 1}(\hat{B}_t^n(k) - B_t(\alpha,k))^2\right] \to 0$$

Proof: The processes $\bar{B}_t^n(\alpha,k) = \int_0^t \alpha(k01(\lambda^n(k) > 0)$ have as martingale estimators the $\hat{B}^n$ of (5.44). According to Theorem 5.12, $E_\alpha[\sup(B_t^n - \bar{B}_t^n(\alpha,k))^2] \to 0$ provided that

$$E_\alpha[\int_0^1 \alpha(k)\lambda^n(k)^{-1}1(\lambda^n(k) > 0)] \to 0$$

(we have omitted some easy calculations), which holds by the law of large numbers and the dominated convergence theorem. To complete the proof it is necessary to show that

$$E_\alpha[\sup(\bar{B}_t^n(\alpha,k) - B_t(\alpha,k))^2] \to 0; \tag{5.45}$$

a stronger statement, (5.48), is established in the proof of Theorem 5.16. □

Theorem 5.16. Assume that for each α and k,

$$\int_0^1 E_\alpha[\lambda_s(k)]^{-1}1(E_\alpha[\lambda_s(k)] > 0)\, ds < \infty \tag{5.46}$$

Then the error processes $n^{1/2}[\hat{B}^n - B(\alpha)]$ converge in P_α-distribution to a continuous K-variate Gaussian martingale $M(\alpha)$ satisfying

$$<M(\alpha,k),M(\alpha,j)>_t = 1(k=j)\int_0^t E_\alpha[\lambda(k)]^{-1}1(E_\alpha[\lambda(k)] > 0)\alpha(k) \tag{5.47}$$

Proof: We verify the hypotheses of Theorem 5.13. Jumps in the martingale $n^{1/2}[\hat{B}^n(k) - \bar{B}^n(\alpha,k)]$ arise only from $\hat{B}^n(k)$, whose jumps are of size $\lambda^n(k)^{-1} \sim n^{-1}$; hence (5.30) holds for $c_n = n^{-1/4}$, for example. To check

(5.41) we note that

$$n\int_0^t \lambda^n(k)^{-2}\lambda^n(\alpha,k) = n\int_0^t \lambda^n(k)^{-1}\alpha \to \int_0^t E_\alpha[\lambda(k)]^{-1}1(E_\alpha[\lambda(k)]>0)\alpha$$

by the law of large numbers and (5.46). It remains to show that

$$n^{1/2}\sup_{t\le 1}|\bar{B}_t^n(\alpha,k) - B_t(\alpha,k)| \to 0 \tag{5.48}$$

in P_α-probability. [This also proves (5.45).] With $I_s = 1(E_\alpha[\lambda_s(k)]>0)$,

$$n^{1/2}\sup|\bar{B}_t^n(\alpha,k) - B_t(\alpha,k)| = n^{1/2}\sup\left|\int_0^t \alpha_s(k)[1(\lambda_s^n(k)>0) - I_s]\,ds\right|$$

$$= n^{1/2}\int_0^1 \alpha_s(k)I_s 1(\lambda_s^n = 0)\,ds;$$

taking expectations with respect to P_α gives

$$E_\alpha[n^{1/2}\sup|\bar{B}_t^n(\alpha,k) - B_t(\alpha,k)|] = \int_0^1 \alpha_s(k)I_s n^{1/2}P_\alpha\{\lambda_s(k)=0\}^n\,ds,$$

which converges to zero by dominated convergence. □

To apply Theorem 5.16, it is necessary to estimate the variances in (5.47), but *this can be done using the martingale method!* The processes

$$C_t(\alpha,k) = n\int_0^t \lambda^n(k)^{-1}1(\lambda^n(k)>0)\alpha(k)$$

which converge to the variance in (5.47), have martingale estimators

$$\hat{C}_t(\alpha,k) = n\int_0^t \lambda^n(k)^{-2}1(\lambda^n(k)>0)\,dN^n(k),$$

which are consistent by a repetition of the proof of Theorem 5.15.

We illustrate for the special cases of Examples 5.7 and 5.8.

Example 5.17. (i.i.d. random variables). Let $X_1, X_2,\ldots$ be i.i.d. random variables in $[0,1]$ with unknown continuous hazard function h. Each point process $N^{(i)} = \varepsilon_{X_i}$ has stochastic intensity $\lambda_t^{(i)} = h(t)1(X_i \ge t)$; consequently, the superposition $N^n = \Sigma_{i=1}^n N^{(i)}$ has stochastic intensity $\lambda_t^n = \Sigma_{i=1}^n \lambda_t^{(i)} = h(t)(n - N_{t-}^n)$. As discussed in Example 5.7, the martingale estimator for $B_t^n(h) = \int_0^t h(s)1(N_s^n < n)\,ds$ is

$$\hat{B}_t^n = \int_0^t (n - N_{s-}^n)^{-1}\,dN_s^n = \sum_{j=1}^{N_t^n}(n-j+1)^{-1} \tag{5.49}$$

Theorem 5.15 does not apply directly, but Theorem 5.12 does, to yield $E_h[\sup(\hat{B}_t^n - B_t^n)^2] \to 0$ for each h, while Theorem 5.13 gives that under P_h, $n^{1/2}[\hat{B}^n - B^n(h)] \xrightarrow{d} M(h)$, where $M(h)$ is a continuous Gaussian martingale with $\langle M(h) \rangle_t = F_h(t)/G_h(t)$, with F_h and $G_h = 1 - F_h$ the distribution and survivor functions, respectively, associated with h. Verification of (5.41) is straightforward: with $b_n = n$ and $Y^n \equiv 1$,

$$n \int_0^t (\lambda^n)^{-2} \lambda^n(h) = n \int_0^t (\lambda^n)^{-1} 1(\lambda^n > 0) h \to \int_0^t \frac{h}{G_h} = \frac{F_h(t)}{G_h(t)}$$

In the absence of censoring, asymptotic behavior of the product limit estimator (5.20) is that of the empirical distribution function (see Section 4.6). For the censored case, see Gill (1980) or Jacobsen (1982). □

For the Markov process model of Example 5.8, while the martingale method is useful for formulating the estimators $\hat{A}$ of (5.22) and $\hat{A}^*$ of (5.23), it is less useful for their analysis, which can be effected directly. For example, since

$$\hat{A}^*(i,j) = \frac{n^{-1} \sum_{\ell=1}^n N_1^{(\ell)}(i,j)}{n^{-1} \sum_{\ell=1}^n \int_0^1 1(X_s(\ell) = i)\, ds}$$

with ℓ referring to the ℓth process, from which—with $P_s(i,j)$ denoting the transition function—it follows simply by the strong law of large numbers that [assuming $X_0(\ell) = i_0$ for all ℓ]

$$\hat{A}^*(i,j) \to \frac{A(i,j) \int_0^1 P_s(i_0,i)\, ds}{\int_0^1 P_s(i_0,i)\, ds} = A(i,j)$$

almost surely; the corresponding central limit theorem is equally easy to derive. Limit behavior of the estimators $\hat{A}$ is treated in Exercise 5.14.

We now take up maximum likelihood estimation given i.i.d. copies of a multiplicative intensity model; for simplicity we take $K = 1$ (no marks). As observed previously, because the index set $I = L_+^1[0,1]$ is infinite-dimensional, the log-likelihood function

$$L_n(\alpha) = \int_0^1 \lambda^n(1 - \alpha) + \int_0^1 (\log \alpha)\, dN^n \tag{5.50}$$

is unbounded above, so the problem must be approached indirectly. We describe one approach, the method of "sieves," due to Grenander and collaborators, and based on the following prescription:

1) For each $a > 0$ devise a subset $I(a)$ of the index set I within which there does exist for each n a maximum likelihood estimator $\hat{\alpha} = \hat{\alpha}(n,a)$.

2) Choose $a_n \downarrow 0$ such that the $I(a_n)$ increase and $\cup_n I(a_n)$ is a dense subset of I.
3) Establish consistency of the resultant sequence of estimators $\hat{\alpha} = \hat{\alpha}(n,a_n)$.

Note the resemblance to methods employed in Sections 4.4 and 4.5 and in particular the simultaneous limit processes as the sample size converges to infinity and as the $I(a_n)$ increase to I. The sets $I(a)$ are said to form a *sieve*.

For each α let $m_s(\alpha) = E_\alpha[\lambda_s]$ (even though λ is functionally independent of α its probability law can depend on α). We assume that $m(\alpha)$ is bounded and bounded away from zero for each α, but not necessarily uniformly in α. The index set I is the family of nonnegative, left-continuous, right-hand limited functions $\alpha \in L^1[0,1]$.

We now introduce the sieves. For each $a > 0$ let $I(a)$ be the set of absolutely continuous functions $\alpha \in I$ satisfying $a \le \alpha \le a^{-1}$ and $|\alpha'| \le \alpha/a$; as $a \downarrow 0$ the $I(a)$ do indeed increase to a dense subset of I. We show that for each n and a there exists a maximum likelihood estimator $\hat{\alpha}(n,a) \in I(a)$. Then with $a = a_n$ converging to zero sufficiently slowly, $\hat{\alpha}(n,a_n) \to \alpha$ in P_α-probability, where the convergence is in $L^1[0,1]$.

Martingale ideas continue to play a central role in proofs, as in the following consistency theorem.

Theorem 5.18. a) For each $a > 0$ and each sample size n there exists a (not necessarily unique) maximum likelihood estimator $\hat{\alpha}(n,a) \in I(a)$ satisfying $L_n(\hat{\alpha}) \ge L_n(\alpha)$ for all $\alpha \in I(a)$.

b) If the "entropy"

$$h(\alpha) = -\int_0^1 m(\alpha)(1 - \alpha + \alpha \log \alpha) \tag{5.51}$$

is finite, then if $n^{1/2}a_n \to \infty$ and $\hat{\alpha} = \hat{\alpha}(n,a_n)$, $\|\hat{\alpha} - \alpha\|_1 \to 0$ in P_α-probability.

Proof: We follow the line of argument in Grenander (1981, Sect. 8.2, Theorem 1). It is shown there that for each $a > 0$,

i) $I(a)$ is uniformly bounded and equicontinuous.
ii) $\{\alpha' : \alpha \in I(a)\}$ is uniformly bounded.

With n and ω fixed, the latter suppressed and $(\alpha(j)) \in I(a)$ such that $L_n(\alpha(j)) \to \sup\{L_n(\alpha) : \alpha \in I(a)\}$, it follows that there is a subsequence, which we continue to denote by $(\alpha(j))$, such that the derivatives $\alpha'(j)$ converge in $L^1[0,1]$ and such that $(\alpha_0(j))$ converges. Hence there is $\alpha(\infty) \in I(a)$ such that $\alpha(j) \to \alpha(\infty)$ uniformly. Since $\alpha(\infty)$ is continuous and bounded away from zero, $L_n(\alpha(\infty)) = \lim L_n(\alpha(j)) = \sup\{L_n(\alpha) : \alpha \in I(a)\}$.

To establish b), we first note that by reasoning in Grenander (1981), given $\alpha \in I$ and $\delta > 0$, for each sufficiently small a there exists $\tilde{\alpha} \in I(a)$, depending on δ and a, such that $\|\alpha - \widetilde{\alpha}\|_1 < \delta$ and

$$\left| \int m(\alpha)(1 - \widetilde{\alpha} + \alpha \log \widetilde{\alpha}) - \int m(\alpha)(1 - \alpha + \alpha \log \alpha) \right| < \delta \quad (5.52)$$

Recalling that (N, λ) is a generic copy of the process, let

$$L(\beta) = \int_0^1 \lambda(1 - \beta) + \int_0^1 (\log \beta)\, dN$$

be the log-likelihood function. By Jensen's inequality the expectation of a log-likelihood function is maximized at the "true" index value:

$$h(\alpha) = -E_\alpha[L(\alpha)] \leq -E_\alpha[L(\beta)] = -\int_0^1 m(\alpha)(1 - \beta + \alpha \log \beta) \quad (5.53)$$

Suppose now that $\alpha \in I$ and $\delta > 0$ are fixed and let $\hat{\alpha} = \hat{\alpha}(n,a)$ be as in a), with n and a variable. Then for n sufficiently large and a sufficiently small,

$$|-\int m(\alpha)(1 - \hat{\alpha} + \alpha \log \hat{\alpha}) - h(\alpha)|$$

$$= -\int m(\alpha)(1 - \hat{\alpha} + \alpha \log \hat{\alpha}) - h(\alpha)$$

[by (5.53)]

$$= -\int m(\alpha)(1 - \hat{\alpha} + \alpha \log \hat{\alpha}) + n^{-1}[\int \lambda^n(1 - \hat{\alpha}) + \int \log \hat{\alpha}\, dN^n]$$

$$- n^{-1}[\int \lambda^n(1 - \hat{\alpha}) + \int \log \hat{\alpha}\, dN^n] + n^{-1}[\int \lambda^n(1 - \widetilde{\alpha}) + \int \log \widetilde{\alpha}\, dN^n]$$

$$- n^{-1}[\int \lambda^n(1 - \widetilde{\alpha}) + \int \log \widetilde{\alpha}\, dN^n] + \int m(\alpha)(1 - \widetilde{\alpha} + \alpha \log \widetilde{\alpha})$$

$$- \int m(\alpha)(1 - \widetilde{\alpha} + \alpha \log \widetilde{\alpha}) - h(\alpha)$$

$$\leq \left| \int m(\alpha)(1 - \hat{\alpha} + \alpha \log \hat{\alpha}) - n^{-1} \int \lambda^n(1 - \hat{\alpha}) + \int \log \hat{\alpha}\, dN^n] \right|$$

$$+ \left| \int m(\alpha)(1 - \widetilde{\alpha} + \alpha \log \widetilde{\alpha}) - n^{-1}[\int \lambda^n(1 - \widetilde{\alpha}) + \int \log \widetilde{\alpha}\, dN^n] \right|$$

$$+ \delta$$

using (5.52) and the definition of $\hat{\alpha}$ as maximum likelihood estimator in $I(a)$.

Assuming for the moment that by proper choice of large n and small a we can dispose of the two remaining terms by showing that they converge to zero in probability, we will have thereby demonstrated that $-\int m(\alpha)(1-\hat{\alpha}+\alpha\log\hat{\alpha}) \to h(\alpha)$ in P_α-probability, which is the same as

$$\int m(\alpha)(\hat{\alpha}-\alpha\log\hat{\alpha}) \to \int m(\alpha)(\alpha-\alpha\log\alpha) \tag{5.54}$$

Following Grenander (1981) one more time, we infer from (5.54) that for $j(y)=y-\log(1+y)$, $\int j(\hat{\alpha}/\alpha-1)\alpha m(\alpha) \to 0$ in probability, which (with a subsequences of subsequences modification of the Grenander argument) shows that $\hat{\alpha}\to\alpha$ in $L^1(m_s(\alpha)\,ds)$ and suffices to complete the proof.

Of those two remaining terms we analyze only the first; the second is handled analogously and more easily. We have

$$\left|\int m(\alpha)(1-\hat{\alpha}+\alpha\log\hat{\alpha}) - n^{-1}[\int \lambda^n(1-\hat{\alpha}) + \int \log\hat{\alpha}\,dN^n]\right|$$

$$\leq \left|\int (1-\hat{\alpha}+\alpha\log\hat{\alpha})\frac{\lambda^n}{n} - m(\alpha)\right|$$

$$+\left|\log\hat{\alpha}_1 \cdot n^{-1}(N_1^n - \int_0^1 \alpha\lambda^n)\right|$$

$$+\left|n^{-1}\int_0^1 \frac{\alpha'_t}{\hat{\alpha}_t}(N_t^n - \int_0^t \alpha\lambda^n)\,dt\right|$$

(with integration by parts in the second term)

$$\leq a^{-1}\int_0^1\left|\frac{\lambda^n}{n} - m(\alpha)\right| + (na)^{-1}\left|N_1^n - \int_0^1 \alpha\lambda^n\right| + a^{-1}\sup_t n^{-1}\left|N_t^n - \int_0^t \alpha\lambda^n\right|$$

By boundedness of λ, for $\varepsilon>0$,

$$P_\alpha\left\{a^{-1}\int_0^1\left|\frac{\lambda^n}{n} - m(\alpha)\right| > \varepsilon\right\} \leq (a\varepsilon)^{-2}E_\alpha\left[\left(\int_0^1\left|\frac{\lambda^n}{n} - m(\alpha)\right|\right)^2\right]$$

$$\leq (a\varepsilon)^{-2}E_\alpha\left[\int_0^1\left(\frac{\lambda^n}{n} - m(\alpha)\right)^2\right]$$

$$= O(\varepsilon^{-2}a^{-2}n^{-1})$$

Since $M_t^n = N_t^n - \int_0^t \alpha\lambda^n$ is a mean zero square integrable martingale, in particular M_1^n is the sum of i.i.d. random variables with finite variances; consequently,

$$P_\alpha\{(an)^{-1}|N_1^n - \int_0^1 \alpha\lambda^n| > \varepsilon\} = O(\varepsilon^{-2}a^{-2}n^{-1})$$

as well. Finally, by Proposition 5.9 applied to $(M^n)^2$ and $<M^n> = \int^{\cdot} \alpha\lambda^n$,

$$P_\alpha\{\sup_t (an)^{-1}|N_t^n - \int_0^t \alpha\lambda^n| > \varepsilon\} = P_\alpha\{\sup(M_t^n)^2 > (\varepsilon an)^2\}$$

$$\leq (\varepsilon an)^{-2}(n^{3/2}) + P_\alpha\{<M^n>_1 > n^{3/2}a_n\}$$

$$\leq (\varepsilon an)^{-2}(n^{3/2}a_n) + (n^{1/2}a_n)^{-1} E_\alpha[\int_0^1 \alpha\lambda^n] \to 0$$

which completes the proof. □

With stronger integrability hypotheses, more sophisticated martingale inequalities, and slower convergence of a to zero, one can obtain strong L^1-consistency (see Karr, 1986b).

For finite-dimensional parametrizations it is possible to achieve strong consistency and asymptotic normality for estimators constructed as solutions of the likelihood equation $L_n'(\theta) = 0$. Let the index set be an open subset Θ of $\mathbf{R}$ (only for simplicity do we take the dimension to be 1) and for each $\theta \in \Theta$ let $\alpha(\theta)$ be a function satisfying (5.16). In what follows, derivatives denoted by primes are all with respect to θ.

Theorem 5.19. Assume that the function $\theta \to \alpha_s(\theta)$ is twice continuously differentiable on Θ for each s and that for each θ

$$0 < I(\theta) = \int_0^1 \frac{\alpha_s'(\theta)^2}{\alpha_s(\theta)} m_s(\theta)\, ds < \infty \tag{5.55}$$

Then

a) There exists a sequence $(\hat{\theta}_n)$ such that almost surely (P_θ) the likelihood equation

$$L_n'(\hat{\theta}_n) = 0 \tag{5.56}$$

is satisfied for all sufficiently large n. Moreover, $\hat{\theta}_n \to \theta$ almost surely.

b) Under P_θ, $n^{1/2}[\hat{\theta}_n - \theta] \xrightarrow{d} N(0, I(\theta)^{-1})$. □

Thus $I(\theta)$ in (5.55) plays the role of the classical Fisher information.

Note the importance throughout the section of martingale techniques, including analysis of maximum likelihood estimators; even if one could avoid them in the latter context (which is uncertain), it would be very inconvenient to do so.

5.3 HYPOTHESIS TESTING

For concreteness we restrict attention to the multiplicative intensity model because it is one case where hypothesis tests can be formulated specifically as well as analyzed. The tests presented here are tests whether a multiplicative intensity process has a prescribed multiplier and whether two multiplicative intensity processes (possibly with different baseline intensities) have the same—but unspecified—multiplier. As expected, results deal mainly with asymptotic distribution theory under the null hypothesis; little is known about behavior under alternative hypotheses. Both martingale and likelihood methods will be explored.

Let I be the set of functions α: $[0,1] \to \mathbf{R}_+^K$ satisfying (5.16); recall that with respect to P_α the point process $N(k)$ has stochastic intensity $\lambda_t(\alpha,k) = \alpha_t(k)\lambda_t(k)$, where λ is predictable and functionally independent of α (but with law depending on α). By P we denote the probability corresponding to $\alpha \equiv 1$, yielding the log-likelihood function

$$L(\alpha) = \sum_{k=1}^{K} \left\{ \int_0^1 \lambda(k)[1 - \alpha(k)] + \int_0^1 (\log \alpha(k))\, dN(k) \right\} \tag{5.57}$$

When we deal with a sequence (N^n, λ^n) the log-likelihood functions are written $L_n(\alpha)$.

Our initial problem is to test the simple hypothesis $H_0: \alpha = \alpha^*$, where α^* is a prescribed element of I, which we address for a sequence (N^n, λ^n) envisioned as partial superpositions of i.i.d. copies of a process (N,λ); for some results this much structure is not necessary, although it may not be easy to construct other specific cases in which the hypotheses are fulfilled.

We begin with the martingale approach. Let $B_t^n(\alpha,k) = \int_0^t \alpha(k) 1(\lambda^n(k) > 0)$, whose martingale estimator is

$$\hat{B}_t^n(k) = \int_0^t 1(\lambda^n(k) > 0)\lambda^n(k)^{-1}\, dN^n(k)$$

Under H_0 the processes $\hat{B}^n(k) - B^n(\alpha^*,k)$ are orthogonal square integrable martingales. For the martingale method the principle of test construction is: *Use test statistics that are martingales*. Thus we can construct a whole family of *martingale test statistics*

$$Z_t^n(k) = \int_0^t Y^n(k)[d\hat{B}_t^n(k) - dB_t^n(\alpha^*,k)], \tag{5.58}$$

where Y^n is a bounded, $\mathcal{H}^n$-predictable process that, by weighting differences between $\hat{B}^n$ and $B^n(\alpha^*)$ at various times, provides great flexibility. Of course, the Y^n must be observable. For each n and k, $Z^n(k)$ is a square integrable martingale (Y^n is bounded) with predictable variation

$$<Z^n(k)>_t = \int_0^t Y^n(k)^2 1(\lambda^n(k) > 0)\lambda^n(k)^{-1}\alpha^*(k),$$

and in addition these martingales are orthogonal.

Theorem 5.13 implies the following result.

Theorem 5.20. Suppose that (b_n) is a sequence of constants increasing to ∞ and that under P_{α^*},

$$b_n \int_0^t Y^n(k)^2 1(\lambda^n(k) > 0)\lambda^n(k)^{-1}\alpha^*(k) \to V_t(k) \tag{5.59}$$

for each t and k, where V satisfies the properties in Theorem 5.13, and assume also that the martingales $b_n^{1/2} Z^n(k)$ fulfill (5.27). Then under P_{α^*}, $b_n^{1/2} Z^n \xrightarrow{d} M$, where the components of M are independent continuous Gaussian martingales with variance functions $V(k)$. □

Given that the processes are observed over [0,1] one should certainly use all the data and base the test on the values $b_n^{1/2} Z_1^n(k)$, which are asymptotically independent and normally distributed with variances

$$V_1(k) = \lim b_n \int_0^1 Y^n(k)^2 1(\lambda^n(k) > 0)\lambda^n(k)^{-1}\alpha^*(k),$$

which can be estimated consistently (see the discussion following Theorem 5.16). Therefore, one arrives at test statistics whose asymptotic distribution does not depend on α^*.

Corollary 5.21. Under H_0, for each k

$$U^n(k) = \frac{Z_1^n(k)}{(<Z^n(k)>_1)^{1/2}} \xrightarrow{d} N(0,1) \quad \square \tag{5.60}$$

The test statistics $U^n(k)$ are the basic martingale tools for testing the hypothesis $H_0: \alpha = \alpha^*$. An advantage of Theorem 5.20 is generality in terms of structure of the sequence (N^n, λ^n). One instance in which its hypotheses are satisfied is superposition of i.i.d. processes: if under each P_α, $(N^{(i)}, \lambda^{(i)})$ are i.i.d. copies of (N, λ), where N has P_α-stochastic intensity $\alpha\lambda$, then Theorem 5.20 holds for $N^n = \Sigma_{i=1}^n N^{(i)}$, $\lambda^n = \Sigma_{i=1}^n \lambda^{(i)}$ and $b_n = n$.

But for this situation we can also employ more classical methods. Given appropriate assumptions the log-likelihood functions

$$L_n(\alpha^*) = \sum_{k=1}^K \left\{ \int_0^1 \lambda^n(k)[1 - \alpha^*(k)] + \int_0^1 (\log \alpha^*(k))\, dN^n(k) \right\}$$

have well-defined limit behavior that follows from asymptotic theory for partial sums of i.i.d. random variables.

Proposition 5.22. If

$$h(\alpha^*) = -\sum_{k=1}^{K} m(\alpha^*,k)[1 - \alpha^*(k) + \alpha^*(k)\log\alpha^*(k)]$$

is finite then under H_0, $n^{-1}L_n(\alpha^*) \to -h(\alpha^*)$ almost surely. If in addition

$$\sigma^2 = \mathrm{Var}_{\alpha^*}\left(\sum_{k=1}^{K}\left\{\int_0^1 \lambda(k)[1 - \alpha^*(k)] + \int_0^1 (\log\alpha^*(k))\, dN(k)\right\}\right) < \infty$$

then $n^{1/2}[n^{-1}L_n(\alpha^*) + h(\alpha^*)] \xrightarrow{d} N(0,\sigma^2)$. □

To use Proposition 5.22, one must estimate the variance σ^2, which can be done using the sample variance.

As in Section 5.2, the martingale method can be used to refine results pertaining to likelihood functions, as we now illustrate. Reverting to a general sequence of pairs (N^n, λ^n) (not necessarily superpositions of i.i.d. processes), the integral $\int_{E\times[0,1]} (\log\alpha^*)\, dN^n$ is martingale estimator of $\Sigma_{k=1}^{K} \int_0^1 (\log\alpha^*(k))\alpha^*(k)\lambda^n(k)$, and one can write the log-likelihood function evaluated at α^* as

$$L_n(\alpha^*) = \sum_{k=1}^{K}\left\{\int_0^1 \lambda^n(k)[1 - \alpha^*(k) \; + \alpha^*(k)\log\alpha^*(k)] + \int_0^1 (\log\alpha^*(k))\, dM^n(k)\right\}$$

where $dM_t^n(k) = dN_t^n(k) - \alpha_t^*(k)\lambda_t^n(k)\, dt$ is an innovation martingale. Whenever the first term can be dealt with, one can generalize Proposition 5.22 by applying a martingale central limit theorem to the second. Indeed, martingale methods can be utilized to describe behavior of log-likelihood functions as random functions of t; even in the i.i.d. case this is difficult using classical techniques. For example, one can derive a theorem on local asymptotic normality.

Theorem 5.23. For each α, n, and t, let

$$L_n(\alpha,t) = \sum_{k=1}^{K}\left\{\int_0^t \lambda^n(k)[1 - \alpha(k)] + \int_0^t (\log\alpha(k))\, dN^n(k)\right\}$$

be the log-likelihood function corresponding to observation of (N^n, λ^n) over $[0,t]$. Assume that there are constants $b_n \uparrow \infty$ and for each k an increasing continuous function $V_t(k)$ such that

$$b_n \int_0^t \lambda^n(\alpha^*,k)^{-1} \to V_t(k) \tag{5.61}$$

in probability for each t. Then for each α the processes $L_n(\alpha^* + b_n^{1/2}\alpha, \cdot) - L_n(\alpha^*, \cdot)$ converge in P_{α^*}-distribution to the process

$$X_t = \sum_{k=1}^{K} \left\{ \int_0^t \frac{\alpha(k)}{\alpha^*(k)} dM(k) - \frac{1}{2} \int_0^t \left[\frac{\alpha(k)}{\alpha^*(k)} \right]^2 dV(k) \right\},$$

where M is a Gaussian martingale with covariance function V.

Proof: Note that by virtue of Theorem 5.13, (5.61) implies that

$$b_n^{-1/2} \left\{ N_t^n - \int_0^t \alpha^*(k)\lambda^n(k) : t \in [0,1] \right\} \xrightarrow{d} \{M_t(k) : t \in [0,1]\}, \tag{5.62}$$

jointly in k, where M is as described above. For each n and t,

$$\begin{aligned}
&L_n(\alpha^* + b_n^{1/2}\alpha,t) - L_n(\alpha^*,t) \\
&\quad = \sum_{k=1}^{K} \left[-b_n^{-1/2} \int_0^t \alpha(k)\lambda^n(k) + \int_0^t \log\left(1 + \frac{b_n^{-1/2}\alpha(k)}{\alpha^*(k)} \right) dN^n(k) \right] \\
&\quad \cong \sum_{k=1}^{K} \int_0^t \frac{\alpha(k)}{\alpha^*(k)} \, [dN^n(k) - \alpha^*(k)\lambda^n(k)] \\
&\qquad\qquad - (2b_n)^{-1} \sum_{k=1}^{K} \int_0^t \left[\frac{\alpha(k)}{\alpha^*(k)} \right]^2 dN^n(k) \\
&\quad \xrightarrow{d} \sum_{k=1}^{K} \int_0^t \frac{\alpha(k)}{\alpha^*(k)} dM(k) - \frac{1}{2} \sum_{k=1}^{K} \int_0^t \left[\frac{\alpha(k)}{\alpha^*(k)} \right]^2 dV(k)
\end{aligned}$$

by (5.61), (5.62), Slutsky's theorem, and the continuous mapping theorem. □

We turn now to two-sample problems. For each n let $N^n = (N^n(1),N^n(2))$ be a marked point process with mark space $E = \{1,2\}$ and suppose that with respect to a history $\mathcal{H}^n$, $N^n(i)$ has stochastic intensity $\alpha_t(i)\lambda_t^n(i)$. The components $N^n(1)$, $N^n(2)$ may but need not be independent; recall that both $N^n(i)$ and $\lambda^n(i)$ are observable. Our concern is the composite null hypothesis $H_0 : \alpha(1) \equiv \alpha(2)$; the results apply with straightforward modification to k-sample problems. Although methods based on likelihoods are desirable, no satisfactory results have been developed because under H_0 the common multiplier α is a nuisance parameter that must be estimated or circumvented. By contrast the martingale method yields asymptotically normal test statistics analogous to those for the one-sample problem, constructed by integration of predictable processes with respect to certain fundamental martingales that we now identify.

Proposition 5.24. Under H_0, for each n

a) The superposition $\bar{N}^n = N^n(1) + N^n(2)$ has $(P_\alpha, \mathcal{H}^n)$-stochastic intensity $\lambda_t^n = \alpha_t \bar{\lambda}_t^n$, where $\bar{\lambda}^n = \lambda^n(1) + \lambda^n(2)$.

b) The processes

$$M_t^n(i) = \int_0^t 1(\lambda^n(i) > 0)[\lambda^n(i)^{-1}\, dN^n(i) - (\bar{\lambda}^n)^{-1}\, d\bar{N}^n] \tag{5.63}$$

are square integrable martingales with

$$<M^n(i), M^n(j)>_t = \int_0^t 1(\lambda^n(i)\lambda^n(j) > 0)\alpha\left[\frac{1(i=j)}{\lambda^n(i)} - 1/\bar{\lambda}^n\right] \tag{5.64}$$

Proof: Part a) is a routine calculation. For b), the first and second equalities in

$$\int_0^t \alpha 1(\lambda^n(i) > 0)(\bar{\lambda}^n)^{-1}\,\bar{\lambda}^n = \int_0^t \alpha 1(\lambda^n(i) > 0)$$

$$= \int_0^t \alpha 1(\lambda^n(i) > 0)\lambda^n(i)^{-1}\lambda^n(i)$$

yield as martingale estimators of the middle term the processes $\int_0^t 1(\lambda^n(i) > 0)(\bar{\lambda}^n)^{-1}\, d\bar{N}^n$ [by a)] and $\int_0^t 1(\lambda^n(i) > 0)\lambda^n(i)^{-1}\, dN^n(i)$, respectively. Consequently, $M^n(i)$, the difference of square integrable martingales, is itself a square integrable martingale. Derivation of (5.64) is straightforward using Theorem B.13 (see Exercise 5.16). □

As in the one-sample problem there results a plethora of test statistics: given bounded predictable processes Y^n, one can define (square integrable) martingale test statistics

$$Z_t^n(i) = \int_0^t Y^n(i)\, dM^n(i), \tag{5.65}$$

where the $M^n(i)$ are given by (5.63). The predictable covariations are

$$<Z^n(i), Z^n(j)>_t = \int_0^t Y^n(i) Y^n(j)\, d<M^n(i), M^n(j)>,$$

so there is one new wrinkle: the $Z^n(i)$ are not orthogonal. Nevertheless, a variant of Theorem 5.13 leads to a central limit theorem for (Z^n), but rather than give the most general result we present instead a specific version that seems useful for applications. Suppose that

$$Y_t^n(i) = \lambda_t^n(i) L_t^n \tag{5.66}$$

where L^n is a predictable process depending only on $(\bar{N}^n, \bar{\lambda}^n)$ and is zero whenever $\bar{\lambda}^n$ is zero. Combining (5.63), (5.65), and (5.66) gives as test statistics the processes

$$Z_t^n(i) = \int_0^t L^n \, dN^n(i) - \int_0^t \frac{\lambda^n(i)}{\bar{\lambda}^n} L^n \, d\bar{N}^n$$

Theorem 5.25. Assume that (5.66) holds for each n and let b_n be constants increasing to infinity. If under H_0

i) $b_n^{1/2} L_t^n \lambda_t^n(i)/\bar{\lambda}_t^n \to 0$ in probability for each t and i;

ii) There exist functions $g(1), g(2) \in L^2[0,1]$ such that

$$b_n (L_t^n)^2 \lambda_t^n(i) \lambda_t^n(j) \frac{\alpha_t}{\bar{\lambda}_t^n} \to g_t(i) g_t(j)$$

in probability for each t, i and j;

iii) Indexed by n, t, i, and j, the random variables in ii) form a uniformly integrable family,

then $b_n^{1/2} Z^n \xrightarrow{d} M$, where

$$M_t(i) = \int_0^t (g(i)\bar{g})^{1/2} \, dW(i) - \sum_{\Omega=1}^{2} \int_0^t g(i) \left(\frac{g(\ell)}{\bar{g}} \right)^{1/2} dW(\ell),$$

$\bar{g} = g(1) + g(2)$, and $W(1)$, $W(2)$ are independent Wiener processes. □

We omit the proof, which is given in Andersen et al. (1982). Conditions ii) and iii) are jointly stronger—and easier to verify—than (5.26) and (5.27). In particular, under the conditions of the theorem, $b_n^{1/2} \times (Z_1^n(1), Z_1^n(2)) \xrightarrow{d} N(0,R)$ under H_0, where $R(i,j) = \int_0^1 g(i)[1(i=j)\bar{g} - g(j)]$, for which the martingale method (one of its beauties!) provides estimators

$$b_n \int_0^1 (L^n)^2 \frac{\lambda^n(i)}{\bar{\lambda}^n} \left[1(i=j) - \frac{\lambda^n(j)}{\bar{\lambda}^n} \right] d\bar{N}^n$$

Given the assumptions of Theorem 5.25 these estimators are weakly consistent under H_0.

We illustrate along the lines of Examples 5.7 and 5.17.

Example 5.26. (i.i.d. random variables). For $i = 1, 2$, let X_{ij} be a sequence of i.i.d. random variables with values in $[0,1]$ and hazard function $h(i)$, and assume that the two sequences are independent. For each n and i

let $N_t^n(i) = \Sigma_{j=1}^n 1(X_{ij} \le t)$; then with respect to the history $\mathcal{H}_t^n = \sigma(X_{ij}1(X_{ij} \le t) : i = 1, 2;\ 1 \le j \le n)$ and with $\lambda_t^n(i) = (n - N_{t-}^n(i))$, $N^n(i)$ has stochastic intensity $h_t(i)\lambda_t^n(i)$. The null hypothesis $H_0 : h(1) \equiv h(2)$ is that both sequences have the same distribution. Suppose that (5.66) holds with $L^n = \ell(\bar{\lambda}^n)$ for a fixed, deterministic function ℓ. Then with the notation $\tilde{\ell}(p) = \ell(p) + \Sigma_{q=p}^{2n} \ell(q)/q$ and since $\lambda_t^n(i) = \int_{[t,1]} dN^n(i)$ and $\bar{\lambda}^n(X_{ij}) = 2n + 1 - R_{ij}$, where R_{ij} is the rank of X_{ij} among $X_{11}, \ldots, X_{1n}, X_{21}, \ldots, X_{2n}$, we have

$$Z_1^n(i) = \sum_{j=1}^{n} \tilde{\ell}(2n + 1 - R_{ij}) \tag{5.67}$$

The choices $\ell(p) = p$ and $\ell(p) \equiv 1$ lead to the rank statistics of Wilcoxon and Savage, respectively. □

Extension to censored data is treated in Andersen et al. (1982).

5.4 FILTERING WITH POINT PROCESS OBSERVATIONS

Stochastic intensities and martingales play important roles in state estimation as well as statistical inference, which is explored in this section. Let N be a point process on $\mathbf{R}_+$ and let $\mathcal{H}$ be a history satisfying Assumptions 2.10. The observations are the (completed) internal history $\mathcal{F}^N$. There is often an $\mathcal{H}$-adapted *state process* Z that is of interest but not directly observable. However, Z is indirectly or partially observable to the extent that $\mathcal{F}^N$ yields information about it, which we seek to utilize in optimal fashion. The underlying probability is completely known, so the problem reduces (see Section 3.3) to computation of conditional expectations $E[J(Z)|\mathcal{F}_t^N]$, where $J(\cdot)$ is a functional of the path $s \to Z_s$. Rather than work at this level of generality, we consider only the "real-time" *filtering* problem of recursive calculation of MMSE state estimators

$$\hat{Z}_t = E[Z_t|\mathcal{F}_t^N], \tag{5.68}$$

Originally, "filtering" pertained to extraction of a signal from additive noise but now connotes, more generally, state estimation of the current state of a process, rather than a previous state (interpolation; smoothing) or a future state (prediction; extrapolation). Filtering is the most fundamental problem of the three because often a solution to it yields rather easily solutions to interpolation and prediction problems.

In what follows the point process N and the state process Z are assumed to possess particular structure. Concerning N we suppose that there exists an $\mathcal{H}$-predictable stochastic intensity (λ_t), which (Exercise 2.16) entails existence of an $\mathcal{F}^N$-stochastic intensity $(\hat{\lambda}_t)$ satisfying $\hat{\lambda}_t =$

$E[\lambda_t|\mathcal{F}_t^N]$ for each t. Let $\hat{M}$ denote the $\mathcal{F}^N$-innovation martingale: $\hat{M}_t = N_t - \int_0^t \hat{\lambda}$. The state process Z is to admit a *semimartingale representation*

$$Z_t = Z_0 + \int_0^t X_s \, ds + M_t, \tag{5.69}$$

where X is an $\mathcal{H}$-progressive process and M is a mean zero $\mathcal{H}$-martingale with $E[\int_0^t |dM_s|] < \infty$ for each $t < \infty$. We further stipulate that Z is *bounded*. Assumptions in this paragraph are in force for the entire general discussion.

The key implication of (5.69) is that one can in effect condition throughout to obtain a semimartingale representation for the estimator process $\hat{Z}$:

$$\hat{Z}_t = E[Z_0] + \int_0^t \hat{X}_s \, ds + \widetilde{M}_t, \tag{5.70}$$

where $\hat{X}$ is the $\mathcal{F}^N$-progressive (and hence computable from the observations) and $\widetilde{M}$ is a mean zero $\mathcal{F}^N$-martingale. In differential form (5.70) becomes the stochastic differential equation

$$d\hat{Z}_t = \hat{X}_t \, dt + d\widetilde{M}_t, \tag{5.70a}$$

the beginning of a method of recursive computation, but one must do more. By Theorem 2.34, the representation theorem for point process martingales, there is an $\mathcal{F}^N$-predictable process H, the *filter gain*, such that $\widetilde{M}_t = \int_0^t H_s \, d\hat{M}_s$, leading to the stochastic differential equation

$$d\hat{Z}_t = \hat{X}_t \, dt + H_t(dN_t - \hat{\lambda}_t \, dt) \tag{5.71}$$

but while a principal step, (5.71) is not yet the end of the story. The filter gain H must be calculated—presumably recursively—if (5.71) is actually to be implemented, which is done in Theorem 5.29, the main result of the section. There is also the issue of computation of $\hat{X}$, as well as the $\mathcal{F}^N$-stochastic intensity $\hat{\lambda}$; the former constitutes one part of Proposition 5.27, in which the semimartingale representation for Z is established, while the latter is considered following Theorem 5.29.

Now for the results, after which examples will be discussed.

Proposition 5.27. Let Z be an $\mathcal{H}$-adapted process with semimartingale representation (5.69). Then the state estimator process $\hat{Z}_t = E[Z_t|\mathcal{F}_t^N]$ has semimartingale representation (5.70), where $\hat{X}$ is the unique $\mathcal{F}^N$-progressive process such that

$$E\left[\int_0^t C_s X_s \, ds\right] = E\left[\int_0^t C_s \hat{X}_s \, ds\right] \tag{5.72}$$

for every bounded $\mathcal{F}^N$-progressive process C, and where $\widetilde{M}$ is a mean zero $\mathcal{F}^N$-martingale.

Proof: In the same way that compensators are (dual) predictable projections of point processes, the process $\hat{X}$ fulfilling (5.72) is the *progressive projection* of X, realized as the Radon-Nikodym derivative, on the σ-algebra on $\mathbf{R}_+ \times \Omega$ generated by the $\mathcal{F}^N$-progressive processes, of the signed measure $X_t(\omega)\,dt\,P(d\omega)$ with respect to the σ-finite measure $dt\,P(d\omega)$. Now let

$$\widetilde{M}_t = E[Z_0|\mathcal{F}_t^N] - E[Z_0] + E[\int_0^t X|\mathcal{F}_t^N] - \int_0^t \hat{X} + E[M_t|\mathcal{F}_t^N]; \qquad (5.73)$$

then provided that $\widetilde{M}$ is a martingale with respect to $\mathcal{F}^N$, (5.70) holds by algebraic manipulation. To show that $\widetilde{M}$ is a martingale it suffices to show that $E[\int_0^t X|\mathcal{F}_t^N] - \int_0^t \hat{X}$ and $E[M_t|\mathcal{F}_t^N]$ are martingales; the latter is nearly immediate (Exercise 5.3). For the former, if $s<t$ and $A \in \mathcal{F}_s^N$, then the process $C_u(\omega) = 1(s<u\leq t)1(\omega \in A)$ is $\mathcal{F}^N$-progressive and hence by (5.72),

$$E[(E[\int_0^t X|\mathcal{F}_t^N] - E[\int_0^s X|\mathcal{F}_s^N]); A] = E[\int_s^t X;A]$$

$$= E[\int CX] = E[\int C\hat{X}] = E[\int_s^t \hat{X};A]$$

as required. □

Corollary 5.28. Under the hypotheses of the proposition there is a bounded, $\mathcal{F}^N$-predictable process H satisfying (5.71). □

By itself (5.72) is not easy to use for computing $\hat{X}$; however, it does imply that for each t,

$$\hat{X}_t = E[X_t|\mathcal{F}_t^N], \qquad (5.72a)$$

and if one can choose versions of these conditional expectations such that a progressive process results (as typically can be done in practice), then (5.72a) is good enough for computation of $\hat{X}$.

We now calculate the filter gain more explicitly.

Theorem 5.29. Let $\widetilde{M}$ be the $\mathcal{F}^N$-martingale defined by (5.73). Let J, K be the unique $\mathcal{F}^N$-predictable processes such that for every bounded, $\mathcal{F}^N$-predictable process C and each $t>0$,

$$E[\int_0^t CZ\lambda] = E[\int_0^t CJ\hat{\lambda}] \qquad (5.74a)$$

and

$$E[\int_0^t C(\Delta M)\,dN] = E[\int_0^t CK\hat{\lambda}], \tag{5.74b}$$

where M is the martingale part of the semimartingale representation of Z. Then almost surely for all t,

$$\widetilde{M}_t = \int_0^t (J_s + K_s - \hat{Z}_{s-})\,d\hat{M}_s \tag{5.75}$$

Proof: The processes J and K are projections onto the $\mathcal{F}^N$-predictable σ-algebra $\mathcal{P}$ on $\mathbf{R}_+ \times \Omega$ and can be realized as Radon-Nikodym derivatives:

i) J is the derivative on $\mathcal{P}$ of the signed measure $Z_t(\omega)\lambda_t(\omega)P(d\omega)$ with respect to the σ-finite measure $\hat{\lambda}_t(\omega)\,dt\,P(d\omega)$ [which on $\mathcal{P}$ agrees with $\lambda_t(\omega)\,dt\,P(d\omega)$, so that absolute continuity holds];

ii) K is the derivative of $\Delta M_t(\omega)\,dN_t(\omega)P(d\omega)$ with respect to $\hat{\lambda}_t(\omega)\,dt\,P(d\omega)$ [which also equals $dN_t(\omega)P(d\omega)$ on $\mathcal{P}$].

Note that the predictable process $\hat{Z}_- = (\hat{Z}_{t-})$ satisfies

$$E[\int_0^t CZ\hat{\lambda}] = E[\int_0^t C\hat{Z}_-\hat{\lambda}] \tag{5.74c}$$

for each C and t; indeed,

$$E[\int_0^t C_s Z_s \hat{\lambda}_s] = \int_0^t E[C_s\hat{\lambda}_s E[Z_s|\mathcal{F}_s^N]]\,ds$$

$$= \int_0^t E[C_s\hat{\lambda}_s\hat{Z}_s] = E[\int_0^t C_s\hat{Z}_{s-}\hat{\lambda}_s]$$

Suppose now that C is an $\mathcal{F}^N$-predictable process such that

a) The martingale $V_t = \int_0^t C\,d\hat{M}$ is *bounded*;

b) With H the filter gain of (5.71), $E[\int_0^t |CH|\hat{\lambda}] < \infty$ for all t.

Then from a), b), and the definition of $\hat{Z}$,

$$E[Z_t V_t] = E[\hat{Z}_t V_t] \tag{5.76}$$

for all t. We now compute both sides of (5.76).

By (2.16) the left-hand side is the expectation of

$$Z_t V_t = \int_0^t Z_{s-}\,dV_s + \int_0^t V_s\,dZ_s$$

$$= \int_0^t Z_{s-}C_s(dN_s - \lambda_s\,ds) + \int_0^t Z_{s-}C_s(\lambda_s - \hat{\lambda}_s)\,ds$$

$$+\int_0^t V_s X_s\,ds + \int_0^t V_{s-}\,dM_s + \int_0^t C_s\,\Delta M_s\,dN_s$$

The first and fourth terms are mean zero $\mathscr{H}$-martingales (the processes Z_- and V_- are predictable) and consequently,

$$E[Z_tV_t] = E[\int_0^t Z_-C(\lambda - \hat{\lambda})] + E[\int_0^t VX] + E[\int_0^t C\,\Delta M\,dN]$$

$$= E[\int_0^t VX] + E[\int_0^t C(J + K - \hat{Z}_-)\hat{\lambda}],$$

where we have used (5.74a) – (5.74c). In the same way,

$$\hat{Z}_tV_t = \int_0^t \hat{Z}_-C\,d\hat{M} + \int_0^t VX + \int_0^t V_-H\,d\hat{M} + \int_0^t CH\,dN,$$

with the first and third terms mean zero $\mathscr{F}^N$-martingales, so that

$$E[\hat{Z}_tV_t] = E[\int_0^t VX] + E[\int_0^t CH\hat{\lambda}],$$

since $\hat{\lambda}$ is the $\mathscr{F}^N$-stochastic intensity of N.

We conclude that

$$E[\int_0^t CH\hat{\lambda}] = E[\int_0^t C(J + K - \hat{Z}_-)\hat{\lambda}] \tag{5.77}$$

for every t and each $\mathscr{F}^N$-predictable C process satisfying a) and b) above. The remainder of the proof is a monotone class argument to deduce from (5.77) that $H = J + K - \hat{Z}_-$ on a large enough set that (5.75) holds. In particular, the conditions are satisfied for $C_s = \widetilde{C}_s 1(s \le S_n)$, where $\widetilde{C}$ is bounded, nonnegative and $\mathscr{F}^N$-predictable and where $S_n = \inf\{t : N_t \ge n$ or $\int_0^t (1 + |H|)\hat{\lambda} \ge n\}$, in which case (5.77) yields

$$E[\int_0^{S_n} \widetilde{C}H\hat{\lambda}] = E[\int_0^{S_n} \widetilde{C}(J + K - \hat{Z}_-)\hat{\lambda}],$$

from which we infer that $H_t 1(t \le S_n) = (J_t + K_t - \hat{Z}_{t-})1(t \le S_n)$ almost everywhere in (t,ω) with respect to the measure $\hat{\lambda}_t(\omega)P(d\omega)$ and hence also (because they agree on $\mathscr{P}$) with respect to $dN_t(\omega)P(d\omega)$. Letting $n \to \infty$ (since $S_n \to \infty$) gives (5.75). □

The representation (5.75) simplifies if the state process Z and point process N have no jumps in common, for then $\Delta M_t\,dN_t \equiv 0$ (jumps in Z arise only from M) and therefore $K \equiv 0$, so that $d\widetilde{M}_t = (J_t - \hat{Z}_{t-})[dN_t - \hat{\lambda}_t\,dt]$. But also, given integrability assumptions, by an argument similar to that leading to (5.74c),

$$J_t = \hat{\lambda}_t^{-1}(Z\lambda)^{\wedge}_{t-} \tag{5.78}$$

Therefore, if $Z\lambda$ can be estimated simultaneously with Z, one obtains the ultimate representation

$$d\widetilde{M}_t = \left[\frac{(Z\lambda)^{\wedge}_t}{\hat{\lambda}_t} - \hat{Z}_{t-}\right](dN_t - \hat{\lambda}_t\,dt), \tag{5.79}$$

but even now we are not finished.

Computation of the $\mathcal{F}^N$-stochastic intensity $\hat{\lambda}$ has been left unresolved; in general, the formula $\hat{\lambda}_t = E[\lambda_t|\mathcal{F}_t^N]$ is not useful directly. Indeed, this calculation is another filtering problem, which may be equally difficult to solve or may engender still further filtering problems (see additional discussion in Section 7.5).

We conclude with two examples. The first, the *disruption problem*, builds upon Exercises 2.20, 3.9, and 3.10 (see also Exercises 7.13, 7.16, and 7.23).

Example 5.30. (Disruption problem). At an unobservable random time T the intensity of N shifts from one known constant to another. The distribution of T is known; the objective is to estimate T itself, given observations only of N. Suppose that T has hazard function h and let $Z_t = 1(T \le t)$ be the state process. Let N be a point process on $\mathbf{R}_+$ and assume that there are known, distinct constants a, $b > 0$ such that with respect to $\mathcal{H}_t = \mathcal{F}_t^Z \vee \mathcal{F}_t^N$, N has stochastic intensity $\lambda_t = a + (b-a)Z_{t-}$ [i.e., $\lambda = a$ on $[0,T]$ and $\lambda = b$ on (T,∞)]. We seek to calculate $\hat{Z}_t = P\{T \le t|\mathcal{F}_t^N\}$.

To start we need a semimartingale representation for Z; the process

$$M_t = Z_t - \int_0^{T\wedge t} h(s)\,ds = Z_t - \int_0^t (1-Z_s)h(s)\,ds \tag{5.80}$$

is the $\mathcal{H}$-innovation martingale for the point process ε_T, so $Z_t = \int_0^t h(s)(1-Z_s)ds + M_t$ is the requisite representation. By Proposition 5.28,

$$\hat{Z}_t = \int_0^t h(s)(1-\hat{Z}_s)\,ds + \widetilde{M}_t;$$

it remains to calculate (5.75) for $\widetilde{M}$. Apparently, $\Delta M\,\Delta N \equiv 0$, so K vanishes, while to calculate J we use (5.78) and the property that $Z^2 = Z$, which implies that $[a+(b-a)Z]Z = bZ$; consequently,

$$J_t = \frac{\{[a+(b-a)Z]Z\}^{\wedge}_{t-}}{\hat{\lambda}_t} = \frac{b\hat{Z}_{t-}}{a+(b-a)\hat{Z}_{t-}}$$

Thus

$$d\hat{Z}_t = h(t)(1 - \hat{Z}_{t-})\,dt + \frac{(b-a)\hat{Z}_{t-}(1-\hat{Z}_{t-})}{a + (b-a)\hat{Z}_{t-}}[dN_t - (a + (b-a)\hat{Z}_{t-})\,dt],$$

a completely self-contained representation. Note the variance factor $\hat{Z}(1 - \hat{Z})$ in the second term; an analog appears in Theorems 6.23 and 7.29. □

In Section 7.2 we solve the second example using different methods applicable on general spaces.

Example 5.31. (Mixed Poisson process). Let N be a mixed Poisson process on $\mathbf{R}_+$ with unobservable directing measure $M(dt) = Y\,dt$, where Y is a positive random variable with known distribution. With respect to $\mathcal{H}_t = \sigma(Y) \vee \mathcal{F}_t^N$, N has time invariant (but random!) stochastic intensity $\lambda_t \equiv Y$. The state process $Z_t \equiv Y$ is a martingale, so computation of MMSE state estimators $\hat{Z}_t = E[Y|\mathcal{F}_t^N]$ is merely a problem of representation of the $\mathcal{F}^N$-martingale $\hat{Z}$. Since the $\mathcal{H}$-stochastic intensity of N is the state process Z, the procedure of this section yields

$$dE[Y|\mathcal{F}_t^N] = \left.\frac{E[Y^2|\mathcal{F}_s^N] - E[Y|\mathcal{F}_s^N]^2}{E[Y|\mathcal{F}_s^N]}\right|_{s=t-}[dN_t - E[Y|\mathcal{F}_s^N]|_{s=t-}\,dt] \quad (5.81)$$

Unfortunately, (5.81) manifests a common flaw of recursive methods applied to computation of moments: the expression for one moment involves that of the next higher order, leading in general to an infinite, or open, system of simultaneous stochastic differential equations that cannot be solved meaningfully. In the case at hand it is easy to show using (5.78) that the equation for $dE[Y^2|\mathcal{F}_t^N]$ involves $E[Y^3|\mathcal{F}_t^N]|_{s=t-}$, and so on. Only in exceptional circumstances (e.g., when conditional distributions are Gaussian, as in the famous Kalman filter, or when the state process satisfies $Z^2 = Z$, as in Example 5.30 and Theorem 6.23) can the system be closed because a conditional moment of some order is a deterministic function of lower-order moments. □

5.5 THE COX REGRESSION MODEL

In 1972 Cox proposed a model for survival times that incorporates *covariates*, that is, additional observable factors influencing the distribution of an individual's lifetime, in such a way that hazard functions for different individuals are mutually proportional. A baseline hazard function is modified via an exponential factor containing the inner product of the covariate vector and a vector of unknown "regression parameters." Known as both the "Cox regression model" and the "proportional hazards model," it can accommodate censored survival times as well. More recently, a point process generalization, the principal topic of the section, has been devised. As will emerge, the point process model is a special case of Model 5.2; however, it is analyzed using a blend of martingale and maximum likelihood techniques, although one maximizes not the full likelihood function but only the "partial likelihood."

We begin by introducing the point process model, but before describing its analysis we discuss the original Cox model.

Model 5.32. (Intensity model with covariates). The index set I consists of all pairs (α,β), where $\alpha:[0,1]\to\mathbf{R}_+$ is left-continuous with right-hand limits and where $\beta\in\mathbf{R}^p$ for some fixed $p\geq 1$. Let N be a marked point process with mark space $E=\{1,\ldots,K\}$. The *intensity model with covariates* has as history $\mathcal{H}_t=\bigvee_{k=1}^{K}\mathcal{F}_t^{N(k)}$ and each component $N(k)$ has $(P_{\alpha,\beta},\mathcal{H})$-stochastic intensity

$$\lambda_t(\alpha,\beta,k)=\alpha_t e^{<\beta,Z_t(k)>}Y_t(k), \tag{5.82}$$

where $<\cdot,\cdot>$ denotes scalar product, where $Z(k)$ is $\mathcal{H}$-predictable with values in $\mathbf{R}^p$, and where $Y(k)$ is $\mathcal{H}$-predictable and assumes only the values 0 and 1. The observations are $(N(k),Z(k),Y(k))$ over $[0,1]$ for each k. □

One should interpret k as an index representing individuals and $Y_t(k)$ as specifying whether individual k is under observation at time t; thus the $Y(k)$ are *censoring* processes. The function α is the *baseline intensity*. For individual k, $Z(k)$ is the *covariate process* depicting time-varying (numerical) factors influencing the event process $N(k)$. The *Cox regression parameter* β determines how covariates alter the baseline intensity; in particular, if some component of β is zero, then the corresponding component of Z has no influence. The functional form $\exp(<\beta,Z_t(k)>)$ matches that of Cox (1972b); it can be generalized (Jacobsen, 1982; Prentice and Self, 1983).

The aim is to estimate α and β from the observations $(N(k), Z(k),Y(k))$, $k=1,\ldots,K$, and to describe asymptotics as $K\to\infty$. Especially in applications, interest focuses on the Cox regression parameter because only through it are individual-to-individual covariate differences manifested in the stochastic intensities. If one thinks of covariates as factors influencing the course of a disease, it is β that determines their relative importance and thus which should be treated first. As we shall see, for both the point process model and the Cox survival time model, one estimates β by maximizing a partial likelihood function not depending on α, then estimates α by martingale methods. [Were β known, (5.82) would be a multiplicative intensity model with $\lambda_t(k)=\exp(<\beta,Z_t(k)>)Y_t(k)$, to which techniques and results from Section 5.2 would apply directly.]

Before treating Model 5.32 we describe briefly the formulation, interpretation, and analysis of the Cox model. Let $T_1, \dots, T_K$ be independent, positive random variables (survival times) such that T_k has hazard function

$$h_t(k) = h(t)e^{<\beta, z_t(k)>}, \tag{5.83}$$

where h is an unknown hazard function common to all individuals, the $z(k)$ are deterministic covariate functions from $\mathbf{R}_+$ to $\mathbf{R}^p$, and $\beta \in \mathbf{R}^p$ is an unknown regression parameter. When covariates do not depend on time, (5.83) becomes $h_t(k) = h(t)\exp(<\beta, z(k)>)$, so that different individuals' hazard functions are indeed mutually proportional. Returning to the general case, the observations are the survival times T_k, but keep in mind that the covariates are known.

The (full) likelihood function for the survival time model is

$$L(\beta,h) = \prod_{k=1}^{K} [h(T_k)\exp(<\beta, z_{T_k}(k)>)\exp(-\int_0^{T_k} h(t)e^{<\beta, z_t(k)>}\,dt)],$$

which when h is unknown is not only too cumbersome but also unbounded above. Moreover, as mentioned, interest often centers on β to the extent that one would like simply to treat h as an infinite-dimensional nuisance parameter. The approach proposed in Cox (1972b, 1975) is to factor the likelihood function into two terms, the first involving only β and hoped to capture most of the dependence on β, and the second dependent on h and β, but not heavily on β, then to estimate β by maximizing the first term alone. Thereafter, h is estimated from the second term, but with β there replaced by the previously derived estimator $\hat{\beta}$.

To carry out this procedure let $\lambda_t(\beta, k) = e^{<\beta, z_t(k)>} 1(T_k \geqslant t)$ and define $\bar{\lambda}_t(\beta) = \sum_{k=1}^{K} \lambda_t(\beta, k)$; then the likelihood function becomes

$$L(\beta,h) = \left[\prod_{k=1}^{K} \frac{\lambda_{T_k}(\beta,k)}{\bar{\lambda}_{T_k}(\beta)}\right]\left[\prod_{k=1}^{K} \bar{\lambda}_{T_k}(\beta)h(T_k)\right]\exp(-\int_0^{\infty} \bar{\lambda}(\beta)h) \tag{5.84}$$

Let $R_k = \{j: T_j \geqslant T_k\}$ be the set of individuals still at risk just before time T_k. Cox proposed that the first term,

$$C(\beta) = \prod_{k=1}^{K} \frac{\lambda_{T_k}(\beta,k)}{\bar{\lambda}_{T_k}(\beta)} = \prod_{k=1}^{K} \frac{\exp(<\beta, z_{T_k}(k)>)}{\sum_{j \in Rk} \exp(<\beta, z_{T_k}(j)>)}, \tag{5.85}$$

now termed the *Cox partial likelihood*, be maximized in order to estimate β. It is not a conditional likelihood, nor does it capture all dependence of the likelihood function on β. One heuristic basis is that at time T_k the overall hazard rate from currently surviving individuals is (by independence) $h(T_k)\Sigma_{R_k} \exp(<\beta, z_{T_k}(j)>)$, of which $h(T_k)\exp(<\beta, z_{T_k}(k)>)$ is contributed by individual k; hence if one thinks of T_k as known only to be the survival time of some unspecified individual still alive at T_k-, then the individual who dies at that time is k with probability $\exp(<\beta, z_{T_k}(k)>)/\Sigma_{R_k}\exp(<\beta, z_{T_k}(j)>)$. Of course, the interpretation is inexact because T_k is already known to arise from individual k; moreover, there is no rigorous justification for the multiplication in (5.85).

The *Cox estimator* $\hat{\beta}$ is any value maximizing the partial likelihood. Then the integrated hazard function $H_t = \int_0^t h$ is estimated by

$$\hat{H}_t = \sum_{T_i \le t} \left[\sum_{j \in R_i} \exp(<\hat{\beta}, z_{T_i}(j)>) \right]^{-1} \tag{5.86}$$

Various interpretations have been advanced, including that sieve-type maximum likelihood estimator relative to the likelihood function $L(\beta, h)$ (Johansen, 1983), but for our purposes it seems best to think of $\hat{H}$ as a martingale estimator, especially since this strengthens the analogy to the point process model. Relative to the history $\mathcal{H}_t = \vee_{k=1}^{K} \mathcal{F}_t^{N(k)}$, the point processes $N(k) = \varepsilon_{T_k}$ have stochastic intensities $h(t)\lambda_t(\beta, k)$; consequently, $\bar{N} = \Sigma N(k)$ has stochastic intensity $h(t)\bar{\lambda}_t(\beta)$. If β were known, the martingale estimator of H would be $H_t = \int_0^t \bar{\lambda}(\beta)^{-1} d\bar{N}$, which with β replaced by $\hat{\beta}$ is precisely (5.86).

With the second and third terms from (5.84) written as

$$\exp[\sum \log(\bar{\lambda}_{T_k}(\beta)h(T_k)) - \int_0^\infty \bar{\lambda}(\beta)h] = \exp[\int_0^\infty \log(\bar{\lambda}(\beta)h)\, d\bar{N} - \int_0^\infty \bar{\lambda}(\beta)\, dH],$$

upon substitution of $\hat{H}$ there ensues a quantity whose value does not depend on β, further argument why one might maximize the partial likelihood in order to estimate β. Note also that for finite t, up to a constant e^t the term $\exp[-\int_0^t h\bar{\lambda}(\beta) + \int_0^t \log(h\bar{\lambda}(\beta))\, d\bar{N}]$ is the likelihood function for $\bar{N}$. Thus the partial likelihood is the ratio of the likelihood for $(N(1), \ldots, N(K))$ to that for the superposition $\bar{N}$, a slightly different interpretation of its allocating unmarked survival times among individuals; this interpretation carries over to the point process model.

Now for Model 5.32. To accord with previous usage we replace K by n (note that the cardinality of the mark space now changes with n). Suppose that there are given processes $(N(k), Y(k), Z(k))$ on $[0,1]$ such that for each n, if $\mathcal{H}_t^n = \vee_{k=1}^{n} \mathcal{F}_t^{N(k)}$, then for $k \le n$, $N(k)$ has $(P_{\alpha,\beta}, \mathcal{H}^n)$-stochastic intensity (5.82). For each n let $\bar{N}(n) = \Sigma_{k=1}^n N(k)$.

Analysis commences with the partial log-likelihood function

$$C(\beta) = \sum_{k=1}^{n} \int_0^1 <\beta, Z(k)>\, dN(k) - \int_0^1 \log\left(\sum_{j=1}^{n} Y(j) e^{<\beta, Z(j)>}\right) d\bar{N}(n) \tag{5.87}$$

With identification of T_k in the Cox model with a point process $N(k)$ having stochastic intensity $h(t)\exp(<\beta,z_t(k)>)1(T_k \geq t)$, the Cox partial likelihood (5.85) satisfies

$$\log C(\beta) = \sum_{k=1}^{K} [\log \lambda_{T_k}(\beta,k) - \log \bar{\lambda}_{T_k}(\beta)]$$
$$= \sum_{k=1}^{K} \left\{ \int <\beta,z(k)> dN(k) - \int \log\left(\sum_j e^{<\beta,z(j)>}\right) d\bar{N}, \right.$$

which is formally identical to (5.87). Returning to the general setting, *a propos* the discussion following derivation of (5.86), by Theorem 2.31 the log-likelihood function associated with $N(1), \ldots, N(n)$ is

$$L(\alpha,\beta) = \sum_{k=1}^{n} \left\{ \int_0^1 [1 - \alpha Y(k) e^{<\beta,\, Z(k)>}] + \int_0^1 (\log \alpha + <\beta,Z(k)>)\, dN(k) \right\},$$

while the log-likelihood function for $\bar{N}(n)$ *alone* is

$$L^0(\alpha,\beta) = \int_0^1 \left(1 - \alpha \sum_j Y(j) e^{<\beta,Z(j)>}\right)$$
$$+ \int_0^1 \left[\log \alpha + \log\left(\sum_j Y(j) e^{<\beta,Z(j)>}\right)\right] d\bar{N}(n)$$

Up to a factor depending on neither α nor β, we have $C(\beta) = L(\alpha,\beta) - L^0(\alpha,\beta)$; hence one can interpret $C(\beta)$ heuristically as a conditional likelihood for $N(1), \ldots, N(n)$ given $\bar{N}(n)$.

The partial likelihood function has gradient (a random p-vector)

$$\nabla C(\beta) = \sum_{k=1}^{n} \int Z(k)\, dN(k) - \int \frac{\Sigma\, Y(j) e^{<\beta,Z(j)>} Z(j)}{\Sigma\, Y(j) e^{<\beta,Z(j)>}}\, d\bar{N}(n);$$

the estimator of β is any solution $\hat{\beta}$ to the likelihood equation

$$\nabla C(\beta) = 0 \tag{5.88}$$

For estimation of integrals $B_t(\alpha) = \int_0^t \alpha 1(\Sigma Y(j) e^{<\hat{\beta},Z(j)>} > 0) \cong \int_0^t \alpha$, one uses what would be martingale estimators if β were known, but with β replaced by $\hat{\beta}$.

$$\hat{B}_t = \int_0^t (\sum Y(j) e^{<\hat{\beta},Z(j)>})^{-1}\, d\bar{N}(n) \tag{5.89}$$

Note the exact analogy to the estimators $\hat{H}$ of (5.86).

The central issue is asymptotic properties of $\hat{\beta}$ and B as $n \to \infty$. For the sake of economy, results are presented below for the case of i.i.d. processes, and only conditions not immediately implied by the i.i.d. hypothesis are stated as assumptions. A more complete set of assumptions is formulated by Andersen and Gill (1982).

Suppose, therefore, that $(N(k),Z(k),Y(k))$ are i.i.d. copies of a process (N,Z,Y). With the true parameter values denoted by α_0, β_0, let

$$s(\beta,t) = E_{\alpha_0,\beta_0}[Y_t e^{<\beta,Z_t>}] \tag{5.90a}$$

The conditions needed for consistency and asymptotic normality are as follows. Here and below, α_0 and β_0 are suppressed from the notation when possible.

Assumptions 5.33. a) $\int_0^1 \alpha_0 < \infty$.

b) There is a neighborhood U of β_0 such that

$$E[\sup\{Y_t|Z_t|^2 e^{<\beta,Z_t>}\} : t \in [0,1], \beta \in U\}] < \infty$$

c) $P\{Y_t = 1 \text{ for all } t\} > 0$.

d) The gradient ∇s and Hessian $\nabla^2 s$ of s satisfy

$$\nabla s(\beta,t) = E[Y_t e^{<\beta,Z_t>} Z_t] \tag{5.90b}$$

and

$$\nabla^2 s(\beta,t) = E[Y_t e^{<\beta,Z_t>} Z_t Z_t^T] \tag{5.90c}$$

[We regard Z as a column vector; the T in (5.90c) denotes transpose. Thus ∇ is a p-vector and $\nabla^2 s$ is a $p \times p$ matrix.]

e) With $u(\beta,t) = \nabla s(\beta,t)/s(\beta,t)$, the matrix

$$R = \int_0^1 \left[\frac{\nabla^2 s(\beta_0,t)}{s(\beta_0,t)} - u(\beta_0,t)u(\beta_0,t)^T \right] \alpha_0(t) s(\beta_0,t)\, dt \tag{5.91}$$

is positive definite. □

These are quite standard. The first has been assumed throughout the chapter, while the third bounds s away from zero, permitting division by it and by estimators as well. Differentiation inside the expectation to obtain (5.90b) and (5.90c) is justified by b), which is also used to verify a Lindeberg condition similar to (5.30). Finally, the matrix R is the inverse of the limit covariance matrix of $n^{1/2}[\hat{\beta} - \beta_0]$ and so must be positive definite; the term "$[\cdot]$" in (5.91) is a logarithmic second derivative, which shows that R plays the role of the Fisher information matrix.

Before treating asymptotic normality, we first consider weak consistency of the $\hat{\beta}$.

Proposition 5.34. If α_0, β_0 satisfy Assumptions 5.33, then $\hat{\beta} \to \beta_0$ in probability.

Proof: (Sketch). For sufficiently large n the function

$$X_n(\beta) = n^{-1}[C(\beta) - C(\beta_0)]$$

$$= n^{-1}\left[\sum_{k=1}^{n}\int_0^1 <\beta - \beta_0, Z(k)> dN(k) - \int_0^1 \log\left(\frac{\Sigma_1^n Y(j)e^{<\beta,Z(j)>}}{\Sigma_1^n Y(j)e^{<\beta_0,Z(j)>}}\right) d\bar{N}(n)\right]$$

is concave and uniquely maximized at $\hat{\beta}$. By the law of large numbers,

$$X_n(\beta) \to \int_0^1\left[<\beta - \beta_0, \nabla s(\beta_0,t)> - \log\frac{s(\beta,t)}{s(\beta_0,t)}s(\beta,t)\right]\alpha_t\,dt$$

in probability, uniformly in a neighborhood of β_0, and evidently this function is concave and uniquely maximized at $\beta = \beta_0$. That $\hat{\beta} \to \beta_0$ follows by the analytical property that maximizers of convergent concave functions on $\mathbf{R}^p$ converge to the maximizer of the limit. □

Given consistency of the $\hat{\beta}$, asymptotic normality of $n^{1/2}[\hat{B} - B]$ ensues from that of $n^{1/2}[\hat{\beta} - \beta_0]$ and results for the multiplicative intensity model. Therefore, the crucial central limit theorem is the next one.

Theorem 5.35. If α_0, β_0 fulfill Assumptions 5.33, then

$$n^{1/2}[\hat{\beta} - \beta_0] \xrightarrow{d} N(0,R^{-1}), \tag{5.92}$$

where R is given by (5.91).

Proof: (Sketch). In form the argument is classical. Under Assumptions 5.33 there exists a Taylor expansion of the gradient $\nabla C(\beta)$: for β near the true value β_0,

$$\nabla C(\beta) - \nabla C(\beta_0) = -J(\beta^*)(\beta - \beta_0), \tag{5.93}$$

where β^* is an intermediate value and

$$J(\beta) = \int_0^1 U(\beta,t)\,d\bar{N}_t(n), \tag{5.94}$$

with $U(\beta,t)$ the obvious empirical estimator of the term "$[\cdot]$" in (5.91): $s(\beta,t)$ is replaced by $n^{-1}\Sigma_1^n Y_t(k)e^{<\beta,Z_t(k)>}$, and so on. Since $\nabla C(\hat{\beta}) = 0$ by definition, (5.93) implies that

$$n^{-1/2}\nabla C(\beta_0) = n^{-1}J(\beta^*)(n^{1/2}[\hat{\beta} - \beta_0]),$$

so that (5.92) holds provided that

$$n^{-1/2}\nabla C(\beta_0) \xrightarrow{d} N(0,R) \tag{5.95}$$

and

$$n^{-1}J(\beta^*) \to R \tag{5.96}$$

in probability.

The latter is direct: since $\hat{\beta} \to \beta_0$ we also have $\beta^* \to \beta_0$, so it is enough to show that $n^{-1}J(\beta_0) \to R$ in probability, which holds by the weak law of large numbers and the very definition of $J(\beta_0)$. The s-ratios in (5.91) are limits of the empirical estimators replacing them in (5.94), while $n^{-1}\,d\bar{N}_t(n) \to E[dN_t] = \alpha_0(t)s(\beta_0,t)\,dt$. Hence (5.96) holds.

For each k, let $M_t(k) = N_t(k) - \int_0^t \alpha_0 Y(k)e^{<\beta_0, Z(k)>}$ be the innovation martingale associated with $N(k)$ and put $\bar{M}(n) = \Sigma_{k=1}^n M(k)$. Then

$$\begin{aligned}
n^{-1/2}\nabla C(\beta_0) &= n^{-1/2}\left[\sum_{k=1}^n \int Z(k)\,dN(k) - \int \frac{\Sigma\, Y(j)e^{<\beta_0,Z(j)>}Z(j)}{\Sigma\, Y(j)e^{<\beta_0,Z(j)>}}\,d\bar{N}(n)\right] \\
&= n^{-1/2}\left[\sum_{k=1}^n \int Z(k)\,dM(k) - \int \frac{\Sigma\, Y(j)e^{<\beta_0,Z(j)>}Z(j)}{\Sigma\, Y(j)e^{<\beta_0,Z(j)>}}\,d\bar{M}(n)\right] \\
&\sim n^{-1/2}\left[\sum_{k=1}^n \int Z(k)\,dM(k) - \int \frac{\nabla s(\beta_0,\cdot)}{s(\beta_0,\cdot)}\,d\bar{M}(n)\right] \\
&= n^{-1/2}\sum_{k=1}^n \int \left\{Z(k) - \frac{\nabla s(\beta_0,\cdot)}{s(\beta_0,\cdot)}\right\} dM(k),
\end{aligned}$$

to which we can apply Theorem 5.11 and associated calculations to derive (5.95). □

Asymptotic behavior of $B_t^* = \int_0^t (\Sigma_{k=1}^n Y(k)e^{<\beta_0, Z(k)>})^{-1}\,d\bar{N}(n)$ (note that here β_0 is fixed) is an immediate consequence of Theorem 5.16.

Proposition 5.36. If α_0, β_0 satisfy Assumptions 5.33, then $n^{1/2}[B^* - B] \xrightarrow{d} M$, where M is a continuous Gaussian martingale with variance function

$$V_t(\alpha_0,\beta_0) = \int_0^t \alpha_0(u)s(\beta_0,u)^{-1}\,du \quad \square \tag{5.97}$$

Taking account of the difference

$$\begin{aligned}
n^{1/2}[\hat{B}_t - B_t^*] &= n^{1/2}\int_0^t [(\,\Sigma Y(j)e^{<\hat{\beta},Z(j)>})^{-1} - (\,\Sigma Y(j)e^{<\beta_0,Z(j)>})^{-1}]\,d\bar{N}(n) \\
&\sim -n^{1/2}\left\langle \hat{\beta} - \beta_0, \int_0^t \frac{\Sigma\, Y(j)e^{<\beta_0,Z(j)>}Z(j)}{(\Sigma\, Y(j)e^{<\beta_0,Z(j)>})^2}\,d\bar{N}(n)\right\rangle
\end{aligned}$$

$$\sim -n^{1/2}\left\langle \hat{\beta} - \beta_0, \int_0^t \frac{\nabla s(\beta_0,u)}{s(\beta_0,u)}\, \alpha_0(u)\, du \right\rangle$$

leads to the final result of the chapter.

Theorem 5.37. If α_0, β_0 satisfy Assumptions 5.33, then the vectors $n^{1/2}[\hat{\beta} - \beta_0]$ and the processes

$$n^{1/2}[\hat{B}_t - B_t(\alpha_0)] + n^{1/2}\left\langle \hat{\beta} - \beta_0, \int_0^t \frac{\nabla s(\beta_0,u)}{s(\beta_0,u)}\, \alpha_0(u)\, du \right\rangle$$

are asymptotically independent; the latter converges to a Gaussian martingale with variance function (5.97). □

We omit the proof of asymptotic independence; the crux of the argument is that the prelimit martingales are orthogonal (Andersen and Gill, 1982), which becomes independence for the Gaussian limits.

EXERCISES

5.1. Let h be the hazard function associated with a distribution function F on $\mathbf{R}_+$. Prove that for each t

$$\int_0^t \frac{h(x)}{1 - F(x)} dx = \frac{F(t)}{1 - F(t)}$$

5.2. Let $T_1, \dots, T_k$ be independent random variables with hazard functions $h_1, \dots, h_k$, respectively. Calculate the hazard function of $T = \min\{T_1, \dots, T_k\}$.

5.3. Suppose that $\mathcal{G}$, $\mathcal{H}$ are histories with $\mathcal{G}_t \subset \mathcal{H}_t$ for each t, and that M is an $\mathcal{H}$-martingale. Prove that the process $(E[M_t|\mathcal{G}_t])$ is a $\mathcal{G}$-martingale.

5.4. Let M be a mean zero Gaussian martingale with $E[M_1^2] = V_t$.

a) Prove that M has independent increments.

b) Calculate the covariance function $E[M_s M_t]$.

c) Suppose that $V_t = \int_0^t v_s ds$. Show that M has the same distribution as the process $(\int_0^t v_s dW_s)$, where W is a Wiener process.

5.5. a) Let X be a Markov process with finite state space S and generator A. Show that for each function f on S the process $M_t = f(X_t) - \int_0^t Af(X_s)ds$ is a martingale, where $Af(i) = \Sigma_j A(Ai, j) f(j)$.

b) Let W be a Wiener process on $\mathbf{R}_+$. Prove that for each $f \in C^2(\mathbf{R}_+)$, the process $M_t = f(W_t) - (1/2)\int_0^t f''(W_s)ds$ is a martingale.

5.6. Prove that the product limit estimator of (5.20) satisfies the differential equation (5.19).

5.7. Devise an analog of (5.19) for the censored data model of Example 5.7 and from it derive the product limit estimator in (5.20a).

5.8. a) Let N be a Poisson process on $\mathbf{R}_+$ with rate λ and let $M_t = N_t - \lambda t$ be the innovation martingale. Prove directly that $(M_t^2) - \lambda t$ is a martingale.

b) Let W be a Wiener process on $\mathbf{R}_+$. Prove that the processes (W_t) and $(W_t^2 - t)$ are $\mathcal{F}^W$-martingales.

c) Relate the properties in a) and b). [*Hint*: Poisson and Wiener processes share the property of independent and stationary increments.]

5.9. Let N be a Poisson process on $\mathbf{R}_+$ with rate 1 and for each n let $M_t(n) = n^{-1/2}(N_{nt} - nt)$, $t \in [0,1]$.

a) Prove that for each n, $M.(n)$ is a martingale. [Be careful! The history must be specified for each n.]

b) Use Theorem 5.10 to prove that $M(n) \xrightarrow{d} W$ in $D[0,1]$, where W is a Wiener process.

5.10. Let N be a Poisson process on $[0,1]$ with unknown mean measure $\mu(dt) = \alpha_t\, dt$, where $\alpha \in L_+^1[0,1]$.

a) Formulate this model as a multiplicative intensity model.

b) Derive the martingale estimator of $B_t(\alpha) = \int_0^t \alpha_s\, ds$.

c) Given i.i.d. copies of N, describe asymptotic behaviour of the resultant sequence of martingale estimators.

5.11. Use Proposition 5.9 to show that (5.39) is sufficient for weak uniform consistency in the setting of Theorem 5.12.

5.12. Formulate and prove analogs of Theorems 5.12 and 5.13 for the case that the baseline stochastic intensities λ^n are *not* observable.

5.13. Let N be a point process on $[0,1]$ that under P_θ, $\theta > 0$, has stochastic intensity $\lambda_t(\theta) = \theta\lambda_t$, where λ is an observable, predictable process.

a) Calculate the martingale estimator of the process $B_t(\theta) = \int_0^t \lambda_s(\theta)ds$.

b) For each t compute the $\mathcal{F}_t^N$-maximum likelihood estimator of θ.

5.14. a) Prove that the estimators $\hat{A}$ of (5.22) are asymptotically normal.

b) Derive asymptotic properties (as $n \to \infty$) for the estimators $\hat{A}^*(i,j)$ of (5.23).

c) Calculate the asymptotic efficiency of $\hat{A}^*$ relative to $\hat{A}$.

5.15. Prove Proposition 5.22.

5.16. a) Prove Proposition 5.24a).

b) Verify (5.64).

5.17. Consider the two-sample multiplicative intensity model $\mathcal{P} = \{P_\theta : \theta \in \mathbf{R}_+^2\}$, where under P_θ each point process $N^n(i)$, $n \geqslant 1$, $i = 1,2$, has stochastic intensity $\theta(i)\lambda_t^n$. Formulate a test of the null hypothesis $H_0 : \theta(1) = \theta(2)$ and—utilizing Theorem 5.25—derive its large-sample properties. Impose assumptions on (N^n) that are reasonabale and defensible.

5.18. Consider the statistical model $\mathcal{P} = \{P_\lambda : \lambda > 0\}$, where under P_λ for each n, N^n is a Poisson process on $[0,1]$ with rate $n\lambda$. To test the null hypothesis $H_0 : \lambda = \lambda^*$ the procedure in Section 5.3 yields martingales

$$Z_t^n = \int_0^t Y_s^n(n^{-1}dN_s^n - \lambda^*\, ds),$$

where the Y^n are bounded, predictable processes, and hence the test statistics U^n in (5.60).

a) Describe explicitly the statistics U^n and their asymptotic behaviour when $Y^n \equiv 1$ for each n.

b) Propose and defend another choice of proceses Y^n, or else argue that within the given model no other choice is more sensible.

5.19. Generalize Exercise 5.18 to the case that for $\alpha \in L^1_+[0,1]$, under the probability P_α, the N^n are Poisson processes with intensity functions $n\alpha$.

5.20. Within the context of Example 5.26,

a) Derive (5.67).

b) Show that the choice $\ell(p) = p$ leads to the Wilcoxon rank statistic $n(2n+1) - 2\sum_{j=1}^{n} R_{ij}$, where R_{ij} is the rank of X_{ij} among $X_{11}, \ldots, X_{1n}, X_{21}, \ldots, X_{2n}$.

c) Show that the choice $\ell(p) \equiv 1$ leads to the Savage rank statistic

$$n - \sum_{j=1}^{n} \sum_{\nu=2n+1-R_{ij}}^{n} 1/\nu$$

5.21. Construct explicitly a point process satisfying the description in Example 5.30. [*Hint*: Use a Cox process.]

5.22. In Example 5.30, give a direct proof that the process (M_t) in (5.80) is a martingale.

5.23. In the context of Example 5.30, suppose that the disruption time T has an exponential distribution with parameter λ. Show that the stochastic differential equation for $\hat{Z}_t = P\{T \leqslant t | \mathcal{F}_t^N\}$ can be solved in closed form in each interval (T_j, T_{j+1}), where the T_j are the points of N, and calculate the solution.

5.24. Let N be a Cox process on $\mathbf{R}_+$ with directing measure $M(A) = \int_A W_t^2 dt$, where W is a Wiener process. Use techniques in Section 4 to develop a stochastic differential equation fulfilled by the MMSE state estimators $E[W_t^2 | \mathcal{F}_t^N]$.

5.25. For the Cox regression model, show that when the covariate vector β is known to be zero the estimator $\hat{H}$ of (5.86) reduces to the estimator derived in Example 5.7.

5.26. Consider the Cox regression model for independent random variables $T_1, \ldots, T_K$ under the stipulation that the common component h of the hazard functions is known.

a) Show that the likelihood function is

$$L(\beta) = \prod_{k=1}^{K} [h(T_k) \exp(<\beta, z_{T_k}(k)>) \exp(-\int_0^{T_k} h(t) e^{<\beta, z_t(k)>} dt)]$$

b) Prove that there exists a maximum likelihood estimator of β.

5.27. Suppose that $\beta \in \mathbf{R}^p$ and that the functions $z_t(1), \ldots, z_t(K)$ from $[0,1]$ into $\mathbf{R}^p$ satisfy $\sum_{k=1}^{K} e^{<\beta, z_t(k)>} = 1$ for all t. Let λ be a positive, integrable function on $[0,1]$ and let N be a Poisson process with intensity function λ. Construct independent Poisson processes $N(1), \ldots, N(K)$ by assigning a point of N at t

to $N(k)$ with probability $e^{<\beta, z_t(k)>}$, with different points allocated independently (see Definition 1.38). Calculate the conditional likelihood function of $(N(1), \ldots, N(K))$ given N and discuss its relationship to the Cox partial likelihood of (5.87).

NOTES

Broad yet detailed presentations are Gill (1980b), Jacobsen (1982), and Rebolledo (1978). The first emphasizes applications to censored data but has a good introduction to the martingale theory of inference for point processes; the second minimizes measure-theoretic technicalities by canonical constructions and algebraic arguments, and does much to reveal the true underlying structure; the third, in the French style, has influenced significantly our formulation here. Aalen (1978) (based on Aalen, 1975) is also a basic reference, while Andersen and Borgan (1985), Grigelionis (1980) and Karr (1984c) are concise surveys.

Section 5.1

The setting and stochastic intensity model (Model 5.1) are fashioned after Rebolledo (1978), while the multiplicative intensity model (Model 5.6), due to Aalen (1978), seems currently to be the best compromise between generality and usefulness. Aalen (1978), presaged in a sense by Nelson (1970), also introduced (in a special case) the martingale estimators of Definition 5.3 and established properties of the error processes that are described in Proposition 5.4. One should not underestimate Corollary 5.5; a sometimes decisive advantage of martingale estimators is that they are unbiased and that variances can be computed and estimated. Another, of course, is the ease with which the estimators themselves are calculated. Perhaps the most apparent disadvantage is that, in general, they estimate other random processes rather than deterministic parameters, but sometimes (see Theorems 5.5 and 5.16) this can be overcome. The likelihood representation in Theorem 5.2 (compare Theorem 2.31), due to Jacod (1975b), is analogous to the famous theorem of Girsanov (1960) for diffusion processes. Concerning the product limit estimators of Example 5.7, see Gill (1980b), Jacobsen (1982), Johansen (1978), and Nelson (1982); the key early paper on nonparametric estimation of distribution functions from censored data is Kaplan and Meier (1958). Additional discussion of the Markov process model in Example 5.8 can be found in Aalen (1978), Aalen and Johansen (1978), and Jacobsen (1982); although derived here by a martingale method the estimators A^* of (5.23) are classical.

Section 5.2

The elegant Proposition 5.9 is due to Lenglart (1977). Theorem 5.10 is not the most general central limit theorem extant for sequences of martingales, but suffices for our development; it is based on Rebolledo (1977a, b). Rebolledo (1980) proves the theorem under an "asymptotic rarefaction of jumps" condition weaker than the

Lindeberg condition (5.30); an analogous functional central limit theorem, but for semimartingales, is given in Liptser and Shiryayev (1980). Helland (1982) contains perhaps the most accessible proofs. Theorems 5.12 and 5.13 are related most directly to Rebolledo (1978), where one also finds L^p-consistency results for arbitrary values of $p > 1$. The choices $p = 2$ and $p = 4$ [the latter not used here, but employed in Karr (1986b) to derive a strong version of Theorem 5.18] give particularly simple error bounds. The sufficiency assertion of Proposition 5.14 is due to Aalen (1978), where there is also a proof of completeness. In Theorems 5.15 and 5.16, new here, one can estimate deterministic integrals rather than random processes. Theorem 5.18, due to Karr (1986b), is modeled after Grenander (1981), which along with German and Hwang (1982) can be consulted concerning related applications to nonparametric density estimation. Ramlau-Hansen (1983) uses kernel techniques from density estimation to convert estimators of integrals of α to estimators of α itself. Concerning finite-dimensional versions such as Theorem 5.19, see Borgan (1984). Jacobsen (1984) discusses an approach to maximum likelihood estimation in the multiplicative intensity model that expands the model to allow processes with multiple jumps. Asymptotics for Example 5.17 are treated in Breslow and Crowley (1974), Gill (1980b), and Jacobsen (1982); the heroic covariance calculations in the first of these have been reduced by the martingale method to one line.

Section 5.3

Except for the elementary Proposition 5.22 and Theorem 5.23, the latter taken from Karr (1986b), all of the section comes from Andersen et al. (1982), but see also Aalen (1978).

Section 5.4

The formulation and main result, Theorem 5.29, are patterned after Brémaud (1981); the theorem is due to Grigelionis (1973). Another expository treatment is given in Liptser and Shiryayev (1978). See also references cited in connection with the martingale representation of Theorem 2.31, the crucial result underlying Theorem 5.29. Applications in latter chapters include Theorems 6.23 and 7.29. Further analysis of the disruption problem of Example 5.30 appears in Chapter 7 and in Brémaud (1981), Galtchouk and Rozovskii (1971), Liptser and Shiryayev (1978); Example 5.31 is taken up again in Section 7.2. Extension to marked point processes is considered in Brémaud (1975b, 1981) and Hadjiev (1978).

Section 5.5

Cox (1972b) formulates the original model of survival times with covariates and shared hazard function in the functional form (5.83), and proposes the partial likelihood (5.85) for estimation of β; Cox (1975) contains further analysis and justification, while Oakes (1981) treats additional aspects. Other properties are treated in Breslow (1975), Naes (1982), Tsiatis (1981) (asymptotics), Andersen (1982) (tests for model fit), and Johansen (1983) (an extended model in which the estimators $\hat{\beta}$, $\hat{H}$ become maximum likelihood estimators). The point process version of Model 5.32 is due to Andersen and Gill (1982), with more general risk functions considered in Jacobsen (1982) and Prentice and Self (1983); however, the fundamental results remain those of Andersen and Gill (1982). The regularity conditions in Assumptions 5.33 and their consequences—Propositions 5.34 and 5.36 and Theorems 5.35 and 5.37—are investigated by Andersen and Gill (1982) in a setting more general than the i.i.d. case presented here.

Exercises

5.5. The principle underlying both parts in Dynkin's formula: see Çinlar (1975a) or Karlin and Taylor (1975) for elementary description and Blumenthal and Getoor (1968) for more advanced treatment. Additional applications are in Theorem 6.23 and Section 7.5.

5.6; 5.7. See Gill (1980a) and Jacobsen (1982).

5.9. Comparison with Exercise 4.22 emphasizes the power of martingale techniques.

5.10. See also Section 6.1, especially Theorem 6.8.

5.12. Even when $\mathcal{H} = \mathcal{F}^N$ the processes λ^n can be unobservable. Although λ_t^n is *some* function of $\{N_u^n : u \leqslant t\}$, without knowledge of the law of N^n it may be impossible to determine *which* function.

5.13. See Rebolledo (1978); different approaches often coincide in special cases.

5.14. Details are given in Jacobsen (1982) (see also Aalen and Johansen, 1978). In the martingale framework temporal homogeneity is unnecessary.

5.16. See Andersen et al. (1982).

5.20. See Andersen et al. (1982); Hajek and Sidak (1967) is a standard general reference on rank tests.

5.24. See Section 7.5, especially Theorems 7.29 and 7.30.

5.27. This is meant to illustrate the narrowness (but nonemptiness) of the conditional likelihood interpretation of the Andersen-Gill partial likelihood (5.87).

6
Inference for Poisson Processes on General Spaces

Beginning with this chapter we examine specific classes of point processes, but the underlying space E remains rather arbitrary (except in the case of renewal processes); the material is a mixture of general results and interesting special cases. Techniques range from martingale methods to strong approximation to sieves. The unifying thread running through Chapters 6 – 9 is *exploitation of stringent structural assumptions*, utilizing concepts and methods developed in Chapters 4 and 5. Justification for the restrictive assumptions is the concomitant increase in precision, specificity, and usefulness of the results.

For Poisson processes we rely principally on the properties of independent increments and conditional uniformity; the former renders many single-realization inference problems identical mathematically to multiple realization problems, while the latter allows one to make use of empirical process ideas and results. Throughout the chapter our point of view is nonparametric, with the underlying mean measure completely unknown. Parametric problems can be addressed using a combination of techniques described in Section 3.5 and *ad hoc* tools depending on the parametric model at hand, or by "general" parametric methods, as in Kutoyants (1979, 1982). Although we will not pursue such cases in detail, we do mention two by way of illustration.

First, material in Section 3.5 applies almost verbatim when the mean measure μ is known up to a scalar multiple (see Exercise 6.3 and also Chapter 9).

For a second example, suppose that the mean measure μ of a Poisson process N on $\mathbf{R}_+$ has the known functional form $\mu(ds) = \lambda s^a\, ds$, with $\lambda > 0$ and $a \geq 0$ unknown. One can apply techniques tailored to this form but based nonetheless on general principles enunciated in Chapters 4 and 5. The log-likelihood function corresponding to observation of a single realization of N over $[0,t]$ is

$$L(\lambda,a) = \int_0^t (1-\lambda s^a)\,ds + \int_0^t \log(\lambda s^a)\,dN_s$$

$$= t - \frac{\lambda t^{a+1}}{a+1} + N_t(\log\lambda) + a\int_0^t (\log s)\,dN_s,$$

which is not intractable. The likelihood equations can be solved numerically but not (it seems) in closed form. On the other hand, the martingale method yields N_t as martingale estimator of $\mu_t = \lambda t^{a+1}/(a+1)$, but is of no help in devising from this estimators of λ and a, nor for all their power do martingale limit theorems discussed in Chapter 5 provide much information about asymptotics as $t \to \infty$).

Specific structure of the model can be utilized to estimate λ and a as follows: by the law of large numbers $N_t/\mu_t = N_t/(\lambda t^{a+1}/(a+1)) \to 1$ almost surely; therefore, $N_{2t}/N_t \to 2^{a+1}$ a.s., and one can form strongly consistent estimators of a:

$$\hat{a} = \log_2 \frac{N_t}{N_{t/2}} - 1 \tag{6.1}$$

Then the estimators

$$\hat{\lambda} = \frac{N_t}{t^{\hat{a}+1}/(\hat{a}+1)} \tag{6.2}$$

are also strongly consistent.

As noted before, our viewpoint in the chapter is quite different: we assume nothing or very little about the unknown mean measure and place most results in the Model 3.1 context of data comprising i.i.d. copies of a process on a compact space. (In view of independent increments the compactness assumption is harmless: if it fails one partitions the space into compact subsets and constructs independent estimators on each, using the techniques to be described.) We shall, however, adopt a dual viewpoint: a sequence of Poisson processes N^n with measures $n\mu$ can be interpreted as partial superpositions of i.i.d. Poisson processes with mean measure μ or just as processes whose mean measures converge to infinity in a certain way. In most instances our rendition is the former, but occasionally it is useful to invoke the latter.

Contents of the sections are as follows. Section 6.1 is devoted to estimation of the mean measure and substitution of estimators of the mean measure to form empirical Laplace functionals and empirical zero-probability functionals specific to Poisson processes and different from those in Sections 4.2 and 4.3 (see Example 4.15 for a particular case). We also discuss an approach based on strong approximation of empirical processes and partial sums of i.i.d. random variables that is especially useful for analysis of set-indexed processes such as empirical zero-probability functionals; the crucial property is conditional uniformity. In Section 6.2 we treat nonparametric likelihood ratio tests of simple hypothesis $H_0 : \mu = \mu_0$; $H_1 : \mu = \mu_1$, with emphasis on structure of likelihood ratios, absolute continuity, and singularity of the probability laws of Poisson processes, and local asymptotic normality of likelihood ratios. Qualitative questions of testing (e.g., for the structure of the mean measure and for equality or ordering of mean measures in the two-sample case) are mentioned only briefly. State estimation given observation of a Poisson process restricted to a subset, because of independent increments, is equally uninteresting on general spaces as on $\mathbf{R}_+$. However, for other kinds of partial observation the state estimation problem does have substance, and three such models are presented in Section 6.3. One is p-thinned Poisson processes as in Section 4.5, another is the stochastic integral process of Examples 3.5 and 3.10 (see also Chapter 7), and the third is a Poisson process on a product space with only one component observable. For the latter, which has applications in positron emission tomography, additional structure is stipulated. Section 6.4 concerns statistical inference and state estimation for Poisson processes given integral data observations. Finally, in Section 6.5 we consider briefly how results in other sections can be employed for inference for random measures with independent increments, on the basis of the Poisson cluster representation of Theorem 1.34, with more abbreviated treatment of Poisson cluster processes. That ideas and techniques from earlier sections apply at all to more general classes of point processes and random measures is an important consequence of our working on general spaces, for then these kinds of extensions can be handled with little further difficulty.

6.1 ESTIMATION

Let E be a compact set and let $N_1, N_2, \ldots$ be point processes on E. Consider the statistical model $\mathcal{P} = \{P_\mu : \mu \in \mathbf{M}_d\}$, where with respect to P_μ the N_i are i.i.d. copies of a Poisson process N with mean measure $\mu_N = \mu$($\mathbf{M}_d$ is the set of diffuse measures on E). Our initial concern is nonparametric estimation of μ_N; thereafter we examine estimators of the Laplace functional $L_N(f) = \exp[-\int(1-e^{-f})\,d\mu]$, the zero-probability functional $z_N(A) = \exp[-\mu(A)]$ and (only in exercises, although see also Chapter 9) the covariance measure $\rho_N(A \times B) = \mu(A \cap B)$, each formed by substituting estimators $\hat{\mu}$ for μ. We present a strong approximation that produces better results for the empirical zero-probability functional and additional results in general. Discussion of estimation in the i.i.d. case concludes with application of sieve methods (see Section 5.2) to maximum likelihood estimation of an unknown intensity function. Finally, we mention some single realization results.

For each n let $N^n = \Sigma_{i=1}^n N_i$; we propose as estimators of μ the scaled partial superpositions

$$\hat{\mu} = \frac{N^n}{n}, \tag{6.3}$$

whose asymptotic properties follow from Propositions 4.28 – 4.31. Evidently, $\hat{\mu}$ is the method of moments estimator of μ given $N_1, \dots, N_n$; before assessing its role as maximum likelihood estimator we verify the plausible property that inference for μ, given observation of $N_1, \dots, N_n$, can be based solely on N^n.

Lemma 6.1. The statistic N^n is sufficient for μ given the data $N_1, \dots, N_n$.

Proof: Let μ be fixed. We may suppose that $N_1, \dots, N_n$ have been constructed as follows. Let $N^n = \Sigma \varepsilon_{Y_k}$ be Poisson with mean $n\mu$ and let U_k be i.i.d. random varibles, independent of N^n, with $P\{U_k = i\} = 1/n$, $i = 1, \dots, n$, and put

$$N_i = \sum_k 1(U_k = i)\varepsilon_{Y_k} \tag{6.4}$$

Then $N_1, \dots, N_n$ are independent Poisson processes (Exercise 1.16) with mean measure μ, by construction $N^n = \Sigma_{i=1}^n N_i$, and by the very nature of (6.4), the conditional distribution of $(N_1, \dots, N_n)$ given N^n is independent of μ. □

The construction in the proof highlights the dual role of N^n, which—at least for results concerning convergence in distribution—can be thought of as a superposition or simply as a single realization of a Poisson process with mean measure $n\mu$, in the same way that processes in Theorems 5.12 and 5.13 may but need not arise as superpositions of i.i.d. components. The crucial asymptotic property in both settings is that "intensities" converge to infinity. Another consequence is that likelihoods (when they exist) may be interpreted as arising from a single realization of one process.

Although there is no direct maximum likelihood interpretation of the estimators $\hat{\mu}$, a rather precise indirect interpretation can be provided in the dominated case; see Theorem 6.8 and related discussion. This interpretation is based on sieves in L^1, but we may also use sieves in $\mathbf{M}_d$ to interpret the $\hat{\mu}$ as maximum likelihood estimators. Assume that we have one observation of a Poisson process N with unknown mean measure, that $E = \Sigma_{j=1}^m A_j$ is a partition and that for each j, η_j is a diffuse probability measure on A_j. If the unknown mean measure is restricted to belong to the set of measures on E of the form $\eta = \Sigma_{j=1}^m a_j \eta_j$ $(a_j \geq 0)$, then the log-likelihood function

$$\begin{aligned} L(\eta) &= L(a_1, \ldots, a_m) \\ &= -\eta(E) + \sum_{j=1}^m N(A_j) \log \eta(A_j) - \sum_{j=1}^m \log(N(A_j)!) \\ &= -\sum_{j=1}^m a_j + \sum_{j=1}^m N(A_j) \log a_j - \sum_{j=1}^m \log(N(A_j)!) \end{aligned} \tag{6.5}$$

is maximized for $a_j = N(A_j)$, $j = 1, \ldots, m$, yielding the estimator

$$\hat{\eta} = \sum_{j=1}^m N(A_j)\eta_j, \tag{6.6}$$

which, unlike the purely atomic estimator $\hat{\mu} = N$ of (6.3), belongs to the space $\mathbf{M}_d$ containing the unknown mean measure. If there are sufficiently many and small A_j that with high probability each $N(A_j)$ is either zero or 1, then (with $N = \Sigma_i \varepsilon_{X_i}$)

$$\hat{\eta} = \sum_{j=1}^m 1(N(A_j) = 1)\eta_j = \sum_i \sum_{j=1}^m 1(X_i \in A_j)\eta_j \cong \sum_i \varepsilon_{X_i} = \hat{\mu}$$

Thus we obtain a sieve justification of $\hat{\mu}$ as maximum likelihood estimator.

We will not pursue the estimators (6.6), even though there is interest attached to them because they are diffuse. In Theorem 6.8 we examine the related situation of estimating a mean measure of the form $\mu(dx) = \alpha(x)\mu_0(dx)$, with μ_0 known and $\alpha \in L^1(d\mu_0)$ unknown, using histogram sieves.

As stated before, limit behavior of the $\hat{\mu}$ has been established already in Section 4.6; for completeness we recapitulate it.

Proposition 6.2. For each $\mu \in \mathbf{M}_d$, under P_μ

a) For each compact subset $\mathcal{K}$ of $C_+(E)$, almost surely

$$\sup\{|\hat{\mu}(f) - \mu(f)| : f \in \mathcal{K}\} \to 0 \tag{6.7}$$

b) The error processes $n^{1/2}[\hat{\mu} - \mu]$, as signed random measures on E, converge in distribution to a Gaussian random measure with covariance function

$$R(f,g) = \mu(fg) \tag{6.8}$$

c) Let $\mathcal{K}$ be a compact subset of $C_+(E)$ having finite metric entropy with respect to the uniform metric. Then the error processes $\{n^{1/2}[\hat{\mu}(f) - \mu(f)] : f \in \mathcal{K}\}$, as random elements of $C(\mathcal{K})$, converge in distribution to a Gaussian process with covariance function (6.8).

d) Almost surely the set $\{n^{1/2}[\hat{\mu} - \mu]/(2 \log \log n)^{1/2}\}$ is relatively compact in the function space $C(E)^*$ with limit set $\{\eta : |\eta(f)| \le \mu(f^2)^{1/2} \text{ for all } f\}$. □

For Poisson processes we define by substitution the *empirical Laplace functionals*

$$\hat{L}(f) = \exp[-\hat{\mu}(1 - e^{-f})], \tag{6.9}$$

with $\hat{\mu}$ given by (6.3). These have the advantage of being tailored to the Poisson model, but the disadvantage of lacking robustness. If the Poisson assumption were suspect, the general empirical Laplace functionals of (4.16) would be preferred. In fact, as Example 4.15 illustrates, for ordinary Poisson processes on $\mathbf{R}_+$ the latter may even be more efficient. Although this does not happen for the estimators (6.9), which are always more efficient than those of (4.16), there may be other parametric settings in which general estimators are (sometimes) preferred to parametrically derived estimators.

We now summarize properties of the empirical Laplace functionals. Let $L_\mu(f) = \exp[-\mu(1 - e^{-f})]$ be the Laplace functional of the N_i under P_μ.

Proposition 6.3. Let $\hat{L}$ be given by (6.9). Then for each μ

a) For each equicontinuous subset $\mathcal{K}$ of $C_+(E)$, almost surely

$$\sup\{|\hat{L}(f) - L_\mu(f)| : f \in \mathcal{K}\} \to 0 \tag{6.10}$$

b) If $\mathcal{K}$ is a compact subset of $C_+(E)$ with finite metric entropy, then the error processes $\{n^{1/2}[\hat{L}(f) - L_\mu(f)] : f \in \mathcal{K}\}$ converge in distribution, as random elements of $C(\mathcal{K})$, to a Gaussian process with covariance function

$$R(f,g) = \mu((1 - e^{-f})(1 - e^{-g})) \exp[-\mu(1 - e^{-f}) - \mu(1 - e^{-g})] \tag{6.11}$$

Proof: a) Equicontinuity of $\mathcal{K}$ implies that of $K' = \{(1 - e^{-f}) : f \in \mathcal{K}\}$, and $\mathcal{K}'$ is uniformly bounded; since $\sup_{\mathcal{K}} |\hat{L}(f) - L_\mu(f)| \le \sup_{\mathcal{K}'} |\hat{\mu}(g) - \mu(g)|$, (6.10) follows from (6.7).

b) For fixed $f \in \mathcal{K}$

$$\begin{aligned} n^{1/2}[\hat{L}(f) - L_\mu(f)] &= n^{1/2}(\exp[-\hat{\mu}(1 - e^{-f})] - \exp[-\mu(1 - e^{-f})]) \\ &= n^{1/2}[\hat{\mu}(1 - e^{-f}) - \mu(1 - e^{-f})]e^{-x^*}, \end{aligned}$$

where x^* lies between $\hat{\mu}(1-e^{-f})$ and $\mu(1-e^{-f})$. In view of a), therefore,

$$\sup_{f\in\mathcal{K}} |n^{1/2}[\hat{L}(f)-L_\mu(f)] - n^{1/2}[\hat{\mu}(1-e^{-f})-\mu(1-e^{-f})]e^{-\mu(1-e^{-f})}|$$

converges to zero in P_μ-probability, and the same reasoning used to show that a continuous image of a compact set is compact confirms that $\mathcal{K}'$ has finite metric entropy. Consequently, Proposition 6.3b) ensues from Proposition 6.2c), with (6.11) obtained from (6.7) by straightforward computation. □

The general empirical Laplace functionals of (4.16), that is

$$\hat{L}^*(f) = n^{-1}\sum_{i=1}^{n} e^{-N_i(f)},$$

by Theorem 4.13 have limit covariance function

$$R^*(f,g) = \exp[-\mu(1-e^{-(f+g)})] - \exp[-\mu(1-e^{-f})]\exp[-\mu(1-e^{-g})]$$

In particular, $n^{1/2}[\hat{L}^*(f)-L_\mu(f)]$ has asymptotic variance

$$R^*(f,f) = \exp[-\mu(1-e^{-2f})] - \exp[-2\mu(1-e^{-f})],$$

while by (6.11), that of $n^{1/2}[\hat{L}(f)-L_\mu(f)]$ is

$$R(f,f) = \mu((1-e^{-f})^2)\exp[-2\mu(1-e^{-f})],$$

so that the asymptotic efficiency of $\hat{L}(f)$ relative to $\hat{L}^*(f)$ is

$$e(\hat{L}^*(f),\hat{L}(f);\mu) = \frac{R^*(f,f)}{R(f,f)} = \frac{\exp[\mu(1-e^{-f})^2]-1}{\mu((1-e^{-f})^2)},$$

which always exceeds 1. Therefore, $\hat{L}$ is preferred to $\hat{L}^*$. (No contradiction with Example 4.15 exists because the specialized Laplace functional there is not $\hat{L}$.)

Based on the relationship $z_N(A) = \exp[-\mu(A)]$ we define *empirical zero-probability functionals*

$$\hat{z}(A) = \exp[-\hat{\mu}(A)], \tag{6.12}$$

which, while nonparametric within the Poisson model, differ nevertheless from general empirical zero-probability functionals of Section 4.3. So far we have viewed the estimators $\hat{\mu}$ as indexed by functions rather than sets, but obviously this will not do for analysis of zero-probability functionals. Results from Chapter 4 concerning asymptotics of set-indexed empirical processes do not apply directly, although we shall discuss presently how they can be made to apply. Before doing so, however, we record elementary results.

Proposition 6.4. For each μ let $z_\mu(A) = \exp[-\mu(A)] = P_\mu\{N(A) = 0\}$. Then with $\hat{z}$ given by (6.12),

a) For every A, $\hat{z}(A) \to z_\mu(A)$ almost surely.

b) For $A_1, \ldots, A_k \in \mathscr{E}$,

$$n^{1/2}[(\hat{z}(A_1), \ldots, \hat{z}(A_k)) - (z_\mu(A_1), \ldots, z_\mu(A_k))] \xrightarrow{d} N(0,R)$$

on $\mathbf{R}^k$, where

$$R(i,j) = \mu(A_i \cap A_j)e^{-\mu(A_i)} \quad \square \tag{6.13}$$

Independence of $n^{1/2}[\hat{z}(A) - z_\mu(A)]$ and $n^{1/2}[\hat{z}(B) - z_\mu(B)]$ for disjoint A and B carries over in (6.13). Moreover, Proposition 6.4 implies that for any sequence of error processes $\{n^{1/2}[\hat{z}(A) - z_\mu(A)] : A \in \mathscr{C}\}$ converging in distribution the limit covariance function must be a restriction of

$$R(A,B) = \mu(A \cap B)e^{-[\mu(A)+\mu(B)]} \tag{6.14}$$

We next explore more general version of Proposition 6.4b) based on strong approximation (see Section 4.6); the starting point is a strong approximation for Poisson processes on [0,1] related to but different from that in Theorem 4.33. The key idea is to invoke conditional uniformity by viewing a Poisson process N^n on [0,1] with mean $\mu_n(dx) = n\,dx$ as (see Exercise 1.1)

$$N^n = \sum_{i=1}^{S_n} \varepsilon_{X_i}, \tag{6.15}$$

where the X_i are i.i.d. random variables, uniformly distributed on [0,1], and $S_n = Y_1 + \cdots + Y_n$, with the Y_i i.i.d. random variables independent of the X_i, each having Poisson distribution with mean 1. That is, (6.15) exhibits the Poisson process N^n as a *uniform empirical process with random sample size*. By approximating separately the empirical error processes

$$\alpha_k(x) = k^{1/2}\left[k^{-1}\sum_{i=1}^{k} 1(X_i \le x) - x\right] \tag{6.16}$$

by a Kiefer process (a particular two-dimensional Gaussian process) and the partial sums $n^{-1/2}(S_n - n)$ by a Wiener process, we construct a strong approximation for the processes $\{n^{-1/2}(N^n(x) - nx) : 0 \le x \le 1\}$. Not only is the result of intrinsic interest, but also one can deduce from it a variety of consequences, including properties of empirical zero-probability functionals.

Theorem 6.5. There exists a probability space on which are defined a sequence (N^n) of Poisson processes on [0,1] represented in the form (6.15), a

Kiefer process K, and a Wiener process W independent of K such that

$$\sup_x |n^{-1/2}[N^n(x) - nx] - n^{-1/2}K(x,n) - n^{-1/2}W(n)x| = O(n^{-1/4}(\log n)^2) \quad (6.17)$$

as $n \to \infty$, almost surely.

A *Kiefer process* $\{K(x,k): 0 \le x \le 1, k \in \mathbf{N}\}$ is constructed from a two-dimensional Wiener process $\widetilde{W}(x,y)$ by the transformation

$$K(x,k) = \widetilde{W}(x,k) - x\widetilde{W}(1,k) \quad (6.18)$$

For fixed k, $B_k(x) = k^{-1/2}K(x,k)$ is a Brownian bridge (Section 4.6); the processes $\widetilde{B}_k(x) = K(x,k) - K(x,k-1)$ are independent Brownian bridges. The Kiefer process is a mean zero Gaussian process with covariance

$$E[K(x_1,k_1)K(x_2,k_2)] = (x_1x_2 - x_1 \wedge x_2)(k_1 \wedge k_2);$$

(see Csörgö and Révész, 1981, for additional properties).

Proof of Theorem 6.5: Let (α_k) be the sequence of empirical error processes given by (6.16). By the Komlós et al. (1975, 1976) strong approximation theorem for empirical processes (see also Csörgö and Révész, 1981, Theorem 4.4.6), there is a probability space supporting the sequence (X_i) and a Kiefer process K such that

$$\sup_x |k^{1/2}\alpha_k(x) - K(x,k)| = O((\log k)^2) \quad (6.19)$$

almost surely. Furthermore, by the Komlós et al. (1975, 1976) strong approximation for partial sum processes (cited in Theorem 4.33) there is a probability space on which are defined a Wiener process W and Poisson-distribued summands Y_i such that

$$|S_n - n - W(n)| = O(\log n) \quad (6.20)$$

almost surely. Independence can be effected by construction, in which case (6.15) defines a Poisson process N^n with rate n. Now let $F_k(x) = k^{1/2}\alpha_k(x) + x$ denote the sequence of empirical distribution functions associated with (X_i). Then with error bounds uniform in x,

$$\begin{aligned} n^{-1/2}(N^n(x) - nx) &= n^{-1/2}(S_nF_{S_n}(x) - nx) \\ &= n^{-1/2}(S_nF_{S_n}(x) - S_nx) + n^{-1/2}(S_n - n)x \\ &= n^{-1/2}S_n(F_{S_n}(x) - x) + n^{-1/2}W(n)x \end{aligned}$$

[within $n^{-1/2}$ log n, by (6.20)]

$$\begin{aligned} &= n^{-1/2}S_n{}^{1/2}\alpha_{S_n}(x) + n^{-1/2}W(n)x \\ &= n^{-1/2}K(x,S_n) + n^{-1/2}W(n)x, \end{aligned}$$

within $S_n^{-1/2}(\log S_n)^2$, by (6.19).

But $S_n^{-1/2}(\log S_n)^2 = n^{-1/2}(\log n)^2$ within the error bound in (6.17), so it remains to analyze the term $n^{-1/2}[K(x,n) - K(x,S_n)]$. We have

$$\begin{aligned}\sup_x n^{-1/2}|K(x,n) - K(x,S_n)| &\\ \leq \sup_x n^{-1/2}|\widetilde{W}(x,n) - \widetilde{W}(x,S_n)| + \sup_x n^{-1/2}|x\widetilde{W}(1,n) - x\widetilde{W}(1,S_n)| &\\ \leq 2\sup_x n^{-1/2}|\widetilde{W}(x,n) - \widetilde{W}(x,S_n)| = O(n^{-1/4}(\log n)^2)&\end{aligned}$$

by (6.18) and Theorem 1.2.1 of Csörgö and Révész (1981) on increments of a Wiener process, and this confirms (6.17). □

Therefore, asymptotic behavior of the processes $\{(n^{-1/2}(N^n(x) - nx): 0 \leq x \leq 1\}$ is that of the Gaussian process

$$G(x) = B(x) + Zx, \tag{6.21}$$

where B is a Brownian bridge, Z has distribution $N(0,1)$, and B, Z are independent (see Theorem 4.18).

We next discuss applications to empirical zero-probability functionals, for which Theorem 6.5 must be extended to more general Poisson processes. First, if N^n has rate λn rather than n, then Theorem 6.5 applies verbatim to yield

$$\{n^{-1/2}(N^n(x) - \lambda nx) : 0 \leqslant x \leqslant 1\} \xrightarrow{d} \{\lambda G(x) : 0 \leq x \leq 1\}, \tag{6.22}$$

where G is as in (6.21). To obtain a result for zero-probability functionals we put $z_\lambda(x) = e^{-\lambda x} = P_\lambda\{N(x) = 0\}$ and [from (6.12)] $\hat{z}(x) = \exp[-\hat{\mu}(x)]$.

Corollary 6.6. For each λ, under P_λ

$$\{n^{1/2}[\hat{z}(x) - z_\lambda(x)] : 0 \leq x \leq 1\} \xrightarrow{d} \{\lambda e^{-\lambda x}G(x) : 0 \leqslant x \leqslant 1\} \quad \square \tag{6.23}$$

To generalize further we employ a device appearing in Kallenberg (1983) that can be viewed as a generalized quantile transformation. Suppose that μ is a diffuse measure on E, assumed without loss of generality to satisfy $\mu(E) = 1$. Then there exists an invertible mapping f_μ (not unique) of (a measure 1 subset of) [0,1] into E such that $\mu = \lambda f_\mu^{-1}$, where λ denotes Lebesgue measure. Hence a Poisson process N^n on E with mean measure $n\mu$ admits the representation

$$N^n = \sum_{i=1}^{S_n} \varepsilon_{f_\mu(X_i)}, \tag{6.24}$$

where (X_i) and (S_n) are *exactly* as in (6.16).

Proposition 6..7 Let N^n be given by (6.24), where f_μ satisfies $\mu = \lambda f_\mu^{-1}$, and let $(A_n)_{0 \leqslant x \leqslant 1}$ be an increasing family of subsets of E. Then with $\hat{z}$ given by (6.12), $z_\mu(A) = \exp[-\mu(A)]$ and $a(x) = \mu(A_x)$,

$$\{n^{1/2}[\hat{z}(A_x) - z(A_x)] : 0 \le x \le 1\} \xrightarrow{d} \{e^{-a(x)}G(a(x)) : 0 \le x \le 1\}$$

under P_μ, where G is the Gaussian process given by (6.21).

Proof: We can assume that $f_\mu^{-1}(A_x) = [0,a(x)]$. Let $\tilde{N}^n = \Sigma_{i=1}^{S_n} \varepsilon_{X_i}$; then $N^n = \tilde{N}^n f_\mu^{-1}$ (see Section 1.5) and in particular $\hat{\mu}(A_x) = n^{-1}\tilde{N}^n(a(x))$. Therefore, by Theorem 6.5,

$$\begin{aligned} n^{1/2}[\hat{z}(A_x) - z_\mu(A_x)] &= n^{1/2}(\exp[-n^{-1}\tilde{N}^n(a(x))] - \exp[-a(x)]) \\ &\cong e^{-a(x)} n^{-1/2}[\tilde{N}^n(a(x)) - na(x)] \\ &\xrightarrow{d} e^{-a(x)}G(a(x)) \quad \square \end{aligned}$$

In some instances an unknown mean measure is stipulated to be dominated by a known measure μ_0, but not further. For example, a Poisson process N on $\mathbf{R}_+$ may be known on physical grounds to admit an *intensity function* $\alpha_t = dE[N_t]/dt$, but α itself may be unknown. By sieve methods analogous to those in Theorem 5.18—but with histogram sieves—we can estimate by maximum likelihood the unknown derivative $\alpha = d\mu_N/d\mu_0$ for the model $\mathcal{P} = \{P_\alpha : \alpha \in L_+^1(d\mu_0)\}$, even on a general space E. Suppose that under each P_α, $N_1, N_2, \ldots$ are i.i.d. copies of a Poisson process N on E with mean measure $\mu(dx) = \alpha(x)\mu_0(dx)$, where $\mu_0 \in \mathbf{M}_d$ is *known*. With $N^n = \Sigma_{i=1}^n \mathrm{N}_i$, which is a sufficient statistic for α given the data $N_1, \ldots, N_n$, and with P the probability under which the N_i are i.i.d. Poisson processes with mean measure μ_0, Proposition 6.11 (in Section 6.2) yields the log-likelihood function

$$L_n(\alpha) = \log \frac{dP_\alpha}{dP} = n\int_E (1 - \alpha)\, d\mu_0 + \int_E \log \alpha \, dN^n \tag{6.25}$$

As with the multiplicative intensity model (Sections 5.1 and 5.2), of which this model is a special case when E is a compact interval in $\mathbf{R}_+$, L_n is unbounded above as a function of α, necessitating an indirect approach to maximum likelihood estimation.

Our approach utilizes histogram sieves (see Grenander, 1981). For each $m \in \mathbf{N}$ let $E = \Sigma_{j=1}^{\ell_n} A_{mj}$ be a partition of E into sets A_{mj} with $\mu_0(A_{mj}) > 0$, and assume that $\max_j \mu_0(A_{mj}) \to 0$. For each m let $I(m)$ be the *histogram sieve* of functions $\alpha \in L_+^1(\mu_0)$ that are constant over each partition set A_{mj}. Then by computations identical to those leading to (6.6), the maximum likelihood estimator $\hat{\alpha}(n,m)$ relative to sample size n and the set $I(m)$ is

$$\hat{\alpha}(n,m) = \sum_{j=1}^{\ell_m} \frac{\hat{\mu}(A_{mj})}{\mu_0(A_{mj})} 1_{A_{mj}}, \tag{6.26}$$

where $\hat{\mu}$ is given by (6.3). The appropriate sense for estimators $\hat{\alpha} = \hat{\alpha}(n,m_n)$ defined momentarily to converge to α is with respect to the norm on $L^1(\mu_0)$, which we denote by $\|\cdot\|_1$. For suitable choice of (m_n) convergence takes place almost surely.

Theorem 6.8. If $m = m_n$ is chosen in such a manner that with $\tilde{\ell}_n = \ell_{m_n}$,

$$\sum_{n=1}^{\infty} \frac{\tilde{\ell}_n^4}{n^2} < \infty \tag{6.27}$$

then for $\hat{\alpha} = \hat{\alpha}(n,m_n)$, $\|\hat{\alpha} - \alpha\|_1 \rightarrow 0$ almost surely with respect to P_α.

Proof: Let n and m both be variable; then by (6.26)

$$\begin{aligned}
|\hat{\alpha}(n,m) - \alpha\|_1 &= \sum_{j=1}^{\ell m} \int_{A_{mj}} \left| \frac{\hat{\mu}(A_{mj})}{\mu_0(A_{mj})} - \alpha(x) \right| \mu_0(dx) \\
&\leq \sum_{j=1}^{\ell m} \int_{A_{mj}} \left| \frac{\hat{\mu}(A_{mj})}{\mu_0(A_{mj})} - \frac{\mu_0(\alpha;A_{mj})}{\mu_0(A_{mj})} \right| d\mu_0 \\
&\quad + \sum_{j=1}^{\ell m} \int_{A_{mj}} \left| \frac{\mu_0(\alpha;A_{mj})}{\mu_0(A_{mj})} - \alpha(x) \right| \mu_0(dx) \\
&\leq \sum_{j=1}^{\ell m} |\hat{\mu}(A_{mj}) - E_\alpha[N(A_{mj})]| \\
&\quad + \sum_{j=1}^{\ell m} \int_{A_{mj}} \left| \frac{\mu_0(\alpha;A_{mj})}{\mu_0(A_{mj})} - \alpha(x) \right| \mu_0(dx),
\end{aligned}$$

where $\mu_0(\alpha;A_{mj}) = \int_{A_{mj}} \alpha \, d\mu_0$. That the second, nonrandom term converges to zero provided only that $m \rightarrow \infty$ is analytical and can be shown by a variety of techniques (see, e.g., Grenander, 1981, pp. 419 – 420). We deal with the first by methods analogous to those in Theorems 4.19 and 4.22. For $\varepsilon > 0$,

$$\begin{aligned}
&P_\alpha\left\{\sum_{j=1}^{\ell m} |\hat{\mu}(A_{mj}) - E_\alpha[N(A_{mj})]| > \varepsilon\right\} \\
&\qquad \leqslant \varepsilon^{-4} E_\alpha\left[\left\{\sum_{j=1}^{\ell m} |\hat{\mu}(A_{mj}) - E_\alpha[N(A_{mj})]|\right\}^4\right]
\end{aligned}$$

With n and m fixed the random variables $\hat{\mu}(A_{mj}) - E_\alpha[N(A_{mj})]$ are independent in j and by brute-force expansion of the fourth power of the summation followed by repeated application of the Cauchy-Schwarz inequality, we obtain

$$P_\alpha\left\{\sum_{j=1}^{\ell m} |\hat{\mu}(A_{mj}) - E_\alpha[N(A_{mj})]| > \varepsilon\right\} = O(\varepsilon^{-4} n^{-2} \ell_m^4);$$

the dominant term comes from the order of ℓ_m^4 summands with all indices distinct. Thus (6.27) suffices to give

$$\sum_n P_\alpha\left\{\sum_{j=1}^{\tilde{\ell}_n} |\hat{\mu}(A_{mj}) - E_\alpha[N(A_{mj})]| > \varepsilon\right\} < \infty$$

for every $\varepsilon > 0$, which completes the proof. □

In general, little useful can be said about estimation of the mean measure of a Poisson process in the Model 3.2 context of observation of a single realization over a noncompact set E; one can do arbitrarily poorly if the mean measure is particularly ill-chosen. However, good results obtain for *stationary* Poisson processes. Suppose that E is a locally compact Abelian group with infinite, diffuse Haar measure μ_0, and for each $\nu > 0$ let N be a stationary Poisson process (Definition 1.58, and Example 1.59) under the probability P_ν, with intensity ν. It is natural to estimate ν with estimators $\hat{\nu} = N(B)/\mu_0(B)$, where $B \uparrow E$, but if $\mu_0(B) \uparrow \infty$ in a completely arbitrary manner, consistency and asymptotic normality fail. For $B_n = \Sigma_{i=1}^n A_i$, where the A_i are disjoint sets with $\mu_0(A_i)$ the same for all i, the independent and stationary increments properties of N yield at once the following result.

Proposition 6.9. Let $A_1, A_2, \ldots$ be disjoint subsets of E with $0 < a = \mu_0(A_i) < \infty$ independently of i, and for each n *let* $\hat{\nu} = N(\Sigma_{i=1}^n A_i)/na$. Then for each ν.

a) $\hat{\nu} \to \nu$ almost surely;
b) $n^{1/2}[\hat{\nu} - \nu] \xrightarrow{d} N(0, \nu a)$. □

The simplest choice of the A_i is as translates of a fixed set A_0. More generally, Theorem 1.68 and the final result of the section enable one to construct a plethora of consistent estimators of ν in the case that E is a Euclidean space (see Chapter 9 for specific choices).

Proposition 6.10. Let N be a stationary Poisson process on $\mathbf{R}^d$. Then N is ergodic.

Proof: Assuming that N is defined on the probability space $(\Omega, \mathcal{F}, P)$, the Palm distribution $\tilde{P}$ (Definition 1.62) of N, by Example 1.54, is just the P-distribution of $N + \varepsilon_0$; here we have assumed for simplicity that $\nu = 1$. Therefore, P and $\tilde{P}$ agree on the invariant σ-algebra $\mathcal{I}$ (Definition 1.67) and hence by Theorem 1.68a) and (1.80), $E[N(f)|\mathcal{I}] = \mu_0(f)$ for all f, which proves that $\mathcal{I}$ is P-degenerate. □

This concludes our general treatment of estimation, but not our consideration of the topic. Additional aspects appear in Section 6.3–6.5.

6.2 EQUIVALENCE AND SINGULARITY; HYPOTHESIS TESTING

This section is more about properties of likelihood ratios and probability distributions of Poisson processes than about hypothesis testing per se, but even less overtly statistical portions can be interpreted as addressing by means of likelihood ratios the simple-vs.-simple hypothesis test

$$\begin{aligned} H_0 &: \mu_N = \mu_0 \\ H_1 &: \mu_N = \mu_1 \end{aligned} \tag{6.28}$$

for a Poisson process N on a general space E with unknown mean measure μ_N. As in Section 6.1 the setting is nonparametric, but here we accentuate the single-realization case, with asymptotics pertaining to behavior of Poisson process likelihood ratios and probability laws associated with bounded sets increasing to a noncompact space E. In particular we examine equivalence and singularity of probability laws P_{μ_0}, P_{μ_1} engendering Poisson processes with mean measures μ_0, μ_1, respectively, especially in the case that μ_0, μ_1 are equivalent: in this case P_{μ_0} and P_{μ_1} must be either equivalent or singular; intermediate cases cannot arise. This dichotomy theorem (Theorem 6.12; see also Proposition 3.24) has analogs for Gaussian processes and diffusions (Liptser and Shiryayev, 1978) and is a main result of the section, although Proposition 6.11, which derives explicitly the form of likelihood ratios, is probably the most useful. General discussion concludes with a local asymptotic normality theorem for Poisson process likelihood ratios over bounded sets.

To the extent that qualitative tests are difficult for ordinary Poisson processes on $\mathbf{R}_+$ they verge on hopeless for Poisson processes on general spaces, especially in nonparametric contexts. We end the section with brief consideration of these kinds of questions, but without pretense of covering what little is known, let alone desirable.

Before proceeding we recall some terminology from measure theory. Measures ν,η (assumed σ-finite) on a measurable space $(E,\mathscr{E})$ are *equivalent* if they are mutually absolutely continuous ($\nu \ll \eta$ and $\eta \ll \nu$), in which case the Radon-Nikodym derivatives satisfy $d\nu/d\eta = (d\eta/d\nu)^{-1}$. By contrast, ν and η are *singular*, denoted by $\nu \perp \eta$, if there is $D \in \mathscr{E}$ such that $\nu(D) = \eta(D^c) = 0$ (i.e., singular measures are concentrated on disjoint sets). For each ν and η there exists the *Lebesgue decomposition* of ν with respect to η: $\nu = \nu_a + \nu_s$, where $\nu_a \ll \eta$ and $\nu_s \perp \eta$. It is important to keep in mind that these properties depend on the σ-algebra $\mathscr{E}$; in general, for larger $\mathscr{E}$ absolute continuity is less likely to hold and singularity more likely.

Much of the section is devoted to analysis of implications of relations between mean measures of Poisson processes concerning probability laws of the processes. To this end, let E be noncompact but locally compact and let $(\Omega,\mathcal{F})$ be a measurable space supporting a (simple) point process N on E; we consider the statistical model $\mathcal{P} = \{P_\mu : \mu \in \mathbf{M}_d\}$, where under P_μ, N is a Poisson process with mean measure μ. For construction of likelihood ratio tests a principal interest is equivalence or singularity of probability measures P_{μ_0}, P_{μ_1} corresponding to the hypotheses (6.28) on σ-algebras $\mathcal{F}^N(B)$, where B is bounded, as well as on $\mathcal{F}^N = \mathcal{F}^N(E)$, as a function of equivalence or singularity of the mean measures μ_0, μ_1. When $P_{\mu_0} \sim P_{\mu_1}$ there exists a positive, finite likelihood ratio, and hence a log-likelihood function usable for testing the hypotheses (6.28). At the other extreme, if $P_{\mu_0} \perp P_{\mu_1}$, then the test is rendered trivial: every realization of N leads without error to μ_0 or μ_1 as the "true" mean measure. Singularity of P_{μ_0} and P_{μ_1} cannot occur unless at least one of $\mu_0(E)$ and $\mu_1(E)$ is infinite; see Exercise 6.9.

Within this setting the most interesting case mathematically and physically is that $\mu_0 \sim \mu_1$. We commence with a criterion for equivalence of P_{μ_0}, P_{μ_1} on $\mathcal{F}^N(B)$ when B is bounded and $\mu_0 \sim \mu_1$ on B; thereafter we develop equivalence criteria for the unbounded case, including the dichotomy theorem.

Proposition 6.11. Suppose that B is bounded and that $\mu_1 \ll \mu_0$ on the σ-algebra $\mathcal{E} \cap B$. Then $P_{\mu_1} \ll P_{\mu_0}$ on $\mathcal{F}^N(B)$, with

$$\frac{dP_{\mu_1}}{dP_{\mu_0}} = \exp\left[\int_B \left(1 - \frac{d\mu_1}{d\mu_0}\right) d\mu_0 + \int_B \log \frac{d\mu_1}{d\mu_0}\, dN\right] \tag{6.29}$$

Proof: For $f \geq 0$ vanishing off B,

$$E_{\mu_0}\left[e^{-N(f)} \exp\left[\int_B \left(1 - \frac{d\mu_1}{d\mu_0}\right) d\mu_0 + \int_B \log \frac{d\mu_1}{d\mu_0}\, dN\right]\right]$$

$$= \exp\left[\int_B \left(1 - \frac{d\mu_1}{d\mu_0}\right) d\mu_0\right] E_{\mu_0}\left[\exp\left[-\int_B \left(f - \log\frac{d\mu_1}{d\mu_0}\right) dN\right]\right]$$

$$= \exp\left[\int_B \left(1 - \frac{d\mu_1}{d\mu_0}\right) d\mu_0\right] \exp\left[-\int_B \left(1 - e^{-f}\frac{d\mu_1}{d\mu_0}\right) d\mu_0\right]$$

$$= \exp\left[-\int_B (1 - e^{-f})\, d\mu_1\right] = E_{\mu_1}[e^{-N(f)}];$$

hence (6.29) holds by Theorem 1.12. □

Note the resemblance of (6.29) to the likelihood function (5.50) for multiplicative intensity processes.

By interchanging μ_0 and μ_1 we establish a criterion for equivalence.

Corollary 6.12. If $\mu_0 \sim \mu_1$ on B, then $P_{\mu_0} \sim P_{\mu_1}$ on $\mathcal{F}^N(B)$ and (6.29) holds. □

In the setting of Proposition 6.11, Neyman-Pearson critical regions have the form $\Omega_c = \{\int_B \log(d\mu_1/d\mu_0)\, dN \geq \delta\}$, with δ a constant depending on B, μ_0, μ_1, and the power bound under H_0 (probability of type I error). The null distribution of the test statistic is difficult to express in closed form; however, its Laplace transform is computable and together for example with Chebyshev's inequality can be used to construct approximate critical regions. Alternatively, if $\mu_0(B)$ is large then the test statistic—by independent increments—is approximately normally distributed, yielding another way to derive approximate critical regions (see M. Brown, 1972, for details).

Remaining a moment longer with the hypotheses (6.28) and observations $\mathcal{F}^N(B)$, even when μ_0, μ_1 are not equivalent, one can formulate a likelihood ratio test by the following rather standard device. Both μ_0 and μ_1 are absolutely continuous with respect to $\mu^* = \mu_0 + \mu_1$ and Proposition 6.11 implies that on $\mathcal{F}^N(B)$

$$\frac{dP_{\mu_i}}{dP_{\mu^*}} = \exp\left[\int_B \left(1 - \frac{d\mu_i}{d\mu^*}\right) d\mu^* + \int_B \log \frac{d\mu_i}{d\mu^*}\, dN\right]$$

for $i = 0, 1$, so we may form the likelihood ratio (not a Radon-Nikodym derivative but a ratio of likelihoods nevertheless)

$$\frac{dP_{\mu_1}/dP_{\mu^*}}{dP_{\mu_0}/dP_{\mu^*}} = \exp\left[\mu_0(B) - \mu_1(B) + \int_B \log \frac{d\mu_1/d\mu^*}{d\mu_0/d\mu^*}\, dN\right]$$

and use it to test the hypotheses (6.28).

For the σ-algebra $\mathcal{F}^N(E)$ corresponding to observation of N over the noncompact set E it is possible that $\mu_0 \sim \mu_1$ on E—and hence that $P_{\mu_0} \sim P_{\mu_1}$ on $\mathcal{F}^N(B)$ for every bounded set B—but that $P_{\mu_0} \perp P_{\mu_1}$ on $\mathcal{F}^N(E)$ (Exercise 6.10). Next we give a necessary and sufficient condition for $P_{\mu_0} \sim P_{\mu_1}$ in the unbounded case when $\mu_0 \sim \mu_1$, but this *dichotomy theorem* yields more: if P_{μ_0} and P_{μ_1} are not equivalent, then they must be singular.

Theorem 6.13. Suppose that $\mu_0 \sim \mu_1$ on E and that $\mu_0(E) = \mu_1(E) = \infty$. Then on $\mathcal{F}^N$, either $P_{\mu_0} \sim P_{\mu_1}$ or $P_{\mu_0} \perp P_{\mu_1}$ according as the integral

$$\int_E \left[1 - \left(\frac{d\mu_1}{d\mu_0}\right)^{1/2}\right]^2 d\mu_0 \tag{6.30}$$

converges or diverges.

Proof: We show first that convergence in (6.30) implies that $P_{\mu_1} \ll P_{\mu_0}$ on $\mathcal{F}^N(E)$; since convergence of (6.30) implies convergence of the corresponding integral with μ_0 and μ_1 interchanged, this provides the "convergence implies equivalence" part of the theorem. By Liptser and Shiryayev (1978, Lemma 19.13) it is enough to show that for bounded sets $B_n \uparrow E$, $\lim(dP_{\mu_1}/dP_{\mu_0})|\mathcal{F}^N(B_n)$ exists and is finite almost surely with respect to P_{μ_1}. From (6.29),

$$\frac{dP_{\mu_1}}{dP_{\mu_0}}|\mathcal{F}^N(B_n) = \exp\left[\int_{B_n}\left(1 - \frac{d\mu_1}{d\mu_0}\right) d\mu_0 + \int_{B_n} \log\frac{d\mu_1}{d\mu_0}\, dN\right],$$

and by the P_{μ_1}-strong law of large numbers for N

$$\lim \frac{\int_{B_n} \log(d\mu_1/d\mu_0)\, dN}{\int_{B_n} \log(d\mu_1/d\mu_0)\, d\mu_1} = 1$$

almost surely with respect to P_{μ_1} (one can choose the B_n in a manner ensuring that this holds); consequently,

$$\frac{dP_{\mu_1}}{dP_{\mu_0}}|\mathcal{F}^N(B_n) \sim \exp\left[\int_{B_n}\left(1 - \frac{d\mu_1}{d\mu_0} + \frac{d\mu_1}{d\mu_0}\log\frac{d\mu_1}{d\mu_0}\right) d\mu_0\right]$$

The proof that $P_{\mu_1} \ll P_{\mu_0}$ is completed by showing that

$$\int_E \left(1 - \frac{d\mu_1}{d\mu_0} + \frac{d\mu_1}{d\mu_0}\log\frac{d\mu_1}{d\mu_0}\right) d\mu_0 < \infty \tag{6.31}$$

That (6.31) and convergence in (6.30) are equivalent is seen by first noting that the function $h(y) = 1 - y + y(\log y)$ in (6.31) can be replaced without affecting convergence of the integral by $\tilde{h}(y) = 1 - y + y\varphi(\log y)$, where $\varphi(x) = x$ for $|x| \le 1$ and $x/|x|$ otherwise, and then using the property that there are constants c, c' such that $c(1 - y^{1/2})^2 \le \tilde{h}(y) \le c'(1 - y^{1/2})^2$, $y \ge 0$ (see Grenander, 1981, Chap. 8, and Liptser and Shiryayev, 1978, Chap. 19, for details).

Conversely, given divergence in (6.30), almost surely with respect to P_{μ_1},

$$\lim \frac{dP_{\mu_0}}{dP_{\mu_1}}|\mathcal{F}^N(B_n) = \left(\lim \frac{dP_{\mu_1}}{dP_{\mu_0}}|\mathcal{F}^N(B_n)\right)^{-1}$$

$$= \frac{1}{\infty} = 0, \tag{6.32}$$

but by Neveu (1975, Proposition III-1-5) the limit in (6.32) is the Radon-Nikodym derivative of the absolutely continuous component in the Lebesgue decomposition of P_{μ_0} with respect to P_{μ_1}, and therefore $P_{\mu_0} \perp P_{\mu_1}$. □

Equivalent means of stating convergence in (6.30) exist. For example, M. Brown (1971, 1972) demonstrates that $P_{\mu_0} \sim P_{\mu_1}$ if and only if for some $c > 0$ (and with $f = d\mu_1/d\mu_0$)

$$\int_{\{|1-f|>c\}} |1-f|\, d\mu_0 + \int_{\{|1-f|\leq c\}} (1-f)^2\, d\mu_0 < \infty \tag{6.33}$$

An advantage of (6.30) as opposed to (6.33) is that it generalizes to point processes with nondeterministic compensators (see Liptser and Shiryayev, 1978).

Here is an extended criterion for singularity, in which it is not assumed that $\mu_0 \sim \mu_1$.

Proposition 6.14. Suppose that

$$\mu_1(dx) = f(x)\mu_0(dx) + \nu(dx) \tag{6.34}$$

is the Lebesgue decomposition of μ_1 with respect to μ_0. Then the following are equivalent:

a) $P_{\mu_0} \perp P_{\mu_1}$ on $\mathcal{F}^N$;

b) Either

$$\nu(E) = \infty \tag{6.35}$$

or

$$\int_E (1 - f^{1/2})^2\, d\mu_0 = \infty \tag{6.36}$$

Proof: a) ⇒ b). If (6.35) and (6.36) both fail, then there exists $A \in \mathcal{E}$ with $\mu_0(A) = \nu(A^c) = 0$ and $P_{\mu_1}\{N(A) = 0\} = \exp[-\nu(A)] > 0$. Conditional on the event $\{N(A) = 0\}$, which has P_{μ_0}-probability 1, N is Poisson with mean μ_0 under P_{μ_0} and Poisson with mean $f(x)\mu_0(dx)$ under P_{μ_1}. Failure of (6.36) implies convergence in (6.30) for these conditional processes; hence by the "convergence implies equivalence" part of Theorem 6.13, $P_{\mu_1} \ll P_{\mu_0}$ on an event having positive probability for both, which makes $P_{\mu_0} \perp P_{\mu_1}$ impossible.

b) ⇒ a). If (6.35) holds, then there is a set A with $\mu_0(A) = 0$ (hence $P_{\mu_0}\{N(A) = 0\} = 1$) but $\nu(A) = \infty$, which implies that $P_{\mu_1}\{N(A) = 0\} = 0$; therefore, a) holds. If (6.36) is fulfilled and P_{μ_2} denotes the probability law of the Poisson process on E with mean measure $\mu_2(dx) = f(x)\mu_0(dx)$, then $P_{\mu_0} \perp P_{\mu_2}$ by Theorem 6.13. By (6.34), P_{μ_1} is the convolution of P_{μ_2} and P_ν, the first of which is singular with respect to P_{μ_0}, as is the second, except possibly on $\{N(E) = 0\}$; therefore, $P_{\mu_0} \perp P_{\mu_1}$. □

From the perspective of hypothesis testing, Theorem 6.13 describes the pleasant situation that from a single realization of N there is either error-free discrimination between P_{μ_0} and P_{μ_1} (the singular case) or a positive, finite log-likelihood ratio

$$\log\left(\lim \frac{dP_{\mu_1}}{dP_{\mu_0}} \,|\mathcal{F}^N(B_n)\right) = \int_E \left(1 - \frac{d\mu_1}{d\mu_0}\right) d\mu_0 + \int_E \log \frac{d\mu_1}{d\mu_0}\, dN$$

that can be used to perform likelihood ratio tests. Proposition 6.14 provides additional conditions for perfect discrimination between μ_0 and μ_1. If either measure is infinite, then there is no error in testing the hypotheses (6.28), while if both are finite, then every nonzero realization of N yields unerring determination of μ_0 or μ_1, but on $\{N(E) = 0\}$ no certain choice is possible. However, on this event, an atom of $\mathcal{F}^N(E)$, one can employ the log-likelihood ratio

$$\log \frac{P_{\mu_1}\{N(E) = 0\}}{P_{\mu_0}\{N(E) = 0\}} = \mu_0(E) - \mu_1(E)$$

in the usual fashion.

We now examine log-likelihood function asymptotics for a sequence of processes on a bounded set with mean measures converging to infinity. As Proposition 3.25 and Theorem 5.23 suggest, the appropriate milieu for such study is local asymptotic normality under contiguous alternatives, formulated in this case as follows. Let $N_1, N_2, \ldots$ be point processes on a compact set E and let μ_0, μ^* be elements of $\mathbf{M}_d$ with $\mu^* \ll \mu_0$. Suppose that under the probability measure P_0 the N_i are i.i.d. Poisson processes with mean measure μ_0 and that there is a sequence of probabilities P^n under which $N_1, \ldots, N_n$ are i.i.d. Poisson processes with mean measure $\mu_0 + n^{-1/2}\mu^*$. The contiguity theorem describes asymptotic behavior of log-likelihood ratios associated with the superpositions $N^n = N_1 + \cdots + N_n$. Since under P_0, N^n is Poisson with mean $n\mu_0$, while under P^n its mean measure is $n\mu_0 + n^{1/2}\mu^*$, by (6.29) the log-likelihood ratio is

$$\begin{aligned} L_n &= \log \frac{dP_1^n}{dP_0} \\ &= n\mu_0(E) - [n\mu_0(E) + n^{1/2}\mu^*(E)] + \int_E \log \frac{d(n\mu_0 + n^{1/2}\mu^*)}{d(n\mu_0)}\, dN^n \\ &= -n^{1/2}\mu^*(E) + \int_E \log\left(1 + n^{-1/2}\frac{d\mu^*}{d\mu_0}\right) dN^n \end{aligned}$$

More generally, allowing for observation over a variable subset A of E we have the log-likelihood function

$$L_n(A) = -\, n^{1/2}\mu^*(A) + \int_A \log\left(1 + n^{-1/2}\frac{d\mu^*}{d\mu_0}\right) dN^n, \qquad (6.37)$$

a signed random measure on E. In Theorem 5.23 we used martingale methods to analyze such processes; here we employ instead the independent increments property of Poisson processes.

Theorem 6.15. Suppose that $\int_E (d\mu^*/d\mu_0)\, d\mu^* < \infty$. Then with notation and hypotheses as in the preceding paragraph, under P_0, $L_n \xrightarrow{d} G$ as random measures on E, where G is a Gaussian random measure with independent increments,

$$E[G(A)] = -\frac{1}{2}\mu^*(A) \tag{6.38}$$

and

$$\mathrm{Var}(G(A)) = \int_A \frac{d\mu^*}{d\mu_0}\, d\mu^* \tag{6.39}$$

Proof: By Theorem 1.21 and the fact that the L_n and G have independent increments (for details of the construction of G, see Neveu, 1965, pp. 84 ff.), it suffices to show that $L_n(A) \xrightarrow{d} G(A)$ for each set A. Continuing from (6.37), we have

$$\begin{aligned} L_n(A) &\cong -\, n^{1/2}\mu^*(A) + n^{-1/2}\int_A \frac{d\mu^*}{d\mu_0}\, dN^n - \frac{1}{2n}\int_A \frac{d\mu^*}{d\mu_0}\, dN^n \\ &= n^{-1/2}\sum_{i=1}^{n}\left(\int_A \frac{d\mu^*}{d\mu_0}\, dN_i - \int_A \frac{d\mu^*}{d\mu_0}\, d\mu_0\right) \\ &\quad - \frac{1}{2n}\sum_{i=1}^{n}\int_A \frac{d\mu^*}{d\mu_0}\, dN_i \end{aligned}$$

The summands in the first term are i.i.d. under P_0 with mean zero and finite variance $\int_A (d\mu^*/d\mu_0)^2\, d\mu_0 = \int_A (d\mu^*/d\mu_0)\, d\mu^*$, while by the strong law of large numbers, the second term converges to $-(1/2)\mu^*(A)$; hence $L_n(A) \xrightarrow{d} G(A)$ by the central limit theorem and Slutsky's theorem. □

The ease with which further testing problems can be posed for Poisson processes on general spaces is matched by the paucity of procedures available for addressing them in nonparametric settings. For nonhomogeneous Poisson processes on $\mathbf{R}_+$ there do exist rigorously derived techniques (some to be mentioned presently) for testing structure of rate functions or comparing rate functions. For planar Poisson processes there are methods, developed mainly by British statisticians, based on distances to nearest neighbors in the point process and distances from points in $\mathbf{R}^2$ to points of N ("distance methods") that have been applied extensively to "real" problems, but not all of them have unambiguously established properties. The reader interested in applications is urged to pursue them; see the chapter notes for some references.

Returning to the context of general spaces, among testing problems of both mathematical and practical interest, but concerning which little is known, are the following.

1) *Structure of the mean measure.* The canonical problem is to test whether μ_N is absolutely continuous with respect to a prescribed measure μ_0. Consistent estimators of $d\mu_N / d\mu_0$ under H_0 can be constructed using Theorem 6.8, but it is unclear what properties of them (intuitively, "largeness") should lead to rejection of H_0.
2) *Equality and ordering of mean measures.* Given independent Poisson processes N_1, N_2 with mean measures μ_1, μ_2, except on $\mathbf{R}_+$ and with further assumptions, few procedures are available for testing even the hypothesis that $\mu_1 = \mu_2$, let alone the hypothesis that $\mu_1 \geq \mu_2$.
3) *Whether a point process is Poisson.* The main difficulty, beyond those already present for ordinary Poisson processes on $\mathbf{R}_+$, is lack of easily specified, reasonably, yet not unreasonably, small classes of point processes to serve as alternative hypotheses.

For Poisson processes on $\mathbf{R}_+$ martingale methods are available. One could use techniques described in Section 5.3 to test, for example, whether a Poisson process N on [0,1] known to have an intensity function, has a prescribed intensity function. Two-sample tests can be effected with martingale test statistics (5.65), which under $H_0: \alpha(1) \equiv \alpha(2)$ [here $N(1)$, $N(2)$ are independent Poisson processes with intensity functions $\alpha(1)$, $\alpha(2)$] have the advantage of removing the nuisance parameter $\alpha(1) = \alpha(2)$ from the problem. Under H_0 the process $M_t = N_t(1) - N_t(2)$ is a mean zero martingale; procedures whose asymptotics are described by Theorem 5.25 test for this property. Boswell (1966), M. Brown (1972), and Saw (1975) examine other aspects of testing for nonhomogeneous Poisson processes on $\mathbf{R}_+$. Brown treats mainly Neyman-Pearson tests of simple hypotheses for models $\mathcal{P} = \{P_\alpha\}$, where under P_α the Poisson process has mean measure $\alpha_t \mu_0(dt)$ with μ_0 known but not necessarily Lebesgue measure; results include normal approximations for test statistics $\int_0^t \log(\alpha(1)/\alpha(0))\, dN$ as $t \to \infty$. Boswell and Saw consider trend and monotonicity properties of several kinds.

6.3 PARTIALLY OBSERVED POISSON PROCESSES

In this section we consider three models of Poisson processes that are only partially observable and for each present results on statistical inference and state estimation. We begin with p-thinned Poisson processes, for which we refine and specialize results appearing in Sections 4.5 and 6.1; in addition, we develop a maximum likelihood interpretation of the estimators $\hat{p}$ of (4.46) and establish consistency under conditions weaker than in Theorem 4.22. Next we examine the stochastic integral process studied already in Examples 3.5 and 3.10 but in greater generality: for some results the integrand is allowed to be a semi-Markov process rather than a Markov process. For the Markov case we give an explicit application of the state estimation procedure culminating in Theorem 5.29. Finally, we treat inference for Poisson processes on product spaces such that only one component of the process is observable. The three models are illustrative rather than exhaustive; in some ways the variety of models and the techniques with which they are analyzed is the real content of the section.

p-Thinned Poisson Processes

We recall the setting established in Section 4.5. Let E be a compact space and let $N_1', N_2', \dots$ be observable point processes. They are thinnings of unobservable Poisson processes N_i, so we retain the "prime" notation from Section 4.5. The statistical model is $\mathcal{P} = \{P_p : p \text{ is a measurable function from } E \text{ to } [0,1]\}$ and under P_p the N_i' are i.i.d. copies of the p-thinning of the Poisson process with *known* mean measure $\mu_0 \in \mathbf{M}_d$; the model is hence a submodel of the model treated in Theorem 6.8, as well as a specialized version of Model 3.3. Our goal is to estimate the thinning function p.

Denote by P the probability measure corresponding to $p \equiv 1$ (i.e., no thinning), under which the N_i' have mean measure μ_0, and for each n let $\overline{N}_n' = \Sigma_{i=1}^n N_i'$. Then from (6.29) we obtain log-likelihood functions

$$L_n(p) = \log \frac{dP_p}{dP}\Big|_{\sigma(N_1', \dots, N_n')}$$

$$= n \int_E (1-p)\, d\mu_0 + \int_E \log p \, d\overline{N}_n' \tag{6.40}$$

By contrast with the situation of Theorem 6.8, there do exist maximum likelihood estimators, but they are *not consistent*. Since $\log p \le 0$, the second term in the log-likelihood function is maximized by taking $\hat{p}$ to be one at each atom of $\overline{N}_n'$, and then the first term is maximized by letting $\hat{p}$ be zero everywhere else. Because $P\{\overline{N}_n'(E) < \infty\} = 1$ and μ_0 is diffuse, $\|\hat{p} - p\|_1 = \|p\|_1$, where $\|\cdot\|_1$ is the norm on $L^1(d\mu_0)$; consequently, the $\hat{p}$ are not consistent. However, sieve estimators of (6.26), which coincide with the estimators defined in (4.46), are strongly consistent under conditions weaker in some ways than those in Theorem 4.22.

For each m, let $E = \Sigma_{j=1}^{\ell_m} A_{mj}$ be a partition; assume that $\mu_0(A_{mj}) > 0$ for each m and j and that $\max\{\mu_0(A_{mj}) : 1 \le j \le \ell_m\} \to 0$. Let $I(m)$ be the sieve of functions p: $E \to [0,1]$ constant on each A_{mj}. Then we have the following counterpart of Theorems 4.22 and 6.8.

Theorem 6.16. a) For each n and m, the n-sample maximum likelihood estimator of p relative to $I(m)$ is

$$\hat{p}(n,m) = \sum_{j=1}^{\ell_m} \min\left\{\frac{\bar{N}_n'(A_{mj})}{n\mu_0(A_{mj})}, 1\right\} 1_{A_{mj}} \tag{6.41}$$

b) Suppose that $0 \le \delta < 1$ and that $m = m_n$ is chosen so that with $\tilde{\ell}_n = \ell_{m_n}$,

$$\sum_n \frac{\tilde{\ell}_n^4}{n^{2-\delta}} < \infty \tag{6.42}$$

Then with $\hat{p} = \hat{p}(n,m_n)$, for each p such that

$$\lim_n n^{\delta/4} \sum_{j=1}^{\tilde{\ell}_n} \int_{A_{mj}} \left| \frac{\mu_0(p; A_{mj})}{\mu_0(A_{mj})} - p(x) \right| \mu_0(dx) = 0, \tag{6.43}$$

we have $n^{\delta/4} \| \hat{p} - p \|_1 \to 0$ almost surely.

Proof: a) For $p = \Sigma a_j 1_{A_{mj}}$ belonging to $I(m)$, (6.40) gives

$$L_n(p) = \sum_{j=1}^{\ell_m} [(1 - a_j)\mu_0(A_{mj}) + \bar{N}_n'(A_{mj})(\log a_j)]$$

Since

$$\frac{\partial L_n(p)}{\partial a_j} = -\mu_0(A_{mj}) + \frac{\bar{N}_n'(A_{mj})}{a_j}, \tag{6.44}$$

maximization with respect to a_j occurs at $\hat{a}_j = \bar{N}_n'(A_{mj})/\mu_0(A_{mj})$ provided that this quantity does not exceed 1. If it does exceed 1, then the derivative in (6.44) is positive everywhere on [0,1], so the maximizing value in [0,1] is $\hat{a}_j = 1$.

b) The argument follows the proof of Theorem 6.8. One splits $n^{\delta/4} \| \hat{p} - p \|_1$ into random and nonrandom terms; for the former

$$P_p\left\{ n^{\delta/4} \sum_{j=1}^{\ell_m} \left| \frac{\bar{N}_n'(A_{mj})}{n} - E_p[N'(A_{mj})] \right| > \varepsilon \right\}$$

$$\le \varepsilon^{-4} n^{-\delta} E_p\left[\left(\sum_{j=1}^{\ell_m} \left| \frac{\bar{N}_n'(A_{mj})}{n} - E_p[N'(A_{mj})] \right| \right)^4 \right] = O(\varepsilon^{-4} n^{2-\delta} \ell_m^4)$$

for each $\varepsilon > 0$, so that (6.42) implies that the random term converges to zero almost surely. That (6.43) suffices to dispose of the nonrandom term follows from the proof of Theorem 6.8. □

Unlike Theorem 4.22, Theorem 6.16 with $\delta = 0$ imposes no continuity requirement on the thinning function p. A Poisson process N with mean measure μ_0 satisfies

$$E[N(A_{mj})^3] = \mu_0(A_{mj}) + 3\mu_0(A_{mj})^2 + \mu_0(A_{mj})^3,$$

so that (4.48) in this case becomes (essentially) $\widetilde{\ell}_n^3 = O(n^\delta)$ or some $\delta < 1$, i.e., $\widetilde{\ell}_n^3 \sim n^{1/3-\varepsilon}$ for some $\varepsilon > 0$, while (6.42) is more restrictive: it requires that $\widetilde{\ell}_n^3 \sim n^{1/4-\varepsilon}$.

If each N_i', under P_p, is the p-thinning of an unobservable Poisson process N_i with mean measure μ_0, with the pairs (N_i, N_i') i.i.d., then state estimation for the underlying processes N_i is nearly trivial. By Exercise 1.16, for each i, N_i' and $N_i'' = N_i - N_i'$ are independent Poisson processes with mean measures $p(x)\mu_0(dx)$ and $(1 - p(x))\mu_0(dx)$, respectively; consequently,

$$E_p[N_i \mid \mathcal{F}^{N_i'}] = E_p[N_i' + N_i'' \mid \mathcal{F}^{N_i'}] = N_i' + (1 - p)\mu_0, \tag{6.45}$$

where $[(1 - p)\mu_0](dx) = (1 - p(x))\mu_0(dx)$. Of course, however, (6.45) can be implemented only if p is known, and otherwise we are in the combined statistical inference and state estimation setting of Section 3.4. With $\hat{p}$ the estimators in Theorem 6.16b) we can form, guided by principles enunciated in Section 3.4 (see also Sections 4.4, 4.5, 7.4, and 8.3), *pseudo-state estimators*

$$\hat{E}[N_{n+1} \mid \mathcal{F}^{N_{n+1}'}] = N_{n+1}' + (1 - \hat{p})\mu_0, \tag{6.46}$$

which have the following consistency property.

Proposition 6.17. Under the conditions of Theorem 6.16b) for each δ and p satisfying (6.42) and (6.43),

$$n^{\delta/4} \| \hat{E}[N_{n+1} \mid \mathcal{F}^{N_{n+1}'}] - E_p[N_{n+1} \mid \mathcal{F}^{N_{n+1}'}] \| \to 0 \tag{6.47}$$

almost surely, where $\| \cdot \|$ denotes the total variation norm on the set of finite signed measures on E.

Proof: From (6.45) and (6.46),

$$(\hat{E}[N_{n+1} \mid \mathcal{F}^{N_{n+1}'}] - E_p[N_{n+1} \mid \mathcal{F}^{N_{n+1}'}])(dx) = (p(x) - \hat{p}(x))\mu_0(dx),$$

which implies that

$$n^{\delta/4} \| \hat{E}[N_{n+1} \mid \mathcal{F}^{N_{n+1}'}] - E_p[N_{n+1} \mid \mathcal{F}^{N_{n+1}'}] \| = n^{\delta/4} \| \hat{p} - p \|_1;$$

therefore, (6.47) follows from Theorem 6.16b). □

Stochastic Integral Process

Let N be a Poisson process on $\mathbf{R}_+$ having unknown rate α under the probability measure P_α, and suppose that (X_t) is a semi-Markov process with state space $\{0,1\}$ that under each P_α is independent of N with semi-Markov kernel (not a function of α)

$$Q = \begin{bmatrix} 0 & G \\ F & 0 \end{bmatrix}$$

Sojourns of X in 0 having distribution G alternate with sojourns in 1 having distribution F, and all sojourns are mutually independent (see also Section 8.4). We assume that F, G are known, although if, as in Examples 3.5 and 3.10, X is a Markov process with generator

$$A = \begin{bmatrix} -a & a \\ b & -b \end{bmatrix}, \tag{6.48}$$

then the transition rates a,b can be unknown as well (see Karr, 1984a). The observations are the stochastic integral process

$$X * N_t = \int_0^t X \, dN,$$

in this case a point process (interpretation: an atom of N at u is observable in $X * N$ if and only if $X_u = 1$), together, perhaps, with X itself. When X is observable, that is, the observed history is

$$\mathcal{H}_t = \mathcal{F}_t^{X*N} \vee \mathcal{F}_t^X, \tag{6.49}$$

inference concerning α and state estimation are both straightforward; some statistical results are given in Exercise 6.15. (Exercises 3.15 – 3.17 treat the Markov case.) After noting elementary properties of $X * N$ we examine state estimation for unobserved portions of N given observations (6.49); thereafter we move on to the more difficult and interesting case that only $X * N$ is observable.

Proposition 6.18. Under each probability P_α,

a) $X * N$ is a Cox process directed by the random measure $M(A) = \alpha \int_A X_u \, du$.

b) If F, G are nonarithmetic (see Chapter 8) with finite means $\bar{F}$, $\bar{G} > 0$, then with $q = \bar{F}/(\bar{F} + \bar{G})$, almost surely

$$\frac{X * N_t}{qt} \to \alpha \tag{6.50}$$

c) If F, G are nonarithmetic with finite variances $\sigma^2(F)$, $\sigma^2(G)$, respectively, then with $\sigma^2 = (\bar{F} + \bar{G})^{-3}[\bar{F}^2\sigma^2(G) + \bar{G}^2\sigma^2(F)]$, as $t \to \infty$

$$[(1 + q\sigma)(\alpha qt)]^{-1/2}(X * N_t - \alpha qt) \xrightarrow{d} N(0,1) \quad \square \tag{6.51}$$

The quantity q is the asymptotic fraction of time X spends in state 1 (i.e., the fraction of time during which N is observable). For proofs, see Karr (1928) and Grandell (1976); independence of N and X is a crucial assumption. In particular, Proposition 6.18a) implies that the $(P_\alpha, \mathcal{H})$-stochastic intensity of $X * N$ is (αX_{t-}), so that by Exercise 2.16 the $(P_\alpha, \mathcal{F}^{X*N})$-stochastic intensity is $(\alpha E_\alpha[X_t | \mathcal{F}_{t-}^{X*N}])$, whose computation is addressed in Theorem 6.23. With X observable we have a multiplicative intensity model, but asymptotics here, as $t \to \infty$, differ from those treated in Chapter 5.

Suppose now that α is known; then state estimation for N given observations $(\mathcal{H}_t)$ can be performed easily.

Proposition 6.19. For each t, with $A_t = A \cap [0,t]$,

$$E_\alpha[N(A)|\mathcal{H}_t] = X * N(A_t) + \alpha \int_{A_t} (1 - X_u)\, du + \alpha|\mathbf{R} \setminus A_t| \tag{6.52}$$

Proof: For fixed t and A

$$E_\alpha[N(A)|\mathcal{H}_t] = E_\alpha[X * N(A_t) + \int_{A_t} (1 - X_u)\, dN_u + N(\mathbf{R} \setminus A_t)|\mathcal{H}_t]$$

$$= X * N(A_t) + \int_{A_t} (1 - X_u) E_\alpha[dN_u|\mathcal{H}_t] + E_\alpha[N(\mathbf{R} \setminus A_t)|\mathcal{H}_t]$$

By independence of N and X and the independent increments property of N, $(1 - X_u)E_\alpha[dN_u|\mathcal{H}_t] = (1 - X_u)\alpha\, du$; similarly, $N(\mathbf{R} \setminus A_t)$ is independent of $\mathcal{F}_t^N \vee \mathcal{F}_t^X$, and hence also of $\mathcal{H}_t$, so that $E_\alpha[N(\mathbf{R} \setminus A_t)|\mathcal{H}_t] = E_\alpha[N(\mathbf{R} \setminus A_t)] = \alpha|\mathbf{R} \setminus A_t|$. $\square$

We now take up inference and state estimation given observation only of $X * N$; for notational simplicity let $\mathcal{F}_t = \mathcal{F}_t^{X*N}$.

Proposition 6.20. Let q and σ^2 be as in Proposition 6.18 and assume that conditions b) and c) there are fulfilled. Define estimators $\hat{\alpha}$ $(= \hat{\alpha}_t)$ by $\hat{\alpha} = X * N_t / qt$. Then under each probability P_α,

a) $\hat{\alpha} \to \alpha$ almost surely;
b) $[\alpha(1 + \alpha\sigma^2/q)]^{-1/2}(qt)^{1/2}[\hat{\alpha} - \alpha] \xrightarrow{d} N(0,1)$. $\square$

The next result not only can be used to construct likelihood ratio tests, but also serves to emphasize importance of the conditional expectations $E_\alpha[X_t|\mathcal{F}_t]$.

Proposition 6.21. For α_0, $\alpha_1 > 0$ and $t < \infty$, $P_{\alpha_0} \sim P_{\alpha_1}$ on $\mathcal{F}_t$ with

$$\log\frac{dP_{\alpha_1}}{dP_{\alpha_0}} = \int_0^t [\alpha_1\hat{X}_u(\alpha_1) - \alpha_0\hat{X}_u(\alpha_0)]\,du$$

$$+ X * N_t\left(\log\frac{\alpha_1}{\alpha_0}\right) + \int_0^t \log\frac{\hat{X}_u(\alpha_1)}{\hat{X}_u(\alpha_0)}\,d(X*N)_u, \tag{6.53}$$

where

$$\hat{X}_t(\alpha) = E_\alpha[X_u|\mathcal{F}_u]|_{u=t-} \tag{6.54}$$

Proof: According to discussion following Proposition 6.18, the $(P_\alpha, \mathcal{F})$-stochastic intensity of $X * N$ is the process $\lambda_u(\alpha) = \alpha\hat{X}_u(\alpha)$, and if P denotes a probability measure with respect to which $X * N$ (ignoring its genesis from N and X) is Poisson with rate 1, then by Theorem 2.31,

$$\log\frac{dP_\alpha}{dP} = \int_0^t (1 - \lambda_u(\alpha))\,du + X * N_t(\log\alpha) + \int_0^t \log\hat{X}_u(\alpha)\,d(X*N)_u,$$

and from this (6.53) holds by easy computations. □

State estimation for unobserved portions of N—as well as X—requires the MMSE state estimators $E_\alpha[X_t|\mathcal{F}_t]$.

Proposition 6.22. For each t and A,

$$E_\alpha[N(A)|\mathcal{F}_t] = X * N(A_t) + \alpha\int_{A_t}(1 - E_\alpha[X_u|\mathcal{F}_u])\,du + \alpha|\mathbf{R}\setminus A_t| \quad \square \tag{6.55}$$

Thus the crucial state estimation issue is reconstruction of the process X, to which we apply methodology developed in Section 5.4. It seems possible to obtain explicit results only when X is a Markov process [with generator (6.48)], and we so assume for the rest of the discussion.

Theorem 6.23. For each α the MMSE state estimators $E_\alpha[X_t|\mathcal{F}_t]$ satisfy the stochastic differential equation

$$dE_\alpha[X_t|\mathcal{F}_t] = \{-bE_\alpha[X_t|\mathcal{F}_t] + a(1 - E_\alpha[X_t|\mathcal{F}_t])\}\,dt$$
$$+ (1 - E_\alpha[X_u|\mathcal{F}_u]|_{u=t-})[d(X*N)_t - \alpha E_\alpha[X_u|\mathcal{F}_u]|_{u=t-}\,dt] \tag{6.56}$$

Proof: We begin with a semimartingale representation for the state process X relative to the history $\mathcal{H}$ of (6.49), furnished by *Dynkin's formula* (see also Theorem 7.29): for $f(x) = x$, the process

$$M_t = X_t - \int_0^t Af(X_u)\,du = X_t - \int_0^t [a1(X_u = 0) - b1(X_u = 1)]\,du$$

is a martingale; hence

$$X_t = \int_0^t [a1(X_u = 0) - b1(X_u = 1)]\,du + M_t \tag{6.57}$$

is the requisite representation. With α fixed and $\hat{X}_t = E_\alpha[X_t|\mathcal{F}_t]$ it follows from Proposition 5.27, (5.71), and especially (5.72a) that there is an $\mathcal{F}$-predictable process H such that

$$\hat{X}_t = \int_0^t [a(1 - \hat{X}_u) - b\hat{X}_u]\,du + \int_0^t H_u[d(X * N)_u - \alpha\hat{X}_{u-}\,du], \tag{6.58}$$

since the binary nature of X implies that

$$P_\alpha\{X_u = 0|\mathcal{F}_u\} = 1 - P_\alpha\{X_u = 1|\mathcal{F}_u\} = 1 - E_\alpha[X_u|\mathcal{F}_u]$$

and where we have also used the property that (αX_{t-}) is the $(P_\alpha, \mathcal{H})$-stochastic intensity of $X * \mathrm{N}$. By Theorem 5.29, $H_u = J_u + K_u - \hat{X}_{u-}$, where J, K satisfy (5.74a) and (5.74b), respectively, relative to the state process X. Because X is Markov and independent of N, almost surely the martingale M in (6.57) has no jumps in common with N; hence $K \equiv 0$ (see discussion subsequent to Theorem 5.29). By (5.78) and the crucial relationship $X^2 = X$,

$$J_u = (\alpha\hat{X}_{u-})^{-1}(\alpha X^2)\hat{}_{u-} = (\alpha\hat{X}_{u-})^{-1}(\alpha X)\hat{}_{u-} = 1;$$

therefore, $H_u = 1 - \hat{X}_{u-}$ and this together with (6.58) completes the proof. □

Using (6.55) and (6.56) it is straightforward to effect MMSE state estimation for unobserved portions of N. Note also that (6.56) provides a recursive method for computation of the state estimators $E_\alpha[X_t|\mathcal{F}_t]$: in order to calculate the increment $dE_\alpha[X_t|\mathcal{F}_t]$ one needs only the current value (so that the amount of data that must be retained in order to calculate the state estimator does not increase with time) and the innovation $dM_t = d(X * N)_t - \alpha E_\alpha[X_u|\mathcal{F}_u]|_{u=t-}$. The property that $X^2 = X$ is central, as will be seen more graphically in Section 7.5; it avoids introducing a state estimator of the second moment, which would have to be calculated with another stochastic differential equation.

Poisson Processes with One Observable Component

Let E_0, E_1 be compact spaces (compactness of E_1 is not essential), let $E = E_0 \times E_1$ and suppose that K is a *known* transition kernel from E_0 into

E_1. We consider a special case of the following situation: there is an *unobservable* point process $N^0 = \Sigma\, \varepsilon_{X_i}$ on E_0 such that each point X_i has associated to it an observable point $Z_i \in E_1$ with distribution $K(X_i, \cdot)$ and the Z_i are conditionally independent given N^0. One observes $N^1 = \Sigma\, \varepsilon_{Z_i}$, the second component of the two-component point process $N = \Sigma\, \varepsilon_{(X_i, Z_i)}$ on E. We consider the statistical model $\mathcal{P} = \{P_\mu : \mu \in \mathbf{M}_d(E_0)\}$, where under P_μ, N^0 is Poisson with mean measure μ.

Here are elementary properties.

Lemma 6.24. Suppose that for each μ the Z_i are P_μ-conditionally independent given N^0 with $P_\mu\{Z_i \in B | N^0\} = K(X_i, B)$ independently of μ. Then under P_μ

a) $N = \Sigma\, \varepsilon_{(X_i, Z_i)}$ is a Poisson process on E with mean measure $\mu(dx)K(x,dy)$.

b) $N^1 = \Sigma\, \varepsilon_{Z_i}$ is Poisson on E_1 with mean measure

$$\nu_\mu(dy) = \int_{E_0} \mu(dx)K(x,dy) \quad \square \tag{6.59}$$

We consider the usual problems: estimation of μ given data $N_1^1, N_2^1, \ldots$ comprising i.i.d. copies of the second component N^1 and state estimation for N^0 given observation of N^1. As will be seen, structure of the kernel K is crucial. Throughout we work only with the full nonparametric model $\mathcal{P}$ specified above.

Application of the model to positron emission tomography is presented in Vardi, et al. (1985). An anatomical structure, for example a human brain, is injected with radioactive material that emits positrons as it decays. (The procedure differs physically and mathematically from transmission tomography, in which an externally generated beam is transmitted by the structure with varying attenuation and observations represent line integrals of the density of the structure. There, the key mathematical issue is inversion of the Radon transform.) Points of N^0 are locations at which disintegrations occur; knowledge of N^0 provides information about differential concentrations of the radioactive material, which in turn yields physical characteristics of the structure. But points of N^0 cannot be observed directly; rather, each emission consists of two positrons oppositely directed at a random angle, which are (in most cases) detected by two of many circularly arrayed detectors; hence what is known about the point of disintegration is only that it lies on a particular line. Thus each point X_i of N^0 gives rise, via a known kernel K incorporating geometry of the detectors and the location X_i, as well as the random direction of positron emission, to an observed point $Z_i \sim K(X_i, \cdot)$. The conditional independence assumption seems well supported; in this simple model temporal effects are ignored. The mean measure μ of N^0 describes the anatomical structure and is to be estimated from observation of N^1. State estimation for N^0 is also of interest. Indeed, one method of estimation of μ in fact solves a problem of combined statistical inference and state estimation.

We begin with estimation of μ. The extent to which observation of i.i.d. processes $N_1^1, N_2^1, \ldots$ with mean measure ν_μ is informative concerning μ is determined by the kernel K. For example, if $K(x,dy) = \eta(dy)$ independently of x, where η is a probability measure on E_1, then $\nu_\mu = \mu(E_0)\eta$ and one can learn nothing about μ other than $\mu(E_0)$. At the opposite extreme, for $K(x,dy) = \varepsilon_{f(x)}(dy)$ with f a bijection of E_0 onto a subset of E_1, $\nu_\mu = \mu f^{-1}$, by which μ is uniquely determined. Given the N_i^1, one can do nothing more than estimate their mean measure ν_μ using, on the basis of Proposition 6.2, the estimators $\hat{\nu} = n^{-1} \Sigma_{i=1}^n N_i^1$. The following result summarizes what is learned thereby concerning μ itself.

Proposition 6.25. For $g \in C_+(E_1)$, let $Kg(x) = \int K(x,dy)g(y)$ and suppose that K fulfills the Feller property: $Kg \in C_+(E_0)$ for each $g \in C_+(E_1)$. Let $V = \{Kg : g \in C_+(E_1)\}$ and for $Kg \in V$ put $\hat{\mu}(Kg) = \hat{\nu}(g)$. Then Proposition 6.2 holds for compact subsets $\mathcal{K}$ of V. □

Of course, Proposition 6.25 fails to confront the crucial issue of how large the set V is, but especially at our level of generality this issue is difficult. Mathematically, the question is identifiability of mixtures; the mean ν_μ is the μ-mixture of the measures $K(x,\cdot)$. A mixture model $\{K(x,\cdot) : x \in E_0\}$ is *identifiable* if the mapping $\mu \to \nu_\mu$ is one-to-one. Some results are available, but we will not pursue them here; see additional comments and references in the chapter notes.

For the special case that $\mu(dx) = \alpha(x)\mu_0(dx)$ with μ_0 a known element of $\mathbf{M}_d(E_0)$ and $\alpha \in L_+^1(d\mu_0)$, and that K has the form $K(x,dy) = k(x,y)\rho(dy)$ with ρ a probability on E_1 and $k \geq 0$ a jointly measurable function, one can show existence of maximum likelihood estimators by sieve methods. However, the estimators are only defined implicitly, although they can be calculated—as discussed momentarily—with the "EM algorithm." Consistency properties, consequently, are difficult and we do not elaborate on them here (see Karr, 1985d). Suppose that for each $\alpha \in L_+^1(d\mu_0)$ there is a probability measure P_α under which point processes $N_1^1, N_2^1, \ldots$ are i.i.d. copies of a Poisson process on E_1 having intensity function $\int \mu_0(dx)\alpha(x)k(x,y)$ and for each n, let $\bar{N}_n^1 = \Sigma_{i=1}^n N_i^1$. Suppose that $E_0 = \Sigma_{j=1}^\ell A_j$ is a partition, which is fixed and dropped from notation, and let I be the histogram sieve consisting of functions constant over each A_j.

Proposition 6.26. For each n there exists a maximum likelihood estimator $\hat{\alpha}_n = \Sigma_{j=1}^{\ell} a_j^* 1_{A_j}$, where the a_j^* satisfy the equations

$$\int_{A_i} \mu_0(dx)\left[-1 + \int_{E_1} \frac{k(x,y)}{\Sigma a_j^* \int_{A_j} \mu_0(dz)k(z,y)} n^{-1}\bar{N}_n^1(dy)\right] = 0, \quad (6.60)$$

for $i = 1, \ldots, \ell$.

Proof: For general $\alpha \in L_+^1(\mu_0)$ the log-likelihood function relative to $\alpha \equiv 1$, by Proposition 6.11, is

$$L_n(\alpha) = n\int_{E_0} \mu_0(dx)\left[1 - \alpha(x)\int_{E_1} k(x,y)\rho(dy)\right]$$

$$+ \int_{E_1} \log\left[\int_{E_0} \mu_0(dx)\alpha(x)k(x,y)\right]\bar{N}_n^1(dy)$$

$$- \int_{E_1} \log\left[\int_{E_0} \mu_0(dx)k(x,y)\right]\bar{N}_n^1(dy);$$

for the special case at hand, with terms not depending on α deleted, this becomes

$$L_n(a) = -n\sum_{j=1}^{\ell} a_j\mu_0(A_j) + \int_{E_1} \log\left[\sum_{j=1}^{\ell} a_j \int_{A_j} \mu_0(dx)k(x,y)\right]\bar{N}_n^1(dy)$$

Division by n and differentiation with respect to a_i give (6.60); that a solution exists and maximizes the likelihood function follows by arguments in Vardi et al. (1985). □

The equations (6.60) are troublesome computationally: they are nonlinear and in applications such as the emission tomography model of Vardi et al. (1985), ℓ may be of the order of 10^4. An iterative method of solution is the EM algorithm of Dempster et al. (1977). In this context it leads to recursive computation of revised valu es a_i' from current values by means of the expression

$$a_i'\mu_0(A_i) = a_i\int_{E_1} \frac{\int_{A_i} \mu_0(dx)k(x,y)}{\Sigma a_j \int_{A_j} \mu_0(dz)k(z,y)} n^{-1}\bar{N}_n^1(dy)$$

Given that the recursion converges, it is evident that the limits a_j^* satisfy (6.60).

The final result of the section examines state estimation for N^0 given observation of N^1 under the assumption that $K(x,dy) = k(x,y)\rho(dy)$ and—of course—that μ is known, but without stipulated special form. An explicit solution can be obtained for state estimation of the Laplace functional of N.

Theorem 6.27. For each μ and f,

$$E_\mu[e^{-N^0(f)}|\mathcal{F}^{N^1}] = \exp\left[-\int_{E_1} -\log\frac{\int \mu(dx)e^{-f(x)}k(x,y)}{\int \mu(dx)k(x,y)} N^1(dy)\right] \quad (6.61)$$

Proof: Let $\mu k(y) = \int \mu(dx)k(x,y)$. Under P_μ, N^1 is Poisson with mean measure $\mu k(y)\rho(dy)$; therefore, for $g \in C_+(E_1)$,

$$E_\mu\left[\exp\left[-\int -\log\frac{\int \mu(dx)e^{-f(x)}k(x,y)}{\mu k(y)} N^1(dy)\right]e^{-N^1(g)}\right]$$

$$= E_\mu\left[\exp\left[-\left(\int -\log\frac{\int \mu(dx)e^{-f(x)}k(x,y)}{\mu k(y)} + g(y)\right) N^1(dy)\right]\right]$$

$$= \exp\left[-\int_{E_1}\left(1 - \frac{\int \mu(dx)e^{-f(x)}k(x,y)}{\mu k(y)} e^{-g(y)}\right)\mu k(y)\rho(dy)\right]$$

$$= \exp\left[-\int_{E_1} [\mu k(y) - \int_{E_0} \mu(dx)e^{-f(x)}k(x,y)e^{-g(y)}]\rho(dy)\right]$$

$$= \exp\left[-\int_{E_0}\int_{E_1} (1 - e^{-f(x)-g(y)})\mu(dx)k(x,y)\rho(dy)\right]$$

$$= E_\mu[e^{-N^0(f)-N^1(g)}],$$

and an appeal to Theorem 1.12 serves to complete the proof. □

From (6.61) it follows that

$$E_\mu[N^0(f)|\mathcal{F}^{N^1}] = -\frac{d}{d\alpha} E_\mu[e^{-N^0(\alpha f)}|\mathcal{F}^{N^1}]|_{\alpha=0}$$

$$= \int_{E_1} \frac{\int \mu(dx)f(x)k(x,y)}{\int \mu(dx)k(x,y)} N^1(dy) \quad (6.62)$$

Comparison of (6.60) and (6.62) reveals the essence of the EM algorithm: it is a procedure for combined inference and state estimation, but with emphasis on inference, whereas in our other results emphasis is on state estimation. The algorithm performs an expectation step, using current parameter estimates to effect a provisional reconstruction of the unobservable Poisson process N^0, uses this reconstruction in a maximum likelihood step to revise parameter estimates, and proceeds until convergence occurs.

6.4 INFERENCE GIVEN INTEGRAL DATA

In this section we discuss inference for Poisson processes when observations are *integral data*

$$\mathcal{F}^N(g_0, \ldots, g_m) = \sigma(N(g_0), \ldots, N(g_m))$$
$$= \sigma(N(g) : g \in sp\{g_0, \ldots, g_m\}), \qquad (6.63)$$

where $g_0 \equiv 1$, $g_1, \ldots, g_m$ are continuous functions on E (assumed compact) and "sp" denotes "linear space spanned by." For an introductory discussion, see Section 3.1. The problems are estimation of the mean measure given observation of i.i.d. copies N_i, and state estimation, now nontrivial, for N given observations (6.63). Moreover, we add a design element absent from other analyses: the functions g_i can be interpreted as modes of measurement of the stochastic system N that are under control of the statistician or experimenter; the design problem is to choose which measurements to perform. Although the more challenging question of choosing the g_i adaptively is not treated, we do examine criteria that should be taken into account when the g_i are chosen. What appears here is not a complete treatment; many techniques and results remain to be developed.

As expounded at greater length in Karr (1985e), a fundamental idea for either inference or state estimation is to choose $g_1, \ldots, g_m$ in such a way that every point measure $\nu = \Sigma_{i=1}^{\ell} \varepsilon_{x_i}$ on E with $\ell \leq m$ is uniquely determined by the integrals $\nu(g_0) = \nu(E)$, $\nu(g_1), \ldots, \nu(g_m)$. Then on the event $\{N(E) = N(g_0) \leq m\}$ the point process N is completely specified by $N(g_0), \ldots, N(g_m)$ and one can in principle reconstruct the value of $N(f)$ for every $f \in C_+(E)$. Additional discussion of such sets is presented following Proposition 6.29. In particular, for $m = 1$, g_1 must be an invertible mapping from E onto a subset of $\mathbf{R}$. For the Poisson case, by virtue of conditional uniformity, such a function g_1 and the function $g_0 \equiv 1$ are all one requires for consistent estimation of the mean measure.

Theorem 6.28. Let $N_1, N_2, \ldots$ be i.i.d. Poisson processes with mean measure μ and assume that g is a bounded, invertible mapping of E onto a subset of $\mathbf{R}$. On $\{N_i(E) = 1\}$, let

$$X_i = g^{-1}(N_i(g)) \qquad (6.64)$$

and for $f \in C_+(E)$ define

$$\hat{\mu}(f) = \frac{\Sigma_{i=1}^n f(X_i) 1(N_i(E) = 1)}{\Sigma_{i=1}^n 1(N_i(E) = 0)} \qquad (6.65)$$

Then for each compact subset $\mathcal{K}$ of $C_+(E)$, almost surely

$$\sup\{|\hat{\mu}(f) - \mu(f)| : f \in \mathcal{K}\} \to 0 \qquad (6.66)$$

Motivation for the estimators $\hat{\mu}$ will emerge during the proof. The key idea is that by invertibility of g, on $\{N_i(E) = 1\}$, X_i is the sole point of N_i and by conditional uniformity has distribution $\mu(\cdot)/\mu(E)$. Note that the estimator $\hat{\mu}$ is computable from the data $(N_1(E),N_1(g)),\ldots,(N_n(E),N_n(g))$.

Proof of Theorem 6.28: With N a generic copy.

$$e^{-\mu(E)} = P\{N(E)=0\} = \lim n^{-1}\sum_{i=1}^{n} 1(N_i(E)=0), \tag{6.67}$$

while by conditional uniformity, for each f,

$$E[N(f)1(N(E)=1)] = E[N(f)|N(E)=1]P\{N(E)=1\} = e^{-\mu(E)}\mu(f)$$

With X_i given by (6.64), invertibility of g implies that

$$N_i 1(N_i(E)=1) = \varepsilon_{X_i} 1(N_i(E)=1); \tag{6.68}$$

in particular $N_i(f)1(N_i(E)=1) = f(X_i)1(N_i(E)=1)$. Since $\mu(E)<\infty$, application of Proposition 4.29 to the observable point processes $\tilde{N}_i = N_i 1(N_i(E)=1)$, together with (6.68), gives

$$\sup_{\mathcal{K}} \left| n^{-1}\sum_{i=1}^{n} f(X_i)1(N_i(E)=1) - e^{-\mu(E)}\mu(f) \right| \to 0 \tag{6.69}$$

almost surely, and (6.66) follows from (6.67) and (6.69). □

Analogs of Propositions 4.30 and 4.31 exist as well; for the former, see Exercise 6.21. When $m>1$, more efficient methods of estimation—Theorem 6.28 uses only realizations for which $N_i(E)\leqslant 1$—can be devised.

Now for state estimation; we assume that the mean measure μ is known. Exact calculation of conditional expectations of the form $E[N(f)\,|\,\mathcal{F}^N(g_0,\ldots,g_m)]$ seems difficult in generality, even for Poisson processes, but other state estimators can be computed rather easily. More precisely, we consider optimal state estimators that are linear as functions of the observed data $N(g_0),\ldots,N(g_m)$; although it is for the moment inessential, we continue to assume that $g_0=1$. The pleasant solution to this problem is that the optimal linear estimator (which of course differs from the MMSE nonlinear state estimator except in degenerate cases) is given by $\hat{N}(f) = a(f) + N(\hat{f})$, where $a(f)$ is a constant and $\hat{f}$ is the projection of f onto $sp\{g_0,\ldots,g_m\}$ in the Hilbert space $L^2(d\mu)$.

Proposition 6.29. Let $V = sp\{g_0,\ldots,g_m\}$ in $L^2(d\mu)$. Then for $f\in C_+(E)$ the solution to the optimization problem

$$\underset{a\in\mathbf{R},\, g\in V}{\text{minimize}}\, E[(N(f)-a-N(g))^2],$$

is $g^* = \hat{f}$, the orthogonal projection of f onto V, and $a^* = \mu(f)-\mu(\hat{f})$.

Proof: We minimize $h(a, a_0, \ldots, a_m) = E[(N(f) - a - \Sigma_{j=0}^m a_j N(g_j))^2]$ by brute force. Setting the derivative $\partial h/\partial a$ equal to zero implies that

$$a^* = \mu(f) - \sum_{j=0}^m a_j \mu(g_j) \tag{6.70}$$

Similarly, setting $\partial h/\partial a_i$ equal to zero gives

$$E[N(f)N(g_i)] = a\mu(g_i) + \sum_{j=0}^m a_j E[N(g_j)N(g_i)],$$

into which we substitute $E[N(f)N(g)] = \mu(fg) + \mu(f)\mu(g)$ [see Example 1.15 and (6.70)] to obtain

$$\mu(fg_i) + \mu(f)\mu(g_i) = \mu(f)\mu(g_i) - \mu(g_i)\sum_{j=0}^m a_j\mu(g_j)$$
$$+ \sum_{j=0}^m a_j\mu(g_jg_i) + \sum_{j=0}^m a_j\mu(g_j)\mu(g_i), \tag{6.71}$$

which simplifies to

$$\mu(fg_i) = \sum_{j=0}^m a_j\mu(g_jg_i), \tag{6.72}$$

$i = 0, \ldots, m$, precisely the "normal equations" characterizing the projection $\hat{f}$. That $\hat{f}$ minimizes $\mu((f-g)^2)$ over $g \in V$ implies that the solution to (6.70) and (6.71) minimizes $E[(N(f) - a - N(g))^2]$. □

Computation of the state estimator

$$\hat{N}(f) = \mu(f) - \mu(\hat{f}) + N(\hat{f}) \tag{6.73}$$

arising from Proposition 6.29 is no more difficult than calculation of the projection $\hat{f}$, which is ordinarily effected by solving the normal equations (6.72). A design aspect is that if $g_0, \ldots, g_m$ are chosen to be an orthonormal subset of $L^2(d\mu)$ [i.e., $\mu(g_jg_k) = 1(k=j)$], then the normal equations are solvable in closed form: $a_i = \mu(fg_i)$ for $i = 0, \ldots, m$.

An alternative approach to state estimation disregards the Poisson structure of N and chooses, if possible, $g_0, \ldots, g_m$ so that each point measure $\nu = \Sigma_{i=1}^{\ell} \varepsilon_{X_i}$ on E with $\ell \le m$ is fully determined by $\nu(g_0), \ldots, \nu(g_m)$. Such sets of functions are known as *Tchebycheff systems*; the simplest example is the functions $g_j(x) = x^j$ on $[0,1]$. In this case, whatever the law of N, the point process $\widetilde{N} = N1(N(E) \le m)$ is reconstructible without error from $\mathcal{F}^N(g_0, \ldots, g_m)$, and the only problem remaining is how to estimate N on $\{N(E) > m\}$. One approach is to suppose that (g_j) is a *complete* Tchebycheff system, that is, $\{g_0, \ldots, g_m\}$ is a Tchebycheff system for each m. For data $\mathcal{F}^N(g_0, \ldots, g_m)$, exact reconstruction of N is then possible on $\{N(E) \le m\}$, while on $\{N(E) = k\}$, $k > m$, one can reconstruct N under the approximation that $N(g_i) = E[N(g_i)]$ for $i = m + 1, \ldots, k$; more sophisticated variations can be formulated.

6.5 RANDOM MEASURES WITH INDEPENDENT INCREMENTS; POISSON CLUSTER PROCESSES

This section contains no new material in that everything in it is a combination of results presented previously; rather, its purpose is to emphasize that inference procedures for Poisson processes on general spaces—this is an important reason for our working in a general context—apply also to larger classes of random measures and point processes that admit Poisson cluster representations. In particular we consider random measures with independent increments, whose Poisson cluster representation is given by Theorem 1.34 and for which reasonably specific results can be developed. For infinitely divisible point processes, with Poisson cluster representation derived in Theorem 1.32, the situation is less tractable, although the third model of Section 6.3, of a Poisson process with an unobservable component, is germane for some problems. Our development is purposely sketchy; as stated before, one principal intention is to stress breadth of applicability of methods developed earlier in the chapter.

We begin with (purely atomic) random measures with independent increments but no fixed atoms. According to Theorem 1.34, such a random measure M on a (compact) space E admits the *Poisson cluster representation*

$$M = \sum_j U_j \varepsilon_{X_j}, \tag{6.74}$$

where $N = \Sigma\, \varepsilon_{(X_j, U_j)}$, the "measure of jumps," is a Poisson process on $E \times (0,\infty)$ with mean measure μ_N satisfying

$$\int_{(0,\infty)} (1 - e^{-u})\mu_N(E \times du) < \infty \tag{6.75}$$

Since observation of M over $A \subset E$ is equivalent to observation of N over $A \times (0,\infty)$, provided the mean measure of N is locally finite, material in Sections 6.1 – 6.4 applies. The key difference from the case of infinitely divisible point processes is that here the Poisson cluster representation is compatible with natural forms of observation. It is important to remember that local finiteness of μ_N is defined relative to $E \times (0,\infty)$—not $E \times [0,\infty)$ — because in many interesting cases, for example most Lévy processes on $\mathbf{R}_+$

(increasing processes with independent and stationary increments), $\mu_N(E \times (0,\varepsilon)) = \infty$ for every $\varepsilon > 0$ (there are many small jumps) even though $\mu_N(E \times [\varepsilon,\infty)) < \infty$ (there cannot be too many large jumps).

For P_μ the probability measure under which N is Poisson with mean μ and the M given by (6.74),

$$E_\mu[e^{-M(f)}] = E_\mu[\exp(-\Sigma\, U_j f(X_j))]$$
$$= E_\mu[\exp(-\int_{E\times(0,\infty)} uf(x)N(dx,du))]$$
$$= \exp[-\int_{E\times(0,\infty)} (1-e^{-uf(x)})\mu(dx,du)]; \qquad (6.76)$$

therefore, the law of M is determined by μ. Estimation of Laplace functionals of truncations of M, given i.i.d. copies, can be performed in the following manner.

Proposition 6.30. Suppose that $\mu \in \mathbf{M}_d(E \times (0,\infty))$ satisfies (6.75) and that P_μ is a probability with respect to which the random measures M_1, $M_2, \ldots$ are i.i.d. with Laplace functional (6.76), and let N_1, $N_2, \ldots$ be the associated Poisson processes on $E \times (0,\infty)$. Given the representation $M_i = \Sigma_j U_{ij}\varepsilon_{X_{ij}}$, for each $\delta > 0$, let M_i^δ be the thinned random measure

$$M_i^\delta = \sum_j U_{ij} 1(U_{ij} > \delta)\varepsilon_{X_{ij}},$$

and let $\hat{\mu} = n^{-1}\Sigma_{i=1}^n N_i$ be the usual estimators of the mean measure of the N_i. Then for the empirical Laplace functionals

$$\hat{L}^\delta(f) = \exp[-\int_{E\times(\delta,\infty)} (1-e^{-uf(x)})\hat{\mu}(dx,du)]$$

and each compact subset $\mathcal{K}$ of $C_+(E)$, almost surely

$$\sup\{|\hat{L}^\delta(f) - E_\mu[\exp(-M^\delta(f))]| : f \in \mathcal{K}\} \to 0 \qquad (6.77)$$

Proof: Observe first that on $E \times (\delta,\infty)$ one can apply Proposition 6.3a) to the Poisson processes N_i, to deduce that for $\mathcal{K}'$ a compact subset of $C_+(E \times (\delta,\infty))$,

$$\sup_{g\in\mathcal{K}'} \left| \exp[-\hat{\mu}(g)] - \exp\left[-\int_{E\times(\delta,\infty)} (1-e^{-g(x,u)})\mu(dx,du)\right] \right| \to 0 \qquad (6.78)$$

almost surely. For $f \in C_+(E)$ define $\tilde{f} \in C_+(E \times (\delta,\infty))$ by $\tilde{f}(x,u) = uf(x)$;

then for each $a>0$, $\{\tilde{f}: f\in\mathcal{H}\}$ is a compact subset of $C_+(E\times(\delta,a))$ and (6.78) yields

$$\sup_{f\in\mathcal{H}}\left|\exp[-\hat{\mu}(\tilde{f}1_{E\times(\delta,a)})]-\exp\left[-\int_{E\times(\delta,a)}(1-e^{-uf(x)})\mu(dx,du)\right]\right|\to 0$$

almost surely. For a sufficiently large (6.75) implies that the difference between $\exp[-\hat{\mu}(\tilde{f}1_{E\times(\delta,a)})]$ and $\exp[-\hat{\mu}(\tilde{f}1_{E\times(\delta,\infty)})]$ is small uniformly in $f\in\mathcal{H}$, as is that between $\exp[-\int_{E\times(\delta,a)}(1-e^{-uf(x)})d\mu]$ and $\exp[-\int_{E\times(\delta,\infty)}(1-e^{-uf(x)})d\mu]$, and hence (6.77) holds. □

For likelihood ratios we obtain a rather precise result by invoking the independent increments property of the Poisson process N that represents M: for (δ_n) a sequence decreasing to zero the point processes $N_n(\cdot)=N(\cdot\cap(E\times(\delta_n,\delta_{n-1}]))$ are independent Poisson processes under P_μ with *finite* mean measures $\mu(\cdot\cap(E\times(\delta_n,\delta_{n-1}]))$. Using Theorem 6.13 we derive a sufficient condition for equivalence, together with an expression for the likelihood ratio. The sample space is presumed to be $\Omega=\mathbf{M}_a(E)$, the set of purely atomic measures of E, with coordinate mapping $M(\omega)=\omega$ and σ-algebra $\mathcal{F}^M=\sigma(M)$.

Theorem 6.31. Suppose that μ_0, μ_1 are Radon measures on $E\times(0,\infty)$, each fulfilling (6.75), and that in addition

a) $\mu_0\sim\mu_1$ on $E\times(0,\infty)$;

b) $\int_{E\times(0,\infty)}[1-(d\mu_1/d\mu_0)^{1/2}]^2\,d\mu_0<\infty$

Then $P_{\mu_0}\sim P_{\mu_1}$ on $\mathcal{F}^M$, and with $\delta_n\to 0$, $\delta_0=\infty$ and $B_n=(\delta_n,\delta_{n-1}]$,

$$\frac{dP_{\mu_1}}{dP_{\mu_0}}=\prod_{n=1}^{\infty}\exp\left[(\mu_1-\mu_0)(E\times B_n)+\sum_{U_j\in B_n}\log\frac{d\mu_1}{d\mu_0}(X_i,U_j)\right]\qquad(6.79)$$

Proof: For the Poisson process N, Theorem 6.13, which applies by assumptions a) and b), gives, with $C_n=[\delta_n,\infty)$,

$$\frac{dP_{\mu_1}}{dP_{\mu_0}}\Big|_{\mathcal{F}^N(E\times C_n)}=\exp\left[(\mu_1-\mu_0)(E\times C_n)+\int_{E\times E_n}\log\frac{d\mu_1}{d\mu_0}\,dN\right]$$

Utilizing the martingale convergence theorem (as in Liptser and Shiryayev, 1978, Theorem 19.6) to take limits, along with invertibility of the mapping $\Sigma\, u_j\varepsilon_{x_j}\to\Sigma\,\varepsilon_{(x_j,u_j)}$ of $\mathbf{M}_a(E)$ into $\mathbf{M}_p(E\times(0,\infty))$, we arrive at (6.79). □

Note that unlike the analog (6.29) for Poisson processes, the log-likelihood ratio (6.79) is not a linear function of M, but it is, of course, linear in N. Theorem 6.31 has application to problems ranging from L^1-maximum likelihood estimation via sieves (as in Theorem 6.8) to likelihood ratio tests to state estimation for random measures with conditionally independent increments; concerning the latter, see Karr (1983).

For infinitely divisible point processes we recall from Theorem 1.32 that a point process N on E is infinitely divisible if and only if N is equal in distribution to a Poisson cluster process (Definition 1.30)

$$N \stackrel{d}{=} \sum_j N_j, \tag{6.80}$$

where $\widetilde{N} = \Sigma\, \varepsilon_{N_j}$ is a Poisson process on $\mathbf{M}_p(E)$ with mean measure $\mu_{\widetilde{N}}$ satisfying

$$\int_{\mathbf{M}_p} (1 - e^{-\nu(E)})\, \mu_{\widetilde{N}}(d\nu) < \infty \tag{6.81}$$

Potentially, methods introduced in the chapter are applicable, but there is a major difficulty: (6.80) gives only equality in distribution, as contrasted with the pointwise representation (6.74). There is no way in general to reconstruct the components N_j from the superposition N, which prevents one's exploiting the Poisson cluster representation (6.80). Empirical methods from Section 4.2 can be applied to estimate from i.i.d. copies of N the Laplace functional

$$E_\lambda[e^{-N(f)}] = \exp[-\int (1 - e^{-\nu(f)})\lambda(d\nu)],$$

where the index λ ranges over the set of measures on $\mathbf{M}_p$ satisfying (6.81), but it is usually impossible to estimate λ directly as the mean of the processes $\widetilde{N}^i = \Sigma_j\, \varepsilon_{N_{ij}}$ because the N_{ij} cannot be calculated from N_i.

Of course, for processes constructed explicitly as Poisson cluster processes in the sense of Definition 1.30, things are more hopeful. A Poisson cluster process N on E has the form (6.80), where there is an underlying Poisson process $\overline{N} = \Sigma\, \varepsilon_{(X_j, N_j)}$ on a space $E_0 \times E$, where $N^0 = \Sigma\, \varepsilon_{X_j}$, the cluster center process, is Poisson on E_0 with mean measure μ_0, and where $\overline{N}$ is obtained from N^0 by position-dependent marking (Example 1.28) using a transition kernel K: $E_0 \rightarrow \mathbf{M}_p(E)$. Hence $\overline{N}$ is Poisson with mean measure $\mu_0(dx)K(x,d\nu)$; while $\mathbf{M}_p(E)$ is not compact, the general theory applies as long as E_0 is compact. If the process $N^1 = \Sigma\, \varepsilon_{N_j}$ is observable, as in some applications it is even though none of the X_i are observable, then methods from Section 6.3 apply provided that the kernel K is known. In particular, Theorem 6.27 allows MMSE reconstruction of the cluster center process N^0 from observation of N^1. The *really* interesting state estimation problem of estimating the cluster center process N^0 from the Poisson cluster process $N = \Sigma\, N_j$ alone has yet to be solved.

EXERCISES

6.1. Let N be a Poisson process on a general space E with mean measure μ. Show that for each k the cumulant measure γ^k (Definition 9.6) is given by $\gamma^k(f_1 \otimes \cdots \otimes f_k) = \mu(\Pi_{i=1}^k f_i)$.

6.2. Let N be a Poisson process on $[1,\infty)$ with mean measure $\mu(ds) = \lambda s^a\, ds$, where $\lambda > 0$ and $a \in \mathbf{R}$ are unknown.

a) Show that if $a < -1$, then consistent estimation of λ and a given one realization of N is impossible.

b) Given that $a > 0$, show that the estimators $\hat{\lambda}$ of (6.2) are strongly consistent as $t \to \infty$.

c) Still with $a > 0$, discuss asymptotic normality of the estimators $\hat{a}$, $\hat{\lambda}$ of (6.1) and (6.2).

6.3. Consider the statistical model $\mathcal{P} = \{P_\lambda : \lambda > 0\}$, where under P_λ, $N_1, N_2, \ldots$ are i.i.d. Poisson processes on a compact set E with mean measure $\mu = \lambda\mu^*$, where μ^* is a *known* element of $\mathbf{M}_d$.

a) Construct estimators $\hat{\lambda}$ of λ, given observations $N_1, \ldots, N_n$, that are strongly consistent and asymptotically normal; verify these properties.

b) Examine asymptotic behavior of the empirical Laplace functionals $\hat{L}(f) = \exp[-\hat{\lambda}\mu^*(1 - e^{-f})]$ and calculate their asymptotic efficency relative to the general empirical Laplace functionals of (4.16).

6.4. Let E be noncompact and let μ^* be a *known* element of $\mathbf{M}_d$ with $\mu^*(E) = \infty$. Suppose that under P_λ, $\lambda > 0$, N is a Poisson process with mean measure $\lambda\mu^*$. Repeat Exercise 6.3 relative to observation of a single realization of N over bounded sets $B_n \uparrow E$. (If necessary, specify the B_n to have additional properties, but impose as few restrictions as possible.)

6.5. Analyze consistency and asymptotic normality of the estimators $\hat{\rho}(A \times B) = \hat{\mu}(A \cap B)$ of the covariance measure ρ of a Poisson process, given as data i.i.d. copies of the process.

6.6. Verify that (6.14) is the correct covariance function in the context of Proposition 6.4.

6.7. Let N be a Poisson process on $[0,1]$ with unknown rate function α.

a) Derive the martingale estimator (Definition 5.3) of the process $B_t(\alpha) = \int_0^t \alpha_s\, ds$.

b) Discuss asymptotic behavior of estimators $\hat{B}^n$ associated with Poisson processes N^n having rate functions $\alpha^n = n\alpha$.

6.8. Let N be a Poisson process on $[0,1]$ with unknown *decreasing* intensity function α further assumed to be continuous and strictly positive. Show that within the statistical model specified by these restrictions the maximum likelihood estimator of α is the derivative of the least concave majorant of the function $t \to N_t$.

6.9. Suppose that μ_0, μ_1 are *finite* measures on E and let N be a Poisson process that under P_i has mean μ_i.

a) Prove that P_0 and P_1 cannot be singular on $\mathcal{F}^N(E)$.

b) Given the Lebesgue decomposition $dP_1 = Z\,dP_0 + dQ$, where $Q \perp P_0$, provide an upper bound on the mass of Q.

c) Discuss implications for hypothesis testing.

6.10. Give an explicit example of a space E (necessarily not compact—why?) and measures μ_0, μ_1 on E such that (for P_μ the law of the Poisson process with mean measure μ), $P_{\mu_0} \sim P_{\mu_1}$ on $\mathcal{F}^N(B)$ for every bounded set B but $P_{\mu_0} \perp P_{\mu_1}$ on $\mathcal{F}^N(E)$.

6.11. Let P_0, P_1 be probabilities with respect to which the Poisson process N on $\mathbf{R}_+$ has equivalent mean measures μ_0, μ_1.

a) Prove directly (from properties of Poisson processes) that the likelihood ratio process

$$Z_t = \exp\left[\int_0^t \left(1 - \frac{d\mu_1}{d\mu_0}\right) d\mu_0 + \int_0^t \log \frac{d\mu_1}{d\mu_0}\, dN\right]$$

is a $(P_0, \mathcal{F}^N)$-martingale.

b) Derive a stochastic differential equation satisfied by Z and discuss its application to recursive computation of likelihood ratios.

6.12. a) Let N be a Poisson process on $\mathbf{R}_+$ with unknown rate λ, let $L_t(\lambda) = -\lambda t + N_t(\log \lambda)$ be the log-likelihood process (with a constant term deleted), and let $\hat{\lambda} = N_t/t$ be the maximum likelihood estimator given by (3.40). Prove that under P_μ, $2[L_t(\hat{\lambda}) - L_t(\lambda)] \xrightarrow{d} \chi_1^2$, a chi-squared distribution with 1 degree of freedom, as $t \to \infty$.

b) Generalize to the case of Exercise 6.3.

6.13. Let $N_1, N_2, \ldots$ be i.i.d. copies of a Poisson process on a compact space E with unknown mean measure μ, let μ^* be a fixed element of $\mathbf{M}_d$, and consider testing the hypothesis $H_0 : \mu \ll \mu^*$. Given observations $N_1, \ldots, N_n$ and a finite partition $E = \Sigma A_j$, under H_0 (6.26) provides a restricted maximum likelihood estimator of $d\mu/d\mu^*$. A plausible (but not rigorously justified) criterion for testing H_0 is to reject if the test statistic $T = \max\{N^n(A_j)/\mu^*(A_j)\}$ is too large, where $N^n = N_1 + \cdots + N_n$. (If H_0 fails, then there exists a component of μ singular with respect to μ^* and if successively finer partitions were employed, then the test statistics would diverge to $+\infty$ almost surely.)

a) Given fixed $\mu \ll \mu^*$, sample size n, and partition (A_j), calculate the distribution of T.

b) Analyze large-sample behavior under the assumptions of Theorem 6.8.

6.14. Let N_1, N_2 be independent Poisson processes on $[0,1]$ with mean measures μ_1, μ_2 satisfying (in distribution function notation) $\mu_1(1) = \mu_2(1) = 1$. Propose, motivate, and analyze a test of the hypothesis $H_0 : \mu_1(t) \geq \mu_2(t)$ for all t. [*Hint*: Use conditional uniformity; see the discussion of stochastic orderings in Section 8.2.]

6.15. Formulate and prove an analog of Proposition 6.21 for the observations $\mathcal{H}_t$ of (6.49).

6.16. Prove Proposition 6.22.

6.17. Prove Lemma 6.24.

6.18. Let $N_1 = \Sigma\, \varepsilon_{T_i}$ and $N_2 = \Sigma\, \varepsilon_{S_i}$ be independent Poisson processes on $\mathbf{R}_+$ with rates λ, μ and suppose that one is able to observe only the Poisson samples $N_1(S_i)$, $i = 1,2,\ldots$. That is, observations are the cumulative numbers of arrivals in N_1 at each arrival time of N_2. Suppose that μ is known but that λ is not.

a) Show that the maximum likelihood estimator of λ given the observations $N_1(S_1), \ldots, N_1(S_n)$ is $\hat{\lambda} = \mu N_1(S_n)/n$.

b) Prove that the estimators are strongly consistent and asymptotically normal.

c) Construct a likelihood ratio test of the hypotheses $H_0 : \lambda = \lambda_0$; $H_1 : \lambda = \lambda_1 > \lambda_0$.

6.19. (Continuation of Exercise 6.18) Suppose that in addition to the values $N_1(S_n)$ the sampling times S_n are observable. Develop asymptotic properties of the estimators $\hat{\lambda} = N_1(S_n)/S_n$ corresponding to the data $(S_1, N_1(S_1)), \ldots, (S_n, N_1(S_n))$ and calculate their asymptotic efficiency relative to the estimators $\hat{\lambda}$ of Exercise 6.18.

6.20. (Continuation of Exercise 6.18) Suppose now that λ and μ are both known. Given as data the pairs $(S_n, N_1(S_n))$, let $\mathcal{H}_t = \sigma(1(S_n \le t), (S_n, N_1(S_n)) : n \ge 1)$ be the σ-algebra describing observations over $[0,t]$.

a) Calculate explicitly the MMSE state estimators $E[N_1(t)|\mathcal{H}_t]$.

b) Derive a stochastic differential equation satisfied by these estimators.

6.21. Formulate and prove a central limit theorem for the estimators $\hat{\mu}$ given by (6.65).

6.22. Let N be a Poisson process on a compact space E with known mean measure μ and let the observations be the integral data $N(g_0) = N(E), N(g_1), \ldots, N(g_m)$, where $g_0 \equiv 1$, $g_1, \ldots, g_m$ are continuous functions on E. Let $V = sp\{g_0, \ldots, g_m\}$. Show that for fixed $f \in C_+(E)$ the MMSE linear predictor of $e^{-N(f)}$ given these observations, that is, the values of $g \in V$ and $a \in \mathbf{R}$ minimizing $E[\,(e^{-N(f)} - a - N(g))^2]$, are g^*, the orthogonal projection of $\exp[-\mu(1 - e^{-f})](e^{-f} - 1)$ onto V, and $a^* = E[e^{-N(f)}] - \mu(g^*)$.

6.23. Let $N_1, N_2, \ldots$ be i.i.d. Poisson processes on $[0,1]$ with unknown rate $\lambda > 0$ and let g be a positive function on $[0,1]$. Suppose that one can observe only the integral data $N_1(g), N_2(g), \ldots$.

a) Develop asymptotic properties of the estimators $\hat{\lambda} = \Sigma_{i=1}^{n} N_i(g)/n \int_0^1 g$.

b) Suppose that g can be chosen (but only a priori) by the experimenter. Propose and defend an optimality criterion and calculate the optimal choice of g relative to it.

6.24. Let $N = \Sigma \varepsilon_{T_i}$ be a Poisson process on $\mathbf{R}$ with known rate λ, let $(S_i)_{i \in \mathbf{Z}}$ be i.i.d. random variables independent of N with distribution function G, and let $\tilde{N} = \Sigma\, \varepsilon_{T_i + S_i}$ be the point process formed by displacing each point T_i of N by the distance S_i. Exercise 1.20 shows that estimation of G from observation of $\tilde{N}$ alone is impossible; here we examine estimation of G given observation of N and $\tilde{N}$ separately. (One does not know the links between the two processes.)

a) Prove that $N^* = \Sigma\, \varepsilon_{(T_i, T_i+S_i)}$ is a Poisson process with

$$E[\exp^{-N^*(h)}] = \exp[-\lambda \int\int (1-e^{-h(x,x+y)})\, dx\, G(dy)].$$

b) Show that for $z \in \mathbf{R}$ and $t,\ \tau > 0$,

$$P\{N((z,t+z]) = 0, \widetilde{N}((z,\tau+z]) = 0\}$$

$$= \exp\left[-\lambda \int_0^t [G(\tau - x) - G(x)]dx\right] \quad (6.82)$$

c) Prove that G is uniquely determined by the integrals on the right-hand side of (6.82).

d) For fixed t and τ devise strongly consistent estimators of the probability $P\{N((0,t]) = 0,\ \widetilde{N}((0,\tau]) = 0\}$ given observation of N and $\widetilde{N}$ over increasingly large sets.

6.25. Suppose that there is an unknown number m of sources transmitting signals according to mutually independent Poisson processes $N_1, \ldots, N_m$ with known rate λ. When a signal is received one can discern from which source it emanated. Let $K(t) = \Sigma_{j=1}^m 1(N_j(t) \geq 1)$ be the number of sources that have transmitted in $[0,t]$ and let $N(t) = \Sigma_{j=1}^m N_j(t)$ be the total number of signals received during the same interval.

a) Show that $K(t)$ is a sufficient statistic for m.

b) Show that the estimators $\hat{m} = K(t)/(1 - e^{-\lambda t})$ are unbiased.

c) Suppose that there is a observation cost $c > 0$ per unit of time and that the loss function is $\ell(t) = E[(\hat{m} - m)^2] + ct$. Show that as $m \to \infty$ the optimal observation time t^* minimizing ℓ satisfies $t^* = \lambda^{-1}\log(m\lambda/c) + O(1/m)$.

6.26. (Continuation of Exercise 6.25) Suppose that the individual transmission rate λ is also unknown.

a) Prove that given observation over $[0,t]$, the statistic $(K(t),N(t))$ is sufficient for (m,λ).

b) Let a be a function on $[1,\infty)$ satisfying

i) $a(x)/(1 - e^{-a(x)}) = x$ for $x \geq 2$;

ii) a has four continuous derivatives on $[1,3]$;

iii) a is bounded away from zero.

Show that as $m \to \infty$ and $\lambda t \to \infty$ the estimators $\hat{m} = K(t)/[1 - e^{-a(N(t)/K(t))}]$ satisfy

$$E[\hat{m} - m] = -\frac{1}{2}\lambda t e^{-\lambda t}[1 + o(1)] + O(m^{-k})$$

and

$$\mathrm{Var}(\hat{m}) = m e^{-\lambda t}[1 + o(1)] + O(m^{-k})$$

for every $k > 0$. [The intuitive basis: $a(N(t)/K(t))$ is an estimator of λt, which one then substitutes into the estimators of Exercise 6.25b).]

6.27. Let $\{K(x,k): 0 \le x \le 1, k \in \mathbf{N}\}$ be the Kiefer process defined by (6.18), where the two-dimensional Wiener process $\widetilde{W}$ is by definition a mean zero Gaussian process with $E[\widetilde{W}(x_1,y_1)\widetilde{W}(x_2,y_2)] = (x_1 \wedge x_2)(y_1 \wedge y_2)$.
a) Prove that K is a mean zero Gaussian process with covariance function $E[K(x_1,k_1)K(x_2,k_2)] = (x_1x_2 - x_1 \wedge x_2)(k_1 \wedge k_2)$.
b) Prove that for each k, $B_k(x) = k^{-1/2}K(x,k)$ is a Brownian bridge.
c) Prove that the processes $\widetilde{B}_k(x) = K(x,k) - K(x,k-1)$ are independent Brownian bridges.

NOTES

References for basic theory of nonhomogeneous Poisson processes on $\mathbf{R}_+$ are Khinchin (1956b, 1960) and Snyder (1975), while for Poisson processes on general spaces, in addition to the still readable early works of Dobrushin (1956) and Doob (1953) there are more modern expositions in Kallenberg (1983) and Matthes et al. (1978). In this book we have omitted the recently emerged central role of Poisson processes as building blocks for construction of Markov (and other) processes; works treating this role include Çinlar and Jacod (1981), El-Karoui and Lepeltier (1977), Watanabe (1977), and Yor (1976). To our knowledge there exists no treatment other than this of inference for Poisson processes on general spaces.

Section 6.1

Without restriction on the form of the mean measure, single-realization inference for Poisson processes can be arbitrarily uninformative even on $\mathbf{R}_+$; thus one is led in the nonparametric case to the multiple-realization formulation of this section. The simplest parametric case, homogeneous Poisson processes on $\mathbf{R}_+$, is treated in Section 3.5; see the associated notes for additional references. In Chapter 9 homogeneous Poisson process on $\mathbf{R}^d$, $d \ge 1$, are analyzed as stationary point processes. In Exercise 6.3 the mean measure is presumed known within a scalar multiplier. Many other sources address special estimation problems or applicability of particular methods. Prominent among them are Kutoyants (1979, 1982), by far the most comprehensive examination of parametric maximum likelihood estimation, Bartoszyński et al. (1981) on penalized maximum likelihood estimation of intensity functions, and Krickeberg (1974, 1982) on estimation of moment measures. Nonparametric Bayesian estimation is treated by Lo (1982); sequential methods are described by Dvoretsky et al. (1953b) and in a more modern setting permitting censored observations by Vardi (1979). "Linear" inference is discussed by Clevenson and Zidek (1977). The problem of estimating the "number of sources," parts of which appear in Exercises 6.25 and 6.26, receives its most detailed study in Hall (1982) (see also Starr, 1974; Vardi, 1980). Ripley (1981) and Diggle (1983) contain expository development of "distance methods" for planar Poisson processes. These techniques, which focus on distances from points in $\mathbf{R}^2$ to the nearest point of the process or from a point of the process to its nearest neighbor, have received widespread and productive application despite as yet incomplete understanding of their theoretical properties.

The main theorems of the section, Theorems 6.5 and 6.8, appear here for the first time. The former, a strong approximation for ordinary Poisson processes that admits numerous useful implications, should be compared to Theorem 4.33 and to Csörgö and Révész (1981, Theorem 7.3.2); Pyke (1968) contains an "in distribution" analog of one of its consequences. Basic properties of the Kiefer process, first used by Kiefer (1972), can be found in Csörgö and Révész (1981). Theorem 6.8 employs ideas based on Grenander (1981, Chapter 8) and Karr (1986b).

Section 6.2

Of results here, Theorem 6.13, taken from Karr (1983), is an extension to general spaces of Liptser and Shiryayev (1978, Theorem 19.7). Related versions are given by M. Brown (1971, 1972)—the latter is perhaps the most detailed treatment available of likelihood ratios and likelihood ratio tests for Poisson processes on $\mathbf{R}_+$, as well as the source of Proposition 6.14. Weiss (1979) shows that in the noncompact case singularity of Poisson process laws on $\mathcal{F}^N$ is equivalent to singularity on an appropriately defined tail σ-algebra. Albeit new, the contiguity property in Theorem 6.15 is not especially deep. Variations and special cases can be found in Boswell (1966), Cox and Lewis (1966), and Saw (1975), all concerning tests of hypotheses of "trend" (see also Section 3.5); in Dvoretsky et al. (1953a), Epstein (1960), and Kiefer and Wolfowitz (1957), which examine sequential testing of hypotheses for Poisson processes on $\mathbf{R}_+$; and in Lewis (1965), which deals with the problem of testing whether a renewal process is Poisson, given asynchronous observations (see also Section 8.2).

Section 6.3

As mentioned in the text, even though each model treated here is of particular interest for applications, an important message of the section is the variety of problems that can be formulated and of techniques available for their solution Theorem 6.16 has not appeared previously; it is a closer relative of Theorem 6.8 in terms of hypotheses and method of proof than of Theorem 4.22 even though the latter also addresses—but in a more general context—estimation of thinning functions.

With the exception of Propositions 6.21 and 6.22, which are elementary, results on the stochastic integral process $X * N$ come from Karr (1982), but proofs here are simplified by use of martingale techniques. Theorem 6.23 is a very special case of Theorem 7.29; the property that $X^2 = X$ not only makes $X * N$ a Cox process (see Karr, 1978) but also yields the single recursive equation (6.56) rather than the infinite system of equations in Theorem 7.30. Konecny (1984) applies filtering techniques to calculate likelihood functions for use in maximum likelihood estimation; see Smith and Karr (1985) and also Section 7.5 for application to models of precipitation. Further analysis of the process as a renewal process (Exercise 3.15 and Theorem 8.24) can be found in Chapter 8. Freed and Shepp (1982) examine calculation of state estimators $P\{X_0 = 1 | \mathcal{F}_t^{X*N}\}$; even this seemingly innocuous problem admits no truly "simple" solution.

Results concerning Poisson processes with one observable component stem from Karr (1985d), motivated by Vardi et al. (1985). In the latter the *EM* algorithm is utilized to calculate the maximum likelihood estimators in the context of Proposition 6.26; good descriptions of the algorithm itself and of applications to inference when there is "missing data" are the original paper of Dempster et al. (1977) and the recent survey paper of Redner and Walker (1984). The mixture identifiability problem arising in Proposition 6.25 is discussed by Barndorff-Nielson (1965) and Teicher (1961) (see also Lindsay, 1983a,b). In view of Theorem 1.32, some problems of inference and state estimation for Poisson cluster processes fall within this setting (see also Section 6.5).

Section 6.4

Karr (1985e) is the source of results presented here; see Section 3.1 for introductory discussion of inference given integral data. Complete Tchebycheff systems are important in approximation theory and description of the structure of compact, convex sets of measures. Karlin and Studden (1966) provide a comprehensive and comprehensible exposition.

Section 6.5

Brief as it is, this section exhibits rather clearly two kinds of problems: easy problems, for example inference for random measures with independent increments, for which the Poisson cluster representation is compatible with natural forms of observation and to which previously derived results apply in a straightforward manner, and difficult problems, whose solution is not facilitated by the point process's admitting a Poisson cluster representation, because the representation cannot be reconstructed from observation of the process. For such point processes other methods must be—and remain to be—developed. Of results here, Theorem 6.31 comes from Karr (1983) (see also Akritas and Johnson, 1981; Kailath and Segall, 1975), while Proposition 6.30 is new. There exist several papers treating special topics in inference for Poisson cluster processes; among them are Baudin (1981) on nearest-neighbor methods in $\mathbf{R}^2$; Brown and Silverman (1978) on tests based on distance methods; Kryscio and Saunders (1983) on nearest neighbors, interpoint distances, and discrimination between Poisson and Poisson cluster processes; and Strauss (1975) on tests for clustering.

Exercises

6.5. See also Section 9.1.

6.8. Grenander (1956) derives analogs in the context of estimation of density functions.

6.12. This property corresponds to a result for the classical case of i.i.d. real data (see Bickel and Doksum, 1977, or Wilks, 1962). Fundamentally a consequence of asymptotic normality of derivatives of log-likelihood functions ("score" functions), it also appears, for example, in Theorem 9.26.

6.13. The suggested technique is related to methods proposed by Saw (1975).

6.17. See the discussion of position-dependent marking in Section 1.5 and also Exercise 1.11.

6.18–6.20. See Kingman (1963) and Section 8.1 concerning Poisson sampling of renewal processes. Additional Poisson sampling problems are examined in Chapter 10.

6.21. See Karr (1985e).

6.24. Random translations of Poisson processes are discussed in M. Brown (1969), Milne (1970), and Thédeen (1964, 1967a); the second of the four is most oriented to statistical inference.

6.25. See Starr (1974), to whom the model seems due, and Vardi (1980).

6.26. See Hall (1982).

6.27. These and deeper properties of the Kiefer process are given in Csörgö and Révész (1981).

7

Inference for Cox Processes on General Spaces

With their conditionally Poisson structure, Cox processes are uniquely suited to state estimation and to application in a variety of situations. Lundberg (1940) used Pólya processes, a class of mixed Poisson processes, to model accident statistics. Cox (1955), a seminal work, presented a model of breakages of thread as it is fed into looms: the thread is composed of long, internally homogeneous segments tied together; within each, weak points causing breakages can be represented as an ordinary Poisson process, but the rate varies randomly from segment to segment. If the rate is modeled as a pure jump random process (X_t), then the point process of breakages is a Cox process directed by the random measure $M(B) = \int_B X_t \, dt$. Snyder (1975) treats a number of communications applications in which the directing intensity is an unobservable information process modulating the observable Cox process. A Cox process model of precipitation occurrences is formulated in Smith and Karr (1983) (see also Rohde and Grandell, 1981) using the stochastic integral $X * N$ of Examples 3.5 and 3.10 and Section 6.3; the 0-1 Markov intensity process represents alternating climatological states, a "dry" state $(X_t = 0)$ during which (for meteorological reasons) precipitation cannot arise, and a "wet" state $(X_t = 1)$ during which precipitation events form a Poisson process.

The unifying element in these applications is a physical interpretation for the directing random measure, that is, the two-stage structure of a Cox process mirrors the physical nature of the system being modeled. Thus, for another example, a binary Markov process X may represent an on/off signal transmitted optically through space or through a fiber. The signal process is not observed directly. Instead (see Snyder, 1975), observations are photon counts, some of them noise, for which there is significant theoretical and empirical evidence substantiating the assumption that they are conditionally Poisson.

Another shared trait, perhaps most evident in the optical communication model, is that typically only the Cox process is observable; the directing measure cannot be observed directly even though it may be of equal or greater physical importance. Thus arises the state estimation problem for Cox processes: optimal reconstruction, realization by realization, of the directing measure from Cox process observations. It will occupy much of the chapter. Even for statistical inference one ordinarily wishes to perform estimation and testing for probability laws of directing measures—because it is they rather than the laws of the Cox processes that are dictated by physical considerations—but must do so with observations solely of the Cox processes. Special structure of Cox processes is exploited for statistical inference through distributional relationships that make possible estimation and hypothesis testing for unobservable directing measures, while for state estimation the key property appears in Theorem 7.6, and expresses MMSE state estimators in terms of Palm distributions of the directing measure, which are expressible in turn using those of the Cox process by means of Proposition 1.55. Throughout the chapter we assume that only Cox processes, not their directing measures, are observable.

We pause to recollect definitions and results from Chapter 1 and to introduce further terminology. Given a point process N and random measure M on the space E and defined over the same probability space, N is a *Cox process directed by* M if conditional on M (i.e., on the σ-algebra $\mathcal{F}^M$), N is a Poisson process with mean measure M:

$$E[e^{-N(f)}|\mathcal{F}^M] = \exp[-M(1-e^{-f})] \tag{7.1}$$

for each $f \geq 0$. We say also that (N,M) is a *Cox pair*. When M has the form

$$M(B) = \int_B X(y)\nu(dy), \tag{7.2}$$

with X a positive, measurable process and ν a diffuse measure on E, we call X the *directing intensity* of N. The following result collects computational relationships derived in Chapter 1.

Lemma 7.1. Let (N,M) be a Cox pair. Then

a) The probability laws $\mathcal{L}(N)$ and $\mathcal{L}(M)$ determine each other uniquely.

b) N is simple if and only if M is diffuse almost surely.

c) The following hold:

$$L_N(f) = L_M(1-e^{-f}); \tag{7.3}$$

$$z_N(A) = L_M(1_A); \tag{7.4}$$

$$\mu_N = \mu_M \tag{7.5}$$

Throughout the chapter we assume—often without explicit mention—that all *directing measures are diffuse* almost surely, so that Cox processes are simple point processes.

It follows that the family of Cox processes is as large as that of diffuse random measures; not surprisingly, for statistical estimation, the topic of Section 7.1, one must rely primarily on the general methods developed in Chapter 4, as opposed to highly specialized techniques such as those in Chapter 6. In some cases martingale methods from Chapter 5 apply, but it seems difficult to exploit Cox and martingale structures simultaneously. Moreover, throughout much of the chapter our concern is with Cox processes on general spaces, to which martingale methods do not apply at all.

One class of Cox processes for which specialized estimation techniques have been developed is the mixed Poisson processes, treated earlier in Examples 4.26 and 5.31. A *mixed Poisson process* is a Cox process with directing measure

$$M = Y\nu \tag{7.6}$$

where $\nu \in \mathbf{M}_d$ and Y is a positive random variable. There are two unknown, infinite-dimensional parameters, ν and the distribution F of Y, and special structure of a mixed Poisson process can be utilized effectively not only for statistical estimation (Section 7.1) and state estimation (Section 7.2) but also for combined statistical inference and state estimation (Section 7.4).

Figuratively and literally, the heart of the chapter is Section 7.2, on state estimation. There we develop a variety of techniques for MMSE reconstruction of unobservable directing measures from Cox process observations, in settings ranging from completely general to Cox processes with directing intensities to mixed Poisson processes. Explicit and recursive computation of MMSE state estimators $E[\exp(-M(f))|\mathcal{F}^N(A)]$ and $E[M(f)|\mathcal{F}^N(A)]$ of the directing measure given observations of the Cox process are addressed. Of course, in general spaces the notion of recursive computation is somewhat nebulous, so only on $\mathbf{R}_+$ is recursive calculation of MMSE state estimators treated meaningfully. On $\mathbf{R}_+$ as well as on general spaces, optimal state estimators are nonlinear functions of the observations $N_A(\cdot) = N(\cdot \cap A)$, which is a principal reason for their being very difficult to calculate; consequently, we also consider minimum error linear state estimators (Theorem 7.17). Applications of state estimation results appear in Sections 7.2 and 7.5.

Properties of likelihood ratios and associated issues of hypothesis testing are examined in Section 7.3, where state estimation techniques are central to representation and calculation of likelihood ratios. We also derive a separation theorem for Cox processes on $\mathbf{R}_+$ that allows recursive computation of likelihood ratios by substitution of state estimators of the directing intensity into Poisson process likelihood ratios.

Section 7.4 treats in detail the problem of combined statistical inference and state estimation for mixed Poisson processes, with briefer analysis of more difficult and less effective techniques available for general Cox processes.

A Cox process N on $\mathbf{R}_+$ with directing intensity (X_t) has $\mathcal{F}^N$-stochastic intensity

$$\lambda_t = E[X_u \mid \mathcal{F}_u^N]|_{u=t-} \tag{7.7}$$

and martingale tools, from Theorem 2.31 for calculation of likelihood ratios to the state estimation procedure of Section 5.4, apply but only (it seems) at the expense of one's more or less ignoring the Cox process structure. Such matters are treated in Section 7.5, which also contains a state estimation theorem for Cox processes on $\mathbf{R}_+$ whose directing intensity is a function of a Markov process.

7.1 GENERAL ASPECTS OF ESTIMATION

Let (N_1,M_1), (N_2,M_2), ... be i.i.d. copies of a Cox pair (N,M) on a compact space E and assume that only the Cox processes N_1, N_2, ... are observable. To estimate the Laplace functional $L_N(f) = E[\exp(-N(f))]$ in the absence of additional assumptions on the law $\mathcal{L}(N)$, one must employ empirical Laplace functionals introduced in (4.16):

$$\hat{L}_N(f) = n^{-1}\sum_{i=1}^{n} e^{-N_i(f)}, \tag{7.8}$$

to which Theorems 4.12 and 4.13 apply directly, with the limiting covariance function given by (4.19). In many cases, however, as examples and applications throughout the chapter confirm, principal interest focuses on the law of the directing measures since typically it is specified initially. Consequently, it is more germane to estimate the Laplace functional $L_M(f) = E[\exp(-M(f))]$. By Lemma 7.1 and Theorem 1.12, L_N and L_M determine each other via (7.3), which in inverted form becomes

$$L_M(g) = L_N(-\log(1-g)), \tag{7.9}$$

provided that $0 \le g < 1$. For statistical estimation this restriction is harmless, but it can cause difficulties for state estimation (see Section 7.4). Amalgamation of (7.8) and (7.9) yields the empirical Laplace functionals

$$\hat{L}_M(g) = \hat{L}_N(-\log(1-g)) = n^{-1}\sum_{i=1}^{n} \exp[-N_i(-\log(1-g))], \tag{7.10}$$

defined only for functions $g \in C_+(E)$ satisfying $0 \le g < 1$. Without additional restrictions, (7.10) is the extent to which special structure of Cox processes is exploited.

Properties of $\hat{L}_M$, all straightforward consequences of Theorems 4.12 and 4.13, are now summarized.

Proposition 7.2. Let $\hat{L}_M$ be given by (7.10), where the N_i are i.i.d. copies of a Cox process N satisfying (7.9). Then

a) For each equicontinuous subset $\mathcal{K}$ of $C_+(E)$ such that

$$\sup\{\|g\|_\infty : g \in \mathcal{K}\} < 1, \tag{7.11}$$

$\sup\{|\hat{L}_M(g) - L_M(g)| : g \in \mathcal{K}\} \to 0$ almost surely.

b) If $\mathcal{K}$ is a compact subset of $C_+(E)$ fulfilling (7.11) and having finite metric entropy, then the error processes $\{n^{1/2}[\hat{L}_M(g) - L_M(g)] : g \in \mathcal{K}\}$ converge in distribution, as random elements of $C(\mathcal{K})$, to a continuous Gaussian process G with covariance function

$$R(g_1,g_2) = L_M(g_1 + g_2 - g_1g_2) - L_M(g_1)L_M(g_2) \quad \square \tag{7.12}$$

The covariance function R of (7.12) differs from the covariance function R^* of the empirical Laplace functionals

$$\hat{L}^*_M = n^{-1}\sum_{i=1}^{n} e^{-M_i(f)},$$

that would be used if the M_i were observable but which are not, however, computable from the data $N_1, N_2, \ldots$; the latter covariance by, (4.19), is

$$R^*(g_1,g_2) = L_M(g_1 + g_2) - L_M(g_1)L_M(g_2)$$

In particular, for $g \in C_+(E)$ with $g < 1$, the asymptotic efficiency of $\hat{L}_M(g)$ relative to $\hat{L}^*_M(g)$ is

$$e(\hat{L}_M(g),\hat{L}^*_M(g)) = \frac{R^*(g,g)}{R(g,g)}$$

$$= \frac{L_M(2g) - L_M(g)^2}{L_M(2g - g^2) - L_M(g)^2} < 1;$$

hence there is, not unexpectedly, loss of efficiency in consequence of one's inability to observe the M_i directly.

By contrast with the Poisson case, superpositions $N^n = N_1 + \cdots + N_n$ are *not* sufficient statistics for the unknown law of the M_i, even for mixed Poisson processes, and therefore statistical inference cannot be based on N^n alone. In Section 7.3 we confirm this assertion by calculation of likelihood functions, but another justification can be deduced from the following result.

Lemma 7.3. For each n the superposition N^n is a Cox process with directing measure $M^n = M_1 + \cdots + M_n$. □

That Lemma 7.3 implies that N^n is not sufficient can be argued heuristically as follows: if it were, then $L_M(f) = \lim L_{M^n}(f)^{1/n}$ could be estimated from the N^n alone, but by applying (7.10) to the Cox pair (N^n, M^n) treated as a sample of size 1 we obtain the estimators $\hat{L}_{M^n}(g) = \exp[-N^n(-\log(1-g))]$, whose n^{th} roots converge not to L_M but only to $\exp[-\mu_N(-\log(1-g))]$.

In view of (7.5), estimation of the mean measure $\mu_M = \mu_N$ is routine; Lemma 7.3 shows that N^n is a sufficient statistic for μ_M and the estimators

$$\hat{\mu}_M = n^{-1}N^n = n^{-1}\sum_{i=1}^{n} N_i \tag{7.13}$$

have the limiting behavior described by Propositions 4.28 – 4.31, provided only that suitable moment hypotheses be fulfilled. As was true for the empirical Laplace functionals of (7.10), variance properties differ from those of the estimators $\hat{\mu}^*_M = n^{-1}\Sigma_{i=1}^n M_i$ that would be utilized if the M_i were observable, and there is concomitant loss of efficiency (Exercise 7.7).

Depending on one's point of view, for estimation in the context of general Cox processes there is little need or little hope to go beyond what has been presented, but given appropriate structural assumptions one can develop effective specialized techniques. The one class of Cox processes (other than Poisson processes, obviously) for which this has been done not only in reasonable detail but also in (relative) generality is the mixed Poisson processes. Given a diffuse (finite) measure ν on E and a probability distribution F on $\mathbf{R}_+$, let $P_{\nu,F}$ be a probability measure under which (N_1, Y_1), $(N_2, Y_2), \ldots$ are i.i.d. copies of a pair (N, Y) such that Y is a random variable with distribution F and N is a Cox process with directing measure $M = Y\nu$. While the methods above apply to estimation of the Laplace functional

$$L_N(g) = \int F(du)\exp[-u\nu(1 - e^{-g})],$$

the goal now is to estimate ν and F separately, assuming that only the N_i are observable.

We assume that

$$F(0) = 0, \tag{7.14a}$$

(purely for convenience in this case), that

$$\nu(E) = 1, \tag{7.14b}$$

and that

$$m_F = \int uF(du) < \infty \tag{7.14c}$$

The assumption (7.14c) implies that the mean measure $\mu_N = \mu_M = m_F\nu$ is finite, while (7.14b)—or an alternative—is needed in order to ensure that the full nonparametric model $\mathcal{P} = \{P_{\nu,F}\}$ be identifiable. Otherwise, a multiplicative factor can be shifted between ν and (the distribution F of) the Y_i while affecting neither the distribution of the M_i nor that of the N_i. Alternatives to (7.14b) have been employed (for estimation of ν alone it is convenient to suppose instead that $m_F = 1$, because then ν is the mean of the N_i), but the form we have chosen permits use of conditional uniformity of mixed Poisson processes (Exercise 7.3), allows straightforward estimation of ν and F, and is well suited to estimation of more complicated objects required in the combined statistical inference and state estimation setting of Section 7.4. We work below only with the full nonparametric model specified by (7.14), with the N_i as the sole observations. If ν is known or the Y_i are observable, further specialization is possible. The latter problem can be approached directly (Exercise 7.8); the former, which ceases to be a point process problem, has been treated in the literature, some references to which are given in the chapter notes. See also the discussion relative to (7.18).

For the mixed Poisson process model as we have formulated it, estimation of ν and F is direct. In view of Exercise 7.2, N_i has the structure

$$N_i = \sum_{j=1}^{N_i(E)} \varepsilon_{X_{ij}}, \tag{7.15}$$

where $N_i(E)$ has the mixed Poisson distribution

$$Q_F(k) = \frac{\int F(du)e^{-u}u^k}{k!} = P_{\nu,F}\{N(E) = k\}, \tag{7.16}$$

which by (7.14b) is not dependent on ν, and where the X_{ij} are i.i.d. random elements of E, independent of $N_i(E)$ and with distribution ν. The latter independence manifests itself as asymptotic independence of the estimators we are about to define. The estimators

$$\hat{\nu} = \frac{\sum_{i=1}^n N_i}{\sum_{i=1}^n N_i(E)} \tag{7.17}$$

are normalizations of empirical processes with random sample sizes. (For similar analysis of Poisson processes see Theorem 6.5 and related discussion.) The distribution F is determined uniquely by the integrals $Q_F(k)$: we estimate the latter simply with empirical averages

$$\hat{Q}(k) = n^{-1} \sum_{i=1}^{n} 1(N_i(E) = k) \tag{7.18}$$

There are disadvantages to this choice, for characteristics of F such as moments are not easily calculated from Q_F and must be estimated independently rather than by substitution as functionals of Q_F. Concerning alternatives, Karr (1984b) presents techniques for estimation of ν and the somewhat more useful Laplace transform of F, and Simar (1976) addresses maximum likelihood estimation of F itself, given i.i.d. mixed Poisson random variables as data; in this context estimation of moments as functionals is direct.

Properties of $\hat{\nu}$ and $\hat{Q}$ can be developed without difficulty.

Theorem 7.4. Let $\hat{\nu}$, $\hat{Q}$ be given by (7.17) and (7.18), respectively, and assume that (7.14) holds. Then almost surely with respect to $P_{\nu,F}$,

a) For each compact subset $\mathcal{K}$ of $C_+(E)$, $\sup\{|\hat{\nu}(f) - \nu(f)| : f \in \mathcal{K}\} \to 0$.

b) $\sup\{|\hat{Q}(k) - Q(k)| : k \in \mathbf{N}\} \to 0$.

Proof: Of course, b) is simply the Glivenko-Cantelli theorem (see Gaenssler, 1984) applied to the empirical distributions Q of the random variables $N_i(E)$, and requires no further comment. As for a),

$$\sup_{\mathcal{K}} |\hat{\nu}(f) - \nu(f)| \le [n^{-1} \sum_{i=1}^{n} N_i(E)]^{-1} \sup_{\mathcal{K}} \left| n^{-1} \sum_{i=1}^{n} N_i(f) - m_F \nu(f) \right|$$
$$+ \left(\sup_{\mathcal{K}} \nu(f) \right) \left| [n^{-1} \sum_{i=1}^{n} N_i(E)]^{-1} - \frac{1}{m_F} \right|$$

By the strong law of large numbers, almost surely $n^{-1} \sum_{i=1}^{n} N_i(E) \to m_F$ and hence the second term converges to zero. By Proposition 4.29, since under $P_{\nu,F}$ the N_i have finite mean measure $m_F \nu$,

$$\sup\{|n^{-1} \sum_{i=1}^{n} N_i(f) - m_F \nu(f)| : f \in \mathcal{K}\} \to 0$$

as well. □

Asymptotic normality is hardly more difficult; however, we present only a sketch of the proof because more complicated results—in particular

Theorem 7.25—are proved using arguments that can easily be adapted to this case.

Theorem 7.5. Assume that (7.14) holds and that $E_{\nu,F}[Y^2] < \infty$. Then under $P_{\nu,F}$,

$$\begin{bmatrix} \{n^{1/2}[\hat{\nu}(f) - \nu(f)] : f \in C_+(E)\} \\ \{n^{1/2}[\hat{Q}(k) - Q(k)] : k \in \mathbf{N}\} \end{bmatrix} \xrightarrow{d} \begin{bmatrix} \{G(f) : f \in C_+(E)\} \\ \{Z(k) : k \in \mathbf{N}\} \end{bmatrix}, \tag{7.19}$$

with the first components viewed as random measures on E and the second as ordinary sequences of random variables, where G is a mean zero Gaussian random measure with covariance function

$$R_G(f,g) = m_F^{-1}[\nu(fg) - \nu(f)\nu(g)], \tag{7.20a}$$

Z is a mean zero Gaussian sequence with covariance function

$$R_Z(k,j) = 1(k=j)Q_F(k) - Q_F(k)Q_F(j), \tag{7.20b}$$

and G, Z are independent.

Proof: (Sketch). For each f,

$$n^{1/2}[\hat{\nu}(f) - \nu(f)] = \frac{n^{1/2}}{n^{-1}\sum_{i=1}^n N_i(E)}\left[n^{-1}\sum_{i=1}^n \{N_i(f) - \nu(f)N_i(E)\}\right]$$

$$\cong m_F^{-1} n^{-1/2} \sum_{i=1}^n \int (f - \nu(f))\, dN_i$$

by Slutsky's theorem. Asymptotic normality is, ultimately, a consequence of the central limit theorem (we view the first components as random measures, not as random elements of $C(\mathcal{H})$ for some $\mathcal{H} \subset C(E)$). Covariance computations are routine: (7.20b) follows from the role of the $\hat{Q}$ as empirical processes on $\mathbf{N}$, (7.20a) comes from

$$\mathrm{Cov}(N(g),N(h)) = m_F\nu(gh) + \mathrm{Var}_{\nu,F}(Y)\nu(g)\nu(h)$$

(Exercise 4.16), and finally independence of G and Z (i.e., the covariance is zero) is a consequence of conditional uniformity:

$$E_{\nu,F}[(\int (f - \nu(f))\, dN)1(N(E) = k)] = 0$$

for each k. □

As promised, more complicated analogs are proved more carefully in Section 7.4, where in order to perform approximate state estimation for mixed Poisson processes whose probability law is unknown, we estimate integrals

$$\int F(du)e^{-u\nu(A)}u^k = \frac{k!}{\nu(A)^k} P_{\nu,F}\{N(A) = k\}$$

for arbitrary subsets A of E. For now, though, Theorem 7.5 concludes our discussion of estimation for mixed Poisson processes.

7.2 STATE ESTIMATION

Let (N,M) be a Cox pair on a locally compact space E. This section is devoted to the state estimation problem of minimum mean squared error (MMSE) reconstruction of the directing measure M from observation of the Cox process N over a subset A of E, under the assumption that the probability law of M is known. We begin with a very general result, which is then specialized to produce MMSE state estimators for specific classes of Cox processes.

The general result is based on Palm distributions, concerning which we recall basic properties (see Section 1.7 for details). Denote by $Q_N(\mu,d\nu)$, $\mu \in \mathbf{M}_p$, the reduced Palm distributions of N, with heuristic interpretation

$$Q_N(\mu,d\nu) = P\{N - \mu \in d\nu | \mu \subset \operatorname{supp} N\},$$

and let $Q_M(\mu,d\nu)$, $\mu \in \mathbf{M}_p$, be the *un*reduced Palm distributions associated with M. The $Q_N(\mu,\cdot)$ are probability measures on $\mathbf{M}_p$, while the $Q_M(\mu,\cdot)$ are probability measures on $\mathbf{M}_d$; for precise definitions, see Section 1.7. For $\mu \in \mathbf{M}_p$ we recall the reduced *Palm process* N_μ with probability law $Q_N(\mu,\cdot)$, and similarly, let M_μ be the unreduced Palm process of M, with law $Q_M(\mu,\cdot)$. Proposition 1.55 relates the N_μ and M_μ: for each μ, N_μ has the distribution of a Cox process directed by M_μ.

With this preparation we can formulate the main theorem on state estimation, in which we calculate explicitly the MMSE state estimators $E[e^{-M(f)}\mathcal{F}^N(A)]$.

Theorem 7.6. Let (N,M) be a Cox pair. Then for each set A with $E[M(A)] < \infty$ and each $f \in C_+(E)$.

$$E[e^{-M(f)}|\mathcal{F}^N(A)] = \frac{E[e^{-M_\mu(A)}e^{-M_\mu(f)}]}{E[e^{-M_\mu(A)}]}\Big|_{\mu=N_A}, \tag{7.21}$$

where $N_A(\cdot) = N(\cdot \cap A)$ is the restriction of N to A (i.e., the observations).

Proof: Let $\mathcal{L}$ denote the probability law of M and let

$$L(\mu,f) = E[e^{-M_\mu(f)}] = \int Q_M(\mu,d\nu)e^{-\nu(f)}$$

denote the Laplace functional of the Palm process M_μ. To prove (7.21) it suffices (Theorem 1.12) to show that for g a positive function vanishing on A^c and each k.

$$E\left[e^{-N(g)}1(N(A)=k)\frac{L(N,f+1_A)}{L(N,1_A)}\right]=E[e^{-N(g)}1(N(A)=k)e^{-M(f)}],$$

which we do by calculating the two sides separately. Since N is conditionally Poisson given M and by conditional uniformity of Poisson processes,

$$\begin{aligned}E&\left[e^{-N(g)}1(N(A)=k)\frac{L(N,f+1_A)}{L(N,1_A)}\right]\\&=\frac{1}{k!}\int\mathcal{L}(dv)e^{-v(A)}\int_A v(dx_1)\cdots\int_A v(dx_k)\exp\left[-\sum_{i=1}^k g(x_i)\right]\\&\quad\times\frac{L(\Sigma_{i=1}^k\varepsilon_{x_i},f+1_A)}{L(\Sigma_{i=1}^k\varepsilon_{x_i},1_A)}\end{aligned}\tag{7.22}$$

Consider now a random measure $\tilde{M}$ with law $\tilde{\mathcal{L}}(dv)=C^{-1}e^{-v(A)}\mathcal{L}(dv)$, where $C=\int\mathcal{L}(dv)e^{-v(A)}$, which is positive and finite; then it is straightforward (Exercise 7.10) to verify that the Palm Laplace functionals $\tilde{L}(\mu,f)=E[e^{-\tilde{M}_\mu(f)}]$ are related to those for M by

$$\tilde{L}(\mu,f)=\frac{L(\mu,f+1_A)}{L(\mu,1_A)}\tag{7.23}$$

Substitution of (7.23) into (7.22) yields

$$\begin{aligned}E&\left[e^{-N(g)}1(N(A)=k)\frac{L(N,f+1_A)}{L(N,1_A)}\right]\\&=\frac{C}{k!}\int\mathcal{L}(dv)\int_A v(dx_1)\cdots\int_A v(dx_k)\exp\left[-\sum_{i=1}^k g(x_i)\right]\tilde{L}\left(\sum_{i=1}^k\varepsilon_{x_i},f\right)\\&=\frac{C}{k!}E\left[\int_A\tilde{M}(dx_1)\cdots\int_A\tilde{M}(dx_k)\exp\left[-\sum_{i=1}^k g(x_i)\right]e^{-\tilde{M}(f)}\right]\\&=\frac{1}{k!}E\left[\int_A M(dx_1)\cdots\int_A M(dx_k)\exp\left[-\sum_{i=1}^k g(x_i)\right]e^{-M(A)-M(f)}\right]\end{aligned}\tag{7.24}$$

On the other hand, suppose that for each v, $\tilde{N}^v$ is Poisson with mean measure v; then once more by conditional uniformity

$$\begin{aligned}E[e^{-\tilde{N}^v(g)}1(\tilde{N}^v(A)=k)]&=P\{\tilde{N}^v(A)=k\}E[e^{-\tilde{N}^v(g)}|\tilde{N}^v(A)=k]\\&=\frac{e^{-v(A)}}{k!}\int_A v(dx_1)\cdots\int_A v(dx_k)\exp\left[-\sum_{i=1}^k g(x_i)\right];\end{aligned}$$

consequently,

$$E[e^{-N(g)}1(N(A)=k)e^{-M(f)}]$$

$$= \int \mathcal{L}(d\nu)e^{-\nu(f)}E[e^{-\tilde{N}^{\nu}(g)}1(\tilde{N}^{\nu}(A)=k)]$$

$$= \frac{1}{k!}E\left[\int_A M(dx_1)\cdots\int_A M(dx_k)\exp\left[-\sum_{i=1}^{k} g(x_i)\right]e^{-M(A)-M(f)}\right] \quad (7.25)$$

Comparison of (7.24) and (7.25) indicates that the proof is complete. □

Alternative forms of (7.21) will be used elsewhere in this section and in other sections; we give them as corollaries.

Corollary 7.7. With the notation and hypotheses of Theorem 7.6, for each set Γ,

$$P\{M \in \Gamma | \mathcal{F}^N(A)\} = \frac{E[e^{-M_\mu(A)}1(M_\mu \in \Gamma)]}{E[e^{-M_\mu(A)}]}\Big|_{\mu=N_A} \quad (7.26)$$

Proof: By (7.21) the two sides of (7.26), viewed with N_A fixed as probability measures on $\mathbf{M}_d$, have the same Laplace functional; hence they are equal by Theorem 1.12. □

Corollary 7.8. Under the hypotheses of Theorem 7.6, for each function $f \in C_+(E)$,

$$E[M(f)|\mathcal{F}^N(A)] = \frac{E[e^{-M_\mu(A)}M_\mu(f)]}{E[e^{-M_\mu(A)}]}\Big|_{\mu=N_A} \quad (7.27)$$

Proof: For f such that $E[M(f)] < \infty$, (7.27) is a direct consequence of (7.26); validity for general $f \geq 0$ holds by approximation arguments. □

Note that (7.26) provides explicitly a regular version of the conditional distribution $P\{M \in (\cdot)|\mathcal{F}^N(A)\}$. The Cox relationship between M_μ and the reduced Palm process N_μ allows one to convert (7.26) to the very intuitive expression

$$P\{M \in \Gamma|\mathcal{F}^N(A)\} = P\{M_\mu \in \Gamma | N_\mu(A)=0\}\big|_{\mu=N_A}; \quad (7.28)$$

conditioning on $\{N_\mu(A)=0\}$ simply assures that N has no points in A other than at the atoms of μ.

One can hardly call Theorem 7.6 really "useful" for calculation of MMSE state estimators; it is too general. However, specialized versions such as Theorem 7.9 for Cox processes admitting directing intensities and

Proposition 7.10 for mixed Poisson processes certainly are useful, among other things for derivation of recursive methods for computation of state estimators and likelihood ratios. As discussed in Chapter 1 and also in Section 3.3 (see in particular Example 3.16) the *raison d'être* for recursive methods is to avoid recalculating state estimators from scratch as additional observations are obtained, by instead computing in some fairly simple fashion incremental changes in the estimators. Only for point processes on $\mathbf{R}_+$ can one unambiguously make sense of observations increasing infinitesimally; in this context Theorem 6.23 and the more general Theorem 7.29 are prome examples. The stochastic differential equation (6.56) is a recursive method for calculation of state estimators $E_\alpha[X_t \,|\mathcal{F}^{X*N}]$: the change at time t in the value of the estimator is determined by previously calculated values and the innovation $d(C*N)_t - \alpha E_\alpha[X_t|\mathcal{F}_{t-}^{X*N}]dt$, and consists of two terms. The first, a predictable change, involves only a dt-differential and results solely from one's having observed the process over a larger set (it is functionally independent of the innovation at time t). The second, expressed in (6.56) in the canonical form "filter gain times innovation," vanishes if there is no innovation, that is, if one's knowledge does not increase beyond what can be predicted from past observations. It is quite feasible to utilize (6.56) for real-time calculations. Related computational methods are developed in Theorems 7.15 and 7.29.

For Cox processes on general spaces there is neither a natural "order" in which observations are gathered, even if they are gathered over linearly increasing subsets of E, nor methodology for solving stochastic differential equations that might be derived. Some works treating observations over linearly increasing sets are mentioned in the chapter notes; however, their flavor, despite a seemingly more general setting, is distinctly one-dimensional. The crucial difficulty, of course, is lack of analogs of martingales and stochastic differential equations.

We now specialize, here by restricting the structure of the directing measure and in Section 7.5 by assuming the underlying space to be $\mathbf{R}_+$. The next result concerns Cox processes with directing intensities; however, rather than introduce explicitly the law of the intensity, we continue to work with that of the directing measure M.

Theorem 7.9. Let (N,M) be a Cox pair and suppose that there is a measure $\nu \in M_d$ such that $P\{M \ll \nu\} = 1$; let $\mathcal{L}$ denote the probability law of M. Then for each set A such that $E[M(A)] < \infty$ and each $f \in C_+(E)$,

$$E[e^{-M(f)}|\mathcal{F}^N(A)] = \frac{\int \mathcal{L}(d\eta)\exp[-\eta(A) + \int_A \log(d\eta/d\nu)\, dN]e^{-\eta(f)}}{\int \mathcal{L}(d\eta)\exp[-\eta(A) + \int_A \log(d\eta/d\nu)\, dN]} \tag{7.29}$$

Proof: In view of (7.21) it is enough to show that for $\mu = \Sigma_{i=1}^k \varepsilon_{x_i}$,

$$E[e^{-M_\mu(g)}] = \frac{E[e^{-M(g)}\Pi_{i=1}^k \{dM/d\nu(x_i)\}]}{E[\Pi_{i=1}^k \{dM/d\nu(x_i)\}]}, \tag{7.30}$$

where the M_μ are the Palm processes of M. Heuristically, this is obvious from the interpretation,

$$E[e^{-M_\mu(g)}] = \frac{E[e^{-M(g)}M(dx_1)\cdots M(dx_k)]}{E[M(dx_1)\cdots M(dx_k)]},$$

and the assumed form of M. To make the reasoning rigorous we use (1.63a): with μ_M^k the k-variate mean measure of M,

$$\int_{A_1} \nu(dx_1)\cdots\int_{A_k}\nu(dx_k)E\left[e^{-M(g)}\prod_{i=1}^k \frac{dM}{d\nu}(x_i)\right]$$

$$= E[e^{-M(g)}M(A_1)\cdots M(A_k)]$$

$$= \int_{A_1\times\cdots\times A_k} \mu_M^k(dx_1,\ldots,dx_k)L_M\left(\sum_{i=1}^k \varepsilon_{x_i}, g\right)$$

$$= \int_{A_1} \nu(dx_1)\cdots\int_{A_k}\nu(dx_k)E\left[\prod_{i=1}^k \frac{dM}{d\nu}(x_i)\right]L_M\left(\sum_{i=1}^k \varepsilon_{x_i}, g\right)$$

for all choices of $A_1,\ldots,A_k$, which proves (7.30). □

We next examine the most important special case of Theorem 7.9, mixed Poisson processes.

Proposition 7.10. Let N be a mixed Poisson process on E with directing measure $M = Y\nu$, where $\nu \in \mathbf{M}_d$ and Y is a positive random variable with distribution F satisfying (7.14c). Then for each set A with $\nu(A) < \infty$ and each f,

$$E[e^{-M(f)}|\mathcal{F}^N(A)] = \frac{\int F(du)e^{-u\nu(A)}u^{N(A)}e^{-u\nu(f)}}{\int F(du)e^{-u\nu(A)}u^{N(A)}} \tag{7.31}$$

Proof: Define h: $[0,\infty)\to\mathbf{M}_d$ by $h(u) = u\nu$, so that the law $\mathcal{L}$ of M satisfies $\mathcal{L} = Fh^{-1}$. By (7.29) and change of variables

$$E[e^{-M(f)}|\mathcal{F}^N(A)] = \frac{\int F(du)\exp[-u\nu(A) + \int_A \log(d[u\nu]/d\nu)dN]e^{-u\nu(f)}}{\int F(du)\exp[-u\nu(A) + \int_A \log(d[u\nu]/d\nu)\,dN]}$$

$$= \frac{\int F(du)\exp[-u\nu(A) + N(A)(\log u)]e^{-u\nu(f)}}{\int F(du)\exp[-u\nu(A) + N(A)(\log u)]}$$

as asserted. □

In particular, we have the following consequence, which is central in Section 4 for analysis of combined inference and state estimation for mixed Poisson processes.

Corollary 7.11. Suppose that (N,M), ν, F satisfy the hypotheses of Proposition 7.10. Then for each set A with $\nu(A) < \infty$,

$$E[M|\mathcal{F}^N(A)] = \frac{\int F(du)e^{-u\nu(A)}u^{N(A)+1}}{\int F(du)e^{-u\nu(A)}u^{N(A)}}\,\nu \quad \square \tag{7.32}$$

Obviously, since $M = Y\nu$, (7.32) implies that

$$E[Y|\mathcal{F}^N(A)] = \frac{\int F(du)e^{-u\nu(A)}u^{N(A)+1}}{\int F(du)e^{-u\nu(A)}u^{N(A)}} \tag{7.33}$$

Integrals

$$K_A(k) = \int F(du)e^{-u\nu(A)}u^k = \frac{k!}{\nu(A)^k}P\{N(A) = k\} \tag{7.34}$$

are used throughout Section 7.4 and will be estimated from i.i.d. copies N_i using estimators analogous to (7.18). Note also in (7.31) – (7.34) the role of $N(A)$ as "sufficient statistic" for state estimation given the observations $\mathcal{F}^N(A)$; this is another manifestation of conditional uniformity and, indeed, characterizes mixed Poisson processes.

When $\nu(E) = \infty$ it is possible given observations over bounded sets B_n increasing to E to recover the multiplier Y—and hence also the directing measure M— exactly. In some ways this is a negative result; it demonstrates that mixed Poisson processes are not ergodic, a sometimes serious limitation in inference and modeling applications.

Proposition 7.12. Let (N,M), ν, F satisfy the hypotheses of Proposition 7.10, suppose that $\nu(E) = \infty$, and let $B_1 \subset B_2 \subset \cdots$ be bounded sets increasing to E such that for $\tilde{N}$ a Poisson process with mean measure ν, $\lim \tilde{N}(B_n)/\nu(B_n) = 1$ almost surely. Then $\lim E[Y|\mathcal{F}^N(B_n)] = Y$ almost surely.

Proof: By the martingale convergence theorem, $E[Y|\mathcal{F}^N(B_n)] \to E[Y|\mathcal{F}^N(E)]$ a.s., so it is enough to show that Y is $\mathcal{F}^N(E)$-measurable, but by the strong law of large numbers and the conditionally Poisson structure of N, $\lim N(B_n)/\nu(B_n) = Y$ almost surely. $\square$

We next solve the state estimation problem for p-thinned Cox processes by reducing it to the principal problem of the section, so that methods already described apply. Comparable results for Poisson processes appear in Section 6.3. Suppose that (N,M) is a Cox pair with known law and that $p : E \rightarrow (0,1]$ is a known function. Neither the directing measure nor the Cox process N is observable; instead, one observes only the p-thinning N' of N, possibly only over a subset A of E (see Definition 1.38 and Section 4.5). By Theorem 1.39, the observed process N' and the random measure $M'(dx) = p(x)M(dx)$ form a Cox pair; this relationship together with the following result enables one to reconstruct the underlying process N from the thinned process N'. (See also Theorem 4.27, a general result formulated using Palm distributions.)

Theorem 7.13. For each A with $E[M(A)] < \infty$ and each $f \in C_+(E)$,

$$E[N(f)|\mathcal{F}^{N'}(A)] = \int_A f\,dN' + E[\int_A f(1-p)\,dM|\mathcal{F}^{N'}(A)]$$
$$+ E[\int_{A^c} f\,dM|\mathcal{F}^{N'}(A)] \tag{7.35}$$

We prove the theorem after interpreting (7.35) and indicating how—in conjunction, for example, with (7.27) or (7.23)—it would be applied. First the interpretation: write N as $N = N'_A + N''_A + N_{A^c}$, where $N'' = N - N'$; then the three terms on the right-hand side of (7.35) are MMSE estimators of $N'_A(f)$, $N''_A(f)$, and $N_{A^c}(f)$, respectively, given the observations $\mathcal{F}^{N'}(A)$. The first is clear: N'_A is observable, while the second results from the property that N'_A and N''_A are conditionally independent given M, with mean measures $1_A(x)p(x)M(dx)$ and $1_A(x)(1-p(x))M(dx)$, so that the best estimator of $N''_A(f)$ is that of its conditional mean $\int_A f(1-p)\,dM$. Similarly, N_{A^c} is conditionally independent of N'_A given M, so $N_{A^c}(f)$ can be estimated only to the extent that one can estimate $\int_{A^c} f\,dM$. To utilize (7.35) we write it as

$$E[N(f)|\mathcal{F}^{N'}(A)] = \int_A f\,dN' + E\left[\int_A \frac{f(1-p)}{p}\,dM'\mathcal{F}^{N'}(A)\right]$$
$$+ E\left[\int_{A^c} \frac{f}{p}\,dM'\ \mathcal{F}^{N'}(A)\right], \tag{7.36}$$

where M' is the directing measure of the observable Cox process N'. Results elsewhere in the section can then be applied to calculate the second and third terms in (7.36).

Proof of Theorem 7.13: We may and do prove (7.35) separately for functions f vanishing on A^c, then for functions f vanishing on A.

Suppose that $f, g \geqslant 0$ vanish on A^c. Then on the one hand, conditional independence of N' and N'' given M implies that

$$E[N(f)e^{-N'(g)}|M=v]$$

$$= E[N'(f)e^{-N'(g)}|M=v] + E[N''(f)|M=v]\,E[e^{-N'(g)}|M=v]$$

$$= v(pfe^{-g})\exp[-v(p(1-e^{-g}))] + v(f(1-p))\exp[-v(p(1-e^{-g}))]$$

(Exercise 7.9); consequently,

$$E[N(f)e^{-N'(g)}] = E[M(fpe^{-g})\exp[-M(p(1-e^{-g}))]$$
$$+ E[M(f(1-p))\exp[-M(p(1-e^{-g}))]] \tag{7.37}$$

On the other hand,

$$E[\{N'(f) + E[\int_A f(1-p)dM|\mathcal{F}^{N'}(A)]\}e^{-N'(g)}]$$

$$= E[N'(f)e^{-N'(g)}] + E[(\int_A f(1-p)dM)e^{-N'(g)}]$$

$$= E[N'(f)e^{-N'(g)}] + E[(\int_A f(1-p)dM)E[e^{-N'(g)}|M]]$$

$$= E[N'(f)e^{-N'(g)}] + E[M(f(1-p))\exp[-M(p(1-e^{-g}))]],$$

which by Exercise 7.14 is identical to (7.37) and establishes the first part of (7.35).

For f vanishing on A and g on A^c the agrument is easier:

$$E[N(f)e^{-N'(g)}] = E[E[N(f)e^{-N'(g)}|\mathcal{F}^M \vee \mathcal{F}^{N'}(A)]]$$

$$= E[M(f)e^{-N'(g)}]$$

$$= E[E[M(f)|\mathcal{F}^{N'}(A)]e^{-N'(g)}],$$

and thus the proof is complete. □

This ends for the moment our study of state estimation on general spaces; we return to the topic at the end of the section, in Section 7.3 for computation of likelihood ratios and in Section 7.4 in the context of combined statistical inference and state estimation. Now we explore consequences of the assumption that the underlying space is $\mathbf{R}_+$, but in a context different from that of Section 7.5. There we assess applicability of martingale methods of inference and state estimation for Cox processes on $\mathbf{R}_+$ with directing measures $M(A) = \int_A X_u\,du$; here, more generally, directing measures have the form

$$M(A) = \int_A X_u \nu(du), \tag{7.38}$$

where ν is a diffuse measure on $\mathbf{R}_+$ and X is a positive, measurable stochastic process. It follows that Theorem 7.9, and for mixed Poisson processes, Proposition 7.10 apply. The latter, especially, yields useful methods for recursive calculation of state estimators. Nonetheless, explicit techniques of computation are also important, one of which we illustrate for a specific case.

Example 7.14. Let N be a Cox process on $\mathbf{R}_+$ with directing intensity a Markov process X with state space $[0,\infty)$. Let (P_t) be the transition function of X and let (Q_t) be the subordinate transition function generated by the multiplicative functional $Z_t = \exp(-\int_0^t X_u\, du)$, that is, $Q_t g(x) = E_x[g(X_t)Z_t]$, where P_x is the probability law of X corresponding to the condition $X_0 = x$ (see Blumenthal and Getoor, 1968, for details). For $r \in \mathbf{N}$ and $0 \le t_1 < \cdots < t_r \le t$ let

$$g_t(t_1, \ldots, t_r) = \int \pi(dx_0)\left(\prod_{i=1}^{r} \int Q_{t_i - t_{i-1}}(x_{i-1}, dx_i) x_i\right) \int Q_{t-t_r}(x_r, dy),$$

where π is the distribution of X_0, and further define

$$\begin{aligned} h_t(t_1, \ldots, t_r; v) \\ &= g_t(t_1, \ldots, t_j, v, t_{j+1}, \ldots, t_r) \qquad \text{if } v \in (t_j, t_{j+1}) \\ &= g_t(t_1, \ldots, t_r, v) \qquad \text{if } v \in (t_r, t) \\ &= \int \pi(dx_0)\left(\prod_{i=1}^{r} \int Q_{t_i - t_{i-1}})x_{i-1} dx_i\right) \int P_{v-tt}(x_r, dx_{r+1})\, x_{r+1} \qquad \text{if } v > t \end{aligned}$$

Theorem 7.9 applies and shows that if on $[0,t]$, $N = \sum_{i=1}^{k} \varepsilon_{T_i}$, then for $f \in C_+(\mathbf{R}_+)$,

$$E[M(f) | \mathcal{F}_t^N] = \frac{\int_0^t f(v) h_t(T_1, \ldots, T_k; v)}{g_t(T_1, \ldots, T_k)}\, dv \tag{7.39}$$

Verifications and computational details are given in Karr (1983). For the special case that X has state space $\{i,j\}$ and transition rates q_i $(i \to j)$ and q_j $(j \to i)$ with $0 < q_i, q_j < \infty$, the transition function (Q_t) can be calculated in closed form:

$$Q_t = \begin{bmatrix} p(t) + (j + q_j)r(t) & q_i r(t) \\ q_j r(t) & p(t) + (i + q_i)r(t) \end{bmatrix}, \tag{7.40}$$

where $p(t)=(r_1-r_2)^{-1}(r_1e^{r_1t}-r_2e^{r_2t})$ and $r(t)=(r_1-r_2)^{-1}(e^{r_1t}-e^{r_2t})$, with r_1,r_2 the real and negative roots of the equation

$$x^2+(i+j+q_i+q_j)x+(iq_i+jq_j+ij)=0 \tag{7.41}$$

When $i=0$, $j=\alpha$, $q_0=b$, we have the process of Theorem 6.23; in this case (7.41) can be solved in closed form, there is a simplified form of (7.40), and one obtains an explicit calculation of state estimators as a complement to the recursive method of Theorem 6.23. □

We next analyze recursive versions of (7.33) for estimation of the directing multiplier of a mixed Poisson process on $\mathbf{R}_+$. We assume that the measure ν is diffuse (but not necessarily Lebesgue measure) and use distribution function notation: $\nu(t)=\nu([0,t])$, $N(t)=N([0,t])$, $N(t-)=N([0,t))$. According to (7.33),

$$E[Y|\mathcal{F}_t^N]=\frac{\int F(du)e^{-u\nu(t)}u^{N(t)+1}}{\int F(du)e^{-u\nu(t)}u^{N(t)}}\,; \tag{7.42}$$

while this expression is satisfactory for many purposes, state estimators can also be calculated recursively.

Theorem 7.15. Let $H(1)$, $H(2)$ be the $\mathcal{F}^N$-predictable processes defined by

$$H_t(1)=[\int F(du)e^{-u\nu(t)}u^{1+N(t-)}]^2$$
$$-[\int F(du)e^{-u\nu(t)}u^{2+N(t-)}][\int F(du)e^{-u\nu(t)}u^{N(t-)}] \tag{7.43a}$$

and

$$H_t(2)=[\int F(du)e^{-u\nu(t)}u^{1+N(t-)}][\int F(du)e^{-u\nu(t)}u^{N(t-)}], \tag{7.43b}$$

respectively. Then the MMSE state estimators $E[Y|\mathcal{F}_t^N]$ satisfy the stochastic differential equation

$$dE[Y|\mathcal{F}_t^N]=\frac{-H_t(1)}{H_t(2)}[dN_t-E[Y|\mathcal{F}_{t-}^N]\nu(dt)] \tag{7.44}$$

Proof: (Sketch). We argue somewhat heuristically that

$$dE[Y|\mathcal{F}_t^N]=[\int F(du)e^{-u\nu(t)}u^{N(t-)}]^{-2}H_t(1)\nu(dt)-\frac{H_t(1)}{H_t(2)}dN_t, \tag{7.45}$$

which in view of (7.42) and (7.43b) is equivalent to (7.44). The first term in

(7.45) arises as follows. For $\Delta t > 0$, let $\Delta N(t) = N(t+\Delta t) - N(t)$ and $\Delta v(t) = v(t = \Delta t) - v(t)$. Then on the event $\{\Delta N(t) = 0\}$,

$$E[Y|\mathcal{F}^N_{t+\Delta t}] - E[Y|\mathcal{F}^N_t]$$

$$= \frac{\iint F(du)F(dv)[e^{-uv(t)}u^{1+N(t)}e^{-vv(t)}v^{N(t)}][e^{-u\Delta v(t)}-e^{-v\Delta v(t)}]}{\iint F(du)F(dv)e^{-v(t)}u^{N(t)}e^{-u\Delta v(t)}e^{-v\Delta v(t)}v^{N(t)}}$$

$$= \frac{\iint F(du)F(dv)[e^{-uv(t)}u^{1+N(t)}e^{-vv(t)}v^{N(t)}](u-v)}{[\int F(du)e^{-uv(t)}u^{N(t)}]^2}\Delta v(t) + o(\Delta v(t));$$

only straightforward calculations are needed in order to obtain the first term. If $N(t) = N(t-) + 1$ (in other words, N has a point at t), then in addition to the predictable change above we must include the discontinuous change

$$E[Y|\mathcal{F}^N_t] - E[Y|\mathcal{F}^N_{t-}]$$

$$= \frac{\iint F(du)F(dv)[e^{-uv(t)}u^{N(t)}e^{-vv(t)}v^{N(t-)}][ue^{-u\Delta v(t)} - ve^{-v\Delta v(t)}]}{\iint F(du)F(dv)e^{-v(t)}u^{N(g)}e^{-u\Delta v(t)}v^{N(t-)}}$$

$$\cong \frac{\iint F(du)F(dv)[e^{-uv(t)}u^{1+N(t-)}e^{-vv(t)}v^{N(t-)}](u-v)}{[\int F(du)e^{-uv(t)}u^{1+N(t-)}][\int F(du)e^{-vv(t)}v^{N(t-)}]},$$

which gives the second term in (7.45). □

By arguments given in Chapter 2, the $\mathcal{F}^N$-compensator of N is $A_t = \int_0^t E[Y|\mathcal{F}^N_{u-}]v(du)$ and therefore (7.44) is in the cannonical form "filter gain times innovation," with no predictable term.

For some applications even recursive methods for computation of state estimators are too cumbersome, in part because—as made manifest in Section 7.5—there often results an infinite, or "open" system of simultaneous stochastic differential equations, each introducing successively higher-order moments of the quantities for which state estimators were sought originally, to which no solution can sensibly be said to exist. (In many ways, Theorem 6.23, with its single equation, is the exceptional case.) Therefore, it is of interest to develop simplified techniques for state estimation. One approach is to impose a priori restrictions on the form of state estimators as functions of the observations, the most obvious of which, of course, is linearity. We end the section with two results: one for Cox processes on $\mathbf{R}_+$, the other for Cox processes on general spaces. Proposition 6.29 is analogous in motivation (both underlying state estimation problems are difficult to solve in full generality) and in the mathematical form of the results.

Let N be a Cox process on $\mathbf{R}_+$ with directing measure (7.38), where v is a diffuse measure on $\mathbf{R}_+$ and (X_t) is a positive process. For each t it is desired to estimate X_t by a *linear* state estimator

$$\hat{X}_t = a + \int_0^t h(u)\, dN_u, \tag{7.46}$$

where a is a constant and h a deterministic function, so that the estimator $\hat{X}_t$ is a *nonrandom* linear functional of the observations. More specifically, we wish to calculate the optimal estimator of the form (7.46), relative to the criterion of mean squared error. Even though in some applications the time t is fixed, it is easier to think of calculating $\hat{X}_t$ for all values of t, in the same way that a recursive computational technique such as that of Theorem 7.15 does not calculate a state estimator for only a single t-value but provides instead a recipe valid for every value. Thus in (7.46) the constant a becomes a function $a(t)$ and h becomes a function of two variables. As might be suspected, the process $dN_t - E[X_t]\nu(dt)$ plays a role analogous to that of the innovation martingale in, for example, Theorems 5.29 and 6.23. Although deterministic, equation (7.49), as discussed further following the proof of Theorem 7.16, is analogous to stochastic differential equations derived previously.

Theorem 7.16. Let N be a Cox process on $\mathbf{R}_+$ admitting directing intensity (X_t) with respect to the diffuse measure ν. Assume that $E[X_t^2] < \infty$ for each t, and let

$$R(t,u) = \mathrm{Cov}(X_t, X_u) \tag{7.47}$$

be the covariance function of X. For each t the MMSE linear state estimator of X_t, given observations $\mathcal{F}_t^N$, is

$$\hat{X}_t = E[X_t] + \int_0^t h^*(t,u)[dN_u - E[X_u]\nu(du)], \tag{7.48}$$

where the function h^* satisfies the integral equation

$$h^*(t,u)E[X_u] + \int_0^t h^*(t,s)R(s,u)\nu(ds) = R(t,u) \tag{7.49}$$

for $0 \le u \le t$. Moreover, the optimal estimation error is

$$E[(\hat{X}_t - X_t)^2] = h^*(t,t)E[X_t] \tag{7.50}$$

Proof: For optimality it is enough to show that with

$$\tilde{X}_t = a(t) + \int_0^t [h^*(t,u) + g(t,u)][dN_u - E[X_u]\nu(du)],$$

the mean squared error $E[(\tilde{X}_t - X_t)^2]$ is minimized by $a(t) = E[X_t]$ and $g \equiv 0$. By straightforward computations making use of (7.49) (and relegated to Exercise 7.15)

$$E[(\tilde{X}_t - X_t)^2] = R(t,t) + (a(t) - E[X_t])^2 - \int_0^t h^*(t,u)R(u,t)\nu(du)$$

$$+ \int_0^t \int_0^t g(t,u)R(u,s)g(t,s)\nu(du)\nu(ds)$$

$$+ \int_0^t E[X_u]g(t,u)^2\nu(du) \tag{7.51}$$

In this expression, only the second term, which is evidently minimized by choosing $a(t) = E[X_t]$, and the fourth and fifth terms depend on a or g. The latter two are nonnegative (the first because as a covariance function R is positive definite) and each is minimized by taking $g \equiv 0$. Consequently, the minimal error is

$$E[(\hat{X}_t - X_t)^2] = R(t,t) - \int_0^t h^*(t,u)R(u,t)\nu(du) = h^*(t,t)E[X_t]$$

by another application of (7.49). □

Solving (7.49) can be easy or difficult depending on the form of the functions $t \to E[X_t]$ and R; for an illustration of the former, see Exercise 7.24. In less fortuitous instances, however, the same escalation as in Theorem 7.29 can arise: one equation may only engender others. However, note one additional advantage of linear state estimation: only first and second moments of the directing intensity are required.

We revert now to Cox processes on general spaces and derive an analog of Theorem 7.16. Let (N,M) be a Cox pair on a locally compact space E, let μ be the mean measure of N and M, and let $f \in C_+(E)$ satisfy $E[M(f)^2] < \infty$. Suppose that N has been observed over the bounded set A. We seek that element of $C_+(E)$ minimizing, by analogy with (7.48), the mean squared error $e(g) = E[(M(f) - \int_A g(dN - d\mu))^2]$; the minimizer $\hat{f}$ leads to (7.52) as MMSE linear state estimator of $M(f)$ given the observations $\mathcal{F}^N(A)$. Not only is the problem analogous to that treated in Theorem 7.16, but so is the solution.

Theorem 7.17. The MMSE linear state estimator of $M(f)$ given observations $\mathcal{F}^N(A)$ is

$$\hat{M}(f) = \mu(f) + \int_A \hat{f}(dN - d\mu), \tag{7.52}$$

where the function $\hat{f}$ satisfies the equation

$$E[M(f)M(h)] + \mu(f)\mu(h) = \mu(\hat{f}h) + E[M(\hat{f})M(h)] - \mu(\hat{f})\mu(h) \tag{7.53}$$

for every nonnegative function h on A such that $E[M(h)^2] < \infty$. Also,

$$E[(\hat{M}(f) - M(f))^2] = \text{Var}(M(f)) - E[M(f)M(\hat{f})] + \mu(f)\mu(\hat{f}) \tag{7.54}$$

Proof: The optimal linear estimator $\hat{M}(f)$ is characterized by the property that $\hat{M}(f) - M(f)$ is orthogonal to $N(h)$ in the Hilbert space $L^2(dP)$ for each function h on A with $N(h) \in L^2(dP)$; that is, $\hat{M}(f)$ is the unique solution of the "normal equations"

$$\begin{aligned} 0 &= E[\{M(f) - (\mu(f) + N(\hat{f}) - \mu(\hat{f}))\}N(h)] \\ &= E[M(f)N(h)] - \mu(f)\mu(h) - E[N(\hat{f})N(h)] + \mu(\hat{f})\mu(h) \\ &= E[M(f)M(h)] - \mu(f)\mu(h) - E[M(\hat{f}h) + M(\hat{f})M(h)] + \mu(\hat{f})\mu(h), \end{aligned}$$

with $\hat{f}$ assumed to vanish on A^c and where we have used the property that $E[N(\hat{f})N(h)|M] = M(\hat{f}h) + M(\hat{f})M(h)$. The last expression above gives (7.53) as characterization of the optimal integrand $\hat{f}$. Then, (7.54) follows by routine Hilbert space reasoning (see also Grandell, 1976). □

The mapping $f \to \hat{f}$ defined by (7.53) is linear, so for estimation of $M(f)$ for functions f belonging to a given class V, taken without loss of generality to be a vector space, it is necessary to estimate only for functions comprising a basis of V. One should regard (7.53) as an equation for measures on A, as the following example illustrates.

Example 7.18. Let N be a mixed Poisson process on E with directing measure $M = Y\nu$, where $\nu \in \mathbf{M}_d$ and $E[Y^2] < \infty$; we assume for simplicity that $E[Y] = 1$, so that $\mu_N = \mu_M = \nu$. Then (7.53) becomes

$$E[Y^2]\nu(f)\nu(h) - \nu(f)\nu(h) = \nu(\hat{f}h) + E[Y^2]\nu(\hat{f})\nu(h) - \nu(\hat{f})\nu(h),$$

which simplifies to

$$\text{Var}(Y)[\nu(f) - \nu(\hat{f})]\nu(\cdot) = \nu(\hat{f} \times \cdot),$$

whose solution is $\hat{f} = \text{Var}(Y)\nu(f)/[1 + \text{Var}(Y)\nu(A)]$, a constant function. □

While challenging and intriguing in its own right, state estimation is also important to calculation of likelihood ratios, the topic of next section.

7.3 LIKELIHOOD RATIOS AND HYPOTHESIS TESTS

Structurally, this section parallels Section 7.2 and, although less clearly, Section 6.2; however, it is in the spirit of the latter in emphasizing structure of likelihood ratios and questions of absolute continuity and singularity rather than details of hypothesis testing. Notwithstanding, at some level much of it can be interpreted as addressing construction of likelihood ratio tests when the data are observations of a Cox process. But whereas in the Poisson case sufficiency of the superposition of i.i.d. processes (Lemma 6.1) allowed us in effect to consider only single-realization data regardless of whether single or multiple realizations were observed, for Cox processes there is no simplification at all. To form the multiple-realization likelihood function, one merely multiplies associated single-realization likelihood functions. Hence in this section—but for reasons different from those in the Poisson case—we still analyze only single-realization likelihoods.

Let (N,M) be a Cox pair on a locally compact space E with only the Cox process N observable, and consider the test

$$\begin{aligned} H_0 &: \mathcal{L}(M) = \mathcal{L}_0 \\ H_1 &: \mathcal{L}(M) = \mathcal{L}_1 \end{aligned} \tag{7.55}$$

of two simple hypotheses concerning the probability law $\mathcal{L}(M)$ of M. Although one could formulate an equivalent problem in terms of the law of N, not only the mathematics of Cox processes but also the physics of systems they are used to model dictate that the law of M be regarded as the fundamental object defining the distribution of the pair (N,M). But since the observations are the Cox process rather than the directing measure, a likelihood ratio test for the hypotheses (7.55) must not be based on the likelihood ratio $(d\mathcal{L}_1/d\mathcal{L}_0)(M)$, which cannot be calculated from the observations, but instead on the N-likelihood ratio

$$\frac{dP_1}{dP_0}\Big|_{\mathcal{F}^N} = \frac{dP_1\{N \in (\cdot)\}}{dP_0\{N \in (\cdot)\}} \tag{7.56}$$

Here under P_i the directing measure has law $\mathcal{L}_i$. Moreover, the likelihood ratio must be calculated either explicitly or recursively; we consider both, beginning with the former.

Concerning the N-likelihood ratio, its key feature is that it is the conditional expectation of the M-likelihood ratio $(d\mathcal{L}_1/d\mathcal{L}_0)(M)$ given the observations of N, and from this, together with Theorems 7.6 and 7.9 and

Proposition 7.10, we develop a number of expressions and computational techniques. A preliminary result establishes this property along with a sufficient condition for equivalence of P_0 and P_1 on $\mathcal{F}^N$, namely that of $\mathcal{L}_0$ and $\mathcal{L}_1$.

Proposition 7.19. Let (N,M) be a Cox pair on E such that under the probability P_i, the directing measure has law $\mathcal{L}_i$, $i=0,1$. If $\mathcal{L}_1 \ll \mathcal{L}_0$, then $P_1 \ll P_0$ on $\mathcal{F}^N$ and for each set A

$$\frac{dP_1}{dP_0}\Big|_{\mathcal{F}^N(A)} = E_0\left[\frac{d\mathcal{L}_1}{d\mathcal{L}_0}(M)|\mathcal{F}^N(A)\right] \tag{7.57}$$

Proof: For each function f vanishing outside A,

$$\begin{aligned} E_1[e^{-N(f)}] &= \int \mathcal{L}_1(d\nu)\exp[-\nu(1-e^{-f})] \\ &= \int \mathcal{L}_0(d\nu)\frac{d\mathcal{L}_1}{d\mathcal{L}_0}(\nu)\exp[-\nu(1-e^{-f})] \\ &= E_0\left[\frac{d\mathcal{L}_1}{d\mathcal{L}_0}(M)E_0[e^{-N(f)}|M]\right] \\ &= E_0\left[E_0\left[\frac{d\mathcal{L}_1}{d\mathcal{L}_0}(M)|\mathcal{F}^N(A)\right]e^{-N(f)}\right], \end{aligned}$$

which together with Theorem 1.12 verifies (7.57). □

Thus we have several explicit expressions for likelihood ratios.

Theorem 7.20. Suppose that under P_i, (N,M) is a Cox pair with the directing measure having law $\mathcal{L}_i$. If $\mathcal{L}_1 \ll \mathcal{L}_0$, then $P_1 \ll P_0$ on $\mathcal{F}^N$ and for each set A with $E_0[M(A)] < \infty$,

$$\frac{dP_1}{dP_0}\Big|_{\mathcal{F}^N(A)} = \frac{E[e^{-M_\mu^0(A)}(d\mathcal{L}_1/d\mathcal{L}_0)(M_\mu^0)]}{E[e^{-M_\mu^0(A)}]}\Big|_{\mu=N_A}, \tag{7.58}$$

where the M_μ^0 are Palm processes of M *relative to the probability* $\mathcal{L}_0$. □

Theorem 7.21. Let (N,M) be a Cox pair such that under the probability P_i, M has law $\mathcal{L}_i$. Suppose that $\mathcal{L}_1 \ll \mathcal{L}_0$ and that there is $\nu \in \mathbf{M}_d$ such that $P_0\{M \ll \nu\} = 1$. Then $P_1 \ll P_0$ on $\mathcal{F}^N$ and for each set A with $E_0[M(A)] < \infty$,

$$\frac{dP_1}{dP_0}\Big|_{\mathcal{F}^N(A)} = \frac{\int \mathcal{L}_1(d\eta)\exp[-\eta(A) + \int_A \log(d\eta/d\nu)\,dN]}{\int \mathcal{L}_0(d\eta)\exp[-\eta(A) + \int_A \log(d\eta/d\nu)\,dN]} \quad \square \tag{7.59}$$

Note the continuing primacy of state estimation: to calculate likelihood ratios one utilizes techniques for state estimation of an unobservable directing measure.

If in the setting of Theorem 7.21, $\mathcal{L}_0 = \varepsilon_{\nu_0}$ and $\mathcal{L}_1 = \varepsilon_{\nu_1}$, so that under P_i, N is Poisson with mean measure ν_i, then with $\nu = \nu_0 + \nu_1$ one obtains by formal substitution into (7.59) the expression

$$\frac{dP_1}{dP_0}\Big|_{\mathcal{F}^N(A)} = \exp\left[(\nu_0 - \nu_1)(A) + \int_A \log \frac{d\nu_1/d\nu}{d\nu_0/d\nu}\, dN\right],$$

proposed in Section 6.2 for likelihood ratio tests for Poisson processes with not necessarily equivalent mean measures. In particular, when $\nu_1 \ll \nu_0$ it assumes the form

$$\frac{dP_1}{dP_0}\Big|_{\mathcal{F}^N(A)} = \exp\left[\int_A \left(1 - \frac{d\nu_1}{d\nu_0}\right) d\nu_0 + \int_A \log \frac{d\nu_1}{d\nu_0}\, dN\right], \tag{7.60}$$

the principal Poisson likelihood ratio of Proposition 6.11. However, none of this is justified in terms of Theorem 7.21 because for $\mathcal{L}_i = \varepsilon_{\nu_i}$ we have $\mathcal{L}_0 \perp \mathcal{L}_1$, the extreme opposite of the absolute continuity assumed in the theorem. Nonetheless, (7.60) is a valid Poisson likelihood ratio and thus for Cox processes it is possible that $P_0 \sim P_1$ on $\mathcal{F}^N$ but $P_0 \perp P_1$ on $\mathcal{F}^M$. Further comments about singularity will be given momentarily, but first we conclude the discussion of absolute continuity by considering mixed Poisson processes.

Given probability distributions F_0, F_1 on $\mathbf{R}_+$ and diffuse measures ν_0, ν_1 on E [to ensure identifiability, assume that $\int uF_0(du) = \int uF_1(du) = 1$] we need a criterion for absolute continuity of the distribution $\mathcal{L}_1$ of $M = Y\nu_1$ under a probability P_1 such that $P_1\{Y \in (\cdot)\} = F_1(\cdot)$ with respect to the corresponding law $\mathcal{L}_0$ constructed from ν_0 and F_0; it is provided by the following result, whose proof is left as Exercise 7.20.

Lemma 7.22. With the notation and hypotheses of the preceding paragraph, the following assertions are equivalent:

a) $\mathcal{L}_1 \ll \mathcal{L}_0$;

b) There is $a \in (0, \infty)$ such that $\nu_1 = a\nu_0$ and such that the measure $B \to F_1(a^{-1}B)$ is absolutely continuous with respect to F_0. □

In view of the lemma there is no loss of generality in supposing that $\nu_0 = \nu_1$, whose common value we denote by ν, and dropping the restriction that $\int uF_i(du) = 1$. From Propositions 7.10 and 7.19 we derive a representation of likelihood ratios for mixed Poisson processes.

Proposition 7.23. Suppose that under P_i, N is a mixed Poisson process on E with directing measure $M = Y\nu$, where $\nu \in \mathbf{M}_d$ does not depend on i, while Y has P_i-distribution F_i. Assume that $E_0[Y] < \infty$. If $F_1 \ll F_0$, then $P_1 \ll P_0$ on $\mathcal{F}^N$ and for each bounded set A,

$$\frac{dP_1}{dP_0}\Big|_{\mathcal{F}^N(A)} = \frac{\int F_1(du) e^{-u\nu(A)} u^{N(A)}}{\int F_0(du) e^{-u\nu(A)} u^{N(A)}} \quad \square \tag{7.61}$$

Inspection of (7.61) confirms even in the mixed Poisson case that the superposition of i.i.d. Cox processes is not a sufficient statistic for an unknown probability law: the n-sample likelihood function obtained by multiplying terms of the form (7.61) cannot be expressed as a function of only the superposition $N_1 + \cdots + N_n$.

Attention so far has focused mainly on absolute continuity relationships between probability laws, and for a reason: singularity of laws of directing measures, as already noted, need not imply that of the associated Cox processes. Thus there is no analog for Cox processes of Theorems 6.13 and 6.15, nor is there a dichotomy theorem. If under P_0, N is Poisson with mean measure μ_0, and under P_1, N is a Cox process whose directing measure M has distribution $P_1\{M = \mu_0\} = P_1\{M = \mu_1\} = 1/2$, where $\mu_0 \perp \mu_1$, then the law of N under P_1 is neither equivalent to nor singular with respect to the law under P_0. It is rather easy (Exercise 7.18) to show that if $P_0 \perp P_1$ on $\mathcal{F}^N$, then $P_0 \perp P_1$ on $\mathcal{F}^M$; except for this and the absolute continuity result given in general form in Theorem 7.20, one can say little more.

Recursive computation is as important for likelihood ratios as for any state estimation problem. In Theorem 7.24 we provide for a large class of Cox processes on $\mathbf{R}_+$ not only a recursive method for calculation but also a *separation theorem* whose content, in effect, is that because a Cox process is conditionally Poisson, one can calculate a likelihood ratio for Cox processes by taking a Poisson likelihood ratio

$$\frac{dP_{\nu_1}}{dP_{\nu_0}} = \exp\left[\int_A \left(\frac{d\nu_0}{d\nu} - \frac{d\nu_1}{d\nu}\right) d\nu + \int_A \log \frac{d\nu_1/d\nu}{d\nu_0/d\nu}\, dN\right]$$

and simply replacing $d\nu_0/d\nu$ and $d\nu_1/d\nu$ by conditional expectations of the directing intensity calculated under the candidate probability laws for M. Thus hypothesis testing (or, in the engineering literature, "detection") is separated from (state) estimation. It does not seem feasible to prove a separation theorem for Cox processes on spaces other than $\mathbf{R}_+$, because—as will become clarified in discussion following the theorem—a separation theorem is equivalent to a method for recursive calculation of likelihood ratios.

We assume a Cox pair (N,M) on $\mathbf{R}_+$ such that

$$M(A) = \int_A X_t \nu(dt) \tag{7.62}$$

with ν stipulated only to be diffuse. The case that ν is Lebesgue measure is pursued further in Section 5 using stochastic intensity methods; there a separation theorem is proved directly from Theorem 2.31.

Theorem 7.24. Let (N,M) be a Cox pair on $\mathbf{R}_+$ with M of the form (7.62), where ν is diffuse, and assume that $\mathcal{L}_1 \ll \mathcal{L}_0$, where $\mathcal{L}_i$ is the P_i-law of M. Then $P_1 \ll P_0$ on $\mathcal{F}^N$ and for each t

$$\frac{dP_1}{dP_0}\Big|_{\mathcal{F}_t^N} = \exp\left[\int_0^t (E_0[X_u|\mathcal{F}_u^N] - E_1[X_u|\mathcal{F}_u^N])\nu(du) + \int_0^t \log\left(\frac{E_1[X_u|\mathcal{F}_u^N]}{E_0[X_u|\mathcal{F}_u^N]}\Big|_{u=s-}\right) dN_s\right] \tag{7.63}$$

Proof: A rigorous proof proceeds by using the property that the $(P_i,\mathcal{F}^N)$-compensator of N is $A_t^i = \int_0^t E_i[X_u|\mathcal{F}_{u-}^N]\nu(du)$, then appeals to representation theorems for likelihood ratios that generalize Theorem 2.31 (see, e.g., Jacod, 1975b, Theorem 4.5; or Liptser and Shiryayev, 1978, Theorem 19.3). Here we give an heuristic argument based on calculation of a stochastic differential. Specifically, with distribution function notation used as convenient,

$$d\left(\log \int \mathcal{L}_0(d\eta)\exp\left[-\eta(t) + \int_0^t \log\frac{d\eta}{d\nu}\,dN\right]\right)$$

$$= -\frac{\int \mathcal{L}_0(d\eta)\exp[-\eta(t) + \int_0^t \log(d\eta/d\nu)\,dN](d\eta/d\nu)(t)}{\int \mathcal{L}_0(d\eta)\exp[-\eta(t) + \int_0^t \log(d\eta/d\nu)\,dN]}\,d\nu(t)$$

$$+ \left\{\log \int \mathcal{L}_0(d\eta)\exp\left[-\eta(t) + \int_0^t \log\frac{d\eta}{d\nu}\,dN\right]\frac{d\eta}{d\nu}(t)\right.$$

$$\left. - \log \int \mathcal{L}_0(d\eta)\exp\left[-\eta(t) + \int_0^t \log\frac{d\eta}{d\nu}\,dN\right] dN_t\right\}$$

$$= -E_0[X_t|\mathcal{F}_t^N]\,d\nu(t) + (\log E_0[X_t|\mathcal{F}_{t-}^N])\,dN_t,$$

since (7.29) implies that

$$E_0[X_t \mid \mathcal{F}_t^N] = E_0\left[\frac{dM}{d\nu}(t)|\mathcal{F}_t^N\right]$$

$$= -\frac{\int \mathcal{L}_0(d\eta)\exp[-\eta(t) + \int_0^t \log(d\eta/d\nu)\,dN](d\eta/d\nu)(t)}{\int \mathcal{L}_0(d\eta)\exp[-\eta(t) + \int_0^t \log(d\eta/d\nu)\,dN]}$$

Hence, by (7.59),

$$d\left(\log \frac{dP_1}{dP_0}\Big|_{\mathcal{F}_t^N}\right) = (E_0[X_t|\mathcal{F}_t^N] - E_1[X_t|\mathcal{F}_t^N])\, dv(t) + \log \frac{E_1[X_t|\mathcal{F}_{t-}^N]}{E_0[X_t|\mathcal{F}_{t-}^N]}\, dN_t, \tag{7.64}$$

which proves (7.63). □

The formula (7.63)—especially its differential form (7.64)—is of computational as well as intuitive value: it provides a technique for recursive calculation of the likelihood ratios $(dP_1/dP_0)|_{\mathcal{F}_t^N}$ as random functions of the observation time t. Indeed, in view of the conditionally Poisson structure of a Cox process, if one could begin with a method for recursive computation of likelihood ratios (e.g., obtained by appeal to Theorem 2.34, which would yield a stochastic integral representation of the likelihood ratio martingale) and then reverse the steps above, the result would be a separation theorem. One should not view Theorem 7.24 as an end in itself: as an instrument for computation it merely reduces calculation of likelihood ratios to that of state estimators for the directing intensity. Even in rigidly structured situations the latter problem is not trivial; we take it up again in Section 7.5.

We turn briefly to testing hypotheses other than the simple pair (7.55). There are two fundamental difficulties (at least). The first is the size of the class of Cox processes: by Lemma 7.1 there are as many Cox processes as diffuse random measures. It is inescapable that restriction of the class of candidate Cox processes in the statistical model must precede any serious attempt to test hypotheses, which leads to the second difficulty, of specifying subclasses of Cox processes that are sufficiently restricted to permit reasonably precise tests, yet sufficiently broad not to render the test meaningless in the sense, for example, that physical grounds suggest that the model itself is inappropriate. Even among Cox processes on $\mathbf{R}_+$ only the following subclasses seem to admit both restrictive structure and breadth:

1) Poisson processes;
2) Mixed Poisson processes;
3) Cox processes that are also renewal processes;
4) Cox processes that are also Poisson cluster processes.

The third class includes the stochastic integral process discussed in Sections 6.3 and 7.5 (see also Chapter 8). Unfortunately (in a sense), in each case the additional structure is exploited more intensively than the Cox structure, which becomes superfluous. One can rarely implement the Neyman-Pearson theory and, especially if parameters are estimated simultaneously, tests are often decided only according as a log-likelihood ratio is positive or negative. Computation of power is usually hopeless.

Even for the simple-vs.-simple test (7.55) the outlook for analysis à la Neyman-Pearson is frequently grim. Consider the mixed Poisson case with likelihood ratio (7.61). Its distribution, even though dependent on the data $\mathcal{F}^N(A)$ only through $N(A)$, is intractable. Normal approximation is impossible since a mixed Poisson process—unless it is actually Poisson—is not ergodic. Hence one cannot construct even approximate critical regions.

Tests with qualitative flavor are even more difficult. Consider the problem of testing whether a mixed Poisson process is Poisson, where the underlying measure ν is unknown. This problem is hopeless in the single-realization case. Suppose even that $E = \mathbf{R}_+$ and that ν is proportional to Lebesgue measure. Given observation of a single realization over $[0,t]$, tests such as uniform conditional tests (Section 3.5) for inherent reasons cannot distinguish between Poisson and mixed Poisson processes: both are conditionally uniform. A single realization of the process over $\mathbf{R}_+$ is not more useful; if one conditions on the value of $\lim N_t/t$ (the intensity in the Poisson case and the random value of the directing multiplier in the mixed Poisson case), then again the two classes are indistinguishable. If the null hypothesis were "Poisson with prescribed intensity ν_0," the outlook would be less bleak, but in applications there is rarely justification for hypotheses this narrow.

For data representing i.i.d. realizations there are more possibilities but few have been worked out in detail. Indeed, so little has been developed that one cannot identify with confidence the principal difficulties that arise.

7.4 COMBINED INFERENCE AND STATE ESTIMATION

Suppose that (N_1,M_1), (N_2,M_2), ... are i.i.d. Cox pairs on a compact space E, with the law $\mathcal{L}$ of the M_i unknown, and that only the N_i are observable. State estimators $E[M_{n+1}|\mathcal{F}^{N_{n+1}}(A)]$ derived in Section 7.2 are not computable because their calculation requires effectively full knowledge of the law $\mathcal{L}$. Nonetheless, there are situations in which state estimation remains the primary goal, lack of knowledge of $\mathcal{L}$ notwithstanding. In this section we solve the problem of combined statistical inference and state estimation for mixed Poisson processes. We also mention, albeit briefly, extensions to more general Cox processes. For mixed Poisson processes the true state estimators given by (7.65) are sufficiently simple in structure, and attributes of $\mathcal{L}$ appearing in them sufficiently facile to estimate, that without further restrictions on E pseudo-state estimators $\hat{E}[M_{n+1}|\mathcal{F}^{N_{n+1}}(A)]$ defined in (7.69) approximate the true state estimators, as $n \to \infty$, at a rate that is best possible given our assumption that the parameters F and ν defining $\mathcal{L}$ are entirely unknown. Moreover, for some purposes the processes N_i may be observed over varying sets A_i. By contrast, results available concerning combined nonparametric inference and state estimation for general Cox processes are much less satisfactory, for reasons indicated in discussion concluding the section.

Thus we establish the following setting. Let (N_1,Y_1), (N_2,Y_2), ... be i.i.d. copies of a pair consisting of a simple point process N on E and a positive random variable Y; for $\nu \in \mathbf{M}_d$ and F a probability distribution on $\mathbf{R}_+$, let $P_{\nu,F}$ be a probability under which Y has distribution F and N is a mixed Poisson process with directing measure $M = Y\nu$. We work with the full nonparametric model $\mathcal{P} = \{P_{\nu,F}\}$, except that throughout ν is assumed to be diffuse and the regularity conditions (7.14) are presumed to hold.

Suppose that $N_1, \dots, N_n$ have been observed previously over all of E, that N_{n+1} is observed more recently over a subset A, and that we seek to reconstruct M_{n+1}. To this end we devise *pseudo-state estimators* approximating the true MMSE state estimators.

$$E_{\nu,F}[M_{n+1} | \mathcal{F}^{N_{n+1}}(A)] = \frac{\int F(du) e^{-u\nu(A)} u^{N_{n+1}(A)+1}}{\int F(du) e^{-u\nu(A)} u^{N_{n+1}(A)}} \nu \tag{7.65}$$

and examine the behavior as $n \to \infty$ of the difference between the pseudo- and the true state estimators. We embark on this path by recalling the integrals

$$K_A(k) = \int F(du) e^{-u\nu(A)} u^k = \frac{k!}{\nu(A)^k} P_{\nu,F}\{N(A) = k\} \tag{7.66}$$

introduced in (7.34) and rephrasing (7.65) as

$$E_{\nu,F}[M_{n+1} | \mathcal{F}^{N_{n+1}}(A)] = \frac{K_A(N_{n+1}(A)+1)}{K_A(N_{n+1}(A))} \nu \tag{7.65a}$$

To create pseudo-state estimators, guided by the principle of separation discussed in the introduction to Chapter 1, we utilize the previous observations $N_1, \dots, N_n$ to construct estimators $\hat{\nu}$, $\hat{K}_A$ and then form the pseudo-state estimator $\hat{E}[M_{n+1} | \mathcal{F}^{N_{n+1}}(A)]$ by substituting them for ν, K_A in (7.65a). The estimator $\hat{\nu}$ is that used in Section 7.1:

$$\hat{\nu} = \frac{\Sigma_{i=1}^n N_i}{\Sigma_{i=1}^n N_i(E)}; \tag{7.67}$$

at the expense of increased variance, since $n^{-1} \Sigma_{i=1}^n N_i(E) \to 1$ almost surely by (7.14b), we could use instead $\hat{\nu} = n^{-1} \Sigma_{i=1}^n N_i$, which we do below when the N_i are observed over differing sets and only the multipliers Y_i are to be estimated. On the basis of (7.66) we take

$$\hat{K}_A(k) = \frac{k!}{\hat{\nu}(A)^k} n^{-1} \sum_{i=1}^n 1(N_i(A) = k) \tag{7.68}$$

and finally the $(n+1)$-sample pseudo-state estimator is

$$\hat{E}[M_{n+1}|\mathcal{F}^{N_{n+1}}(A)] = \frac{\hat{K}_A(N_{n+1}(A)+1)}{\hat{K}_A(N_{n+1}(A))}\hat{\nu} \tag{7.69}$$

Note that "current" observations N_{n+1} are *not* used in forming the estimators $\hat{\nu}$, $\hat{K}_A$. In practice they should be used, at least in (7.68), in order not to discard additional information they contain, but for proving theorems it is more convenient to exclude them, for then in (7.69) the estimators $\hat{\nu}$, $\hat{K}_A$ and the random variable $N_{n+1}(A)$ are independent. Asymptotically, it is irrelevant which option one selects.

The main result concerning asymptotic behavior of differences

$$\hat{E}[M_{n+1}|\mathcal{F}^{N_{n+1}}(A)] - E_{\nu,F}[M_{n+1}|\mathcal{F}^{N_{n+1}}(A)],$$

viewed as signed random measures on E, is Theorem 7.26, to the effect that $n^{1/2}$ times the difference converges in distribution to a mixture of Gaussian random measures. In view of Theorems 7.5 and 7.25 this rate of convergence is best possible because the same rate applies to the estimation error processes $\hat{\nu} - \nu$ and $\hat{K}_A - K_A$, and there is no possibility that pseudo-state estimators can approximate true at a rate faster than the law of the directing measures can be estimated.

While Theorem 7.26 is the *main* result, the next theorem is the *key* result. Not only does it yield Theorem 7.26 by a direct, albeit laborious, transformation argument, but it also provides the details needed to flesh out the skeletal proof given for Theorem 7.5. The $Q_F(k)$ and $\hat{Q}(k)$ of (7.16) and (7.18) are intimately related to (indeed, simpler than) the $K_A(k)$, $\hat{K}_A(k)$ of this section. We stress that for now the set A is fixed.

Theorem 7.25. In addition to the hypotheses set forth above, assume that $\nu(A) > 0$ and that $E_{\nu,F}[Y^2] < \infty$. Then

$$\begin{bmatrix} \{n^{1/2}[\hat{\nu}(f) - \nu(f)] : f \in C(E)\} \\ \{n^{1/2}[\hat{K}_A(k) - K_A(k)] : k \in \mathbf{N}\} \end{bmatrix} \xrightarrow{d} \begin{bmatrix} \{G(f) : f \in C(E)\} \\ \{Z(k) : k \in \mathbf{N}\} \end{bmatrix}, \tag{7.70}$$

under $P_{\nu,F}$, the first components as random measures on E and the second as sequences of random variables, where G is a mean zero Gaussian random measure and X is a mean zero Gaussian sequence. The process (G,Z) is Gaussian with covariance function (where $m = E_{\nu,F}[Y]$)

$$\mathrm{Cov}(G(f),G(g)) = m^{-1}[\nu(fg) - \nu(f)\nu(g)] \tag{7.71a}$$

$$\mathrm{Cov}(Z(k),Z(j)) = 1(i=j)k!\frac{K_A(k)}{\nu(A)^k} - K_A(k)K_A(j)$$

$$+ \frac{kjK_A(k)K_A(j)}{\nu(A)}(1-\nu(A))\left(2+\frac{1}{m}\right) \tag{7.71b}$$

$$\begin{aligned}\mathrm{Cov}(G(f),Z(k)) = kK_A(k)\frac{\nu(f;A)}{\nu(A)} &+ \nu(f;A^c)K_A(k+1) \\ &- \nu(f)[kK_A(k)+\nu(A^c)K_A(k+1)] \\ &+ m^{-1}kK_A(k)[\nu(f;A)-\nu(f)\nu(A)]\end{aligned} \tag{7.71c}$$

Proof: Let [compare (7.18)] $\hat{Q}(k) = n^{-1}\Sigma_{i=1}^{n} 1(N_i(A)=k)$ and (with dependence on ν and F suppressed) let $Q(k) = P_{\nu,F}\{N(A)=k\} = \nu(A)^k K_A(k)/k!$. We show first the central limit theorem

$$\begin{bmatrix} \{n^{1/2}[\hat{\nu}(f)-\nu(f)] : f \in C(E)\} \\ n^{1/2}[\hat{\nu}(A)-\nu(A)] \\ \{n^{1/2}[\hat{Q}(k)-Q(k)] : k \in \mathbf{N}\} \end{bmatrix} \xrightarrow{d} \begin{bmatrix} \{G^0(f) : f \in C(E)\} \\ W^0 \\ \{Z^0(k) : k \in \mathbf{N}\} \end{bmatrix}, \tag{7.72}$$

with G^0 a Gaussian random measure, W^0 a normally distributed random variable, and Z^0 a Gaussian sequence. By Theorem 1.12 and Billingsley (1968, Theorem 2.2) it is enough to show that for fixed f and J the random vectors

$$n^{1/2}([\hat{\nu}(f)-\nu(f)], [\hat{\nu}(A)-\nu(A)], [\hat{Q}Z(0)-Q(0)], \ldots, [\hat{Q}(J)-Q(J)]) \tag{7.73}$$

have asymptotically a joint normal distribution, and by the Cramér-Wold device this follows from asymptotic normality of summations

$$n^{1/2}\{[\hat{\nu}(f)-\nu(f)] + a[\hat{\nu}(A)-\nu(A)] + \sum_{\ell=0}^{J} a_\ell[\hat{Q}(\ell)-Q(\ell)]\}, \tag{7.74}$$

where $a, a_0, \ldots, a_J$ are constants. As in Theorem 7.5, within error converging in probability to zero,

$$n^{1/2}[\hat{\nu}(f)-\nu(f)] = m^{-1}n^{-1/2}\sum_{i=1}^{n}\int (f-\nu(f))\,dN_i$$

and in the same way

$$n^{1/2}[\hat{\nu}(A)-\nu(A)] = m^{-1}n^{-1/2}\sum_{i=1}^{n}[N_i(A)-\nu(A)N_i(E)],$$

which enables us to write (7.74) as

$$n^{-1/2}\sum_{i=1}^{n}\left\{(N_i(f-\nu(f))+a[N_i(A)-\nu(A)N_i(E)]+\sum_{\ell=0}^{J}a_\ell[1(N_i(A)=\ell)-Q(\ell)]\right\} \tag{7.75}$$

But (7.75) is just a sum of i.i.d. random variables with mean zero and finite variance, so asymptotic normality holds by the central limit theorem. (A simplified version of the foregoing argument is adequate to prove Theorem 7.5.)

It remains at the moment to calculate covariances associated with the limit process in (7.72). The computations are straightforward and more boring than interesting; hence some of them are given as exercises! Here we simply record the results:

$$\mathrm{Cov}(G^0(f),G^0(g)) = m^{-1}[\nu(fg)-\nu(f)\nu(g)] \tag{7.76a}$$

[compare (7.20a)]

$$\mathrm{Var}(W^0) = m^{-1}\nu(A)(1-\nu(A)) \tag{7.76b}$$

[take $f=g=1_A$ in (7.76a)]

$$\mathrm{Cov}(Z^0(k),Z^0(j)) = 1(k=j)Q(k)-Q(k)Q(j) \tag{7.76c}$$

[see (7.20b) ,and (4.2)]

$$\mathrm{Cov}(G^0(f),W^0) = m^{-1}[\nu(f;A)-\nu(f)\nu(A)] \tag{7.76d}$$

(which is zero if $A=E$)

$$\mathrm{Cov}(G^0(f),Z^0(k)) = kQ(k)\frac{\nu(f;A)}{\nu(A)} + \frac{\nu(f;A^c)\nu(A)^k K_A(k+1)}{k!} - \nu(f)\frac{kQ(k)+\nu(A^c)\nu(A)^k K_A(k+1)}{k!} \tag{7.76e}$$

(also zero for $A=E$)

$$\mathrm{Cov}(W^0,Z^0(k)) = kQ(k)(1-\nu(A)) \tag{7.76f}$$

[take $f=1_A$ in (7.76e)].

To proceed from (7.72) to (7.70) we use transformation theory (the "delta method") and another liberal dose of covariance computations. By reasoning similar to that just gone through, (7.70) holds if for each f and J the asymptotic distribution of the random vectors

$$\begin{bmatrix} n^{1/2}[\hat{\nu}(f) - \nu(f)] \\ n^{1/2}[\hat{K}_A(0) - K_A(0)] \\ n^{1/2}[\hat{K}_A(1) - K_A(1)] \\ \cdot \\ \cdot \\ \cdot \\ n^{1/2}[\hat{K}_A(J) - K_A(J)] \end{bmatrix} = \begin{bmatrix} n^{1/2}[\hat{\nu}(f) - \nu(f)] \\ n^{1/2}[\hat{Q}(0) - Q(0)] \\ n^{1/2}\left[\dfrac{\hat{Q}(1)}{\hat{\nu}(A)} - \dfrac{Q(1)}{\nu(A)}\right] \\ \cdot \\ \cdot \\ \cdot \\ n^{1/2}\left[\dfrac{J!}{\hat{\nu}(A)^J}\hat{Q}(J) - \dfrac{J!}{\nu(A)^J}Q(J)\right] \end{bmatrix}$$

is normal, but this follows from applying the delta method to (7.73) and the transformation $(a,b,c_0,\dots,c_J) \to (a,c_0,c_1/b,\dots,(J!/b^J)c_J)$. The covariance function (7.71) follows from (7.76) by standard techniques. □

With one more transformation argument we obtain the main result.

Theorem 7.26. Under the hypotheses of Theorem 7.25,

$$n^{1/2}(\hat{E}[M_{n+1}|\mathcal{F}^{N_{n+1}}(A)] - E_{\nu,F}[M_{n+1}|\mathcal{F}^{N_{n+1}}(A)]) \xrightarrow{d} H \tag{7.77}$$

as signed random measures on E, where the distribution of the random measure H is a mixture, with mixing distribution $P_{\nu,F}\{N(A) \in (\cdot)\}$, of those of the sequence (G_k) of Gaussian random measures whose covariance function is given by (7.78).

Proof: By Slutsky's theorem, for fixed k and f,

$$n^{1/2}\left[\frac{\hat{K}_A(k+1)}{\hat{K}_A(k)}\hat{\nu}(f) - \frac{K_A(k+1)}{K_A(k)}\nu(f)\right]$$

$$\cong \nu(f)n^{1/2}\left[\frac{\hat{K}_A(k+1)}{\hat{K}_A(k)} - \frac{K_A(k+1)}{K_A(k)}\right] + \frac{K_A(k+1)}{K_A(k)}n^{1/2}[\hat{\nu}(f) - \nu(f)]$$

$$\cong \frac{\nu(f)}{K_A(k)}n^{1/2}[\hat{K}_A(k+1) - K_A(k+1)] - \frac{\nu(f)K_A(k+1)}{K_A(k)^2}n^{1/2}[\hat{K}_A(k) - K_A(k)]$$

$$+ \frac{K_A(k+1)}{K_A(k)}n^{1/2}[\hat{\nu}(f) - \nu(f)];$$

hence by reasoning similar to that in the proof of Theorem 7.25 and by the conclusion to that theorem, the sequence of signed random measures

$$\left\{\left\{n^{1/2}\left[\frac{\hat{K}_A(k+1)}{\hat{K}_A(k)}\hat{\nu}(f) - \frac{K_A(k+1)}{K_A(k)}\nu(f)\right] : f \in C(E)\right\} : k \in \mathbf{N}\right\}$$

converges in distribution to a jointly Gaussian sequence (G_k) of Gaussian random measures whose covariance function, by (7.71) and routine calculations, is

$$\begin{aligned}
&\mathrm{Cov}(G_k(f),G_j(g)) \\
&= \nu(f)\nu(g)\left[\frac{R(k+1,j+1)}{K_A(k)K_A(j)} - \frac{K_A(k+1)R(k,j+1)}{K_A(k)^2K_A(j)} \right.\\
&\quad \left. - \frac{K_A(k+1)R(k+1,j)}{K_A(k)K_A(j)^2} + \frac{K_A(k+1)K_A(j+1)R(k,j)}{K_A(k)^2K_A(j)^2}\right] \\
&\quad + \frac{\nu(f)}{K_A(j)^2}[K_A(j+1)R(f,j+1) - K_A(j+1)^2R(f,j)] \\
&\quad + \frac{\nu(g)}{K_A(k)^2}[K_A(k+1)R(g,k+1) - K_A(k+1)^2R(g,k)] \\
&\quad + \frac{K_A(k+1)K_A(j+1)}{K_A(k)K_A(j)}R(f,g),
\end{aligned} \tag{7.78}$$

where R is the covariance function defined by (7.71) [i.e., $R(f,g) = \mathrm{Cov}(G(f),G(g))$, $R(f,k) = \mathrm{Cov}(G(f),Z(k))$, and $R(k,j) = \mathrm{Cov}(Z(k),Z(j))$].

For each n, $N_{n+1}(A)$ is independent of $\{N_1, \dots, N_n\}$ and hence of the n-sample estimators $\hat{\nu}$, $\hat{K}_A$ and from this, the continuous mapping theorem and Billingsley (1968, Theorem 3.2), (7.77) ensues. □

In particular, taking the suppressed function in (7.77) to be $f = 1$ and using the relationship $E_{\nu,F}[M_{n+1}(E)|\mathcal{F}^{N_{n+1}}(A)] = E_{\nu,F}[Y_{n+1}|\mathcal{F}^{N_{n+1}}(A)]$, we infer that for the pseudo-state estimators

$$\hat{E}[Y_{n+1}|\mathcal{F}^{N_{n+1}}(A)] = \frac{\hat{K}_A(N_{n+1}(A)+1)}{\hat{K}_A(N_{n+1}(A))} \tag{7.79}$$

of the directing multipliers the following limit behavior obtains.

Proposition 7.27. Under the hypotheses of Theorem 7.25,

$$n^{1/2}(\hat{E}[Y_{n+1}|\mathcal{F}^{N_{n+1}}(A)] - E_{\nu,F}[Y_{n+1}|\mathcal{F}^{N_{n+1}}(A)]) \xrightarrow{d} Y,$$

where Y has the mixed normal distribution of $H(1)$ in Theorem 7.26. □

Note that computation of the pseudo-state estimator (7.79) requires knowledge only of $N_1(A), \dots, N_n(A)$ and $N_1(E), \dots, N_n(E)$, and that as mentioned previously, even the latter would be dispensable if instead of

(7.67) one utilized the estimators $\hat{\nu}(A) = n^{-1}\sum_{i=1}^{n} N_i(A)$. An analog of Proposition 7.27 holds, with a different covariance function (see Karr, 1984b). But the distribution of the $N_i(A)$ depends on A only through $\nu(A)$ (Exercise 1.3), raising hope that a result comparable to Proposition 7.27 might hold if each process N_i were observed over a set A_i depending deterministically on i. Indeed, this is so, provided that the values $\nu(A_i)$ satisfy a fairly stringent stability condition. We now approximate the true state estimators

$$E_{\nu,F}[Y_{n+1}|\mathcal{F}^{N_{n+1}}(A_{n+1})] = \frac{K_{A_{n+1}}(N_{n+1}(A_{n+1})+1)}{K_{A_{n+1}}(N_{n+1}(A_{n+1}))}$$

by pseudo-state estimators

$$\hat{E}[Y_{n+1}|\mathcal{F}^{N_{n+1}}(A_{n+1})] = \frac{\hat{K}(N_{n+1}(A_{n+1})+1)}{\hat{K}(N_{n+1}(A_{n+1}))}$$

where

$$\hat{K}(k) = k!\left[n^{-1}\sum_{j=1}^{n} N_j(A_j)\right]^{-k} n^{-1}\sum_{i=1}^{n} 1(N_i(A_i) = k), \tag{7.80}$$

which can be calculated from only the data $N_1(A_1), \ldots, N_n(A_n)$, $N_{n+1}(A_{n+1})$. In Karr (1984b) the following limit theorem is proved.

Theorem 7.28. Assume that

a) There is $c \in (0,1)$ such that $n^{1/2}\sum_{i=1}^{n} |\nu(A_i) - c| \to 0$;

b) $\int F(du)u^k < \infty$ for each k.

Then under $P_{\nu,F}$,

$$n^{1/2}(\hat{E}[Y_{n+1}|\mathcal{F}^{N_{n+1}}(A_{n+1})] - E_{\nu,F}[Y_{n+1}|\mathcal{F}^{N_{n+1}}(A_{n+1})]) \xrightarrow{d} Z,$$

where Z has a mixed normal distribution. □

The proof is little more difficult than those above; where before appeal was made to the central limit theorem for i.i.d. random variables, one uses instead the Lindeberg-Feller theorem, whose applicability is justified by a) and b). These conditions, of course, are quite restrictive and, moreover, one cannot verify a priori whether they are satisfied. See Karr (1984b) for additional discussion.

An altogether different procedure applies to extremely general—but still i.i.d.—Cox pairs (N_i, M_i), with no structure other than diffuseness imposed on the directing measures. We omit a detailed presentation, but inasmuch as the methodology combines several diverse topics, an outline is merited.

1) For simplicity suppose that E is compact and that $A = E$; assuming that $E[M(E)^k] < \infty$ for each k, the true state estimators of (7.21):

$$E[e^{-M_{n+1}(f)}|\mathcal{F}^{N_{n+1}}] = \frac{E[e^{-M_\mu(1+f)}]}{E[e^{-M_\mu(1)}]}\Big|_{\mu = N_{n+1}} \qquad (7.81)$$

are expressed via the Laplace functionals $L_M(\mu,f) = E[\exp(-M_\mu(f))]$ of the Palm processes M_μ of a generic directing measure M. To form pseudo-state estimators, one must estimate these Palm Laplace functionals. Were the M_i observable this could be done directly, but only the Cox processes are observable, so additional analysis is needed.

2) By Proposition 1.55 for each μ the reduced Palm process N_μ of N, with Palm Laplace functional $L_N(\mu,f)$, has the distribution of a Cox process directed by M_μ, so that for functions g satisfying $0 \le g < 1$,

$$L_M(\mu,g) = L_N(\mu, -\log(1-g)) \qquad (7.82)$$

Consequently, one can estimate the $L_M(\mu,\cdot)$ by applying (7.82) to estimators of the $L_N(\mu,\cdot)$. Since the N_i are observable i.i.d. point processes, methods of Section 4.4 can be used to construct estimators $\hat{L}_N(\mu,g)$ that are asymptotically normal, but only in the integrated sense of Theorem 4.20.

3) There is further complication: neither of the functions $g = 1+f$ and $g \equiv 1$ in (7.81) satisfies $0 \le g < 1$ and therefore (7.82) does not apply. One can circumvent this difficulty by passing to moment-generating functionals rather than Laplace functionals, but there is a further penalty: one must assume that $E[\exp(tM(E))] < \infty$ for some $t > 0$. Granted this assumption, for functions f satisfying $0 < f \le 1$ the true state estimators

$$\begin{aligned} E[e^{M_{n+1}(f)}|\mathcal{F}^{N_{n+1}}] &= \frac{E[e^{-M_\mu(1-f)}]}{E[e^{-M_\mu(1)}]}\Big|_{\mu = N_{n+1}} \\ &= \frac{L_M(N_{n+1}, 1-f)}{L_M(N_{n+1},1)} \\ &= \frac{L_N(N_{n+1},-\log f)}{L_N(N_{n+1},\infty)} \end{aligned}$$

(where "∞" denotes the function identically equal to $+\infty$, but $L_N(\mu,\infty)$ is really just shorthand for $P\{N_\mu(E) = 0\} = \int Q_N(\mu,dv)\exp[-\infty v(E)]$) can be approximated by pseudo-state estimators

$$\hat{E}[e^{M_{n+1}(f)}|\mathcal{F}^{N_{n+1}}] = \frac{\hat{L}_N(N_{n+1},-\log f)}{\hat{L}_N(N_{n+1},\infty)}$$

However, the (hitherto proved) degree of approximation is much worse than in the mixed Poisson case, and the culprit is the integrated convergence in distribution in Theorem 4.20. It can be shown that for f a function satisfying $0<f\leq 1$ and $\delta>0$ (and given the moment generating function assumption)

$$\lim n^{1/2-\delta}E[(\hat{E}[e^{M_{n+1}(f)}|\mathcal{F}^{N_{n+1}}]-E[e^{M_{n+1}(f)}|\mathcal{F}^{N_{n+1}}])^2]=0$$

While this is L^2-convergence, the rate of convergence is $n^{-1/4+\delta}$, $\delta>0$, as opposed to the exact and optimal rate $n^{-1/2}$ for mixed Poisson processes in Theorem 7.26.

7.5 MARTINGALE INFERENCE FOR COX PROCESSES ON $\mathbf{R}_+$

Suppose that N is a Cox process on $\mathbf{R}_+$ with directing measure

$$M(A)=\int_A X_t\,dt \tag{7.83}$$

With respect to the history $\mathcal{H}_t=\sigma(X)\vee\mathcal{F}_t^N$, N has stochastic intensity (X_t) (this process is $\mathcal{H}$-predictable because $X_t\in\mathcal{H}_0$ for every t) and hence N has $\mathcal{F}^N$-stochastic intensity $(\hat{X}_{t-})$, where

$$\hat{X}_t=E[X_t|\mathcal{F}_t^N] \tag{7.84}$$

In this section we examine applicability of martingale methods from Chapters 2 and 5 to inference and state estimation for such Cox processes. That the methods do apply in principle is evident; the key issue is to make use, within the martingale context, of the Cox process structure, which is manifested only in that $X_t\in\mathcal{H}_0$ for each t. It seems difficult to do so; indeed, because X is not observable one is often better off simply using either only methods developed elsewhere in this chapter or only martingale techniques.

In connection with statistical estimation, if the directing intensity X is neither restricted to have special structure nor at least partially observable, use of martingale methods seems precluded and one is left with the Laplace functional techniques described in Section 7.1 and Chapter 4. However, the consequences are less than ideal in terms of what can be inferred about the directing intensity viewed as a stochastic process rather than as the random measure of (7.83). For (N,M) a Cox pair with M given by (7.83), Lemma 7.1 yields

$$E[\exp(-\int g(t)X_t\,dt)]=L_N(-\log(1-g)),$$

provided that $0\leq g<1$, but whereas in principle from this one can recover information or estimate finite-dimensional distributions, moments, ... of X, in practice computation seems difficult. Of course, integrals

$\mu_M(f) = \int f(t)E[X_t]\,dt$, the mean measure of M, can be estimated by empirical averages $n^{-1}\sum_{i=1}^{n} N_i(f)$, but provide only "integrated" information about the function $t \to E[X_t]$. For statistical estimation given i.i.d. copies, there seems to be no way (with or without martingale methods) to exploit systematically the special structure inherent in (7.83). Of course, if N is a mixed Poisson process, then Theorems 7.5 and 7.15 apply for statistical and state estimation, respectively, but then there is so much special structure that martingale techniques are unnecessary.

On the other hand, with additional specialization beyond (7.83) and modification of the observability structure, one achieves a multiplicative intensity model to which martingale methods apply but for which it is hard to exploit the Cox process structure. Specifically, if N is a Cox process on $[0,1]$ directed by the random measure $M(A) = \int_A \alpha_t X_t dt$, with α an unknown function, *and* if the process X is observable, then we have a special case of the multiplicative intensity model of Section 5.1. Methods and results from Section 5.2 apply to estimation of α; in particular, given data comprising i.i.d. pairs (N_i, X_i), then with $N^n = \sum_{i=1}^{n} N_i$ and $X^n = \sum_{i=1}^{n} X_i$, Theorems 5.12 and 5.13 apply to the martingale estimators $\hat{B}_t^n = \int_0^t [1\,(X_u^n > 0)/X_u^n]\,dN_u^n$ of the processes $B_t^n(\alpha) = \int_0^t \alpha_u 1\,(X_u^n > 0)\,du$. However, in this setting one cannot refine or sharpen martingale results by using Cox process structure.

Thus for statistical estimation one seemingly cannot have it both ways; it is possible to exploit Cox process structure or stochastic intensity structure, but not both simultaneously.

For state estimation the situation is similar. In the martingale context, Cox structure is usable to the extent that it implies that the $\mathcal{F}^N$-stochastic intensity is the conditional expectation of the directing intensity, but not much further. Given a Cox process N with directing measure (7.83), the central state estimation problem is to reconstruct the directing intensity; we now explore the consequences of Theorem 5.29 concerning it. The main difficulty is derivation of a semimartingale representation [see (5.69)] for the directing intensity X. If anything the Cox structure is a hindrance because it is more difficult to construct a useful representation for the history $\mathcal{H}_t = \sigma(X) \vee \mathcal{F}_t^N$ than for the smaller history $\mathcal{G}_t = \mathcal{F}_t^X \vee \mathcal{F}_t^N$, with respect to which N still has stochastic intensity (X_{t-}). Once one passes to $\mathcal{G}$, the Cox structure of N is lost. However, if that structure is restricted sufficiently, useful conclusions can be obtained. While valid more generally than for Cox processes, the next result is given here because its principal applications have been to Cox processes and because in the Cox case its hypotheses are easily seen to be fulfilled. It treats the case that the stochastic intensity is a function of a Markov process; in the engineering literature such processes are termed Markov-modulated Poisson processes. Special cases are the stochastic integral of Section 6.3 and the Cox processes of Example 7.14.

Theorem 7.29. Let N be an integrable point process on $\mathbf{R}_+$ admitting $\mathcal{H}$-stochastic intensity $\lambda_t = f(Y_{t-})$, where $(Y,\mathcal{H})$ is a standard Markov process with state space E and generator A, and f is a bounded, positive function on E belonging to the domain of A. Then the MMSE state estimators $\hat{\lambda}_t = E[f(T_t)|\mathcal{F}_t^N]$ satisfy the stochastic differential equation

$$d\hat{\lambda}_t = E[Af(Y_t)|\mathcal{F}_t^N]\,dt + (\hat{\lambda}_{t-})^{-1}[(\lambda^2)_{t-} - (\hat{\lambda}_{t-})^2]\,dM_t \qquad (7.85)$$

where $dM_t = dN_t - \hat{\lambda}dt$ is the $\mathcal{F}^N$-innovation martingale and (λ^2) is the process of MMSE state estimators of (λ_t^2). The initial condition is $\hat{\lambda}_0 = E[f(Y_0)]$.

Proof: Formally, the proof is identical to that of the special case in Theorem 6.23, except that the latter contains simplifications (because there $\lambda^2 = \lambda$) not valid in general: one works through the computations associated with Theorem 5.29 and its antecedents. The semimartingale representation of λ is

$$\lambda_t = \lambda_0 + \int_0^t Af(Y_u)\,du + \tilde{M}_t \qquad (7.86)$$

by Dynkin's formula, where $\tilde{M}$ is a mean zero $\mathcal{H}$-martingale and hence by Proposition 5.27, (5.71), and (5.72a), $\hat{\lambda}$ satisfies the stochastic diffenential equation

$$d\hat{\lambda}_t = E[A\,f(Y_t)|\mathcal{F}_t^N]\,dt + H_t dM_t \qquad (7.87)$$

By Theorem 5.29 the filter gain H has the form $H_t = J_t + K_t - \hat{\lambda}_{t-}$, where J, K are predictable processes that remain to be calculated. Since the martingale $\tilde{M}$ and the point process N have no jumps in common almost surely (this uses the Markov property of Y), $K \equiv 0$ by discussion following Theorem 5.29, while by (5.78), $J_t = (\hat{\lambda}_t)^{-1}(\widehat{\lambda^2})_{t-}$. Substitution into (7.87) yields (7.85); the initial condition is given by (5.70). □

Several comments are in order. First, note that the filter gain in (7.85) is the conditional variance $E[\lambda_t^2|\mathcal{F}_t^N] - E[\lambda_t|\mathcal{F}_t^N]^2$ divided by the conditional mean. Second, save in special cases such as Theorem 6.23, the equation (7.86) does not involve $(\hat{\lambda}_t)$ alone and hence is not self-sufficient for recursive computation of $\hat{\lambda}$. Indeed, (7.86) introduces two additional terms, $E[A\,f(Y_t)|\mathcal{F}_t^N] = Af(Y)_{t-}$ and $(\lambda^2)_{t-}$, neither of which is expressible in general as a function of $\hat{\lambda}$. Estimation of them using two equations analogous to (7.86) [see (7.89)] introduces four further equations, and so on, resulting in an infinite system of stochastic differential equations that can be "solved" only by recourse to (almost always unjustified) approximations or simplifications. For example, an odd-order moment (usually the third) can be declared to be zero, permitting the system to be "closed" and solved recursively; however, justification may be lacking.

Alternatively, instead of estimating moments one can allow measure-valued state processes and estimate distributions themselves, as we describe briefly. We remain in the setting of Theorem 7.29; in this discussion important technical details are ignored. The measure-valued state process $\eta_t = \varepsilon_{Y_t}$, given appropriate assumptions, has semimartingale representation

$$\eta_t = \int_0^t A(Y_u)\,du + \widetilde{M}_t, \tag{7.88}$$

where $\widetilde{M}$ is a martingale taking values in the space of signed measures on E and $A(y)$ is the measure against which a function f belonging to the domain of the generator A integrates to $Af(y)$. Note that (7.88) is the crucial restriction. It follows by arguments in the proof of Theorem 7.29 that the state estimators $\hat{\eta}_t = E[\eta_t|\mathcal{F}_t^N]$, where the conditional expectations are also random measures on E (Karr, 1976b), have semimartingale representation

$$\hat{\eta}_t(g) = \int_0^t Ag(Y)_u^{\wedge}\,du + \int_0^t H_t(g)(dN_u - \hat{\lambda}_{u-}du),$$

with $\hat{\lambda}$ as in Theorem 7.29. But

$$\int_0^t Ag(Y)_u^{\wedge}\,du = \int_0^t\int_E P\{Y_u \in dy|\mathcal{F}_t^N\}Ag(y)\,du = \int_0^t\int_E \hat{\eta}_u(dy)Ag(y)\,du,$$

while since $\lambda_t = \int \eta_t(dy)f(y)$, $\hat{\lambda}_t = \int \hat{\eta}\,(dy)f(y)$. Two more computations suffice to give the following result.

Theorem 7.30. Assume that (7.88) holds. Then the measure-valued process $\hat{\eta}_t = E[\eta_t|\mathcal{F}_t^N]$ satisfies the stochastic diffenential equation

$$d\hat{\eta}_t = \int_E \hat{\eta}_t(dy)A(y)\,dt + (\hat{\lambda}_{t-})^{-1}[\hat{\eta}_{t-}(f\times(\cdot)) - \hat{\lambda}_{t-}\hat{\eta}_{t-}]\,dM_t, \tag{7.89}$$

where $dM_t = dN_t - \hat{\lambda}_{t-}dt$ as in Theorem 7.29.

Proof: By Theorem 5.29 the measure-valued filter gain H is $H_t = (\hat{\lambda}_{t-})^{-1}[(\lambda\eta)^{\wedge}_{t-} - \hat{\lambda}_{t-}\hat{\eta}_{t-}]$ (the K-term again vanishes), but for g in the domain of the generator, $\lambda_t\eta_t(g) = f(Y_t) = \eta_t(fg)$, and consequently $(\lambda\eta)^{\wedge}_{t-} = \hat{\eta}_{t-}(f\times(\cdot))$, which gives (7.89). □

We have digressed from Cox processes on $\mathbf{R}_+$, although it should be borne in mind that a Cox process with directing intensity is one case in which the hypotheses of Theorem 7.29 are known to be verified. Concerning likelihood ratios, Theorem 2.31 yields the following separation theorem, which contains a special case of Theorem 7.24.

Proposition 7.31. Suppose that under the probability P, (N,M) is a Cox pair with M of the form (7.83) and let $\widetilde{P}$ be a probability measure under which N is a Poisson process with rate 1. Then for each t, $P \ll \widetilde{P}$ on $\mathcal{F}_t^N$ with

$$\frac{dP}{d\widetilde{P}}\bigg|_{\mathcal{F}_t^N} = \exp\left[\int_0^t (1 - \hat{X}_u)\, du + \int_0^t (\log \hat{X}_{u-})\, dN_u\right], \tag{7.90}$$

where $\hat{X}_u = E[X_u | \mathcal{F}_u^N]$. □

In general, $P \perp \widetilde{P}$ on $\mathcal{F}^M$, so Proposition 7.31 furnishes another example in which $P \ll \widetilde{P}$ on $\mathcal{F}^N$ despite their being singular on $\mathcal{F}^M$.

In particular, if P_0, P_1 are probabilites with respect to each of which N is a Cox process with directing measure (7.83), then (7.90) implies that

$$\frac{dP_1/d\widetilde{P}}{dP_0/d\widetilde{P}}\bigg|_{\mathcal{F}_t^N} = \exp\left[\int_0^t (\hat{X}_u(0) - \hat{X}_u(1))\, du + \int_0^t \log\frac{\hat{X}_{u-}(1)}{\hat{X}_{u-}(0)}\, dN_u\right], \tag{7.91}$$

where $\hat{X}_u(i) = E_i[X_u | \mathcal{F}_u^N]$. Whenever $P_1 \ll P_0$ on $\mathcal{F}^N$, (7.91) gives the likelihood ratio dP_1/dP_0 and—provided that state estimators $\hat{X}_u(i)$ can be calculated—is usable for likelihood ratio tests. The log-likelihood function

$$L_t(P_0,P_1) = \int_0^t (\hat{X}_u(0) - \hat{X}_u(1))\, du + \int_0^t \log\frac{\hat{X}_{u-}(1)}{\hat{X}_{u-}(0)}\, dN_u \tag{7.92}$$

can be used more generally (although without absolute continuity or equivalence it may assume infinite values), but still the state estimators $X_u(i)$ must be computed. Theorem 7.29 presents a set of conditions when this might be possible as well as a potential method. We conclude the section with a specific application in which computations have actually been done.

Contained in Smith and Karr (1985) is a statistical analysis of some point process models for times of summer season precipitation events at a single site. Three Cox process models of rainfall are considered:

1) Poisson processes (with constant intensity);
2) The stochastic integral $X * N$ of Section 6.3, with X a Markov process;
3) A class of Neyman-Scott cluster processes.

The first two have already been analyzed in detail; the third will be described momentarily. Each entails unknown parameters that must be estimated before likelihood ratio tests can be performed to decide which of two of them better describes the data, so the procedure is not a straightforward application, for example, of (7.92). Specifics are discussed following description of the cluster processes.

The Neyman-Scott cluster processes considered as models of rainfall are Poisson cluster processes in the sense of Definition 1.30, but are more easily understood via an informal construction. Let $\tilde{N} = \Sigma\, \varepsilon_{T_i}$, the *cluster center process*, be a Poisson process on $\mathbf{R}_+$ with rate c, and to each cluster center T_i let there be associated a random number X_i of secondary points with locations T_{ij}, $j = 1, \ldots, X_i$, in such a manner that

a) X_i has a Poisson distribution with mean a;
b) The "offsets" $T_{ij} - T_i$ are independent and identically exponentially distributed with mean b^{-1};
c) X_i and the $T_{ij} - T_i$ are independent of one another, of $\tilde{N}$, and of the X_k and $T_{k\ell}$ for $k \neq i$.

The *Neyman-Scott cluster process* is the Poisson cluster process $N = \Sigma_i \Sigma_j\, \varepsilon_{T_{ij}}$ consisting of all the secondary points, with the general interpretation that the cluster center T_i engenders succeeding events at times T_{ij}; a specific interpretation will be given in a moment. First, here are basic properties.

Proposition 7.32. Let $\tilde{N}$, N be as above. Then the Neyman-Scott cluster process N is a Cox process with directing intensity

$$X_t = ab \int_0^t e^{-b(t-u)}\, d\tilde{N}_u \tag{7.93}$$

Moreover, X is a Markov process. □

The proof is straightforward (Exercise 7.25); intuitively, the Markov property of X results from the stochastic differential equation

$$dX_t = -bX_t\, dt + ab\, d\tilde{N}_t,$$

a consequence of (7.93), and from the independent increments property of $\tilde{N}$.

As models of summer season (July-October) rainfall in the Potomac River basin in the eastern United States, each class of processes carries a distinctive physical interpretation. The Poisson model represents a constant rate of occurrence of rainfall events. For the stochastic integral the directing intensity X represents a climatological process, with $X = 0$ a "dry" state (cool, high-pressure conditions) during which there is no precipitation and $X = 1$ a "wet" state during which events occur in Poisson fashion. Not only is this interpretation consistent qualitatively with climatological data but also (Exercise 3.15), $X * N$ is a renewal process, which accords with data analysis in Smith (1981) but lacks clear physical justification. In the Neyman-Scott cluster process, cluster centers correspond to passages of cold and warm fronts, each having associated with it some (or possibly no) subsequent rainfall events (see Kavvas and Delleur, 1981, for details).

Statistical analysis was performed as follows. By Proposition 7.32, Proppoition 7.31 applies to the Neyman-Scott cluster model and to the stochastic integral $X*N$ for calculation of log-likelihood functions. However, it was necessary first to estimate unknown parameters: for Poisson processes the rate, for $X*N$ the rate c of events during periods when $X=1$ and the transition rates a, b in the generator (6.48) of X, and for the cluster process the rate c of the cluster center process, the mean number a of cluster members per cluster center, and b, the exponential parameter of the cluster center-cluster member offsets. For the Poisson model this was done using (3.40), while for the other two classes it was done by maximizing numerically the likelihood functions (7.90) over a grid of a, b, c values using a nonlinear optimization algorithm, choosing as maximum likelihood estimators $\hat{a}$, $\hat{b}$, $\hat{c}$ (the parameters have different interpretations in the two models) the values for which the likelihood was greatest. To compute state estimators of the directing intensity, (6.56) was utilized for $X*N$ and an approximation to (7.86)—third moments were declared to be zero—for the cluster process. For computational purposes, time discretization was used to convert stochastic differential equations to difference equations.

Likelihood ratio tests were performed using the log-likelihood functions (7.92), but with the probability measures P_0, P_1 there replaced by measures associated with previously calculated maximum likelihood estimators. Thus, given two classes, one was deemed better if the maximum value of the likelihood function was greater for it. In terms of (7.92) and likelihood ratio tests, this amounts to choosing P_1 over P_0 if $L_t(P_0,P_1)>0$ and choosing P_0 over P_1 otherwise. The hypotheses H_0: $P=P_0$, H_1: $P=P_1$ are thus treated symmetrically.

Computational difficulties precluded complete analysis of the Neyman-Scott alternative. Between the Poisson model and the stochastic integral, for small-sample-size values of t the Poisson hypothesis was accepted, but for larger values the stochastic integral was the better model. For simulated realizations of $X*N$, convergence of maximum likelihood estimators to true parameter values was exceedingly slow, especially for ill-chosen but physically reasonable combinations of the parameters. See Smith and Karr (1985) for details and numerical results.

In addition to its other properties, the stochastic integral $X*N$ is a renewal process, so it forms a fitting bridge to the next chapter, where it will be analyzed yet again.

EXERCISES

7.1. Let (N,M) be a Cox pair. Prove that if N is simple, then M is diffuse almost surely.

7.2. Let N be a mixed Poisson process with parameters (ν,F) satisfying (7.14). Prove that N admits a representation of the form (7.15): $N=\Sigma_{i=1}^{N(E)}\varepsilon_{X_i}$, where $N(E)$ has the mixed Poisson distribution (7.16) and the X_i are i.i.d. with distributoion ν and independent of $N(E)$.

7.3. *Deduce for* (7.15) *that a mixed Poisson process* N satisfying (7.14) is conditionally uniform with respect to ν, that is, given that $N(E)=k$, N has the same distribution as the empirical process $\tilde{N}=\Sigma_{i=1}^{k}\varepsilon_{Z_i}$, where the Z_i are i.i.d. random elements of E, each with distribution ν.

7.4. Let N be a mixed Poisson process on $\mathbf{R}_+$. Show that if N is a renewal process, then N is in fact Poisson.

7.5. Derive the covariance function R of (7.12).

7.6. Prove Lemma 7.3.

7.7. Let (N_i,M_i) be i.i.d. copies of a Cox pair on a compact space E. Calculate the asymptotic efficiency of the estimators $\hat{\mu}_M=n^{-1}\Sigma_{i=1}^{n}N_i$ of the mean measure M, which are used if only the N_i are observable, relative to the estimators $\hat{\mu}_M^*=n^{-1}\Sigma_{i=1}^{n}M_i$ that would be used if the M_i were observable.

7.8. Let (N_i,Y_i) be i.i.d. copies of the pair (N,Y) in which N is a mixed Poisson process directed by the random measure $M=Y\nu$, where the Y_i satisfy $E[Y_i]=1$. Assume that the underlying space is compact and suppose that both the N_i and the Y_i are observable.

a) Develop asymptotic properties of the estimators $\hat{\nu}^*=\Sigma_{i=1}^{n}N_t/\Sigma_{i=1}^{n}Y_i$ of the mean measure ν of the N_i.

b) Compare with those of the estimators $\hat{\nu}=\Sigma_{i=1}^{n}N_i$ that would be used if the Y_i were not observable; in particular, calculate the symptotic relative efficiency.

7.9. Let N be a Poisson process with mean measure μ. Show that for $g,h\geqslant 0$, $E[e^{-N(g)}N(h)]=\exp[-\mu(1-e^{-g})]\,\mu(he^{-g})$.

7.10. Let M be a diffuse random measure with probability law $\mathcal{L}$ and let $\tilde{M}$ have law $\tilde{\mathcal{L}}(d\nu)=C^{-1}e^{-\nu(A)}\mathcal{L}(d\nu)$, where A is a fixed set and $C=\int\mathcal{L}(d\nu)e^{-\nu(A)}$. Show that the Palm Laplace functionals $L(\mu,f)=E[e^{-M_\mu(f)}]$ and $\tilde{L}(\mu,f)=E[e^{-d\tilde{M}_\mu(f)}]$ satisfy $\tilde{L}(\mu,f)=L(\mu,f+1_A)/L(\mu,1_A)$.

7.11. Let N be a mixed Poisson process on $\mathbf{R}_+$ with directing measure $M=Y\nu$.

a) Prove that for each t,

$$E[Y^2|\mathcal{F}_t^N]=\frac{\int F(dx)e^{-xt}x^{N_t+2}}{\int F(dy)e^{-yt}y^{N_t}}.$$

b) Derive (5.81) directly from this expression.

7.12. Consider the Markov-modulated Poisson process of Example 7.14.

a) Solve (7.41) for the case $i=0$, $j=\alpha$, $q_0=a$, $q_\alpha=b$, where $a,b>0$.

b) Obtain the corresponding explicit version of (7.40).

c) Contrast this method of calculation with that of Theorem 6.23.

7.13. Let a, $b>0$ be known and distinct, let $T>0$ be an unobservable random variable with known, continuous distribution G, let X be the process $X_t=a1(t<T)+b1(t\geq T)$, and suppose that the observations are a Cox

process N with directing intensity X. [This is the "disruption" problem (see also Exercise 5.23).]

a) Derive a stochastic differential equation satisfied by the MMSE state estimators $P\{T \le t | \mathcal{F}_t^N\} = P\{X_t = b | \mathcal{F}_t^N\}$.

b) Solve the equation explicitly for the case that G is an exponential distribution.

7.14. Provide details missing from the proof of Theorem 7.13.

7.15. Give a detailed derivation of (7.51).

7.16. For the disruption problem of Exercise 7.13, for each t calculate the MMSE linear estimator of $1(T \le t)$ given the observations $\mathcal{F}_t^N$.

7.17. Work through the following heuristic derivation of (7.53).

a) Argue that $\hat{M}(f) - M(f)$ is characterized by its being orthogonal to $N(dx)$ for every $x \in A$.

b) Deduce from a) that for $x \in A$, $\hat{f}$ satisfies

$$E[M(f)M(dx)] + \mu(f)\mu(dx) = \hat{f}(x)\mu(dx) + E[M(\hat{f})M(dx)] - \mu(\hat{f})\mu(dx),$$

the differential version of (7.53).

7.18. Let P_0, P_1 be probabilities under each of which (N,M) is a Cox pair on E. Prove that if $P_0 \perp P_1$ on $\mathcal{F}^N$, then $P_0 \perp P_1$ on $\mathcal{F}^M$.

7.19. Let E be a compact set, let $\mathcal{L}_0, \mathcal{L}_1$ be equivalent probabilities on $\mathbf{M}_d$, let M be the coordinate random measure, and let M_μ^i be the Palm processes of M under $\mathcal{L}_i$. Prove that

$$\frac{E[e^{-M_\mu^0(E)}(d\mathcal{L}_1/d\mathcal{L}_0)(M_\mu^0)]}{E[e^{-M_\mu^0(E)}]} = \frac{E[e^{-M_\mu^1(E)}]}{E[e^{-M_\mu^1(E)}\{(d\mathcal{L}_1/d\mathcal{L}_0)(M_\mu^1)\}^{-1}]}$$

7.20. Prove Lemma 7.22.

7.21. Consider once more the disruption problem of Exercise 7.13. For fixed $\alpha > 0$, develop for each t a level-α likelihood ratio test of the "hypothesis" $H_0 : T \le t$, given the observations $\mathcal{F}_t^N$. That is, define a critical region $\Gamma \in \mathcal{F}_t^N$ such that on the event Γ, $P\{T > t | \mathcal{F}_t^N\} \leqslant \alpha$ almost surely.

7.22. In the context of Theorem 7.25,

a) Calculate the covariances in (7.76).

b) Compute from them those in (7.71).

7.23. (Continuation of Exercise 7.13) Consider the disruption problem of Exercise 7.13 with a and b known but G an exponential distribution with *unknown* parameter λ.

a) Show that consistent estimation of λ from a single realization of a generic copy N, even observed over $[0,\infty)$, is not possible.

b) Devise strongly consistent and asymptotically normal estimators $\hat{\lambda}$ (as $n \to \infty$) given as data i.i.d. copies N_i, each observed over $[0,1]$. [*Hint*: One choice uses the identity $E[N_i(1)] = b + \lambda^{-1}(b-a)(1-e^{-\lambda})$.]

c) Construct pseudo-state estimators $\hat{P}\{T_{n+1} \le t | \mathcal{F}_t^{N_{n+1}}\}$ by replacing λ in the stochastic differential equation derived in Exercise 7.13 by $\hat{\lambda}$ (based on $N_1, \ldots, N_n$) and examine asymptotics as $n \to \infty$.

7.24. Work explicitly through the calculations associated with Theorem 7.16 under the assumption that the directing intensity is a Markov process with finite state space and generator A.

7.25. a) Prove Proposition 7.32.

b) Calculate the Laplace functional of the Neyman-Scott cluster process N of that Proposition.

7.26. Let N be a Cox process on $\mathbf{R}_+$ with directing intensity X, a Markov process with state space $\{0, \alpha\}$, where $\alpha > 0$ is unknown, and with known generator

$$A = \begin{bmatrix} -a & a \\ b & -b \end{bmatrix},$$

with a and b both positive. Construct strongly consistent and asymptotically normal estimators of α given the observations $\mathcal{F}_t^N$, $t \geq 0$, and verify these properties.

NOTES

Expository accounts of Cox processes are given by Grandell (1976), Krickeberg (1972), Matthes et al. (1978), and Snyder (1975); only the latter contains applications. As described in Chapter 1, the fundamental theoretical role of Cox processes concerns thinnings (see Kallenberg, 1975a; Mecke, 1968). A plethora of modeling applications—central to each is a physical interpretation of the directing measure or intensity—includes aerosol pollutants (Rohde and Grandell, 1981), optical signal transmission (Karp and Clark, 1970; Snyder, 1975), and precipitation (Smith and Karr, 1983). Previous treatments of inference in some generality are Grandell (1976) and Krickeberg (1982); each concentrates on state estimation.

Section 7.1

As intimated above, there are no previous works on statistical estimation for Cox processes on general spaces, in part because Proposition 7.2 [which merely combines (7.9) with Theorems 4.12 and 4.13] is as far as Cox process structure can be exploited without introducing additional assumptions. Except for Poisson processes, only mixed Poisson processes have been studied in detail; much remains to be done to identify and analyze worthwhile special cases. Theorems 7.4 and 7.5, taken from Karr (1984b), are the most general results available in the i.i.d. case; the estimators have been modified slightly to yield easier proofs and less complicated covariances. Albrecht (1982b), Simar (1976), and Tucker (1964) treat special cases of estimation for mixed Poisson processes. Lemma 7.3 highlights a key difference between Poisson and Cox processes.

Section 7.2

State estimation, motivated originally by applications, is *the* fundamental inference problem for Cox processes from a theoretical viewpoint as well. The main result,

Theorem 7.6, is due to Karr (1985a); along with Theorem 4.27 it emphasizes the central role of Palm distributions for state estimation. However, despite its elegance it is in some sense too general to be *really* useful. The special cases of Theorems 7.9 and 7.15, Propositions 7.10 and 7.12, and Example 7.14, which are more useful, appear in Karr (1983). There, Theorem 7.9 is the most general result and is proved by appeal to the Bayes' theorem of Kallianpur Streiber (1968) (see also Krickeberg, 1982). Special versions of Proposition 7.10 and Theorem 7.15 are in Grandell (1976), Liptser and Shiryayev (1978), and Snyder (1975). Maziotto and Szpirglas (1980) analyze recursive state estimation for planar mixed Poisson processes. Theorem 7.16 is due to Grandell (1976); the formally new Theorem 7.17 uses exactly the same proof. An infinitesimal/recursive version for general spaces is Karr (1983, Theorem 3.11). Other special cases of state estimation are considered in Boel and Beñes (1980) (diffusion process intensities); Freed and Shepp (1982) (estimation of X_0 for the process $X * N$ of Section 6.3); Grandell (1976), Karp and Clark (1970), and Macchi and Picinbono (1972) (photon counts); and Snyder (1972a, 1975) (engineering applications). Snyder (1972a) is the seminal early work.

Section 7.3

Even though none of the results here is especially deep, to our knowledge this is the first general study of likelihood ratios and absolute continuity/singularity for Cox processes on general spaces. Analogs of Theorem 7.24, the separation theorem, appear in Brémaud (1981), Brémaud and Jacod (1977), and Snyder (1975); in the first and third one can locate interpretations and implications of "separation" (see also Proposition 7.31). The intractable problem of discriminating between Poisson and mixed Poisson processes is discussed by Albrecht (1982a) and Sundt (1982); the root of the difficulties is nonergodicity of mixed Poisson processes (see Proposition 7.12).

Section 7.4

All of the results pertaining to mixed Poisson processes are due to Karr (1984b), but—as in Theorems 7.4 and 7.5—estimators and covariances have been simplified. In the proof of Theorem 7.25 are details needed to flesh out that of Theorem 7.5. The only extant treatment of combined inference and state estimation for general Cox processes is Karr (1985a).

Section 7.5

As observed in the text, the message of the section is negative: it seems impossible to exploit Cox structure within the context of martingale inference. The relevant characterization of Cox processes, $\mathcal{H}_0$-measurability of the stochastic intensity or compensator (Brémaud, 1975a, 1981; see also Exeercise 2.10) simply does not help. Indeed, historical development of the theory confirms this. Results originally proved for Cox processes using specialized methods (e.g., Snyder, 1972a) have been extended to point processes with much more general stochastic intensities; one cannot hope to exploit what has become an inessential hypothesis. As a consequence, Section 5.4, particularly Theorem 5.29, is the main antecedent of this section. Theorem 7.29, in some ways the culmination of the pre-martingale era, is due to Snyder (1972a). Dynkin's formula, which yields the semimartingale representation (7.86), is an important result for Markov processes and is explained, for example, in Karlin and Taylor (1975). That Theorem 6.23 is a special case of Theorem 7.29 is apparent; just *how* special is initially less so. Closing an infinite system of stochastic differential equations by writing it as a single equation for a measure-valued process, as in Theorem 7.30, is elegant mathematically but of dubious proctical value. Kunita (1971) and Fujisaki et al. (1972) are related works. Neyman and Scott (1972) contains a discussion of the cluster process bearing their names; applications to modeling of precipitation is studied in Kavvas and Delleur (1981) and Smith and Karr (1985).

Exercises

7.2-7.3. For more on this characteristic feature of mixed Poisson processes, see Feigin (1979), Kallenberg (1975b, 1983) (where "conditionally uniform" is termed "symmetrically distributed"), Matthes et al. (1978), Mecke (1976), and Puri (1982).

7.4. See Theorem 8.24 and Kingman (1963), the original source.

7.7-7.8. Lost efficiency is an inevitable consequence of having "inappropriate" Observations.

7.10. See Karr (1985a).

7.12. See Karr (1983, 1984a) for additional aspects.

7.13; 7.23. Among references on this venerable problem are Brémaud (1980), Davis et al. (1975), Galtchouk and Rozovskii (1971), Liptser and Shiryayev (1978), and Wan and Davis (1977).

7.26. See also exercise 3.17.

8

Nonparametric Inference for Renewal Processes

Although of considerable intrinsic interest, renewal processes are important mathematically because of regenerative processes, that is, continuous-time stochastic processes with embedded random times forming a renewal process. Statistical inference for renewal processes is based on the property that interarrival times are i.i.d. and requires few tools beyond those developed in earlier chapters. For state estimation, as will emerge, central roles are played by the backward and forward recurrence time processes. Usefulness of renewal processes stems principally from asymptotic properties, not only of their own but also of solutions to renewal equations: hence in this chapter there is some emphasis on functionals of renewal processes (e.g., recurrence time processes) and of interarrival time distributions (e.g., renewal measures). We begin with a prefatory discussion of key concepts and results, but it is in no sense complete (nothing is proved); consult the chapter notes for references.

Let $N = \Sigma_{n=0}^{\infty} \varepsilon_{T_n}$ be a simple point process on $\mathbf{R}_+$ with arrival times $T_1 < T_2 < \cdots$ and with $T_0 = 0$ not considered an arrival time, and let $U_i = T_i - T_{i-1}$, $i \geq 1$, be the interarrival times. From Definition 1.6, N is a *renewal process* if the U_i are i.i.d.; that is, at each arrival time T_i the arrival counting process

$$N_t = N((0,t]) = \sum_{i=1}^{\infty} 1(T_i \leq t) \tag{8.1}$$

renews itself probabilistically: its future increments form another renewal process (Therefore, the T_i are sometimes called renewal times.) The distribution F of the interarrival times is termed the *interarrival time distribution* and is the sole object needed to specify the law of N. Because N is simple, $F(0) = 0$, and we assume throughout that $F(\infty) = 1$; the reader should be aware that often neither of these assumptions is imposed when probabilistic properties of renewal processes are studied. The mean and variance of F are denoted by m and σ^2, respectively, and when it exists the hazard function $F'/(1-F)$ by h.

Given the i.i.d. structure of the interarrival times, it is reasonable to surmise that asymptotic properties of the arrival time sequence (T_k) are manifested as well in the counting process (N_t), which is indeed true.

Proposition 8.1. Let N be a renewal process with interarrival time distribution F.

a) If $m < \infty$, then $N_t/t \to 1/m$ almost surely.

b) If $\sigma^2 < \infty$, then

$$t^{1/2}\left(\frac{N_t}{t} - \frac{1}{m}\right) \xrightarrow{d} N\left(0, \frac{\sigma^2}{m^3}\right) \quad \square \tag{8.2}$$

In particular, Proposition 8.1 provides apparatus to estimate the mean interarrival time from observation of the counting process (N_t), although if (8.2) were used, for example, to construct hypothesis tests or confidence intervals, it would also be necessary to estimate the variance (Exercise 8.9).

Associated with a renewal process are many functionals, the two most fundamental of which are the backward and forward recurrence time processes, whose values at t are, respectively, the time elapsed since the last arrival before t and the time from t until the first arrival thereafter. More precisely, the *backward recurrence time* at t is

$$V_t = t - T_{N_t}, \tag{8.3}$$

which is zero if $N_t = 0$; the *forward recurrence time* at t is

$$W_t = T_{N_t+1} - t \tag{8.4}$$

Pictorially, $t \to V_t$ is an "ascending sawtooth" and $t \to W_t$ a "descending sawtooth." For state estimation their roles are crucial but different: to predict the future of a renewal process one needs concerning the past only the cumulative number of arrivals and the current value of the backward recurrence time (i.e., N_t and V_t), while once the forward recurrence time is predicted, the remainder of the future can be predicted deterministically.

The function (also viewed as a measure)

$$R(t) = 1 + E[N_t] \tag{8.5}$$

is the *renewal function*, expressible in terms of the interarrival time distribution F as

$$R(t) = \sum_{k=0}^{\infty} F^k(t) = \sum_{k=0}^{\infty} P\{T_k \le t\}, \tag{8.6}$$

where F^k denotes the k-fold convolution of F with itself and $F^0 \equiv 1$ is the distribution function of T_0. The renewal function is analogous to the potential measure of a Markov process, but its most important role is in the solution of renewal equations (Theorem 8.4). Here are its elementary properties.

Proposition 8.2. Let R be the renewal function associated with the renewal process N. Then $R(t) < \infty$ for all $t < \infty$ and $R(t)/t \to 1/m$ as $t \to \infty$, even if $m = \infty$, in which case $1/m = 0$. □

Denote by B the set of functions f on $\mathbf{R}_+$ that are (measurable and) bounded over bounded intervals. Given a locally finite measure ν on $\mathbf{R}_+$ and $f \in B$, the convolution $\nu * f(t) = \int_0^t f(t-u)\nu(du)$ also belongs to B. A *renewal equation* associated with F is an equation

$$f = g + F * f, \tag{8.7}$$

where $g \in B$ is known and $f \in B$ is unknown. The following example not only illustrates the typical genesis of a renewal equation but also points the way to the general solution.

Example 8.3. Let (W_t) be the forward recurrence time process defined by (8.4); we wish to calculate the distribution of W_t for each t. Let y be fixed and put $f(t) = P\{W_t > y\}$; then

$$\begin{aligned} P\{W_t > y\} &= P\{W_t > y, T_1 > t\} + P\{W_t > y, T_1 \le t\} \\ &= P\{T_1 > t + y\} + E[P\{W_t > y | T_1\} 1(T_1 \le t)] \end{aligned}$$

since once $T_1 > t$, $W_t > y$ if and only if $T_1 > t + y$. Now we make the crucial *renewal argument*: $\widetilde{N} = \Sigma_{k=1}^{\infty} \varepsilon_{T_k - T_1}$ is a renewal process independent of T_1, with the same law as N; therefore (with $\widetilde{W}$ the forward recurrence time for $\widetilde{N}$), $P\{W_t > y | T_1\} = P\{\widetilde{W}_{t-T_1} > y\} = f(t - T_1)$, and hence f satisfies the renewal equation (8.7) with $g(t) = 1 - F(t + y)$. But we may calculate f directly: using (8.6),

$$\begin{aligned} P\{W_t > y\} &= P\{N_{t+y} - N_t = 0\} \\ &= \sum_{k=0}^{\infty} P\{T_k \le t, T_{k+1} > t + y\} \\ &= \sum_{k=0}^{\infty} \int_0^t F^k(du)[1 - F(t + y - u)] \\ &= \int_0^t R(du)[1 - F(t + y - u)] = R * g(t), \end{aligned} \tag{8.8}$$

that is, for this specific case the renewal equation is solved by $f = R * g$. □

The same is true in general.

Theorem 8.4. Given $g \in B$ the unique solution in B to the renewal equation (8.7) is $f = R * g$. □

For a Poisson process with rate λ, $R(t) = 1 + \lambda t$ and substitution in (8.8)—in such computations one must not neglect to include the atom of the renewal measure at the origin—gives $P\{W_t > y\} = e^{-\lambda y}$ regardless of t, a property that characterizes Poisson processes among renewal processes (Exercise 8.24). In other cases the renewal function is difficult or impossible to calculate explicitly, so one is led to hope—based also on "nice" asymptotic behavior of renewal processes—that solutions of renewal equations might be similarly well behaved in that their limits do not depend strongly on the interarrival time distribution. (By Propositions 8.1 and 8.2 there is dependence at least on the mean interarrival time, but one can hope for no other dependence.) Neither hope is unfounded, as the *key renewal theorem* demonstrates.

Theorem 8.5. Assume that $g \in B$ is directly Riemann integrable (Feller, 1971) and that the distribution F is nonarithmetic. Then

$$\lim_{t \to \infty} R * g(t) = m^{-1} \int_0^\infty g(u)\, du; \tag{8.9}$$

the possibility $m = \infty$ is not excluded. □

[The distribution F is *arithmetic* if there exists $a > 0$ such that $F(\{na : n \in \mathbf{N}\}) = 1$ and *nonarithmetic* otherwise. An analog of (8.9) holds in the arithmetic case.]

Choosing $g = 1_{[0,h]}$ in (8.9) yields the *Blackwell renewal theorem*:

$$\lim_{t \to \infty} [R(t+h) - R(t)] = \frac{h}{m}, \tag{8.10}$$

a much strengthened form of the second statement in Proposition 8.2. The equations (8.9) and (8.10) are in fact equivalent.

The function $g(t) = 1 - F(t + y)$ in (8.8) is directly Riemann integrable provided that $m < \infty$, and if in addition F is nonarithmetic, then

$$\lim_{t \to \infty} P\{W_t > y\} = m^{-1} \int_y^\infty [1 - F(u)]\, du \tag{8.11}$$

This distribution arises (Exercise 1.31) in connection with stationary renewal processes. If the renewal process N is modified so that U_1 has distribution

(8.11), then $N^* = \Sigma_{n=0}^{\infty} \varepsilon_{T_n}$ (a special case of a *delayed* renewal process) is a stationary point process on $\mathbf{R}_+$, additional aspects of which are treated in Chapter 9.

Finally, we introduce the notion of a regenerative process.

Definition 8.6. Let X be a (right-continuous, left-hand limited) process taking values in a locally compact Hausdorff space E. Then X is a *regenerative process* if there exist $\mathcal{F}^X$-stopping times $0 = T_0, T_1, T_2, \ldots$ such that

a) $N = \Sigma\, \varepsilon_{T_n}$ is a renewal process;

b) For each $n, k, t_1, \ldots, t_k \geq 0$ and (bounded, continuous) function f on E^k,

$$E[f(X_{T_n+t_1}, \ldots, X_{T_n+t_k}) | \mathcal{F}^X_{T_n}] = E[f(X_{t_1}, \ldots, X_{t_k})] \tag{8.12}$$

At each *regeneration time* T_n, the future of X becomes a probabilistic replica of X that is independent of the past. Given a suitable definition the excursions $Z_{n+1}(t) = X_{T_n+t}, 0 \leq t \leq U_{n+1}$, are such that the pairs (U_n, Z_n) are i.i.d. and hence we can study X using the marked point process $\bar{N} = \Sigma\, \varepsilon_{(T_n, Z_n)}$. Other than a renewal process the simplest example is an irreducible Markov process with finite state space, which is regenerative with respect to the sequence of times of returns to its initial state (Exercise 8.6).

Renewal process results apply to regenerative processes. For example, with A a fixed subset of E, the function $t \to P\{X_t \in A\}$ satisfies the renewal equation

$$P\{X_t \in A\} = P\{X_0 \in A, T_1 > t\} + \int_0^t F(du) P\{X_{t-u} \in A\},$$

where F is the distribution of the times between regenerations; consequently, if F fulfills the hypotheses of Theorem 8.5, then

$$\lim_{t \to \infty} P\{X_t \in A\} = m^{-1} \int_0^{\infty} P\{X_0 \varepsilon A, T_1 > u\}\, du, \tag{8.13}$$

a limit theorem of remarkable power and generality (Exercise 8.6).

This concludes our survey of renewal and regenerative processes. The remainder of the chapter pertains to inference and is organized in the following manner. Section 8.1 is devoted to estimation of the interarrival time distribution and also—by substitution—of functionals such as the renewal function and the distribution (8.11), given as data different kinds of observations of a single realization of a renewal process. For synchronous observations $\mathcal{F}^N_{T_n}$ empirical process methods from Section 4.6 apply; we do not dwell on this case. Mainly, we treat asynchronous data $\mathcal{F}^N_t$ representing "real-time" observation; here empirical process techniques can be utilized with minor modifications. Finally, partly leading to Chapter 10, we consider Poisson samples of a renewal process: observations are the values of the counting process (N_t) at arrival times in a Poisson process independent of N. We also examine martingale methods; when the hazard function h exists, N admits $\mathcal{F}^N$-stochastic intensity

$$\lambda_t = h(V_{t-}) \tag{8.14}$$

Section 8.2 is a study of hypothesis testing, beginning with analysis of absolute continuity, likelihood functions, and local asymptotic normality. Thereafter we discuss some fairly specific tests concerning structural characteristics of the interarrival time distribution. Tests whether a renewal process is a Poisson or Cox process are also treated; they lead to interesting characterizations.

State estimation and combined statistical inference and state estimation form the topic of Section 8.3, the latter treated somewhat sketchily. Finally, Section 8.4 describes martingale techniques for statistical estimation for Markov renewal processes (a generalization of renewal processes with many important applications) based on complete or possibly heavily censored observations.

8.1 ESTIMATION OF THE INTERARRIVAL DISTRIBUTION

Let N be a point process that under the probability P_F is a renewal process with interarrival distribution F, which we take to be completely unknown, yielding the full nonparametric model $\mathcal{P} = \{P_F\}$. Parametric models, for example where the interarrival times have a gamma distribution (Examples 3.4 and 3.9), are ordinarily analyzed using specialized techniques. The reliability literature (see the chapter notes) is a good (perhaps the best) source for specifics; because interarrival times are nonnegative, classical procedures for normally distributed data do not apply. Here our primary focus is on nonparametric estimation of the interarrival distribution F given various kinds of observations of the renewal process.

The case of synchronous data, represented by σ-algebras

$$\mathcal{F}^N_{T_n} = \sigma(T_1, \ldots, T_n) = \sigma(U_1, \ldots, U_n), \tag{8.15}$$

and corresponding to observation of N until the nth arrival, is easy: the empirical distribution functions

$$\hat{F}(x) = n^{-1} \sum_{i=1}^{n} 1(U_i \leq x) \tag{8.16}$$

are nonparametric maximum likelihood estimators of F and standard results for empirical distribution functions (see, e.g., Gaenssler, 1984) apply. In particular, we have the following asymptotic properties.

Proposition 8.7. Let $\hat{F}$ be given by (8.16). Then under each P_F,
a) $\sup\{|\hat{F}(x) - F(x)| : x \in \mathbf{R}_+\} \to 0$ almost surely;
b) As random elements of $D[0,\infty)$,

$$\{n^{1/2}[\hat{F}(x) - F(x)] : x \in \mathbf{R}_+\} \xrightarrow{d} \{G(x) : x \in \mathbf{R}_+\}, \tag{8.17}$$

where G is a Gaussian process with covariance function

$$R(x,y) = F(x)[1 - F(y)] \tag{8.18}$$

for $0 \leq x \leq y$. □

In most instances, however, one does not observe synchronous data; instead, the arrival counting process N is observed in "real time," resulting in the asynchronous data

$$\mathcal{F}_t^N = \sigma(N_u : 0 \leqslant u \leqslant t) \tag{8.19}$$

The obvious analogs of the estimators (8.16) are the empirical distribution functions

$$\hat{F}(x) = N_t^{-1} \sum_{i=1}^{N_t} 1(U_i \leq x), \tag{8.20}$$

defined to be identically 1 on $\{N_t = 0\}$. By contrast with the similar representation (6.15) for Poisson processes, here the random sample size N_t is a deterministic function of (U_n). This extreme dependence is less troublesome than intermediate forms and we can without difficulty establish an analog of Proposition 8.7. First, though, we remark on some additional aspects of (8.20). Note that the "sample size" t, following our custom, is suppressed; x is the argument of $\hat{F}$ as a distribution function. Not all of the information contained in the data $\mathcal{F}_t^N$ is used in forming the estimator $\hat{F}$: the ongoing interarrival interval is known to exceed the current backward recurrence time V_t, but this is ignored in (8.20). Therefore, in particular, the $\hat{F}$ cannot be maximum likelihood estimators. Partial information about U_{N_t+1} *is* incorporated in the modified estimators

$$\begin{aligned} \hat{\hat{F}} &= \frac{N_t}{N_t + 1} \hat{F}(x), \qquad x \leq V_t \\ &= \hat{F}(x), \qquad x > V_t, \end{aligned} \tag{8.20a}$$

whose asymptotic behavior is identical to that of the $\hat{F}$. (In Section 8.4 partial observations of Markov renewal processes are incorporated similarly.) A more serious—and inescapable—shortcoming is the *length bias* in the estimators $\hat{F}$; since N has been observed over an interval of length t, there are no observed U_i with values exceeding t, so that in particular $\hat{F}(x) = 1$ for all $x \geq t$. Only if F has compact support, and then only for t sufficiently large, does asynchronous observation yield good information about its tail.

Concerning asymptotics, strong uniform consistency is nearly immediate given Proposition 8.7a); for a central limit theorem one must account for the "random" sampling rate of the U_i imposed by asynchronous observation.

Proposition 8.8. Let $\hat{F}$ be given by (8.20). Then almost surely under each P_F,

$$\sup\{|\hat{F}(x) - F(x)| : x \in \mathbf{R}_+\} \to 0 \quad \square \tag{8.21}$$

To prove the central limit theorem, we note that

$$\hat{F} = n^{-1}\sum_{i=1}^{n} 1(U_i \leq (\cdot)), \qquad t \in [T_n, T_{n+1});$$

consequently, asymptotic behavior of the $\hat{F}$ of (8.20) is that of the synchronous data estimators (8.16), except that one must reconcile the relative rates of growth of t and n. This is done using the strong law of large numbers, which entails an assumption that the mean interarrival time be finite.

Theorem 8.9. Assume that $m = E_F[U_i] < \infty$. Then under P_F

$$\{t^{1/2}[\hat{F}(x) - F(x)] : x \in \mathbf{R}_+\} \xrightarrow{d} \{G(x) : x \in \mathbf{R}_+\} \tag{8.22}$$

as random elements of $D[0,\infty)$, where G is a Gaussian process with covariance function

$$R(x,y) = mF(x)[1 - F(y)] \tag{8.23}$$

for $0 \leq x \leq y$.

Proof: In view of finiteness of m and Proposition 8.1,

$$\begin{aligned} t^{1/2}[\hat{F} - F] &= \left(\frac{t}{N_t}\right)^{1/2} N_t^{-1/2}\sum_{i=1}^{N_t}[1(U_i \leq (\cdot)) - F] \\ &\sim m^{1/2} N_t^{-1/2}\sum_{i=1}^{N_t}[1(U_i \leq (\cdot)) - F] \end{aligned} \tag{8.24}$$

On the interval $[T_n, T_{n+1})$ the process

$$A_t = N_t^{-1/2}\sum_{i=1}^{N_t}[1(U_i \leq (\cdot)) - F]$$

satisfies

$$A_t = n^{-1/2} \sum_{i=1}^{n} [1(U_i \leq (\cdot)) - F]$$

and hence (Serfozo, 1975), utilizing also the fact that $T_n/n \to m$,

$$\lim_{t\to\infty} A_t \overset{d}{=} \lim_{n\to\infty} n^{-1/2} \sum_{i=1}^{n} [1(U_i \leq (\cdot)) - F],$$

in the sense of convergence in distribution. Appealing to Proposition 8.7b) and remembering the factor $m^{1/2}$ in (8.24), we arrive at (8.23). □

If F is stipulated to admit hazard function h, then martingale concepts are useful for interpretation of the estimators even though martingale results from Chapter 5 do not apply in our single-realization setting. Let

$$L_t = V_{t-} \tag{8.25}$$

be the *left*-continuous backward recurrence time process, which is therefore $\mathcal{F}^N$-predictable. The functions $t \to L_t$ and $t \to V_t$ differ only at the points T_n, where $L_t = U_n$ but $V_t = 0$. In some contexts the difference is immaterial but in others, for example Theorem 8.17, it is crucial. The stochastic intensity of N is then derived easily.

Proposition 8.10. If F admits hazard function h, then the $(P_F, \mathcal{F}^N)$-stochastic intensity of N is $\lambda_t = h(L_t)$. □

Consider now the identity

$$F(x) = E\left[\int_0^{U_i \wedge x} h(y)\, dy\right] \tag{8.26}$$

(Exercise 8.11); with x fixed the process $(\hat{F}_t(x))_{t\geq 0}$ satisfies

$$N_t \hat{F}_t(x) = \sum_{i=1}^{N_t} 1(U_i \leq x) = \int_0^t 1(\lambda_u \leq x)\, dN_u, \tag{8.27}$$

and by Definition 5.4 is the martingale estimator of the process

$$B_t(F) = \int_0^t \lambda_u 1(\lambda_u \leq x)\, du = \int_0^{L_t \wedge x} h(u)\, du + \sum_{i=1}^{N_t} \int_0^{U_i \wedge x} h(u)\, du$$

Therefore, $\hat{F}_t(x)$ estimates $N_t^{-1} B_t(F)$; by (8.26) the latter converges almost surely to $F(x)$, and hence we have a martingale interpretation of the estimators $\hat{F}$.

Returning to properties of the $\hat{F}$, we consider estimation of functionals of F by corresponding functionals of $\hat{F}$. The mean of the empirical distribution function $\hat{F}$ is

$$\hat{m} = N_t^{-1} \sum_{i=1}^{N_t} U_i = N_t^{-1}(t - V_t); \tag{8.28}$$

its asymptotic properties are those of the estimators t/N_t, which are in turn straightforward consequences of Proposition 8.1.

For estimation of the renewal function the natural choice is

$$\hat{R} = \sum_{k=0}^{\infty} \hat{F}^k, \tag{8.29}$$

with $\hat{F}$ given by (8.20); these estimators have the following properties.

Proposition 8.11. Let $\hat{R}$ be given by (8.29). Then for each $x_0 < \infty$, $\sup\{|\hat{R}(x) - R(x)| : x \leq x_0\} \to 0$ almost surely with respect to P_F.

Proof: Proposition 8.8 reduces this proposition to the analytical assertion that if F_n, F are distribution functions on $(0, \infty)$ with renewal functions R_n, R and if $F_n \to F$ uniformly, then $R_n \to R$ uniformly on compact sets. We outline one of several arguments. Let $\tilde{F}_n(\alpha) = \int e^{-\alpha x} F_n(dx)$, $\tilde{F}(\alpha)$, $\tilde{R}_n(\alpha)$, $\tilde{R}(\alpha)$ be the Laplace transforms. Then by (8.6) and the convergence $F_n \to F$, for each α

$$\tilde{R}(\alpha) = [1 - \tilde{F}(\alpha)]^{-1} = \lim\, [1 - \tilde{F}_n(\alpha)]^{-1} = \lim \tilde{R}_n(\alpha);$$

therefore, $R_n \to R$ vaguely. One then applies standard reasoning used to deduce uniform convergence of monotone functions from pointwise convergence. □

Unfortunately, this line of attack does not yield a useful central limit theorem; thus solutions to renewal equations can be estimated consistently, but little can be said about the estimation error. A compromise is to estimate limits of solutions to renewal equations. Under the assumptions of Theorem 8.5, if f is the solution of the renewal equation (8.7), then $\lim_{x\to\infty} f(x) = m^{-1} \int g$, so since m can be estimated using Proposition 8.1 or (8.28), whenever g—which usually depends on F—can be estimated, so can the functional $\alpha(F) = m^{-1} \int g$. Because of the variety of functions involved it is difficult to develop a comprehensive theory; instead, we illustrate with a specific case that arises again in Section 8.3.

Let H be the P_F-limit distribution of the forward recurrence time W_t of N. By (8.11), provided that m is finite and F nonarithmetic,

$$H(y) = m^{-1} \int_y^{\infty} [1 - F(u)]\, du, \tag{8.30}$$

which given the observations $\mathcal{F}_t^N$ we estimate by substitution:

$$\hat{H}(y) = \hat{m}^{-1} \int_y^\infty [1 - \hat{F}(u)]\, du, \tag{8.31}$$

with $\hat{F}$ given by (8.20) and $\hat{m}$ by (8.28). Note that (8.31) makes sense as an estimator of (8.30) as long as $m < \infty$; whether F is nonarithmetic affects only the role of H as limit distribution of the forward recurrence time. In the following result we impose rather stringent hypotheses on F in order to simplify the proof; the practical impact is minor, however.

Theorem 8.12. Assume that F is continuous and that there is $x_0 < \infty$ such that $F(x_0) = 1$. Then

a) Almost surely with respect to P_F, $\sup\{|\hat{H}(y) - H(y)| : y \in \mathbf{R}_+\} \to 0$.

b) As $t \to \infty$,

$$\{t^{1/2}[\hat{H}(y) - H(y)] : 0 \le y \le x_0\} \xrightarrow{d} \{J(y) : 0 \le y \le x_0\}, \tag{8.32}$$

where

$$J(y) = m^{-2} \int_y^{x_0} [1 - F(u)]\, du \int_0^{x_0} G(v)\, dv - m^{-1} \int_y^{x_0} G(u)\, du, \tag{8.33}$$

and G is a continuous Gaussian process with covariance (8.23).

Proof: a) By Proposition 8.1 and (8.28), $\hat{m} \to m$ almost surely, while

$$\sup_{y \in \mathbf{R}_+} \left| \int_y^\infty [1 - \hat{F}(u)]\, du - \int_y^\infty [1 - F(u)]\, du \right|$$

$$= \sup_{y \le x_0} \left| \int_y^{x_0} [1 - \hat{F}(u)]\, du - \int_y^{x_0} [1 - F(u)\, du \right|$$

$$\le x_0 \sup_{x \in \mathbf{R}_+} |\hat{F}(x) - F(x)|,$$

which converges to zero by Proposition 8.8.

b) In view of a), within error converging in probability to zero uniformly in y,

$$t^{1/2}[\hat{H}(y) - H(y)] = t^{1/2}\{(\hat{m}^{-1} - m^{-1}) \int_y^{x_0} (1 - F) + \hat{m}^{-1} \int_y^{x_0} (\hat{F} - F)\}$$

$$\cong m^{-2} \int_y^{x_0} (1 - F) \int_0^{x_0} t^{1/2}(\hat{F} - F) + m^{-1} \int_y^{x_0} t^{1/2}(\hat{F} - F)$$

$$\to m^{-2}\int_y^{x_0}(1-F)\int_0^{x_0}G - m^{-1}\int_y^{x_0}G,$$

by Theorem 8.9 and the continuous mapping theorem. It follows by continuity of F that G is continuous [in fact, $G(x) = B(F(x))$, where B is a Brownian bridge] and hence the integrals in (8.33) are well defined. □

Analysis of asynchronous samples concludes with application of marked point processes and empirical process methods to regenerative processes. Let X be a regenerative process (Definition 8.6) with state space E and renewal process $N = \Sigma\, \varepsilon_{T_n}$ of regeneration times. Let a death state Δ be adjoined to E in the usual manner (Blumenthal and Getoor, 1968) and for each n define a random element Z_{n+1} of the function space $D = D(E \cup \{\Delta\}, [0,\infty))$ by

$$\begin{aligned} Z_{n+1}(t) &= X_{T_n+t} && \text{if } t < U_{n+1} \\ &= \Delta && \text{if } t \geq U_{n+1} \end{aligned} \tag{8.34}$$

The function Z_{n+1} is the *excursion* of X between the regeneration times T_n and T_{n+1}. The Z_n determine X completely; in particular, $U_{n+1} = \inf\{t: Z_{n+1}(t) = \Delta\}$. That the Z_n are i.i.d. random elements of D, which is certainly plausible, is confirmed by the following result.

Proposition 8.13. Let X be regenerative, with excursions Z_n. Then the Z_n are i.i.d. random elements of D whose distribution determines the law of X. Conversely, given i.i.d. random elements Z_n of D such that the death times $U_n = \inf\{t: Z_n(t) = \Delta\}$ are finite almost surely, then with $T_n = \Sigma_{i=1}^n U_i$ and

$$X_t = Z_{n+1}(t - T_n), \qquad t \in [T_n, T_{n+1}),$$

X is regenerative with regeneration times T_n. □

We study the regenerative process X via the marked point process $\bar{N} = \Sigma_{n=0}^{\infty}\, \varepsilon_{(T_n, Z_n)}$, in which the regeneration time T_n is marked by the just-completed excursion Z_n. Observation of X over $[0,t]$ is essentially equivalent to that of

$$\bar{N}_t = \sum_{n=0}^{\infty} 1(T_n \leq t)\, \varepsilon_{(T_n, Z_n)}$$

and a plausible estimator of the law $\mathcal{L}$ of the excursions is

$$\hat{\mathcal{L}} = N_t^{-1} \sum_{n=1}^{N_t} \varepsilon_{Z_n}, \tag{8.35}$$

although, as for the estimators (8.20), information about the "in progress" excursions has been discarded. The space D is a complete, separable metric space and empirical process results apply to the $\hat{\mathcal{L}}$; again in the central limit theorem one must reconcile different sampling rates.

Theorem 8.14. Let X be a regenerative process with state space E and excursion sequence (Z_n), let $\mathcal{L}$ be the law of the Z_n, and let $\hat{\mathcal{L}}$ be given by (8.35). Then

a) If $\mathcal{C}$ is a Vapnik-Chervonenkis class of (Borel) subsets of D, then $\sup\{|\hat{\mathcal{L}}(C) - \mathcal{L}(C)|: C \in \mathcal{C}\} \to 0$ almost surely.

b) If $m = E_F[U_i]$ is finite, and if $\mathcal{C}$ is a class of subsets of D satisfying appropriate regularity conditions (see Gaenssler, 1984), then there is a continuous Gaussian process $\{G(C): C \in \mathcal{C}\}$ such that $\{t^{1/2}[\hat{\mathcal{L}}(C) - \mathcal{L}(C)]: C \in \mathcal{C}\} \xrightarrow{d} G$ as random elements of the metric space V described in Section 4.1; G has covariance function $R(C_1, C_2) = \mathcal{L}(C_1 \cap C_2) - \mathcal{L}(C_1)\mathcal{L}(C_2)$. □

Our discussion of estimation concludes with *Poisson sampling* of renewal processes, a special case of which appeared in Exercise 6.18. Suppose that under each probability P_F, $N = \Sigma_{n=0}^{\infty} \varepsilon_{T_n}$ is a renewal process with interarrival distribution F, that $\tilde{N} = \Sigma_{n=0}^{\infty} \varepsilon_{S_n}$ is a Poisson process with (known) rate 1, and that N, $\tilde{N}$ are independent. The observations are the values of the counting process (N_t) at the arrival times S_n of $\tilde{N}$, that is, the *Poisson samples* $N_{S_1}, N_{S_2}, \ldots$ or equivalently the numbers

$$Y_k = N_{S_k} - N_{S_{k-1}} \tag{8.36}$$

of arrivals in N during interarrival intervals in $\tilde{N}$ (additional models of this ilk are analyzed in Chapter 10.) The goal is to estimate the interarrival distribution F from the (Y_k), observed in real time, so that the σ-algebra corresponding to observation over $[0,t]$ is $H_t = \sigma((S_k, Y_k)1(S_k \le t): k \ge 0)$; at S_k, itself observable, one receives the data Y_k. But since $\tilde{N}$ is independent of N and has rate 1, the real-time sampling rate is equal to 1, and properties established for estimators based on $\mathcal{G}_k = \sigma(Y_1, \ldots, Y_k)$ carry over with *no* change to analogous estimators based on $\mathcal{H}_t$; consequently, we pursue only estimators based on the $\mathcal{G}_k$, beginning with a description of the probabilistic structure of (Y_k).

Proposition 8.15. Let (Y_k) be given by (8.36) with $N, \tilde{N}$ as described; for each k let

$$Q(k) = \int \frac{F(du)e^{-u}u^k}{k!} \tag{8.37}$$

Then under P_F, $(Y_k) \stackrel{d}{=} (\delta_k X_k)$, where

a) The (X_k) are i.i.d. with geometric distribution $P\{X_k = n\} = Q(0)^{n-1}(1 - Q(0))$;

b) For each k, $\delta_k = 1(R_n = k \text{ for some } n)$, where (R_n) is a discrete-time delayed renewal process: R_1, $R_2 - R_1$, $R_3 - R_2$, . . . are independent, $P\{R_1 = \ell\} = Q(\ell - 1)$, $\ell \geqslant 1$; for $j \geqslant 2$, $P\{R_j - R_{j-1} = \ell\} = Q(\ell)/(1 - Q(0))$, $\ell \geqslant 1$;

c) (X_k) and (δ_k) are independent.

One can interpret the construction using point processes: if $N^* = \Sigma\, \varepsilon_{R_n}$ is viewed as a point process on $\mathbf{N}$, then $\delta_k = N^*(\{k\})$ and Y is the integer-valued random measure obtained by integrating X with respect to N^*.

Proof of Proposition 8.15. Introduce the dual process

$$\widetilde{Y}_k = \widetilde{N}_{T_k} - \widetilde{N}_{T_{k-1}}, \tag{8.38}$$

the number of points of $\widetilde{N}$ in the interarrival interval $(T_{k-1}, T_k]$ of N; because $\widetilde{N}$ is Poisson and independent of N, the $\widetilde{Y}_k$ are i.i.d. with the mixed Poisson distribution Q. The sequences (Y_k) and $(\widetilde{Y}_k)$ determine each other pathwise (a picture is recommended!); in particular, given positive integers q, $r_1, \dots, r_q$ and $k_1, \dots, k_q$, and with $R_\ell = \Sigma_{i=1}^{\ell} r_i$ and $K_\ell = \Sigma_{i=1}^{\ell} k_i$,

$$\begin{aligned}
&P\{Y(R_\ell) = k_\ell, \ell = 1, \dots, q;\ Y_j = 0, \text{ other } j < R_q\} \\
&= P\{\widetilde{Y}_1 = r_1 - 1;\ \widetilde{Y}(K_\ell + 1) = r_{\ell+1}, \ell = 1, \dots, q-1;\ \widetilde{Y}(K_q + 1) \geq 1;\ \widetilde{Y}_j = 0, \\
&\qquad\qquad \text{other } j \leqslant K_q\} \\
&= Q(r_1 - 1)Q(r_2) \cdots Q(r_q)(1 - Q(0))Q(0)^{K_q - q}
\end{aligned}$$

Summed over $k_1, \dots, k_q$ this becomes

$$P\{Y(R_\ell) \neq 0, \ell = 1, \dots, q;\ Y_j = 0, \text{ other } j < R_q\} = Q(r_1 - 1)\prod_{j=1}^{q} \frac{Q(j)}{1 - Q(0)},$$

which shows that $\delta_k = 1(Y_k \neq 0)$ has the structure indicated in b). These two formulas further imply that

$$\begin{aligned}
&P\{Y(R_\ell) = k_\ell, \ell = 1, \dots, q \,|\, \delta_j = 1, j \in \{R_1, \dots, R_q\};\ \delta_j = 0, \text{ all other } j < R_q\} \\
&\qquad = (1 - Q(0))^q Q(0)^{K_q - q} = \prod_{j=1}^{q} Q(0)^{k_j - 1}(1 - Q(0)),
\end{aligned}$$

which proves both a) and c). □

To estimate F we estimate the mixed Poisson distribution (8.37), by which F is uniquely determined (see the discussion preceding Theorem 7.4).

Given observations $Y_1, \ldots, Y_n$, the estimator of $Q(k)$ is

$$\hat{Q}(k) = N(S_n)^{-1} \sum_{\ell=1}^{N(S_n)} 1(\widetilde{Y}_\ell = k), \tag{8.39}$$

since having observed $Y_1, \ldots, Y_n$ we can reconstruct $\widetilde{Y}_\ell$ for $\ell = 1, \ldots, N(S_n)$. Relative to $(\widetilde{Y}_k)$, (8.39) is a sequence of asynchronous samples; reasoning used already in the chapter, along with parts of Theorems 7.4 and 7.5, produces the final result of the section.

Theorem 8.16. Within the Poisson sampling scheme let Q, $\hat{Q}$ be given by (8.37), (8.39), respectively, with $\widetilde{Y}$ defined by (8.38). Then

a) Almost surely, $\sup\{|\hat{Q}(k) - Q(k)| : k \in \mathbf{N}\} \to 0$.

b) If $m = E_F[U_i]$ is finite, then as sequences of random variables $\{n^{1/2}[\hat{Q}(k) - Q(k)] : k \in \mathbf{N}\} \xrightarrow{d} \{G(k) : k \in \mathbf{N}\}$, where G is a Gaussian sequence with covariance function $R(k,j) = m[1(k=j)Q(k) - Q(k)Q(j)]$. □

This ends our discussion of estimation for renewal processes; a central point is that no new tools were needed, even for Poisson samples. What is new are the renewal process context, the need to adjust for sampling rate differentials caused by asynchronous or Poisson sampling, and interest in estimation of functionals such as renewal functions and forward recurrence-time distributions.

8.2 LIKELIHOOD RATIOS AND HYPOTHESIS TESTS

Like its counterparts in earlier chapters, this section does not emphasize specifics of hypothesis testing, although there are more details here than in the others. We do consider structure and asymptotic properties of likelihood ratios, structural characteristics of the interarrival distribution, and characterizations of Poisson and Cox processes among renewal processes, which may be useful for purposes of testing. Most of the tests mentioned concerning the interarrival distribution address specific qualitative characteristics and stochastic orderings that have arisen and been studied mainly in the reliability literature. These include hypotheses that the interarrival times U_i (which in reliability are lifetimes of i.i.d. systems, each, when it fails, replaced instantaneously) are stochastically larger than those in a renewal process with interarrival distribution F_0 and hypotheses concerning structure of the hazard function of F, for example that it is increasing (i.e., the systems deteriorate with age) or constant (i.e., the process is Poisson).

We begin with analysis of likelihood functions and ratios. As in earlier chapters, the results, in practical terms, are directed at construction of likelihood ratio tests for simple hypotheses

$$\begin{aligned} H_0 &: F = F_0 \\ H_1 &: F = F_1 \end{aligned} \tag{8.40}$$

about the interarrival distribution, but also have inherent interest. Our statistical model consists of those P_F for which F admits density F'. The hazard function $F'/(1-F)$ is denoted by h. Only likelihood ratios associated with asynchronous observations $\mathcal{F}_t^N$ are treated.

Theorem 8.17. Let P denote the probability with respect to which N is Poisson with rate 1. Then for each F admitting density F' and each $t < \infty$, $P_F \ll P$ on $\mathcal{F}_t^N$ with

$$\frac{dP_F}{dP}\Big|_{\mathcal{F}_t^N} = e^t \left[\prod_{i=1}^{N_t} F'(U_i) \right] [1 - F(L_t)], \tag{8.41}$$

where (L_t) is the left-continuous backward recurrence time process.

Proof: We apply Theorem 2.31 even though direct arguments are available, because the need to use predictable stochastic intensities is made strikingly clear. By Proposition 8.10, the $(P_F, \mathcal{F}^N)$-stochastic intensity of N is $\lambda_t = h(L_t)$; hence by Theorem 2.31, $P_F \ll P$ on $\mathcal{F}_t^N$ for each t and by (2.49) the derivative is

$$\begin{aligned} \frac{dP_F}{dP} &= \exp\left[\int_0^t (1 - \lambda_u)\, du + \int_0^t (\log \lambda_u)\, dN_u \right] \\ &= \exp\left[t - \int_0^t h(L_u)\, du + \int_0^t (\log h(L_u))\, dN_u \right] \\ &= \exp\left[t - \sum_{i=1}^{N_t} \int_0^{U_i} h(u)\, du + \int_0^{L_t} h(u)\, du + \sum_{i=1}^{N_t} \log h(U_i) \right] \end{aligned}$$

(since $L_{T_i} = U_i$)

$$= \exp\left[t + \sum_{i=1}^{N_t} \log F'(U_i) + \log[- \mathrm{F}(\mathrm{L}_t)] \right]; \tag{8.42}$$

consequently, (8.41) holds. □

Note what would have happened had we attempted to use $h(V_t)$ as stochastic intensity in (8.42): since $V_{T_i} = 0$ for each i the dN-integral would have diverged to $-\infty$ and no likelihood function would have arisen; singularity of P_F and P on $\mathcal{F}_t^N$ might have been concluded erroneously.

To test the hypotheses (8.40) given observations $\mathcal{F}_t^N$ one can use the likelihood ratio

$$L_t(F_0,F_1) = \frac{1 - F_1(L_t)}{1 - F_0(L_t)} \prod_{i=1}^{N_t} \frac{F_1'(U_i)}{F_0'(U_i)}, \tag{8.43}$$

additional properties of which will be derived momentarily. First we observe that because there exist strongly consistent estimators of F, for renewal processes singularity on $\mathcal{F}_\infty^N$ always occurs for different interarrival distributions.

Proposition 8.18. Given probability distributions F_0, F_1, either $P_{F_0} = P_{F_1}$ on $\mathcal{F}_\infty^N$ or $P_{F_0} \perp P_{F_1}$ on $\mathcal{F}_\infty^N$ according as $F_0 = F_1$ or $F_0 \neq F_1$. □

Hence interesting behavior of likelihood ratios as $t \to \infty$ occurs only locally. Intuitively, it is clear that one should deal with a fixed null distribution F and t-dependent alternatives converging to it at rate $t^{-1/2}$, but it is not clear in what sense the convergence should take place. The following contiguity theorem reveals that the difference between the hazard functions should converge to zero at rate $t^{-1/2}$, which is not altogether surprising in view of Theorems 2.31 and 8.17. As it turns out, (8.43) is not convenient in this context and we employ a modification of (8.42) instead.

Theorem 8.19. Let F, F^* be distributions on $\mathbf{R}_+$ admitting hazard functions h, h^*, respectively, such that $h^*(x) = 0$ whenever $h(x) = 0$ (which implies that $F^* \ll F$). For each $t > 1$, let F_t have hazard function $h_t = h + t^{-1/2}(h^* - h)$, and assume that

$$\sigma^2 = E_F\left[\left(\frac{h^*(U_i)}{h(U_i)} - 1\right)^2\right] < \infty \tag{8.44}$$

and that $m = E_F[U_i]$ is finite as well. Then for each t, $P_{F_t} \ll P_F$ on $\mathcal{F}_t^N$; moreover, under P_F

$$\log \frac{dP_{F_t}}{dP_F}\Big|_{\mathcal{F}_t^N} \xrightarrow{d} N\left(-\frac{\sigma^2}{2m}, \frac{\sigma^2}{m}\right) \tag{8.45}$$

Proof: Formally, by dividing dP_{F_t}/dP as given by (8.42) by dP_F/dP and using Proposition 8.10 (more rigorously, by mimicking the proof of Theorem 8.17 with Theorem 2.31 replaced by an analog of Theorem 5.2), we infer that absolute continuity holds, with

$$\log \frac{dP_{F_t}}{dP_F}\Big|_{\mathcal{F}_t^N} = \int_0^t [h(L_u) - h_t(L_u)]\, du + \int_0^t \log \frac{h_t(L_u)}{h(L_u)}\, dN_u$$

$$= t^{-1/2} \int_0^t [h(L_u) - h^*(L_u)]\, du + \int_0^t \log\left(1 + t^{-1/2}\left[\frac{h^*(L_u)}{h(L_u)} - 1\right]\right) dN_u$$

$$\cong t^{-1/2}\left\{ \int_0^t [h(L_u) - h^*(L_u)]\, du + \int_0^t \left[\frac{h^*(L_u)}{h(L_u)} - 1\right] dN_u \right\}$$

$$- (2t)^{-1} \int_0^t \left[\frac{h^*(L_u)}{h(L_u)} - 1\right]^2 dN_u$$

$$= t^{-1/2} \int_0^t \left[\frac{h^*(L_u)}{h(L_u)} - 1\right][dN_u - h(L_u)\, du]$$

$$- (2t)^{-1} \int_0^t \left[\frac{h^*(L_u)}{h(L_u)} - 1\right]^2 dN_u$$

By the strong law of large numbers,

$$t^{-1} \int_0^t \left[\frac{h^*(L_u)}{h(L_u)} - 1\right]^2 dN_u = t^{-1} \sum_{i=1}^{N_t} \left[\frac{h^*(U_i)}{h(U_i)} - 1\right]^2$$

$$= \frac{N_t}{t} N_t^{-1} \sum_{i=1}^{N_t} \left[\frac{h^*(U_i)}{h(U_i)} - 1\right]^2$$

$$\to \frac{\sigma^2}{m}$$

This expression, Proposition 8.10, and Theorem B.21, together with ergodicity of renewal processes, imply that

$$t^{-1/2} \int_0^t \left[\frac{h^*(L_u)}{h(L_u)} - 1\right][dN_u - h(L_u)\, du] \xrightarrow{d} N(0, \frac{\sigma^2}{m}),$$

which completes the proof. □

We now take up some specific questions of hypothesis testing. The methods below are based without exception on the i.i.d. structure of the interarrival times, regardless of whether observations are synchronous or asynchronous; moreover, as we have seen, asymptotics in the latter case differ from more directly deduced properties of the former only in the ubiquitous variance adjustment factor $m = E_F[U_i]$ appearing, for example, in (8.23) and (8.45). Hence for simplicity we shall work only with synchronous data. Given observations $U_1, \ldots, U_n$, except that the U_i are nonnegative and hence cannot be normally distributed, the mathematics is classical. Therefore, the contribution of renewal processes must lie principally in the *kinds of questions* that are posed. As intimated previously, reliability theory is the source of many such questions; here the renewal process describes failures of i.i.d. objects tested sequentially, with instantaneous replacements at times of failures.

Consider, for example, the following question. Suppose that a newly developed system is compared with an extant system in order to ascertain whether the new system produces a "longer lifetime"; the data are lifetimes of i.i.d. copies of the new system. This might in some cases be viewed as a two-sample problem, but often there are enough data concerning the old system that its lifetime distribution F_0 is taken as known. Of several interpretations of "longer lifetime," one is the relatively strong condition that the F-distributed lifetimes of the new system be *stochastically larger* than F_0-lifetimes in the sense that

$$F(x) \leq F_0(x) \tag{8.46}$$

for every $x \geq 0$. (One says also that F is stochastically larger than F_0; for elementary properties, see Exercise 8.15.) If (8.46) holds, then for each x the old system is more likely than the new to fail before x; at least in terms of longevity the new system is superior.

The hypothesis

$$H_0 : F \leq F_0, \quad F \neq F_0 \tag{8.47}$$

can be tested nonparametrically with the one-sided Kolmogorov-Smirnov statistic $D^- = \sup\{[F_0(x) - \hat{F}(x)] : x \in \mathbf{R}_+\}$, where $\hat{F}$ is given by (8.16); like the Kolmogorov-Smirnov statistic (its two-sided analog) D^- is distribution-free provided that F_0 is continuous and in this case has known and rather simple limit distribution.

Suppose that F_0, F_1 are distribution functions with F_1 stochastically larger than F_0. If (N_t^0), (N_t^1) are associated renewal processes, then it seems plausible that for each t, N_t^0 should be stochastically larger than N_t^1 (*the process with shorter lifetimes should have more failures*). *Were it possible to construct* N^0 and N^1 on the same probability space in such a manner that $N_t^0 \geq N_t^1$ for all t, this and other distributional relationships would hold simply as expectations of pointwise inequalities. We can achieve this by a coupling construction.

Proposition 8.20. Let F_0, F_1 be distribution functions on $\mathbf{R}_+$ such that F_0 is stochastically *smaller* than F_1. Then there exists a probability space on which are defined counting processes (N_t^0), (N_t^1) such that

a) (N_t^i) is a renewal process with interarrival distribution F_i;
b) Almost surely, $N_t^0 \geq N_t^1$ for all t.

Proof: It suffices to construct random variables (U^0, U^1) such that U^i has distribution F_i and $U^0 \leq U^1$ almost surely, for then i.i.d. copies of

(U^0, U^1) can be taken as the interarrival times for N^0, N^1 and it is evident that a) and b) hold. But construction of (U^0, U^1) is easy: let Z be uniformly distributed on $[0,1]$ and let $(U^0, U^1) = (F_0^{-1}(Z), F_1^{-1}(Z))$. □

An immediate consequence is that N_t^0 is indeed stochastically larger than N_t^1 for each t. Proposition 8.20 is mainly a tool for study of renewal processes and derivation of approximations and bounds for quantities that are difficult to compute explicitly but are important in inference nevertheless. For example, given a renewal process N the distribution of N_t, needed, for example, to calculate critical regions for hypothesis tests and to compute power, is in principle expressible in terms of the interarrival distribution F:

$$P\{N_t < n\} = P\{T_n > t\} = 1 - F^n(t), \tag{8.48}$$

but except in a handful of cases cannot be calculated explicitly. If F^* is a more tractable distribution than F satisfying $F^* \leqslant F$, then by proposition 8.20, for each n and t, $P_F\{N_t < n\} \leqslant P_{F^*}\{N_t < n\}$, which can be used to derive bounds. a specific example appears in Exercise 8.19.

In reliability problems it is often of interest to study aging properties of systems. One would expect a typical physical system to undergo a "burn-in" period during which failure is relatively more likely, that systems surviving the burn-in would exhibit an extended period during which the failure rate is constant, and that finally, systems would age, becoming more and more likely to fail. We introduce three formalizations of the simpler notion of systems that merely grow increasingly less reliable with age, and discuss briefly some procedures that have been developed to test for them, given i.i.d. samples $U_1, \ldots, U_n$, against the null hypothesis that the U_i are exponentially distributed. The tests, therefore, examine aspects of testing whether a renewal process is Poisson; in this case specializations are the motivation for the alternative hypotheses and their qualitative form. First we define the properties to be tested.

Definition 8.21. Let F be a distribution function on $\mathbf{R}_+$ and let $\bar{F}(x) = 1 - F(x)$.

a) F has *increasing failure rate* if the function $-\log \bar{F}(x)$ is convex on the support of F.

b) F has *increasing failure rate average* if the function $-x^{-1} \log \bar{F}(x)$ is increasing on the support of F.

c) F has the property *new better than used* if $\bar{F}(x + y) \geq \bar{F}(x)\bar{F}(y)$ for all x, $y \geq 0$.

It is conventional to write $F \in IFR$, $F \in IFRA$, $F \in NBU$ when F has the indicated property. Analogs such as decreasing failure rate, ..., which describe systems that improve with age, can be formulated and studied using similar methods.

If F admits hazard function h (in reliability, the *failure rate function*), then $F \in IFR$ if and only if h is increasing, and $F \in IFRA$ if and only if the function $x \to x^{-1} \int_0^x h$ is increasing. The less obviously motivated class of $IFRA$ distributions, which contains the IFR distributions, is important because it arises in the reliability of multicomponent systems and systems that fail from the cumulative effect of randomly occurring shocks (Exercise 8.22). Weaker still than $IFRA$ is NBU, whose interpretation is that a single system is less likely to remain in operation at time $x + y$ than if it had been replaced at time x with an independent copy. With respect to each class exponential distributions are the borderline case: if F is exponential, then $-\log \bar{F}$ is linear, $-x^{-1} \log \bar{F}$ is constant, and $\bar{F}(x + y) = \bar{F}(x)\bar{F}(y)$ for all x, y. (In fact, each of these properties characterizes exponential distributions.) The exponential case, consequently, is the natural null hypothesis to be tested against the increasingly broad alternatives

$$H_1 : F \in IFR,\ F \text{ not exponential} \tag{8.49a}$$

$$H_1 : F \in IFRA,\ F \text{ not exponential} \tag{8.49b}$$

$$H_1 : F \in NBU,\ F \text{ not exponential}, \tag{8.49c}$$

which represent particular non-Poisson renewal processes.

The literature on tests and maximum likelihood estimation for distributions with monotone failure rate is extensive; here we only mention selected aspects. For the IFR alternative (8.49a) tests are based (see also Example 3.14) on normalized spacings

$$D_i = (n - i + 1)[U_{(i)} - U_{(i-1)}], \tag{8.50}$$

where $U_{(0)} = 0$, $U_{(1)}, \ldots, U_{(n)}$ are the order statistics engendered by $U_1, \ldots, U_n$. Behavior under the null hypothesis of exponentiality has been noted in Section 3.3 and Excerise 3.34; behavior under the alternative (8.49a) has also been characterized: the spacings decrease in the sense that each is stochastically smaller than its predecessor.

Proposition 8.22. If $F \in IFR$, then the spacings D_i are stochastically decreasing in i. □

Test statistics based on spacings include the *time on test* $(\Sigma_{i=1}^{n}\Sigma_{j=1}^{i}D_j)/(\Sigma_{i=1}^{n}D_i)$, which has asymptotic minimax properties (Barlow et al., 1972) and rank statistics $\Sigma_{i=1}^{n} i\log[1-R_i/(n+1)]$, where R_i is the rank of D_i among $D_1, \ldots, D_n$. The latter are locally most powerful (Bickel and Doksum, 1968). Similar methods have been applied to the *IFRA* alternative (8.49b) (see the chapter notes).

The *NBU* alternative (8.49c) has been examined by Hollander and Proschan (1972). Deviation of $F \in NBU$ from exponentiality is described using the functional

$$\gamma(F) = \int\int [\bar{F}(x)\bar{F}(y) - \bar{F}(x+y)]\, F(dx)F(dy), \tag{8.51}$$

which is zero for F an exponential distribution and strictly positive for all other $F \in NBU$. The test statistic is a U-statistic approximation to the corresponding functional $\gamma(\hat{F})$ of the empirical distribution function $\hat{F}$ of (8.16):

$$[2n(n-1)(n-2)]^{-1}\sum 1(U_i > U_j + U_k), \tag{8.52}$$

where the summation is over all triples of distinct integers i, j, k with $j<k$. Well-developed limit theory for U-statistics (see Serfling, 1980) yields consistency, asymptotic normality, and characteristics of the null distribution.

Alternatives narrower than (8.49) are usually formulated parametrically; see Exercise 8.26 for the most important case, Weibull distributions. In the opposite direction of alternatives broader than (8.49) the situation is less understood. There exist "nonparametric" characterizations of Poisson processes among renewal processes, each reducing ultimately to a characterization of exponential distributions, but none seems to have been investigated as a tool for inference. Nonetheless, they are interesting in their own right, so we present one of the simplest.

Proposition 8.23. Let N be a renewal process with interarrival distribution F and forward recurrence time process (W_t). Then the following are equivalent for each $c>0$:

a) N is a Poisson process with rate $1/c$;
b) For each $t>0$

$$E[W_t] = c \tag{8.53}$$

Proof: Integration with respect to y of the renewal equation

$$P\{W_t > y\} = 1 - F(t+y) + \int_0^t F(du)P\{W_{t-u} > y\}$$

satisfied by the function $t \to P\{W_t > Y\}$ (see example 8.3) and application of (8.53) yield

$$c = E[W_t] = \int_t^\infty [1 - F(u)]\, du + cF(t),$$

so that F is differentiable, with hazard function $h(t) \equiv 1/c$. □

Related characterizations of Poisson processes among renewal processes are mentioned in Exercises 8.24 and 8.25 and in the chapter notes.

There exists also a characterization of Cox processes that are renewal processes. The stochastic integral $X * N$ introduced in Example 3.5 and studied at length in Section 6.3 is a renewal process provided that the distribution governing sojourns in state 1 is exponential. The content of the following theorems is that every Cox process that is also a renewal process has this form. In attempting to determine whether a Cox process or a renewal process provides a better description of a set of data, it is essential to understand that the intersection of the two classes is nonempty and to characterize processes it contains.

The crucial properties are two. Let N be a renewal process with interarrival hazard function h and suppose that N is also a Cox process with directing intensity (X_t). Then first of all, X can assume only one nonzero value, by the following argument. In consequence of Proposition 8.10, N has $\mathcal{F}^N$-stochastic intensity $\lambda_t = h(L_t)$ and therefore by Exercise 2.16 and Theorem 7.9,

$$h(L_t) = E[X_t | \mathcal{F}_t^N] = \frac{E[\exp(-\int_0^t X)(\Pi_{i=1}^n X_{t_i})X_t]}{E[\exp(-\int_0^t X)\Pi_{i=1}^n X_{t_i}]}$$

evaluated at $n = N_t$, $t_1 = T_1, \ldots, t_n = T_n$. That is, for each n and for $t_1 < \ldots < t_n < t$,

$$h(t - t_n) = \frac{E[\exp(-\int_0^t X)(\Pi_{i=1}^n X_{t_i})X_t]}{E[\exp(-\int_0^t X)\Pi_{i=1}^n X_{t_i}]},$$

and letting all the t_i increase to t gives

$$h(0) = \frac{E[\exp(-\int_0^t X)\, X_t^{n+1}]}{E[\exp(-\int_0^t X)h(0)^n]},$$

or

$$E[\exp(-\int_0^t X)\, X_t^n] = E[\exp(-\int_0^t X)]h(0)^n,$$

which implies that the only nonzero value that X can assume is $\lambda = h(0)$. Second, sojourns of X in λ must be exponentially distributed, intuitively because unless the hazard function is constant, times between arrivals in N will not be independent: short interarrival times will cluster as X remains in λ, to be followed by a longer interarrival time during which X spends some time in zero.

Theorem 8.24. Let N be a point process on $\mathbf{R}_+$ that is simultaneously a stationary renewal process and a Cox process with stationary directing intensity. Then the Laplace transform $\ell(\alpha) = \int e^{-\alpha x} F(dx)$ has the form

$$\ell(\alpha) = \frac{\lambda}{\lambda + \alpha + \int (1 - e^{-\alpha z}) K(dz)}, \tag{8.54}$$

where $\lambda > 0$ and K is a measure on $(0,\infty)$ satisfying $\int zK(dz) < \infty$.

Proof: (Sketch). Let $\Lambda_t = \int_0^t X_u \, du$ and let Λ^{-1} denote the right-continuous inverse. By Exercise 2.11 N admits the representation $N = \sum_{n=0}^{\infty} \varepsilon_{T_n}$, where

$$T_n = \Lambda^{-1}(S_n) \tag{8.55}$$

with $\widetilde{N} = \sum \varepsilon_{S_n}$ a Poisson process with rate 1 and independent of X. It follows that

$$E[\exp(-\alpha(\Lambda^{-1}(\alpha)))] = \frac{1 - \ell(\alpha)}{m\alpha\ell(\alpha)} \tag{8.56a}$$

(here m is the mean interarrival time) and that for $s, t > 0$

$$E[\exp(-\alpha[\Lambda^{-1}(t+s) - \Lambda^{-1}(t)])] = \exp\left[\frac{-s(1 - \ell(\alpha))}{\ell(\alpha)}\right]; \tag{8.56b}$$

these are consequences of a theorem of Kingman (1963) concerning the way in which the law of the Poisson sample process (T_n) determines that of the process Λ^{-1} being sampled via (8.55) (see Proposition 10.1). Also by Kingman (1963), the independent increments property of (T_n) and stationarity of the increments $T_2 - T_1$, $T_3 - T_2, \ldots$ are shared by Λ^{-1}. Application of the Lévy-Khinchin representation theorem (see Blumenthal and Getoor, 1968, and Theorem 1.34) provides existence of $\beta \geqslant 0$ and a measure H with $\int zH(dz) < \infty$ such that

$$E[\exp(-\alpha[\Lambda^{-1}(t+s) - \Lambda^{-1}(t)])] = \exp\left[-s\left(\beta\alpha + \int_0^\infty (1 - e^{-\alpha z}) H(dz)\right)\right];$$

together with (8.56b) this implies that

$$\ell(\alpha) = \left[1 + \alpha\beta + \int_0^\infty (1 - e^{-\alpha z}) H(dz)\right]^{-1} \tag{8.57}$$

and that $m = \beta + \int zH(dz)$. But β cannot be zero, for if it were, then (8.57) would imply that

$$P\{\Lambda^{-1}(0) = 0\} = \lim_{\alpha \to \infty} E[\exp(-\alpha\Lambda^{-1}(0))] = \frac{\beta}{m} = 0$$

and hence that $\Lambda \equiv 0$, an obvious contradiction. Thus, (8.55) holds with $\lambda = 1/\beta$ and $K(dz) = \lambda H(dz)$. □

When the measure K in Theorem 8.24 is finite, so that Λ^{-1} is a compound Poisson process, then the directing intensity has the structure described before the theorem.

Theorem 8.25. In the context of Theorem 8.24, suppose that $k = K(\infty) - K(0)$ is finite. Then X is a semi-Markov process with state space $\{0,\lambda\}$, sojourns in λ are exponentially distributed with parameter k, and sojourns in 0 have distribution $G(z) = k^{-1}[K(z) - K(0)]$. □

It may also be wondered which renewal processes are infinitely divisible (i.e., are Poisson cluster processes). Haberland (1975) shows that only Poisson processes are simultaneously renewal processes and Poisson cluster processes.

8.3 STATE ESTIMATION; COMBINED INFERENCE AND STATE ESTIMATION

Let N be a renewal process with known interarrival distribution F; we first consider the prediction problem of estimating future behavior of N given (complete) observations $\mathcal{F}_t^N$ over $[0,t]$. Intuitively, the forward recurrence time W_t should be a "sufficient statistic" for state estimation of the "future" process

$$N_u^t = N_{t+u} - N_t; \tag{8.58}$$

once W_t is estimated the remainder of the future is straightforward, because after $T_{N_t+1} = t + W_t$ the renewal process is a probabilistic replica of itself. Confirmation is contained in the following result, which shows that the situation is symmetric: once the backward recurrence time V_t is known, additional information in $\mathcal{F}_t^N$ does not improve predictions of N^t.

Theorem 8.26. Let $t > 0$ be fixed and let (N_u^t) be given by (8.58). Then for each set Γ,

$$P\{N^t \in \Gamma | \mathcal{F}_t^N\} = P\{N^t \in \Gamma | V_t\} = [1 - F(V_t)]^{-1} \int F(V_t + dy) P\{N \in \Gamma | T_1 = y\} \tag{8.59}$$

Proof: It suffices to establish equality of the first and third members in (8.59) when $\{N^t \in \Gamma\} = \{N^t_{u_1} = j_1, \ldots, N^t_{u_k} = j_k\} = \{N_{t+u_1} - N_t = j_1, \ldots, N_{t+u_k} - N_t = j_k$ for $0 \leqslant u_1 < \ldots < u_k$ and $0 \leqslant j_1 \leqslant \ldots \leqslant j_k$. Given $\Lambda = \{N_{t_1} = i_1, \ldots, N_{t_\ell} = i_\ell\} \in \mathcal{F}^N_t$, with $0 \leqslant t_1 < \ldots < t_\ell \leqslant t$ and $0 \leqslant i_1 \leqslant \ldots \leqslant i_\ell$,

$$P\{\Gamma \cap \Lambda\} = \sum_{q=0}^{\infty} P\{\Lambda, N_t = i_\ell + q, N_{t+u_1} = i_\ell + q + j_1, \ldots, N_{t+u_k} = i_\ell + q + j_k\}$$

$$= \sum_{q=0}^{\infty} \int_0^t P\{\Lambda, T_{i_\ell+q} \in dz\} \int_{(t-z,\infty)} F(dx) P\{N \in \Gamma | T_1 = z + x - t\}$$

$$= \sum_{q=0}^{\infty} \int_0^t P\{\Lambda, T_{i_\ell+q} \in dz\} \int F(dy + t - z) P\{N \in \Gamma | T_1 = y\}$$

$$= \int_0^t P\{\Lambda, V_t \in t - dz\}[1 - F(t-z)]^{-1} \int F(dy + t - z) P\{N \in \Gamma | T_1 = y\}$$

$$= E\left[[1 - F(V_t)]^{-1} \int F(dy + V_t) P\{N \in \Gamma | T_1 = y\}; \Lambda\right];$$

this confirms (8.59). □

In particular, we have the following consequence.

Proposition 8.27. For each t and z,

$$P\{W_t > z | V_t\} = \frac{1 - F(V_t + z)}{1 - F(V_t)} \tag{8.60}$$

Proof: In (8.59) choose Γ so that $\{N^t \in \Gamma\} = \{N^t_z = 0\} = \{N_{t+z} - N_t = 0\} = \{W_t > z\}$; since $P\{N \in \Gamma | T_1 > y\} = P\{T_1 > z | T_1 > y\} = 1(y > z)$, (8.59) gives

$$P\{W_t > z | V_t\} = [1 - F(V_t)]^{-1} \int F(V_t + dy) 1(y > z),$$

which is (8.60). □

Therefore,

$$P\{N^t \in \Gamma | \mathcal{F}^N_t\} = \int P\{W_t \in dy | V_t\} P\{N \in \Gamma | T_1 = y\}, \tag{8.61}$$

with all dependence on $\mathcal{F}^N_t$ contained in the conditional distribution $P\{W_t > z | V_t\}$. For many calculations the explicit form of (8.60) is adequate, but it is also possible to derive a recursive method of computation in the canonical form of predictable part plus filter gain times innovation.

Proposition 8.28. Let N be a renewal process with interarrival hazard function h. Then for each z the MMSE state estimators $P\{W_t > z | V_t\} = P\{W_t > z | \mathscr{F}_t^N\}$ satisfy the stochastic differential equation

$$dP\{W_t > z | V_t\} = \{[1 - F(z)]h(V_t) - P\{W_t > z | V_t\} h(V_t + z)\} dt + [(1 - F(z)) - P\{W_s > z | V_s\}|_{s=t-}](dN_t - h(L_t)\, dt) \quad (8.62)$$

Proof: Recall that (L_t) is the left-continuous backward recurrence time process. We first observe that from (8.60)

$$P\{W_t > z | V_t\} = \frac{1 - F(V_t + z)}{1 - F(V_t)} = \exp[-\int_{V_t}^{V_t + z} h(s)\, ds]$$

Between points of N, $t \to V_t$ is increasing linearly; if $N_{t+\Delta - N_t} = 0$, then

$$dP\{W_t > z | V_t\} = \exp\left[-\int_{V_t+\Delta}^{V_t+z+\Delta} h(s)\, ds] - \exp[-\int_{V_t}^{V_t+z} h(s)\, ds\right]$$

$$= \exp\left[-\int_{V_t}^{V_t+z} h(s)\, ds\right]\left\{\exp[\int_{V_t}^{V_t+\Delta} h(s)\, ds - \int_{V_t+z}^{V_t+z+\Delta} h(s)\, ds] - 1\right\}$$

$$\cong P\{W_t > z | V_t\}[h(V_t) - h(V_t + z)]\, \Delta$$

as $\Delta \to 0$. Hence between points of N the "ordinary" differential equation

$$dP\{W_t > z | V_t\} = P\{W_t > z | V_t\}[h(V_t) - h(V_t + z)]\, dt \quad (8.63)$$

holds. If t is a point of N, then there occurs the discrete correction

$$dP\{W_t > z | V_t\} = [1 - F(z)] - P\{W_u > z | V_u\}|_{u=t-}$$

$$= [1 - F(z)] - \frac{1 - F(L_t + z)}{1 - F(L_t)} \quad (8.64)$$

To obtain (8.62), combine (8.63) and (8.64), add and subtract $\{[1 - F(z)] - [1 - F(L_t + z)]/[1 - F(L_t)]\} h(L_t)\, dt$, and finally appeal to the property that almost surely $L_t = V_t$ for almost every t (with respect to Lebesgue measure) in order to replace L_t by V_t in the predictable part of (8.62). □

For fixed z the equation (8.62) is entirely recursive: it does not—by contrast with some equations derived in Chapter 7—involve other state estimators. Combined with (8.61) it provides a recursive method for calculation of $P\{N' \in (\cdot) | \mathscr{F}_t^N\}$.

Our analysis of state estimation per se concludes with p-thinned renewal processes (Definition 1.38), where p is a constant. Let $N = \Sigma_1^\infty \varepsilon_{T_n}$ be a renewal process with interarrival distribution F and let $Z_1, Z_2, \ldots$ be i.i.d. random variables, independent of N, with $P\{Z_n = 1\} = p \in (0,1]$, $P\{Z_n = 0\} = q = 1 - p$. The p-thinning of N, the point process $N' = \Sigma Z_n \varepsilon_{T_n}$, can be characterized in the following manner.

Proposition 8.29. The p-thinning N' is a renewal process with interarrival distribution

$$\widetilde{F} = p \sum_{k=1}^{\infty} q^{k-1} F^k \quad \square \tag{8.65}$$

For the proof, see Exercise 8.28; some aspects of statistical inference are examined in Exercises 8.29 and 8.30.

The state estimation problem of interest is to reconstruct N from asynchronous observations $\mathcal{F}_t^{N'}$ of N' by calculating MMSE state estimators $E[N_t|\mathcal{F}_t^{N'}]$. General methods developed in Section 5.4 yield the solution.

Theorem 8.30. Suppose that F admits hazard function h. Then the state estimators $E[N_t|\mathcal{F}_t^{N'}]$ satisfy the stochastic differential equation

$$dE[N_t|\mathcal{F}_t^{N'}] = p^{-1}\tilde{h}(L_t')\,dt$$

$$+\left\{1 + \left(\frac{pE[N_u h(L_u)|\mathcal{F}_u^{N'}]}{\widetilde{h}(L_u')} - E[N_u|\mathcal{F}_u^{N'}]\right)\Big|_{u=t-}\right\}(dN_t' - \tilde{h}(L_t' dt), \tag{8.66}$$

where $\widetilde{h}$ is the hazard function of the distribution of (8.65), and L, L' are the left-continuous backward recurrence time processes for N, N', respectively.

Although it is easy to show that the hazard function $\widetilde{h}$ exists, computation is quite another matter; in particular, it is not related in a simple manner to the hazard function of F.

Proof of Theorem 8.30: We work through calculations associated with Theorem 5.29. Notation matches that of Section 5.4 only in part, however: in particular, the observed point process denoted there by N is not the underlying process of this section but rather the p-thinned process N'. By Proposition 8.10 the state process $Z_t = N_t$, with respect to the history $\mathcal{H}_t = \mathcal{F}_t^N \vee \mathcal{F}_t^{N'}$, has semimartingale representation

$$Z_t = \int_0^t h(L_u)\,du + M_t, \tag{8.67}$$

where M is an $\mathcal{H}$-martingale. The $\mathcal{H}$-stochastic intensity of N' is

$$\lambda_t = ph(L_t); \tag{8.68}$$

hence there exists an $\mathcal{F}^{N'}$-stochastic intensity $\hat{\lambda}$ satisfying $\hat{\lambda}_t = pE[h(L_t)|\mathcal{F}_t^{N'}]$ almost surely for each t. By Proposition 8.29 and another appeal to Proposition 8.10, $\hat{\lambda}_t = \tilde{h}(L_t')$, and therefore

$$E[h(L_t)|\mathcal{F}_t^{N'}] = p^{-1}\tilde{h}(L_t') \tag{8.69}$$

almost surely for each t. From Proposition 5.27, (5.72a), Theorem 5.29, (8.68), and (8.69),

$$dE[N_t|\mathcal{F}_t^{N'}] = p^{-1}\tilde{h}(L_t')\,dt + H_t[dN_t' - \tilde{h}(L_t')\,dt], \tag{8.70}$$

where the filter gain H is given by

$$H_t = J_t + K_t - E[N_s|\mathcal{F}_s^{N'}]\,|_{s=t-},$$

with J, K the $\mathcal{F}_t^{N'}$-predictable processes defined by (5.74a) and (5.74b). We compute the former using (5.78) and (8.69):

$$J_t = \hat{\lambda}_t^{-1} pE[\,N_u h(L_u)|\mathcal{F}_u^{N'}]\,|_{u=t-} = [\tilde{h}(L_t')]^{-1}E[N_u\,h(L_u)|\mathcal{F}_u^{N'}]\,|_{u=t-}$$

By contrast with previous cases the process K is not zero. With M the martingale in (8.67), $\Delta M_t = 1$ whenever $dN_t' \neq 0$ and thus K must satisfy

$$E[\int_0^t CK\hat{\lambda}\,ds] = E[\int_0^t C(\Delta M)\,dN'] = E]\int_0^t C\,dN'] = E[\int_0^t C\hat{\lambda}\,ds]$$

for each bounded, $\mathcal{F}^{N'}$-predictable process C and each $t > 0$. Therefore, $K \equiv 1$ [the solution to (5.74b) is unique] and now only substitution of various expressions into (8.70) remains in order to complete the proof. □

We conclude the section with brief analysis of combined statistical inference and state estimation. In view of Proposition 8.27 the key state estimators for which approximating pseudo-state estimators must be produced are the conditional distributions $P\{W_t > z|V_t\} = [1 - F(V_t + z)]/[1 - F(V_t)]$. Our setting (it differs from Sections 3.4 and 7.4) is asynchronous observation of a single realization of a renewal process N with unknown interarrival distribution F, governed by the law P_F. Invoking once again the principle of separation, given observation $\mathcal{F}_t^N$ we construct the estimator $\hat{F}$ of F given by (8.20) and from it pseudo-state estimators

$$\hat{P}\{W_t > z|V_t\} = \frac{1 - \hat{F}(V_t + z)}{1 - \hat{F}(V_t)}, \tag{8.71}$$

where $0/0 = 0$. [The denominator on the right-hand side of (8.71) cannot vanish unless the numerator does also.] These pseudo-state estimators converge to the true state estimators $P_F\{W_t > z|X_t\}$ at an optimal rate, provided only that the mean interarrival time be finite.

Theorem 8.31. Assume that $m = E_F[U_i]$ is finite and let the pseudo-state estimators $\hat{P}\{W_t > z \mid V_t\}$ be given by (8.71). Then under P_F,

$$\{t^{1/2}[\hat{P}\{W_t > z|V_t\} - P_F\{W_t > z|V_t\}] : z \in \mathbf{R}_+\}$$

$$\xrightarrow{d} \left\{ \frac{G(V_\infty + z)}{1 - F(V_\infty)} - \frac{[1 - F(V_\infty + z)]\, G(V_\infty)}{[1 - F(V_\infty)]^2} : z \in \mathbf{R}_+ \right\}, \qquad (8.72)$$

where

i) G is a Gaussian process with covariance function R given by (8.23);

ii) $P\{V_\infty \le y\} = m^{-1} \int_0^y [1 - F(x)]\, dx$;

iii) G and V_∞ are independent.

Proof: With $\hat{F}$ given by (8.20) it is readily shown that the error process $\{t^{1/2}[\hat{F}(x) - F(x)] : x \in \mathbf{R}_+\}$ and the backward recurrence time V_t are asymptotically independent (Exercise 8.32); therefore,

$$(t^{1/2}[\hat{F} - F], V_t) \xrightarrow{d} (G, V_\infty), \qquad (8.73)$$

where G, V_∞ satisfy i) – iii). From (8.73), (8.72) follows by the "delta method" and the continuous mapping theorem. □

Note that Theorems 8.31 and 7.26 are similar structurally as well as in their relationships to Theorems 8.9 and 7.5, respectively, even though one deals with single-realization asymptotics as $t \to \infty$ and the other with asymptotics of i.i.d. copies as $n \to \infty$. To cope with more general state estimators

$$P_F\{\mathrm{N}^t \in \Gamma|\mathcal{F}_t^N\} = [1 - F(V_t)]^{-1} \int_0^t F(V_t + dy) P_F\{N \in \Gamma|T_1 = y\}$$

one must also estimate $P_F\{N \in \Gamma|T_1 = y\}$; whether this is feasible depends on Γ. When it is, in the sense that one can devise asymptotically normal estimators $\hat{P}\{N \in \Gamma|T_1 = y\}$, an analog of Theorem 8.31 holds, but with more complicated covariance relationships. In other cases there are directly defined pseudo-state estimators $\hat{P}\{N^t \in \Gamma|V_t\}$ (see Exercise 8.33 for an example).

8.4 ESTIMATION FOR MARKOV RENEWAL PROCESSES

A Markov renewal process describes by times of jumps and states visited the evolution of a finite-state, continuous-time process (a semi-Markov process, defined momentarily) whose structure generalizes that of a Markov process: the sequence of states visited remains a Markov chain, but time spent in a state need not be exponentially distributed and may depend on the next state entered. A Markov renewal process is also a collection of simultaneously evolving renewal processes, representing times of entrances to various states, with a particular sort of interdependence, and tools developed for statistical inference for renewal processes—empirical and martingale—can be applied, even in the presence of severely censored observations. In this section we treat estimation for two models of Markov renewal processes, one with complete observation of a single realization and the second with i.i.d. copies of a heavily censored process. Asymptotics are as $t \to \infty$ in the first case and as $n \to \infty$ in the second.

We begin with a short description of Markov renewal and semi-Markov processes. Let S be a finite set and let $(X,T) = ((X_n, T_n))_{n \geq 0}$ be a process with $X_n \in S$ for each n and $0 = T_0 < T_1 < \cdots$; then (X,T) is a *Markov renewal process* if

$$P\{X_{n+1} = j, T_{n+1} - T_n \leqslant x | X_0, \ldots, X_n; T_0, \ldots, T_n\}$$
$$= P\{X_{n+1} = j, T_{n+1} - T_n \leq x | X_n\}$$

for each n, j, and x. We assume homogeneity: the transition probabilities

$$Q(i,j,x) = P\{X_{n+1} = j, T_{n+1} - T_n \leq x | X_n = i\} \tag{8.74}$$

do not depend on n. The function Q is the *semi-Markov kernel* of (X,T). The Markov renewal process (X,T) is a Markov chain with state space $S \times \mathbf{R}_+$, but with additional structure.

Proposition 8.32. Let (X,T) be a Markov renewal process with semi-Markov kernel Q. Then

a) $X = (X_n)$ is a Markov chain with transition matrix

$$P(i,j) = Q(i,j,\infty) = \lim_{x \to \infty} Q(i,j,x) \tag{8.75}$$

b) $T_1, T_2 - T_1, \ldots$ are conditionally independent given X.

c) Times of entrances to a fixed state constitute a (possibly delayed or terminating) renewal process. □

The process $Y_t = X_n$, $t \in [T_n, T_{n+1})$, is the *semi-Markov process* engendered by (X,T). As noted previously, its structure generalizes that of a Markov process; indeed, if

$$Q(i,j,x) = P(i,j)(1 - e^{-\lambda(i)x}), \tag{8.76}$$

where $0 < \lambda(i) < \infty$ and $P(i,i) = 0$ for each i, then Y is a Markov process with generator $A(i,j) = \lambda(i)P(i,j)$, $i \neq j$. At the opposite extreme, if $S = \{i\}$ is a singleton, then (T_n) is an ordinary renewal process with interarrival distribution $F(\cdot) = Q(i,i,\cdot)$. One can describe the Markov renewal process

(X,T) not only by the semi-Markov process Y but also (and more completely) by the marked point process

$$N^0 = \sum \varepsilon_{(T_n, X_{n-1}, X_n)} X \tag{8.77}$$

with mark space $E = S \times S$. The time T_n is marked by the states X_{n-1}, X_n from and to which the semi-Markov process then jumps. Our statistical inference procedures are formulated in terms of N^0.

The law of a Markov renewal process is fully defined by the semi-Markov kernel Q, although typically [as in (8.76)] it is specified by giving the transition probabilities $P(i,j)$ for the embedded Markov chain and the conditional sojourn distributions

$$G(i,j,x) = P\{T_{n+1} - T_n \le x | X_n = i, X_{n+1} = j\}, \tag{8.78}$$

with Q then *defined* by

$$Q(i,j,x) = P(i,j)G(i,j,x), \tag{8.79}$$

so that P and G are the fundamental parameters. Below we estimate Q by first estimating unconditional sojourn distributions

$$H(i,x) = P\{T_{n+1} - T_n \le x | X_n = i\} = \sum_j Q(i,j,x) \tag{8.80}$$

Our statistical inference problem is: given either complete, asynchronous observations of the semi-Markov process Y (equivalently, of the marked point process N^0) or partial observation of i.i.d. copies, estimate the semi-Markov kernel Q. We consider only the full nonparametric model $\mathcal{P} = \{P_Q\}$. Below we examine a model involving i.i.d. copies of N^0, each only partially observed, with a censoring mechanism similar to that in Example 5.7. Martingale methods are used to derive estimators; limit theorems are demonstrated by direct appeal to the i.i.d. structure, but with martingale ideas employed to deduce covariance structure of Gaussian limit processes. For complete observation of a single realization, although one must use limit theorems for Markov renewal processes, the results parallel those in Section 8.1.

Asymptotic behavior of a Markov renewal or semi-Markov process is determined by the renewal structure together with limit properties of the Markov chain (X_n). If X is irreducible, so that (since S is finite) there exists a probability distribution ν on S satisfying $\nu(j) = \Sigma_i \nu(i)P(i,j)$ for all $j \in S$, where P is given by (8.75), and if the Markov renewal process (X,T) is aperiodic [this is unrelated to aperiodicity of X and connected instead with the distribution functions $G(i,j,.)$] and if $m(i) = E[\, T_1 | X_0 = i]$ is finite for each i, then

$$\lim_{t\to\infty} P\{Y_t = k\} = \frac{v(k)m(k)}{\Sigma_{j\in S} v(j)m(j)}; \tag{8.81}$$

see Çinlar (1975a,b) for details. Under the same hypotheses there hold strong laws of large numbers and central limit theorems (Pyke and Schaufele, 1964) for functionals $Z_t = \Sigma f(X_{n-1}, X_n, T_n - T_{n-1}) 1(t_n \leqslant t)$.

If for all i, j, x,

$$G(i,j,x) = H(i,x), \tag{8.82}$$

that is, although while not necessarily exponential, the distribution of a sojourn in state i does not depend on the next state to be entered [this assumption always holds for the Markov renewal proxess $(\tilde{X}, T)$ defined by $\tilde{X}_n = (X_n, X_{n+1})$] then $P(i,j)$ and $H(i,x)$ can be estimated from single realizations as follows. Let (compare Examples 2.33 and 5.8)

$$N_t^0(i,j) = \sum_n 1(T_n \le t, X_{n-1} = i, X_n = j) \tag{8.77a}$$

be the number of i-to-j transitions in $[0,t]$ and let

$$N_t^0(i) = \sum_{j\in S} N_t^0(i,j) \tag{8.77b}$$

be the number of sojourns in state i completed before time t. The estimators

$$\hat{H}(i,x) = N_t^0(i)^{-1} \sum_{n=1}^{N_t^0(i)} 1(X_{n-1} = i, T_n - T_{n-1} \le x) \tag{8.83a}$$

are analogous to the empirical distribution functions $\hat{F}$ of (8.20) and have similar asymptotic properties, while

$$\hat{P}(i,j) = \frac{N_t^0(i,j)}{N_t^0(i)}, \tag{8.83b}$$

as estimators of the transition matrix of the embedded Markov chain, except for the random sample size, are nonparametric maximum likelihood estimators and are strongly consistent and asymptotically normal. Finally, by substitution into (8.79), one arrives at

$$\hat{Q}(i,j,x) = \hat{P}(i,j)\hat{H}(i,x), \tag{8.83c}$$

for which the following properties hold.

Theorem 8.33. Assume that the Markov renewal process (X,T) is irreducible and aperiodic that $E[\,T_1 \mid X_0 = i] < \infty$ for each i, and that (8.82) holds. Then with $\hat{H}$, $\hat{P}$, $\hat{Q}$ given by (8.83), for each i and j

a) $\sup\{|\hat{Q}(i,j,x) - Q(i,j,x)| : x \in \mathbf{R}_+\} \to 0$ almost surely;

b) The error processes $\{t^{1/2}[\hat{Q}(i,j,x) - Q(i,j,x)] : x \in \mathbf{R}_+\}$ converge in distribution to a Gaussian process.

Proof: (Sketch). a) That $\sup|\hat{H}(i,x) - H(i,x)| \to 0$ almost surely follows by the argument used to prove Proposition 8.8, while $|\hat{P}(i,j) - P(i,j)| \to 0$, for example, by Basawa and Prakasa Rao (1980, Sect. 4.2).

b) Let $\tilde{m}(i)$ be the mean recurrence time for state i. Then by Basawa and Prakasa Rao (1980, Sect. 4.2) and the sampling rate reconciliation made throughout Section 8.1,

$$t^{1/2}[\hat{P}(i,j) - P(i,j)] \xrightarrow{d} N(0, \tilde{m}(i)P(i,j)[1 - P(i,j)]),$$

while Theorem 8.9 establishes that

$$\{t^{1/2}[\hat{H}(i,x) - H(i,x)] : x \in \mathbf{R}_+\} \xrightarrow{d} \{W(i,x) : x \in \mathbf{R}_+\},$$

where $W(i)$ is Gaussian with covariance $R(x,y) = \tilde{m}(i)H(i,x)[1 - H(i,y)]$, $0 \leqslant x \leqslant y$. These limit processes are independent, and hence b) holds. □

Moreover, the processes $W(i)$ are mutually independent and independent of the process $Z(i,j) = \lim t^{1/2}[\hat{P}(i,j) - P(i,j)]$, which has covariance function

$$R((i,j),(k,\ell)) = 1(i = k)\tilde{m}(i)[1(j = \ell)P(i,j) - P(i,j)P(k,\ell)];$$

thus further covariance relationships can be calculated.

Both the model for partially observed Markov renewal processes we are about to present and results concerning it are due to Gill (1980a). Let (X,T) be a Markov renewal process, described by the marked point process N^0 of (8.77), for each t let

$$N_t^0 = \sum_{i \in S} N_t^0(i) = \sum_n 1(T_n \leq t), \tag{8.77c}$$

and let (L_t^0) be the left-continuous backward recurrence time process of (N_t^0). The underlying process is not completely observable: each sojourn interval $T_{n+1} - T_n$ is subject to (perhaps total) censoring of the kind introduced in Example 5.7. More precisely, there are censoring variables V_n satisfying $T_n \leq V_n \leq T_{n+1}$ (and additional restrictions) such that for each n and each t in the possibly empty interval $(T_n, V_n]$, the values of Y_{t-}, L_t^0, and all $\Delta N_t^0(i,j)$ are observed. Thus about the sojourn of Y in the state X_n one learns nothing, partial information, or everything according as V_n is equal to T_n, strictly between T_n and T_{n+1}, or equal to T_{n+1}. In detail,

1) If $T_n = V_n$, the observation interval $(V_n, T_n]$ is empty and the sojourn—both its length $T_{n+1} - T_n$ and the state X_n occupied—is entirely invisible.
2) If $T_n < V_n < T_{n+1}$, the sojourn is partially observable: the state X_n is known, it is known that $T_{n+1} - T_n$ exceeds the observation span $V_n - T_n$, and it is known when observability ceases at V_n that V_n differs from T_{n+1}. (In Example 5.7 the indicator variables δ_j convey corresponding information.)
3) If $V_n = T_{n+1}$, the length $T_{n+1} - T_n$ of the sojourn, the current state X_n, *and* the state X_{n+1} entered at T_{n+1} are all known.

It is assumed that observation, to the extent possible, is over $[0,\infty)$. However, assumptions below imply that almost surely $V_n = T_n$ for all but finitely many values of n, so that only a finite number of sojourns are even partially observable. By analogy with Example 5.7 and the description of the underlying semi-Markov process using the marked point process N^0, observations are also described by marked point processes; note that in these expressions x is not a time variable, but instead, the argument of distribution functions. Let

$$N(i,j,x) = \sum_n 1(X_n = i, X_{n+1} = j, T_{n+1} - T_n \le x, V_n = T_{n+1}) \quad (8.84a)$$

be the number of completely observed sojourns in i that last no longer than x and terminate with a jump of the semi-Markov process to j, and let

$$K(i,x) = \sum_n 1(X_n = i, T_{n+1} - T_n \ge x, V_n \ge T_n + x) \qquad (8.84b)$$

be the number of sojourns in i *observed* to last at least as long as x. Also put $N(i,x) = \Sigma_j N(i,j,x)$ and $K(x) = \Sigma_i K(i,x)$.

Before constructing estimators of the semi-Markov kernel Q we give the compensators of the point processes $N^0(i,j)$ of (8.77a). They are not in general integrated stochastic intensities because the $Q(i,j,\cdot)$ need not be absolutely continuous; nonetheless, the following result is related closely to Proposition 8.10.

Proposition 8.34. For each i and j, the $\mathcal{F}^N$-compensator of $N^0(i,j)$ is

$$A_t(i,j) = \int_0^t \frac{1(Y_{u-} = i)}{[1 - H_-(i, L_u^0)]} \, dQ(i,j,L_u^0) \qquad (8.85)$$

where H is given by (8.80) and $H_-(i,x) = H(i,x-)$. □

The observability process can depend on the Markov renewal process, but only through the initial state, with all additional randomness conditionally independent of N^0 given X_0. In order that quantities in (8.84) be finite, most of the semi-Markov process must be unobservable. The following assumptions are in force for the remainder of the section.

Assumptions 8.35. a) There exists a σ-algebra $\mathcal{H}$ (containing all null sets) such that $\mathcal{H}$ and $\mathcal{F}^{N^0}(\mathbf{R}_+ \times E)$ are conditionally independent given X_0 and such that the V_n are stopping times of the history $\mathcal{G}_t = \mathcal{H} \vee \sigma(X_0) \vee \{\mathcal{F}_t^{N(i,j)} : i,j \in S\}$.

b) $E[\Sigma_n 1(V_n > T_n)] < \infty$.

Suppose now that we are given i.i.d. copies of both Markov renewal process and observability process, each defined on $\mathbf{R}_+$ and fulfilling Assumptions 8.35. For each n, let $N^n(i,j,x)$, $K^n(i,x)$, $N^n(i,x)$, $K^n(x)$ denote the sums (*un*normalized) over the first n processes of the functionals defined by (8.84). We finally construct estimators of the unconditional sojourn distributions H and the semi-Markov kernel Q. For the former we use product limit estimators, as in (5.20a):

$$\hat{H}(i,x) = 1 - \prod_{u \leqslant t} \left[1 - \frac{\Delta N^n(i,u)}{K^n(i,u)} \right]$$

$$= \int_0^x \frac{1 - \hat{H}(i,u-)}{K^n(i,u)} dN^n(i,u); \tag{8.86}$$

the second expression is the recursive method of calculation. For estimation of Q we unravel $\hat{H}$:

$$\hat{Q}(i,j,x) = \int_0^x \frac{1 - \hat{H}(i, u-)}{K^n(i,u)} dN^n(i,j,u) \tag{8.87}$$

A substitution justification is given in Exercise 8.36.

Note that full use is made of partially observed sojourns. If desired, one can estimate the transition probabilities $P(i,j)$ by $\hat{P}(i,j) = Q(i,j,\infty)$, but the censoring may prevent these estimators from being consistent.

As the number of partially observed copies approaches infinity, the estimators $\hat{H}, \hat{Q}$ are weakly uniformly consistent and asymptotically normal. We sketch a proof of the first assertion, but omit that of the latter. Of course, one must restrict attention to x-values for which the censoring does not prohibit observed sojourns of length x or longer. For each i, let $\tau_i = \sup\{x : E[K(i,x)] > 0\}$; estimation of $H(i,x)$ for $x > \tau_i$ is impossible.

Theorem 8.36. Suppose that Assumptions 8.35 are satisfied. Then for each i and j,

$$\sup\{|\hat{Q}(i,j,x) - Q(i,j,x)| : x < \tau_i\} \to 0 \tag{8.88}$$

in probability; consequently, $\sup\{|\hat{H}(i,x) - H(i,x)| : x < \tau_i\} \to 0$ in probability.

Proof: (Sketch). Introduce processes

$$Z^n(i,j,x) = N^n(i,j,x) - \int_0^x \frac{K^n(i,u)}{[1 - H(i,u-)]} Q(i,j,du) \quad (8.89)$$

(which cannot be calculated from the observations), put $Z^n(i,x) = \Sigma_j Z^n(i,j,x)$, and let $Z(i,j,x)$, $Z(i,x)$ be associated with a generic copy of the process.

By the weak law of large numbers,

$$\sup|n^{-1}N^n(i,j,x) - E[N(i,j,x)]| \to 0 \quad (8.90a)$$

and

$$sup\left|\int_0^x n^{-1}K^n(i,u)[1 - H(i,u-)]^{-1}Q(i,j,du) - \int_0^x E[K(i,u)][1 - H(i,u-)]^{-1}Q(i,j,du)\right| \to 0, \quad (8.90b)$$

in the sense of convergence in probability. Laborious but not deep calculations reveal that for each i, j, x,

$$\begin{aligned}\hat{Q}(i,j,x) - Q(i,j,x) &= \int_0^x [1 - \hat{H}(i,u-)][n^{-1}K^n(i,u)]^{-1}n^{-1}Z^n(i,j,du) \\ &- Q(i,j,x)\int_0^x \frac{1 - \hat{H}(i,u-)}{[1 - H(i,u)]n^{-1}K^n(i,u)} n^{-1}Z^n(i,du), \\ &+ \int_0^x Q(i,j,u)\frac{1 - \hat{H}(i,u-)}{[1 - H(i,u)]n^{-1}K^n(i,u)} n^{-1}Z^n(i,du), \end{aligned} \quad (8.91)$$

in view of which it suffices to show that $\sup_x|n^{-1}Z^n(i,j,x)| \to 0$ for each i and j. By (8.90),

$$\sup\left|n^{-1}Z^n(i,j,x) - \left\{E[N(i,j,x)] - \int_0^x E[K(i,u)][1 - H(i,u-)]^{-1}Q(i,j,du)\right\}\right| \to 0$$

in probability, so it remains only to establish the identity

$$E[N(i,j,x)] = \int_0^x E[K(i,u)][1 - H(i,u-)]^{-1}Q(i,j,du)$$

for all i, j, x, which is done using Proposition 8.34 and martingale techniques; see Gill (1980a) for details. □

Theorem 8.37. Suppose that in addition to Assumption 8.35a),

$$E[(\sum_j 1(V_j > T_j))^{7+\varepsilon}] < \infty$$

for some $\varepsilon > 0$ and for each i, let τ_i be such that $E[K(i,\tau_i)] > 0$. Then as random elements of the function space $\Pi_{i\in S}(D[0,\tau_i])^{|S|}$ the error processes $\{\{n^{1/2}[\hat{Q}(i,j,x) - Q(i,j,x)]: x \leq \tau_i\} : i,j \in S\}$ converge in distribution to a Gaussian process $G = (G(i,j,x))$ with representation

$$\begin{aligned} G(i,j,x) = &\int_0^x [1 - H(i,u-)]E[K(i,u)]^{-1} M(i,j,du) \\ &- Q(i,j,x) \int_0^x \frac{1 - H(i,u-)}{[1 - H(i,u)]E[K(i,u)]} M(i,du) \\ &+ \int_0^x Q(i,j,u) \frac{1 - H(i,u-)}{[1 - H(i,u)]E[K(i,u)]} M(i,du), \end{aligned} \tag{8.92}$$

where

i) For each i and j, $M(i,j,\cdot)$ is a mean zero Gaussian martingale;
ii) The families $\{M(i,j) : j \in S\}$ are independent as i varies;
iii) $M(i) = \Sigma_j M(i,j)$;
iv) The following covariance relationships hold:

$$<M(i,j)>_x = \int_0^x E[K(i,u)] \frac{1 - \Delta Q(i,j,u)}{1 - H(i,u-)} Q(i,j,du),$$

while for $j \neq j'$,

$$<M(i,j),M(i,j')>_x = -\int_0^x E[K(i,u)]\, \Delta Q(i,j,u)[1 - H(i,u-)]^{-1} Q(i,j',du). \quad \square$$

In fact, the error processes $n^{1/2}[\hat{Q} - Q]$ and $n^{1/2}[\hat{H} - H]$ are jointly asymptotically Gaussian since

$$n^{1/2}[\hat{H}(i,x) - H(i,x)] = n^{1/2}\Big[\sum_j \hat{Q}(i,j,x) - \sum_j Q(i,j,x)\Big]$$

The key point of the proof (Gill, 1980a) is to show asymptotic normality of the processes $n^{1/2}Z^n$ introduced in (8.89). Except for tightness this follows from the ordinary central limit theorem, although it must be verified that the limit covariance is correct. In Gill (1980a) tightness is established by a Skorohod embedding construction and the covariance function is computed using martingale techniques. Note that Theorems 8.36 and 8.37 resemble strongly results that would ensue from Theorems 5.12 and 5.13 if the processes Z^n were martingales; unfortunately, this is not the case.

EXERCISES

8.1. Prove that an ordinary Poisson process on $\mathbf{R}_+$ (in the sense of Definition 1.2) is a renewal process.

8.2. Verify (8.6).

8.3. Let W and V be the forward and backward recurrence time processes in a renewal process with interarrival distribution F.

a) For fixed y, x, exhibit $P\{W_t > y, V_t > x\}$ as the solution of a renewal equation.

b) Assuming F to be nonarithmetic, calculate the limit of this probability as $t \to \infty$.

8.4. (Continuation of Exercise 8.3) Let $I_t = V_t + W_t$ be the length of the interarrival interval containing t.

a) Derive a renewal equation for $P\{I_t > x\}$.

b) Calculate $\lim_{t\to\infty} P\{I_t > x\}$.

8.5. Let N be a renewal process with nonarithmetic interarrival distribution F and let h be a bounded, continuous function satisfying $h(x) = \int h(x-y)F(dy)$ for all $x \in \mathbf{R}_+$.

a) Let $X_n = h(x - T_n)$, where x is fixed. Prove that $(X_n, \mathcal{F}^N_{T_n})$ is a martingale.

b) With the aid of a) deduce that h must be constant. [*Hint*: Use the martingale convergence theorem and the Hewitt-Savage zero-one law.]

8.6. Let (X_t) be an irreducible Markov process with finite state space S and generator

$$A(i,i) = -\lambda(i)$$

$$A(i,j) = \lambda(i)Q(i,j), \qquad i \neq j,$$

where $\lambda(i)$ is the exponential parameter for sojourns in state i and Q is the transition matrix of the embedded Markov chain.

a) Verify that X is regenerative with respect to the sequence (T_n) of times of returns to its initial state X_0.

b) Use (8.13) to deduce that for each j, is

$$\lim_{t\to\infty} P_j\{X_t = j\} = \frac{\nu(j)/\lambda(j)}{\sum_{k\in S} \nu(k)/\lambda(k)}$$

where ν is the unique invariant distribution of Q and P_j is the law of X under the condition $X_0 = j$.

8.7. Prove Proposition 8.8.

8.8. Let N be a renewal process with unknown interarrival distribution F, observed asynchronously. Calculate the bias of the estimators $\hat{F}$, $\hat{\hat{F}}$ of (8.20) and (8.20a).

8.9. Let N be a renewal process with unknown interarrival distribution F, observed asynchronously. Propose $\mathcal{F}^N_t$-estimators of the interarrival variance σ^2 and establish their asymptotic properties as $t \to \infty$.

8.10. Let N be an asynchronously observed renewal process with unknown interarrival distribution F and backward recurrence time process V.
a) Prove that $V_t/N_t \to 0$ almost surely.
b) Deduce from this a strong law of large numbers for the estimators $\hat{m}$ of (8.28).
c) Repeat b) for the mean $\hat{\hat{m}}$ of the estimators $\hat{\hat{F}}$ of (8.20a).

8.11. Verify the identity (8.26).

8.12. A *Weibull* distribution with parameters $\lambda, \alpha > 0$ has distribution function $F_{\lambda,\alpha}(x) = 1 - \exp(-\lambda x^{\alpha})$, $x > 0$.
a) Calculate the hazard function and describe its monotonicity properties.
b) Let N be a renewal process with Weibull—but otherwise unknown—interarrival distribution F. Analyze the problem of maximum likelihood estimation of λ and α from asynchronous observations $\mathcal{F}_t^N$.
c) In the context of b), derive $\mathcal{F}_t^N$-estimators that are strongly consistent and asymptotically normal; verify these properties.

8.13. Provide a direct (i.e., non-martingale) proof of Theorem 8.17.

8.14. Generalize Theorem 8.17 to probabilities P_0, P_1 under which N is a renewal process with interarrival distributions F_0, F_1 satisfying $F_1 \ll F_0$.

8.15. a) Let T, S be random variables with distributions F, G, respectively. Prove that F is stochastically larger than G if and only if $E[h(T)] \geqslant E[hS)]$ for every positive, increasing function h.
b) Prove that the property of "stochastically larger" is not preserved under addition; that is, there exist random variables T, $\tilde{T}$, S, $\tilde{S}$ with T ($\tilde{T}$) stochastically larger than S ($\tilde{S}$) for which $T + \tilde{T}$ is *not* stochastically larger than $S + \tilde{S}$.

8.16. Let F be a distribution function on $\mathbf{R}_+$ with finite mean m and let $H(t) = m^{-1} \int_0^t [1 - F(u)]\, du$ be the associated limiting forward recurrence time distribution [see (8.11)].
a) Show that $H = F$ if and only if F is an exponential distribution.
b) Explore stochastic order relationships, if any, between F and H.

8.17. Let F be a distribution function on $\mathbf{R}_+$ with finite mean m and let $G(dv) = m^{-1} v F(dv)$, which according to Exercise 1.31 is the distribution,in a stationary renewal process with interarrival distribution F, of the interarrival interval U_1 that is constrained to contain the origin. Analyze stochastic order relationships between F and G.

8.18. Let F, G be distribution functions on $\mathbf{R}_+$ with hazard functions h_F and h_G, let λ be a function satisfying

$$\lambda(t) \geq \max\{\sup_{s \leq t} h_F(s), \sup_{s \leq t} h_G(s)\},$$

and let $\widetilde{N} = \Sigma\, \varepsilon_{T_n}$ be a Poisson process with intensity function λ. Construct indicator variables δ_k, $k \geq 1$, as follows:

$$P\{\delta_1 = 1\} = \frac{h_G(T_1)}{\lambda(T_1)},$$

while for $k > 1$,

$$P\{\delta_k = 1 | \delta_1, \dots, \delta_{k-1}\} = \frac{h_G(T_k)}{\lambda(T_k)} \quad \text{if } \delta_1 = \cdots = \delta_{k-1} = 0$$

$$= \frac{h_F(T_k - T_j)}{\lambda(T_k)} \quad \text{otherwise}$$

where $j = \max\{i < k : \delta_i = 1\}$.

a) Prove that $N = \Sigma_{k=1}^{\infty} \delta_k \varepsilon_{T_k}$ is a delayed renewal process in which the first arrival time $T_1 = U_1$ has distribution G and the remaining interarrival times have distribution F.

b) Under the further assumption that h_F is bounded and bounded away from zero, use this construction (with $G(dv) = m^{-1}[1 - F(v)]dv$) to prove the Blackwell renewal theorem (8.10) for F.

8.19. Discuss how the construction in Exercise 8.18 can be used to produce computable upper bounds on probabilities $P_F\{N_t \geq k\}$, which would be used, for example, to construct hypothesis tests.

8.20. Prove the inclusions $IFR \subset IFRA \subset NBU$ (see Definition 8.21).

8.21. Let $T \geq 0$ have distribution F and for each t, let $X_t = (T - t)^+$ be the remaining life at t of an object with lifetime T. Prove that $F \in IFR$ if and only if X_t is stochastically decreasing in t.

8.22. Consider a device subject to shocks that constitute a Poisson process $N = \Sigma\, \varepsilon_{T_n}$ with rate $\lambda > 0$; the shock at T_n causes a random damage X_n. Assume that the X_i are i.i.d. with distribution F and independent of N. The device fails when the cumulative damage $Y_t = \Sigma_{i=1}^{N_t} X_i$ exceeds a threshold a. Prove that distribution of the failure time T of the device belongs to the class $IFRA$.

8.23. (Continuation of Exercise 8.22) Suppose that F and a are known but that λ is not. Discuss estimation of λ from observation only of failure times T_j of i.i.d. devices.

8.24. Let N be a renewal process with interarrival distribution F whose forward recurrence time process satisfies $P\{W_t > y\} = 1 - F(y)$ for all t and y. Prove that N is a Poisson process.

8.25. Let N be a renewal process with interarrival distribution F and backward recurrence time process V. For each t, let $F_t(x)$ be $F(x)$ if $x \leq t$ and 1 otherwise. Prove that if V_t has distribution F_t for all t, then N is Poisson.

8.26. Let N be a renewal process with Weibull interarrival distribution F.

a) Propose and analyze a test of the hypotheses $H_0 : \alpha = 0$; $H_1 : \alpha > 0$ given synchronous observations $\mathcal{F}_{T_n}^N$.

b) Repeat a) for asynchronous observations $\mathcal{F}_t^N$.

8.27. Derive (8.66) and (8.67).

8.28. Prove Proposition 8.29.

8.29. Let N' be the p-thinning of a renewal process N with known, absolutely continuous interarrival distribution F. Only N' is observable; only p is unknown.

a) Show that the $\mathcal{F}_t^{N'}$-estimators $\hat{p} = N_t'/R(t)$ are unbiased, where R is the renewal function of F.

b) Describe the large-sample behavior of these estimators.

8.30. (Continuation of Exercise 8.29) Suppose now that N is also observable.

a) Prove that (N_t, N_t') is a sufficient statistic for p given observation over $[0, t]$.

b) Compute the asymptotic efficiency of the estimators $\hat{p}$ of Exercise 8.29 relative to the estimators $\hat{p}^* = N_t'/N_t$, where $0/0 = 0$.

8.31. Work through the calculations in Theorem 8.30 for the case that N is a Poisson process.

8.32. Verify carefully the asymptotic independence used in the proof of Theorem 8.31.

8.33. Let N be a renewal process with interarrival distribution F and backward recurrence time process V. For each t, let $N_u^t = N_{t+u} - N_t$.

a) Show that $P\{N_u^t = k \mid V_t\} = F^k(u + V_t) - F^{k+1}(u + V_t)$ for each k.

b) Suppose now that F is not known. Construct pseudo-state estimators $\hat{P}\{N_u^t = k \mid v_t\}$ given observations $\mathcal{F}_t^N$ and examine their properties as $t \to \infty$.

8.34. Let (Y_t) be a semi-Markov process with finite state space $S = \{1, \dots, K\}$ and semi-Markov kernel $Q(i,j,x) = P(i,j)G(i,j,x)$ as in (8.79). Let N^0 be the marked point process given by (8.77a). Assuming that $G(i,j,\cdot)$ has hazard function $h(i,j,\cdot)$, show that $N^0(i,j)$ has $\mathcal{F}^Y$-stochastic intensity $\lambda_t(i,j) = 1(Y_{t-} = i)P(i,j)h(i,j,V_{t-})$, where V_t is the time since the most recent transition of Y before t.

8.35. Prove Proposition 8.34.

8.36. In the notation of Section 8.4, prove that for each i and t,

$$H(i,t) = \int_0^t [1 - H(i,s-)]E[K(i,s)]^{-1}\, dE[N(i,s)]$$

NOTES

For a number of years, Feller (1971, Chap. XI) was *the* treatment of renewal theory, but because it omits regenerative processes it must be considered incomplete. More modern accounts [e.g., Çinlar (1975a) and Ross (1982)] do include regenerative processes; the former also presents semiregenerative processes. Other general sources are Cox (1962), which contains distributional calculations possibly appearing nowhere else, and Smith (1958). Applications of renewal processes are too numerous to describe. Perhaps the most important is reliability (see Section 8.2); Daley and Milne (1973) provides access to others.

The introduction surveys the fundamental results; proofs can be found, for example, in Çinlar (1975a) and Feller (1971); despite its being in some ways dated, as a compact, self-contained development of the theory, the latter remains preeminent. Theorem 8.5, the main result, is due in the form stated to Smith (1958); the simpler but equivalent form (8.10) was proved by Blackwell (1948). Feller (1971) seems to be the first expository treatment of renewal equations per se and their application to computations involving functionals of renewal processes. As (8.8) and the proof of Theorem 10.7 show, however, "sample path" arguments often render renewal equations superfluous. Concerning Proposition 8.1, see Cox (1962) or Feller (1971).

Section 8.1

Synchronous observation of renewal processes neither introduces new problems nor requires tools beyond those discussed (in greater generality) in Chapter 4. Transition to asynchronous observation entails only having to reconcile real- and discrete-time sampling rates, as in Theorem 8.9, although it must also be assumed that the mean interarrival time is finite. Our approach utilizes results of Serfozo (1975) that deduce asymptotic properties—including functional limit theorems—of a continuous-time process Y_t from those of an embedded sequence $Y(T_n)$, where the T_n are stopping times satisfying a weak law of large numbers. The key additional assumption, that fluctuations of Y between successive T_n not be too wild, is trivially satisfied in our case. Theorem 8.12, although possibly new, is not deep and is but one of many variations. The Poisson sampling model and Proposition 8.15 are from Kingman (1963) (see also Exercise 6.18 and Chapter 10). Theorem 8.16 is a variant of Theorems 7.4 and 7.5.

Parametric inference for renewal processes amounts in practice to inference for gamma- or Weibull-distributed i.i.d. data and is discussed, for example, in Barlow and Proschan (1975) and Cox and Lewis (1966). Sources with more emphasis on survival analysis and estimation of hazard functions are Kalbfleisch and Prentice (1980), Lawless (1982), and Nelson (1982). Vardi (1982a,b) examines length bias in samples of i.i.d. renewal processes over a fixed, finite time interval, possibly with additional censoring.

Section 8.2

The "theoretical" properties of likelihoods, Theorems 8.17 and 8.19, are the main results. Evidently, the former is not new, although the martingale proof—with its striking confirmation of the need to use predictable stochastic intensities—may be; direct proofs are also possible (Exercise 8.13). Theorem 8.19 appears here for the first time.

Except for queueing theory, reliability is the oldest area of application of point processes. In its literature one can often find specifics of inference that are glossed over in statistics books. Previously mentioned sources, such as Barlow and Proschan (1975), Kalbfleisch and Prentice (1980), and Nelson (1982), along with Mann et al. (1974), provide entry to the voluminous literature. Barlow et al. (1972) is a nice treatment of inference problems, many but not all of reliability origin, under order restrictions. Qualitative concepts such as stochastic orderings and monotonicity of hazard functions arose mainly in reliability.

Ross (1982) and Stoyan (1983) are expository developments of stochastic orderings, not only the "stochastically larger than" ordering of (8.46) but also others involving hazard functions and likelihood ratios. The former includes simple coupling constructions such as that in Proposition 8.20; the latter explores extensively implications concerning comparison of stochastic processes. Lindvall (1981, 1982) uses different couplings in order to prove renewal theorems.

The families *IFR, IFRA, NBU* are increasingly broad interpretations of the intuitive idea of "older is worse"; recently, even weaker notions have been devised. Tests for them (and "older is better" counterparts) constitute a central part of reliability theory. In addition to the books mentioned above, one can consult key papers, among them Barlow (1968), Bickel and Doksum (1968), Bickel (1969), and Proschan and Pyke (1967) on tests for *IFR*; Barlow and Proschan (1969) on tests for *IFRA*; and Hollander and Proschan (1972) and Chen et al. (1983) on tests for *NBU*. All treat i.i.d. lifetimes, but some allow censoring. Testing for "*U*-shaped" hazard functions has also been considered; this problem is akin to testing for unimodality of a density function.

Proposition 8.23, possibly the simplest of several related characterizations of Poisson processes among renewal processes, is due to Çinlar and Jagers (1973) (see also Exercises 8.24 and 8.25). Kingman (1963) is the source of Theorems 8.24 and 8.25. In none of these cases have testing implications been probed.

Testing whether a point process is a renewal process has been addressed using serial correlation coefficients and spectra of counts and interarrival times (see, e.g., Cox and Lewis, 1966). Immense difficulties arise with alternative hypotheses. For example, some procedures take as an alternative the point process of Wold (1948), whose interarrival times form a Markov chain, even though this model has never found plausible application.

Section 8.3

Nearly all of this section is new; Theorem 8.30 is the "most novel" result, whereas Theorem 8.26 should really be regarded as part of the folklore and Theorem 8.31 is an amalgamation of previous results. Thinned renewal processes have also been studied by Mogyorodi (1971, 1972, 1973a,b) and Räde (1972a,b). Poisson limit theorems for superpositions of independent, sparse renewal processes are given in Çinlar (1972); the renewal structure of the components leads to simplified conditions for validity of general theorems.

Section 8.4

Estimation for Markov renewal processes divides into pre- and post-martingale eras, each exemplified by a single paper. Moore and Pyke (1968), the source of Theorem 8.33, is the pre-martingale paper. Under (8.82), which holds without loss of generality by expansion of the state space and redefinition of the X_n, one simply combines results for empirical distribution functions (see Chapter 4) and those for Markov chains (see, e.g., Basawa and Prakasa Rao, 1980, or Billingsley, 1961a,b). The post-martingale results are from Gill (1980a). General exposition of Markov renewal and semi-Markov processes is given by Çinlar (1975a, b); Pyke (1961a,b), and Pyke and Schaufele (1964) are key early works, although the idea stems from Lévy (1954).

Exercises

8.5. This property, known in analysis as the Choquet-Deny theorem, is a key step in the proof of the renewal theorem given by Feller (1971).

8.6. See Çinlar (1975a).

8.12. Cox and Lewis (1966) contains this and additional aspects, but with a different parametrization.

8.15. Ross (1982) presents related properties.

8.17. Waiting-time "paradoxes" have been inferred in this context.

8.19. M. Brown (1980) presents related properties.

8.22. See Barlow and Proschan (1975).

8.25. Chung (1972) establishes the result under the weaker hypothesis that V_t has distribution F_t only for $t \leq t_0$.

8.26. This is a parametric test whether F has increasing failure rate.

8.34. See also Proposition 8.34.

8.36. See Gill (1980a); this is the substitution motivation for (8.86).

9
Inference for Stationary Point Processes

In earlier chapters we emphasized "multiple realization" inference (Model 3.1), in which setting neither the structure of the underlying space nor—except broadly—that of the point processes nor that of the characteristics being estimated had to be severely restricted. By contrast, in this chapter we address only single-realization inference (Model 3.2) in a context entailing stronger assumptions on the space and related, but still fairly broad, assumptions on the point processes. Stationarity replaces the "identically distributed" part of "i.i.d."; to derive asymptotic properties of estimators we substitute for "independent," hypotheses of ergodicity and finiteness of reduced cumulant measures. Of course, in order that stationarity even make sense the underlying space must admit a group structure; for concreteness and simplicity we shall suppose that $E = \mathbf{R}^d$ for some integer $d \geq 1$. The discrete case $E = \mathbf{Z}^d$ can be dealt with more easily, typically by random field rather than point process methods. That is, one analyzes directly the stationary random field $\{N(z) : z \in \mathbf{Z}^d\}$, where $N(z) = N(\{z\})$; point process structure of N plays only an incidental role.

Perhaps the most severe restriction imposed by the stationary point process setting concerns *what* can be estimated easily and effectively. Whereas given i.i.d. multiple realizations of a point process N, one can estimate in a straightforward manner almost any aspect of the probability law $\mathcal{L}_N$ (see Chapter 4), here we focus on estimation of moment measures, which even though infinite-dimensional, specify $\mathcal{L}_N$ only in special cases. However, techniques used to estimate moment measures can be used as well to estimate Palm measures, which do determine $\mathcal{L}_N$, but the estimators we construct cannot easily be converted to estimators of $\mathcal{L}_N$ itself. Even for estimation of moment measures, asymptotics, pertaining to observation over compact sets increasing to E, are constrained in that the sets cannot have arbitrary structure; we assume convexity.

We pause to establish notation, to recall concepts and results from Section 1.8, and to introduce some refinements and extensions; in particular we present a limit theorem that serves as basis for all the consistency results in the chapter. Let $E = \mathbf{R}^d$ and for $x \in E$, let τ_x be the translation operator $y \to \tau_x y = y - x$. Lebesgue measure on E is denoted simply by dx, or if necessary by λ. From Section 1.8 we recollect that a point process N on E, defined over the probability space $(\Omega, \mathcal{F}, P)$, is *stationary* with respect to shift operators $(\theta_x)_{x \in E}$ satisfying conditions set forth there, provided that

$$N\tau_x^{-1} = N \circ \theta_x \tag{9.1}$$

for each x and that (θ_x) be measure-preserving for P: $P\theta_x^{-1} = P$ for each x. In consequence, N is *stationary in law*:

$$N\tau_x^{-1} \stackrel{d}{=} N \tag{9.2}$$

for every x; conversely, if (9.2) holds, then there exists a canonical realization of N that is stationary. By (9.2) for each x the mean measure μ_N satisfies $\mu_N \tau_x^{-1} = \mu_N$ and hence has the form $\mu_N = \nu_N \lambda$; the constant ν_N is the *intensity* of N. We assume throughout the chapter, rarely with explicit mention, that ν_N is *finite and positive*; often we impose stronger moment conditions.

The Palm measure P^* of N is defined by (1.72): for $H: \Omega \times E \to \mathbf{R}$ bounded and measurable,

$$E\left[\int N(\omega, dx) H(\theta_x \omega, x)\right] = E^*\left[\int H(\omega, x)\, dx\right] \tag{9.3}$$

Recall the interpretation $P^*(\Gamma)/P^*(\Omega) = P\{N \in \Gamma | N(\{0\}) = 1\}$. Palm measures play a central role in estimation procedures developed in Section 9.1, mainly through a special case of (9.3): for H a function from $\mathbf{M}_p$ to $\mathbf{R}$ and h a function on E satisfying $\lambda(h) = 1$,

$$E^*[H(N)] = E\left[\int N(dx) H(N\tau_x^{-1}) h(x)\right] \tag{9.4}$$

On the right-hand side of (9.4), h serves as a weight function; integrals $\int_K N(dx) H(N\tau_x^{-1}) h_K(x)$, with normalization incorporated in the requirement $\lambda(h_K) = 1$ and with $K \uparrow E$, are used to estimate $E^*[H(N)]$.

In Section 1.8 we gave one ergodic theorem for a stationary point process N; another, in some ways more useful, version is given momentarily in Theorem 9.2, but only after an additional result, a "zero-infinity" law for stationary point processes: for a stationary point process any finitely defined configuration of points (e.g., pairs of points within a prescribed distance of one another) occurs (modulo a set of probability zero) either not at all or infinitely often. Exercise 1.28 contains a special case: for N a stationary point process on $\mathbf{R}$, $P\{N = 0\} + P\{N((-\infty, 0)) = N((0, \infty)) = \infty\} = 1$. Such a result is required in order to ensure that for sets $K \uparrow E$ there are infinitely many data (in particular, infinitely many points of N), without which consistency and asymptotic normality cannot be established. Before giving the result we recall the notation $N^k[dx_1, \ldots, dx_k) = N(dx_1) \ldots N(dx_k)$; for $A \in \mathscr{E}^k$, $(k!)^{-1}N^k(A)$ is the number of k-configurations of N falling in A.

Proposition 9.1. Let N be a stationary point process on E and let A be a measurable subset of E^k invariant under the transformation $(y_1, \ldots, y_k) \to (y_1 - x, \ldots, y_k - x)$ for each $x \in E$. Then

$$P\{N^k(A) = 0\} + P\{N^k(A) = \infty\} = 1 \tag{9.5}$$

Proof: Let $\Gamma = \{0 < N^k(A) < \infty\}$. if $P\{\Gamma\} > 0$, then the formula $Q\{\cdot\} = P\{\cdot|\Gamma\}$ defines a probability with respect to which N remains stationary, but under which the E-valued functional

$$Y = N^k(A)^{-1} \int_A \left(k^{-1} \sum_{i=1}^{k} x_i\right) N^k(dx)$$

has a translation invariant distribution, which would have to be the nonexistent uniform distribution on E. □

We next present a generalization of Theorem 1.68 forming the foundation of the consistency properties derived in Section 9.1. Recall (Definition 1.67) that an event Γ is *invariant* if $\theta_x^{-1}\Gamma = \Gamma$ for all $x \in E$ and that P is *ergodic* if the σ-algebra $\mathscr{I}$ of invariant events is P-degenerate: $P\{\Gamma\} = 0$ or 1 for all $\Gamma \in \mathscr{I}$. With f in (1.83) the indicator function of the unit hypercube, Theorem 1.68 implies in particular that if N is ergodic, then (with inequalities and convergence interpreted coordinatewise)

$$\lim_{x \to \infty} \frac{N(\{y: 0 < y \leq x\})}{\lambda(\{y: 0 < y \leq x\})} = \nu_N \tag{9.6}$$

almost surely as well as in $L^1(P)$. In applications and theoretical analyses it is of interest to permit more general sets K increasing to E or even just satisfying $\lambda(K) \to \infty$, as well as more general functionals of N; the following spatial ergodic theorem of Nguyen and Zessin (1979a) does this and more.

Denote by $\mathscr{K}$ the family of bounded, convex subsets of E and for $K \in \mathscr{K}$ let $\delta(K)$ be the supremum of the radii of Euclidean balls that are subsets of K. If $\delta(K) > 0$, then in particular the interior of K is nonempty.

We consider random measures

$$M(A) = \int_A N(dx) H(N\tau_x^{-1}) = \int_A N(dx) H(N \circ \theta_x) \tag{9.7}$$

where $H: \mathbf{M}_p \to \mathbf{R}$, which are *covariant* with respect to (θ_x) in the sense that almost surely

$$M(\tau_y A) \circ \theta_y = M(A) \tag{9.8}$$

for each A and y (Exercise 9.5). The choice $H \equiv 1$ yields $M = N$, so that the following result not only gives—as we shall see—consistency of estimators of moment measures but also extends Theorem 1.68.

Theorem 9.2. Let N be a stationary, ergodic point process and let H be a function on $\mathbf{M}_p$ satisfying

$$E^*[H(N)] < \infty, \tag{9.9}$$

where P^* is the P-Palm measure of N, and let M be the random measure defined by (9.7). Then

$$\lim_{\delta(K)\to\infty} \frac{M(K)}{\lambda(K)} = E[M([0,1]^d)] \tag{9.10}$$

almost surely and in $L^1(P)$.

Proof: (Sketch). Necessary background is a multidimensional ergodic theorem of Fritz (1970), which asserts that for $X \in L^1(P)$ and (K_α) a linearly ordered subclass of $\mathcal{K}$ with $\delta(K_\alpha) \to \infty$,

$$\lim \lambda(K_\alpha)^{-1} \sum_{x \in K_\alpha} X \circ \theta_{-x} = E[X|\mathcal{I}]$$

almost surely; here and below summations are restricted to the lattice $\mathbf{Z}^d$ of points in E with integral coordinates. In particular, if P is ergodic, then

$$\lim \lambda(K_\alpha)^{-1} \sum_{x \in K_\alpha} X \circ \theta_{-x} = E[X] \tag{9.11}$$

Let $J = (0,1]^d$; then since M is a random measure and since as x varies over $\mathbf{Z}^d$ the sets $J + x$ constitute a partition of E,

$$\begin{aligned} M(K_\alpha) &= \sum M(K_\alpha \cap (J + x)) \\ &= \sum_{x \in K_\alpha} M(J) \circ \theta_{-x} + \sum_{x \notin K_\alpha : (J+x) \cap K_\alpha \neq \emptyset} M((K_\alpha - x) \cap J) \circ \theta_{-x} \\ &\quad + \sum_{x \in K_\alpha : (J+x) \cap K_\alpha^c \neq \emptyset} M((K_\alpha^c - x) \cap J) \circ \theta_{-x}, \end{aligned} \tag{9.12}$$

where we have also used (9.8). By (9.9) and (9.11)

$$\lim \lambda(K_\alpha)^{-1} \sum_{x \in K_\alpha} M(J) \circ \theta_{-x} = E[M(J)] = E\left[\int_J N(dx) H(N \circ \theta_x)\right] = E^*[H(N)];$$

this also confirms that (9.9) is the integrability hypothesis needed in order to apply (9.11).

The crucial geometric property of convex sets with nonempty interiors is that their boundaries are small relative to their volumes, vanishingly so as $\delta(K)$ becomes infinite. In (9.12) the second and third terms are in effect summations over the boundary (somewhat loosely defined) of K_α and hence both, divided by $\lambda(K_\alpha)$, converge to zero. More precisely, the second term, which accounts for points $x \notin K_\alpha$ such that $J+x$ intersects K_α nonetheless, may be bounded as follows:

$$\left|\lambda(K_\alpha)^{-1} \sum_{x \notin K_\alpha : (J+x)\cap K_\alpha \neq \emptyset} M((K_\alpha - x)\cap J)\circ\theta_{-x}\right| \leq \lambda(K_\alpha)^{-1} \sum |M(J)|\circ\theta_{-x}$$

$$\leq \lambda(K_\alpha)^{-1} \sum_{x \notin K_\alpha : d(x,\partial K_\alpha)\leq 1} |M(J)|\circ\theta_{-x} \to 0$$

where $d(x,\partial K_\alpha)$ is the distance from x to the boundary of K_α; convergence to zero ensues from the property that $\lambda(K_\alpha)^{-1}\lambda(\{x : d(x,\partial K_\alpha) \leq 1\}) \to 0$ (Exercise 9.6). The third term in (9.12) is handled analogously, and hence almost sure convergence in (9.10) is established. Convergence in $L^1(P)$ is shown either directly (Nguyen and Zessin, 1979a) or by combining almost sure convergence with uniform integrability. □

Rectangles $C_x = \{y : 0 < y \leq x\}$ with $x \to \infty$ satisfy the conditions of Theorem 9.2, so the choice $H = 1$ does provide an extension of Theorem 1.68.

Corollary 9.3. For N a stationary, ergodic point process with intensity ν, almost surely and in $L^1(P)$,

$$\lim_{\delta(K)\to\infty} \frac{N(K)}{\lambda(K)} = \nu \quad \square \tag{9.13}$$

We conclude this background discussion by describing a particular disintegration of Radon measures with certain translation invariance properties; these ideas are used to analyze moment measures of stationary point processes. For $k \geq 1$ define the *diagonal group* $(\tau_x^{(k)})$ of transformations on E^k by $\tau_x^{(k)} y = (y_1 - x, \ldots, y_k - x)$. Given $z \in E^{k-1}$, let λ_z denote the image of Lebesgue measure under the mapping $x \to (z_1 + x, \ldots, z_{k-1} + x, x)$ of E into E^k. It seems plausible that a measure μ on E^k invariant under the diagonal group should be a mixture (in a generalized sense; the mixing measure need not be a probability) of the measures λ_z, $z \in E^{k-1}$, each of which is evidently invariant, and indeed this is so.

Proposition 9.4. Given a measure μ on E^k invariant under the diagonal group there is a unique measure μ_* on E^{k-1} such that

$$\mu = \int_{E^{k-1}} \lambda_z \mu_*(dz) \quad \square \tag{9.14}$$

We omit the proof (see Krickeberg, 1982). The choice $k = 1$, with the convention $E^0 = \{0\}$, shows that every translation invariant measure on E is a scalar multiple of Lebesgue measure.

The remainder of the chapter concentrates on statistical inference for stationary point processes and is organized in the following manner. Section 9.1 concerns estimation of Palm integrals (9.4), where the weight function h satisfies $\lambda(h) = 1$, using substitution estimators $\int_K N(dx) H(N \circ \theta_x) h_K(x)$; in order to satisfy $\lambda(h_K) = 1$, the function h_K must incorporate the scaling factor $\lambda(K)^{-1}$. Consistency and asymptotic normality are investigated for classes of functions. Among the Palm integrals $E^*[H(N)]$ are the intensity of N as well as reduced moment measures and cumulant measures. There are assumptions of stationarity and ergodicity but the setting is nonparametric, with emphasis on results of general applicability for estimation of first and second moment measures.

Likelihood ratios and parametric maximum likelihood estimation are examined in Section 9.2, along with the structure of the stochastic intensity of a stationary point process on **R**. Concepts and techniques of spectral analysis appear in Section 9.3; here one exploits stationarity of the increments of a stationary point process in the frequency domain rather than the time domain. A spectral representation theorem is derived, spectral measures and density functions are defined, and methods for estimation are described. Finally, in Section 9.4 we discuss state estimation for stationary point processes, concentrating exclusively on optimal linear state estimators but considering both time- and frequency-domain approaches.

9.1 ESTIMATION OF MOMENT AND PALM MEASURES

Let N be a stationary point process on $E = \mathbf{R}^d$, defined over a measurable space $(\Omega, \mathcal{G})$ admitting shift operators $(\theta_x)_{x \in E}$ such that (9.1) holds identically in x and ω; these objects are fixed throughout the section. We examine techniques for estimation of reduced moment measures of N as well as other Palm integrals from single realizations observed over increasing (bounded, convex) subsets of E, within the full nonparametric model $\mathcal{P}$ of probability measures P on $(\Omega, \mathcal{G})$ with respect to which N is simple and stationary with nondegenerate intensity. That is, P must satisfy

$$P\theta_x^{-1} = P \tag{9.15}$$

for all $x \in E$ and the intensity $\nu = \nu_P$ must be positive and finite. The intensity and other moment measures are now (since N is immutable) functionals of P rather than N. The Palm measure, also a functional of P, is denoted by P^*, with associated Expectation E^*.

As in analysis of stationary stochastic processes or time series, emphasis is on first and second moments [i.e., the intensity ν and the covariance measure ρ of (1.12)]. However, the theory is no more difficult to formulate and establish in generality, provided only that the moment measures in question exist, and this is the path we follow. The appropriate form of existence for moment measures is not finiteness (the mean measure of a stationary point process on $\mathbf{R}^d$ cannot be finite without vanishing entirely) but instead *local* finiteness. (Reduced cumulant measures can be—and are often assumed to be—finite.)

Denote by $C(T)$ $[C_+(T)]$ the set of [positive] continuous functions on the space T whose support is compact. For each k, let

$$N^k(dx_1, \dots, dx_k) = N(dx_1) \cdots N(dx_k) \tag{9.16}$$

on E^k.

We now introduce moment measures.

Definition 9.5. The point process N admits a *moment of order k* with respect to P if $E[N^k(g)] < \infty$ for every $g \in C_+(E^k)$, in which case the measure

$$\mu^k = E[N^k] \tag{9.17}$$

is the *moment measure of order k* (or kth-order moment measure).

When μ^k exists, $\mu^k(A)$ is interpreted as the mean number of k-configurations of points of N falling in A.

Cumulant measures are important because finiteness assumptions on reduced cumulant measures have the same kinds of implications concerning near independence of distantly separated parts of a stationary point process as do summability or integrability hypotheses on covariance functions of stationary processes on $\mathbf{R}$ (see Section 10.2).

Definition 9.6. Let N admit a moment of order k with respect to P; then the measure

$$\gamma^k = \sum (-1)^{|\mathcal{I}|-1}(|\mathcal{I}|-1)! \prod_{\ell} [\mu^{|J_\ell|}(\otimes f_j : j \in J_\ell)] \tag{9.18}$$

is the *cumulant measure of order k*. In (9.18) the summation is over all partitions $\mathcal{I} = \{J_1, \dots, J_{|\mathcal{I}|}\}$ of $\{1, \dots, k\}$. (For a discussion of the relationship to cumulants of ordinary random variables, see the proof of Theorem 9.15.)

The most important cumulant measure is the covariance measure. Taking $k=2$ in (9.18) yields the partitions $\mathcal{I}=\{\{1,2\}\}$ and $\mathcal{I}=\{\{1\},\{2\}\}$ and thence the measure

$$\gamma^2(f_1\otimes f_2)=\mu^2(f_1\otimes f_2)-\mu^1(f_1)\mu^1(f_2)$$
$$=\mu^2(f_1\otimes f_2)-\nu^2\lambda(f_1)\lambda(f_2),$$

previously introduced (Definition 1.14) as the *covariance measure* ρ, terminology and notation we retain. Note that the covariance measure, like other cumulant measures, is a signed measure.

With $(\tau_x^{(2)})$ the diagonal group on E^2, stationarity of N implies that for each x,

$$\rho(\tau_x^{(2)})^{-1}(f_1\otimes f_2)=E[N(f_1\circ\tau_x)N(f_2\circ\tau_x)]-\nu^2\lambda(f_1)\lambda(f_2)$$
$$=E[N(f_1)N(f_2)]-\nu^2\lambda(f_1)\lambda(f_2)=\rho(f_1\otimes f_2)$$

and hence ρ admits a disintegration (9.14). More generally, each extant moment and cumulant measure of N is invariant under the appropriate diagonal group (Exercise 9.9) and can be disintegrated similarly. The "mixing" measures are known as reduced moment and cumulant measures.

Definition 9.7. Suppose that N admits a moment of order k with respect to P. Then the measure μ_*^k on E^{k-1} satisfying

$$\mu^k=\int_{E^{k-1}}\lambda_z\mu_*^k(dz) \tag{9.19a}$$

is the *reduced moment measure* of order k. (The measures λ_z are defined just before Proposition 9.4.) The measure γ_*^k satisfying

$$\gamma^k=\int_{E^{k-1}}\lambda_z\gamma_*^k(dz) \tag{9.19b}$$

is the *reduced cumulant measure* of order k.

In particular, $\mu_*^1=\nu\varepsilon_0$, so estimation of μ_*^1 is tantamount to that of the intensity ν. We also note the following property, which is crucial to our estimation procedures: reduced moment measures of N with respect to P, which solely on a principle of economy should be estimated rather than moment measures, are themselves moment measures of N with respect to the Palm measure.

Lemma 9.8. Let N be a stationary point process admitting moment of order $k\geq 2$ with respect to P. Then

$$\mu_*^k=E^*[N^{k-1}]\quad\square \tag{9.20}$$

The proof is left as Exercise 9.7.

Finally, we take up estimation per se; our goal is to estimate integral functionals of the unknown probability P of the form $P^*(H) = \int H\,dP^* = E^*[H(N)]$, where H: $\mathbf{M}_p \to \mathbf{R}$ satisfies $H(N) \in L^1(P^*)$. By (9.4) for any function h on E with $\lambda(h) = 1$,

$$P^*(H) = E[\int N(dx)H(N \circ \theta_x)h(x)],$$

and thus one is led to a family of substitution estimators

$$\hat{P}^*(H) = \int N(dx)H(N \circ \theta_x)h(x)$$

These estimators have a very nice interpretation: only points x that are also points of N contribute to the integrals above, and for x a point of N, $N\tau_x^{-1}$ has an atom at the origin; thus the estimator is just a weighted—via the function h—average of H evaluated at translates of N that place each point in turn at the origin.

However, observation of N is possible only over bounded sets, so given observation over B conditions must be placed on H and h in order that the estimator

$$\hat{P}^*(H) = \int_B N(dx)H(N \circ \theta_x)h(x) = \int_B N(dx)H(N\tau_x^{-1})h(x) \qquad (9.21)$$

can be calculated, if not from the observations $\mathcal{F}^N(B)$, then from $\mathcal{F}^N(K)$ for some compact set K related in a natural manner to B. In particular, whenever $x \in B$ and $h(x) \neq 0$, one must be able to determine $H(N\tau_x^{-1}) = H(N(\cdot + x))$ from $\mathcal{F}^N(K)$.

When such conditions are fulfilled, the estimators (9.21) are unbiased.

Proposition 9.9. Suppose that there exists a bounded set A with $0 \in A$ such that $H(\mu) = H(\mu_A)$ for all $\mu \in \mathbf{M}_p$, where μ_A is the restriction of μ to A, and that

$$A + \operatorname{supp} h \subset B, \qquad (9.22)$$

where $\operatorname{supp} h$ is the support of h and $A_1 + A_2 = \{x + y : x \in A_1, y \in A_2\}$, and let $\hat{P}^*(H)$ be given by (9.21); then $\hat{P}^*(H) \in \mathcal{F}^N(B)$ and

$$E[\hat{P}^*(H)] = P^*(H) \qquad (9.23)$$

for every $P \in \mathcal{P}$ such that $H(N) \in L^1(P^*)$.

Proof: When (9.22) holds, $\operatorname{supp} h \subset B$ since $0 \in A$ and hence

$$\hat{P}^*(H) = \int_B N(dx)H(N \circ \theta_x)h(x),$$

from which (9.23) follows by means of (9.4). Moreover, for $x \in B$, $H(N \circ \theta_x) = H((N \circ \theta_x)_A)$ depends only on N_B by (9.22). □

Here are the important specific instances; we shall pursue them throughout the chapter.

Example 9.10. (First moments). For $H = 1$, $P^*(H) = \nu_P$, and we may take $A = \{0\}$ in (9.22), which becomes simply the requirement that $\operatorname{supp} h \subset B$. Thus as $\mathcal{F}^N(B)$-estimators of the intensity, we have

$$\hat{\nu}_B(h) = N(h), \tag{9.24}$$

where h vanishes outside B and satisfies $\lambda(h) = 1$. The simplest and most logical choice is $h = 1_B/\lambda(B)$, which engenders the "standard" estimator

$$\hat{\nu}_B = \frac{N(B)}{\lambda(B)}, \tag{9.24a}$$

whose strong consistency, provided that P is ergodic, is an immediate consequence of Corollary 9.3. The variance of $\hat{\nu}_B(h)$ is

$$\mathrm{Var}_P(\hat{\nu}_B(h)) = \mathrm{Var}_P(N(h)) = \rho(h \otimes h),$$

but is of course unknown, which leads to the second (if not *the*) most important estimation problem for stationary point processes. □

Example 9.11. (Second moments). Appealing to (9.19a) we estimate not the second moment measure but instead the reduced second moment measure μ_*^2, which by (9.20) satisfies $\mu_*^2(f) = E^*[N(f)]$ for each $f \in C_+$, and hence the techniques described above apply. If h is such that $\operatorname{supp} f + \operatorname{supp} h \subset B$, then (9.21) with $H(N) = N(f)$ yields the $\mathcal{F}^N(B)$-estimator

$$\hat{\mu}_*^2(f) = \int N(dx) h(x) \int_B N(dy) f(y - x) \tag{9.25}$$

But (and equally well for moment measures of order greater than two) one can generalize somewhat. For g a function on E^2 with compact support and satisfying

$$\int g(x, x + y)\, dx = 1 \tag{9.26}$$

for all y, the estimator

$$\hat{\mu}_*^2(f) = \int \int N(dx) N(dy) f(y - x) g(x, y) \tag{9.27}$$

is unbiased within the family of probabilities with respect to which N admits a moment of order 2. Indeed, with

$$\widetilde{H}(\mu) = \int \mu(dy) f(y) g(x, x+y),$$

the fundamental relation (9.3) implies that

$$E[\hat{\mu}_*^2(f)] = E[\int N(dx)\widetilde{H}(N \circ \theta_x, x)]$$

$$= E^*[\int \widetilde{H}(N,x)\, dx]$$

$$= E^*[\int N(dy) f(y) \int g(x, x+y)\, dx] = E^*[N(f)]$$

For extension to moment measures of higher order, see Exercise 9.10. □

We turn now to asymptotic behavior of the estimators $\hat{P}^*(H)$, where the bounded, convex sets K, over which a single realization of N is observed, satisfy $\delta(K) \to \infty$. Consistency requires only ergodicity and existence of the moments being estimated, followed by appeal to Theorem 9.2. Asymptotic normality, by contrast, is rather more difficult and has been proved only by resort to the classical device of establishing that all cumulants of orders higher than 2 converge to zero. While the hypotheses are not especially restrictive, the amount of calculation involved in the proof *is*, so many details are omitted.

The assumption in (9.21) and (9.22) that the weight function h have compact support is not tenable if one is to have the "infinitely much" data necessary for consistency. Neither, then, can the hypothesis $\lambda(h) = 1$ be retained, and although intermediate options are available, we take the drastic step of dropping h altogether, leading to estimators

$$\hat{P}^*(H) = \lambda(K)^{-1} \int_K N(dx) H(N\tau_x^{-1}), \tag{9.28}$$

where in anticipation of Theorem 9.12, K is a bounded, convex subset of E. In (9.28) the empirical nature of our estimators is manifest: $\hat{P}^*(H)$ is an average of H-values evaluated at translations $N\tau_x^{-1}$ of N that—because only points x of N contribute to (9.28)—have a point at the origin.

Here, then, is the principal consistency result.

Theorem 9.12. If P is ergodic and if $H(N) \in L^1(P^*)$, then with $\hat{P}^*(H)$ given by (9.28),

$$\lim_{\delta(K)\to\infty} \hat{P}^*(H) = E^*[H(N)] \tag{9.29}$$

almost surely and in $L^1(P)$.

Proof: We apply Theorem 9.2 to the random measure M defined from N and H via (9.7); since $H(N) \in L^1(P^*)$,

$$\begin{aligned}\lim \frac{M(K)}{\lambda(K)} &= E[M([0,1]^d)] \\ &= E[\int_{[0,1]^d} N(dx)H(N \circ \theta_x)] \\ &= E^*[\int_{[0,1]^d} H(N)\,dx] = E^*[H(N)]\end{aligned}$$

by yet another application of the ubiquitous relationship (9.4). □

To continue Example 9.10, we observe that if P is ergodic, then $\lim N(K)/\lambda(K) = \nu$ almost surely and in $L^1(P)$. Combined with (9.29), this implies that

$$\lim_{\delta(K)\to\infty} N(K)^{-1} \int_K N(dx)H(N\tau_x^{-1}) = \nu^{-1}E^*[H(N)] = E_0[H(N)] \tag{9.30}$$

almost surely and in $L^1(P)$. Except in trivial cases the estimators in (9.30) are biased, but they are strongly consistent notwithstanding. Taking H to be the indicator function of a subset of $\mathbf{M}_p$ produces consistent estimators of the Palm distribution P_0 (uniformity in H is resolved in part below), but these are of only indirect utility for estimating the distribution P itself.

For estimation of the reduced second moment measure, the choice $h \equiv 1$ in (9.25) and appeal to (9.29) give that for $f \in C_+$, almost surely

$$\lim_{\delta(K)\to\infty} \lambda(K)^{-1} \int_K N(dx) \int N(dy)f(y-x) = \mu_*^2(f), \tag{9.31}$$

provided, of course, that N admit a second moment with respect to P. Consistent estimation of the reduced covariance measure is then not difficult (Exercise 9.11).

Using straightforward reasoning (see Proposition 4.1), we infer from (9.31) that almost surely the estimators

$$\hat{\mu}_*^2 = \left\{\lambda(K)^{-1} \int_K N(dx) \int N(dy)f(y-x) : f \in C_+\right\},$$

viewed as random measures on E, converge vaguely to μ_*^2, but as in Section 4.1, uniform versions are of greater interest. Here is one such result, pertaining for simplicity only to reduced moment measures; Karr (1985c) and Krickeberg (1982) contain alternatives. Let k be fixed and for B a bounded subset of E^{k-1} and $K \in \mathcal{K}$ define

$$\hat{\mu}_*^k(B) = \lambda(K)^{-1} \int_K N(dx) N^{k-1} \circ \theta_x(B)$$

$$= \lambda(K)^{-1} \int_K N(dx) N^{k-1}(B + (x, \ldots, x)); \qquad (9.32)$$

the "sample size" K is once again suppressed. The $\hat{\mu}_*^k$ are estimators of the reduced moment measure μ_*^k. For each m and each set $C_0 \in \mathcal{K}(E^{k-1})$ with $\delta(C_0) > 0$, let $\mathcal{K}_m(C_0)$ denote the class of convex polyhedra (intersections of finitely many half-spaces) contained in C_0 with at most m faces. The following theorem provides uniform convergence of the $\hat{\mu}_*^k$ over each class $\mathcal{K}_m(C_0)$.

Theorem 9.13. Suppose that P is ergodic and that the moment measure μ^k exists. Then with C_0 and m fixed

$$\lim_{\delta(K)\to\infty} \sup_{C \in \mathcal{K}_m(C_0)} |\hat{\mu}_*^k(C) - \mu_*^k(C)| = 0 \qquad (9.33)$$

almost surely with respect to P.

Proof: (Sketch). The key is embodied in the proof of Theorem 4.5: a topology is introduced in which $\mathcal{K}_m(C_0)$ is compact, then Theorem 9.12 is employed to establish convergence on finite subsets of $\mathcal{K}_m(C_0)$, and finally, this convergence is extended to uniform convergence on $\mathcal{K}_m(C_0)$ by compactness/approximation arguments. The appropriate topology is generated by the *Hausdorff metric*:

$$r(C_1, C_2) = \inf\{\varepsilon > 0 : C_1 \subset C_2^\varepsilon, C_2 \subset C_1^\varepsilon\},$$

where C^ε is the set of points whose Euclidean distance to C is ε or less. For compactness of $\mathcal{K}_m(C_0)$ and other classes of sets, to which (9.33) applies verbatim (see Sendov, 1969). □

Similar reasoning—once more the proof of Theorem 4.5 is the prototype—leads to a function-indexed analog of Theorem 9.13.

Theorem 9.14. Under the hypotheses of Theorem 9.13, for each compact subclass $\mathcal{H}$ of $C_+(E^{k-1})$, all of whose elements are supported in a common compact subset K_0 of E^{k-1}, almost surely

$$\lim_{\delta(K)\to\infty} \sup_{f \in \mathcal{H}} |\hat{\mu}_*^k(f) - \mu_*^k(f)| = 0 \quad \square \qquad (9.34)$$

As remarked previously, in terms of establishing a central limit theorem corresponding to Theorem 9.12, stationarity has thus far been exploited only in the rather brute-force manner of showing that cumulants of orders higher than 2 converge to zero. The key assumption, finiteness of reduced cumulant measures, entails near independence of remotely located parts of the point process, and hence makes asymptotic normality possible. To our knowledge no explicit mixing conditions have been formulated or investigated for spatial point processes.

Theorem 9.15. Assume that P is ergodic, that with respect to P, N admits moments of all orders, and that for each $k \geq 2$ the reduced cumulant measure γ_*^k has finite total variation. Then for fixed $k \geq 1, f \in C_+(E^{k-1})$, with $\hat{\mu}_*^k(f)$ given by (9.32),

$$\lambda(K)^{1/2}[\hat{\mu}_*^k(f) - \mu_*^k(f)] \rightarrow N(0,\sigma^2) \tag{9.35}$$

in distribution as $\delta(K) \rightarrow \infty$, where the (not easily computed) variance σ^2 depends only on the moment measures $\mu^1, \dots, \mu^{2k}$.

Proof: (Sketch). For a random variable Y admitting moment-generating function as $m(t) = E[e^{tY}]$, the coefficients c_ℓ in the expansion

$$\log m(t) = \sum_{\ell=0}^{\infty} \frac{c_\ell t^\ell}{\ell!}$$

are the *cumulants* of Y. They are also realized by application of (9.18) (with $f_j \equiv 1$) to the random measure $M = Y\varepsilon_0$ on $E^0 = \{0\}$; consequently,

$$c_\ell = \sum (-1)^{|\mathcal{J}-1|}(|\mathcal{J}| - 1)! \prod E[Y^{|J_q|}],$$

where notation and domain of the summation are as in Definition 9.6. The distribution of Y is normal if and only if $c_2 \neq 0$ but $c_\ell = 0$ for every $\ell > 2$; moreover, convergence of cumulants implies convergence in distribution of random variables (Exercise 9.12). Given this, the proof is simple conceptually. One shows that as $\delta(K) \rightarrow \infty$, the ℓth-order cumulant c_ℓ of $\int_K N(dx) N^{k-1} \circ \theta_x(f)$ satisfies

$$c_\ell \sim \lambda(K) \tag{9.36}$$

independently of ℓ, from which it follows that the ℓth-order cumulant of $\lambda(K)^{1/2}[\hat{\mu}_*^k(f) - \mu_*^k(f)]$ behaves asymptotically as $\lambda(K)^{1-\ell/2}$, implying not only the convergence (9.35) but also that the limit is normal.

The difficult (albeit mostly tedious) part (Jolivet, 1981) is calculation of the cumulants c_ℓ. With $g_K(x_1, \dots, x_k) = 1(x_k \in K) f(x_1 - x_k, \dots, x_{k-1} - x_k)$, we have $\int_K N(dx)N^{k-1} \circ \theta_x(f) = N^k(g_K)$ and therefore for each ℓ,

$$E\left[\left(\int_K N(dx)N^{k-1} \circ \theta_x(f)\right)^\ell\right] = E[(N^k(g_K))^\ell] = \mu^{k\ell}[\otimes g_K]$$

By stationarity (and lengthy calculations), $\mu^{k\ell}[\otimes g_K] \sim \lambda(K)$; this is certainly plausible even if it is difficult to verify. The method used by Jolivet (1981) decomposes cumulant measures into factorial moment measures [not introduced here (see Krickeberg, 1982)] and then applies computational techniques of Leonov and Shiryayev (1959). □

Calculation of the variance σ^2 in (9.35) seems to be difficult. That, as asserted in Theorem 9.15, it depends only on the moment measures $\mu^1, \dots, \mu^{2k}$ does not appear to help, nor do the calculations in Jolivet (1981).

In addition to Poisson processes, certain Neyman-Scott cluster processes (Section 7.5) are known to fulfill the hypotheses of the theorem.

We conclude with one "process" analog of Theorem 9.15; for others, see Karr (1985c). Consider a family $\{f_y : y \in [0,1]^d\}$, each function f_y defined on E^{k-1} and having compact support. Define the error processes $\alpha_K(y) = \lambda(K)^{1/2}[\hat{\mu}_*^k(f_y) - \mu_*^k(f_y)]$, which have the following asymptotic normality property.

Theorem 9.16. Suppose that the hypotheses of Theorem 9.15 are satisfied and that, in addition, for each $z \in E^{k-1}$ the mapping $y \to f_y(z)$ is Lipschitz continuous, nondecreasing in each component of y and constant over the set $\{y : y_i = 0 \text{ for some } i\}$. Then as $\delta(K) \to \infty$ the error processes $\{\alpha_K(y) : y \in [0,1]^d\}$ converge in distribution to a Gaussian process. □

9.2 LIKELIHOOD-BASED INFERENCE

In this section we restrict attention to stationary point processes on **R** observable over intervals $[0,t]$, with asymptotics arising as $t \to \infty$. The principal results, Theorems 9.24 and 9.26, pertain to maximum likelihood estimation for finite-dimensional statistical models; while the formulation and hypotheses are reminiscent of the classical case of finite-dimensional families of distribution functions on **R** and i.i.d. data, the techniques, based in part on stochastic intensities and martingale methods, are nonclassical. The model is not a multiplicative intensity model and results such as Theorem 5.13 would not apply even if it were, since we are concerned with single-realization inference; thus different martingale limit theorems must be invoked. As a preliminary step we describe the structure of the stochastic intensity of a stationary point process on **R**; the conjecture that it is a strictly stationary process is confirmed in Theorem 9.18. Thereafter we take up the main material of the section.

To begin, we present a general singularity theorem for ergodic, stationary point processes.

Theorem 9.17. Let N be a point process on $E = \mathbf{R}^d$ and let P_1, P_2 be probability measures with respect to each of which N is stationary and ergodic. Then the following assertions are equivalent:

a) $P_1 \neq P_2$ on $\mathcal{F}^N$;
b) $P_1 \perp P_2$ on $\mathcal{F}^N$;
c) $P_1 \perp P_2$ on the invariant σ-algebra $\mathcal{I}$.

Proof: That c) implies b), which in turn implies a), is obvious, so it suffices to show that a) implies c). If a) holds, then because the Palm measure P^* determines the probability P, there is $\Gamma \in \mathcal{F}^N$ such that $P_1^*\{\Gamma\} \neq P_2^*\{\Gamma\}$. By taking $H = 1_\Gamma$ in (9.10) and applying Theorem 9.2, we conclude that

$$\lim_{r\to\infty} \lambda(B_r)^{-1} \int_{B_r} N(dx)1(N\circ\theta_x \in \Gamma) = P_i^*\{\Gamma\}$$

almost surely with respect to each P_i, where B_r is the ball of radius r centered at the origin. The limit is an invariant random variable (Exercise 9.20) and therefore c) holds. □

Hence perfect discrimination is possible in principle within the model of ergodic probabilities. Optimality considerations manifested in the Neyman-Pearson lemma dictate that simple hypotheses $H_0: P = P_0$; $H_1: P = P_1\,(\neq P_0)$ be tested using likelihood ratios. Unfortunately, few results of generality seem known except for stationary point processes on $\mathbf{R}_+$ that admit stochastic intensities, for which Theorems 2.31 and 5.2 provide likelihood functions and ratios. Issues such as computation of power remain unresolved, in part because stationarity is not a hypothesis conducive to distributional calculations. Nevertheless, the structure of the stochastic intensity can be described quite precisely.

Let N be a stationary point process on $\mathbf{R}$ and for each t let

$$\overline{\mathcal{F}}_t^N = \mathcal{F}^N((-\infty,t]) \tag{9.37a}$$

be the entire history, including the "infinitely remote" past, of N at time t; for $t > 0$, let $N_t = N([0,t])$ and put

$$\mathcal{F}_t^N = \mathcal{F}^N([0,t]) \tag{9.37b}$$

We then have the following characterization.

Theorem 9.18. Suppose that (N_t) admits $(P,\overline{\mathcal{F}}^N)$-stochastic intensity $(\bar{\lambda}_t)_{t\in\mathbf{R}_+}$. Then almost surely for each t and s,

$$\bar{\lambda}_t \circ \theta_s = \bar{\lambda}_{t+s} \tag{9.38}$$

In particular, $(\bar{\lambda}_t)$ is stationary in law:

$$(\bar{\lambda}_{s+t})_{t\geq 0} \stackrel{d}{=} (\bar{\lambda}_t)_{t\geq 0} \tag{9.39}$$

for each $s > 0$. Note the analogy of (9.38) to (1.69) and that (9.39) is conceptually identical to the stationarity in law for point processes introduced in Section 1.8.

Proof of Theorem 9.18: For fixed $s > 0$, consider the point process $\widetilde{N}_t = N_{s+t} - N_s$ together with the history $\widetilde{\mathcal{F}}_t = \overline{\mathcal{F}}^N_{s+t}$. On the one hand, $\widetilde{N}$ has $\widetilde{\mathcal{F}}$-stochastic intensity $\tilde{\lambda}_t = \bar{\lambda}_{s+t}$; since the stochastic intensity is unique (Definition 2.11), if it also has stochastic intensity $(\bar{\lambda}_t \circ \theta_s)$, then (9.38) is established. For $0 < u < t$ and $\Gamma \in \widetilde{\mathcal{F}}_u$,

$$E[\widetilde{N}_t - \widetilde{N}_u; \Gamma] = E[(N_t - N_u)\circ\theta_s; \Gamma]$$

$$= E[N_t - N_u; \theta_s^{-1}(\Gamma)]$$

(since $P\theta_s^{-1} = P$)

$$= E\left[\int_0^t \bar{\lambda}_v \, dv; \theta_s^{-1}(\Gamma)\right]$$

[because $\theta_s^{-1}(\Gamma) \in \overline{\mathcal{F}}^N_u$]

$$= E\left[\int_0^t \bar{\lambda}_v \circ \theta_s \, dv; \Gamma\right],$$

which completes the proof. □

Under the hypothesis of Theorem 9.18 there exists a $(P,\mathcal{F}^N)$-stochastic intensity (λ_t) satisfying $\lambda_t = E[\bar{\lambda}_t|\mathcal{F}^N_t]$ for each t, but (λ_t) need satisfy neither (9.38) nor (9.39) (Exercise 9.21). This complicates analysis of log-likelihood functions (see Theorem 2.31)

$$L_t(P) = \int_0^t (1 - \lambda_s)\, ds + \int_0^t (\log \lambda_s)\, dN_s$$

corresponding to the observations $\mathcal{F}^N_t$, which would be used, for example, to construct likelihood ratio tests. The "pseudo"-log-likelihood function

$$\bar{L}_t(P) = \int_0^t (1 - \bar{\lambda}_s)\, ds + \int_0^t (\log \bar{\lambda}_s)\, dN_s$$

is neither computable from the observations $\mathcal{F}^N_t$ nor the likelihood function associated with $\overline{\mathcal{F}}^N_t$ since in particular, by Theorem 9.17, one can have $P \perp \bar{P}$ on $\overline{\mathcal{F}}^N_t$, where with respect to $\bar{P}$, N is Poisson with intensity 1, so that no likelihood function exists. However, $\bar{L}_t(P)$ is the more amenable to analysis because of stationarity of $(\bar{\lambda}_t)$ and hence is a useful tool for deducing properties of $L_t(P)$ under assumptions (as below) that the difference between the two converges to zero as $t \to \infty$.

We move on to maximum likelihood estimation for statistical models of stationary point processes admitting a finite-dimensional parametrization. Let N be a simple point process on $\mathbf{R}$ defined over a measurable space $(\Omega, \mathscr{G})$ supporting shift operators (θ_t) such that $N\tau_t^{-1} = N \circ \theta_t$ identically in t and ω, and let $\bar{\mathscr{F}}_t^N$, $\mathscr{F}_t^N$ be given by (9.37). The parameter space is a compact subset Θ of $\mathbf{R}^d$, $d \geq 1$, and we suppose given for each $\alpha \in \Theta$ a probability P_α and an $\bar{\mathscr{F}}^N$-predictable process $(\bar{\lambda}_t(\alpha))$ such that

1) $\bar{\lambda}_{s+t}(\alpha) = \bar{\lambda}_t(\alpha) \circ \theta_s$ identically in s, t, ω [see (9.38)];
2) $P_\alpha \theta_t^{-1} = P_\alpha$ for each $t \in \mathbf{R}$ (so that N is P_α-stationary);
3) N has $(P_\alpha, \bar{\mathscr{F}}^N)$-stochastic intensity $(\bar{\lambda}_t(\alpha))$;
4) P_α is ergodic.

We shall deal only with models satisfying these assumptions.

Our goals are maximum likelihood estimation of α given the observations $\mathscr{F}_t^N$ and analysis of asymptotic properties of maximum likelihood estimators. For simplicity we assume that Θ is one-dimensional; only standard techniques are needed to effect extension to the multidimensional case. We assume also that for each α, N has finite P_α-intensity. Associated with the observations $\mathscr{F}_t^N$ is the log-likelihood function (with a constant term deleted)

$$L_t(\alpha) = -\int_0^t \lambda_s(\alpha)\, ds + \int_0^t (\log \lambda_s(\alpha))\, dN_s, \tag{9.40}$$

where $\lambda_t(\alpha)$ is the $(P_\alpha, \mathscr{F}^N)$-stochastic intensity of N.

For analysis of maximum likelihood estimators $\hat{\alpha}$ several assumptions are necessary; they are conveniently segregated into regularity conditions concerning dependence of the $\bar{\lambda}(\alpha)$ on α and approximation hypotheses to the effect that for large values of t, $\lambda_t(\alpha) \cong \bar{\lambda}_t(\alpha)$ for each α (and similarly for derivatives). The latter permit replacement of the log-likelihood functions $L_t(\alpha)$ by the more tractable functions

$$\bar{L}_t(\alpha) = -\int_0^t \bar{\lambda}_s(\alpha)\, ds + \int_0^t (\log \bar{\lambda}_s(\alpha))\, dN_s \tag{9.41}$$

Assumptions 9.19. (Regularity conditions). a) For each t and ω the function $\alpha \to \bar{\lambda}_t(\alpha, \omega)$ is continuous; $P_\alpha\{\bar{\lambda}_0(\alpha) > 0\} = 1$ for each α.

b) For $\alpha_1, \alpha_2 \in \Theta$, $\bar{\lambda}_0(\alpha_1) = \bar{\lambda}_0(\alpha_2)$ almost surely with respect to the dominating Poisson law if and only if $\alpha_1 = \alpha_2$.

c) For each t and ω the function $\alpha \to \bar{\lambda}_t(\alpha, \omega)$ is three times continuously differentiable on (the interior of) Θ; for each t and α the random variables $\bar{\lambda}'_t(\alpha)$ and $\bar{\lambda}''_t(\alpha)$ have finite second moments. (Here and below all derivatives are with respect to α.)

d) For each α there exists a neighborhood $U(\alpha)$ of α in Θ such that

$$E_\alpha[\sup\{\bar{\lambda}_0(\beta) : \beta \in U(\alpha)\}] < \infty$$

and

$$E_\alpha[\sup\{|\log \bar{\lambda}_0(\beta)| : \beta \in U(\alpha)\}] < \infty$$

e) For each $\alpha \in \Theta$

$$0 < I(\alpha) = E_\alpha\left[\frac{\bar{\lambda}''_0(\alpha)^2}{\bar{\lambda}_0(\alpha)}\right] < \infty \tag{9.42}$$

f) For each $\alpha \in \Theta$ there exists a neighborhood $U(\alpha)$ such that

$$E_\alpha[\sup\{|\bar{\lambda}'''(\beta)| : \beta \in U(\alpha)\}] < \infty$$

and

$$E_\alpha[\bar{\lambda}_0(\alpha)^2 \sup\{|(\log \bar{\lambda}_0)'''(\beta)| : \beta \in U(\alpha)\}^2] < \infty$$

Each of these implies a corresponding property of the $\mathcal{F}^N$-stochastic intensities $\lambda_t(\alpha)$. Many of them, which are in some cases substitutes rather than complements, are apparent analogs of conditions imposed in analysis of maximum likelihood estimators given i.i.d. data and a finite-dimensional parameter set. They apply to the model as a whole, whereas the approximation hypotheses apply individually to each α.

Assumptions 9.20. (Approximation hypotheses). a) There is a neighborhood $U(\alpha)$ such that

$$\lim_{t\to\infty} \sup\{|\lambda_t(\beta) - \bar{\lambda}_t(\beta)| : \beta \in U(\alpha)\} = 0$$

in P_α-probability, and such that for some $\delta > 0$,

$$\sup_{t\in\mathbf{R}_+} E_\alpha[\sup\{|\log \lambda_t(\beta)|^{2+\delta} : \beta \in U(\alpha)\}] < \infty$$

b) As $t \to \infty$, $\lambda_t(\alpha) - \bar{\lambda}_t(\alpha) \to 0$, $\lambda'_t(\alpha) - \bar{\lambda}'_t(\alpha) \to 0$ and $\lambda''_t(\alpha) - \bar{\lambda}''_t(\alpha) \to 0$ in P_α-probability; there exists $\delta > 0$ such that

$$\sup_{t\in\mathbf{R}_+} E_\alpha\left[\left|\frac{\bar{\lambda}_t(\alpha)}{\lambda_t(\alpha)}\right|^{2+\delta}\right] < \infty,$$

$$\sup_{t\in\mathbf{R}_+} E_\alpha\left[\left|\frac{\bar{\lambda}'_t(\alpha)}{\bar{\lambda}_t(\alpha)}\right|^{2+\delta}\right] < \infty,$$

$$\sup_{t\in\mathbf{R}_+} E_\alpha[|\bar{\lambda}''_t(\alpha)|^{2+\delta}] < \infty$$

c) As $t \to \infty$,

$$t^{-1/2} E_\alpha\left[\int_0^t |\lambda'_u(\alpha) - \bar{\lambda}'_u(\alpha)| \, du \right] \to 0$$

and

$$t^{-1/2} E_\alpha\left[\int_0^t |\lambda_u(\alpha) - \bar{\lambda}_u(\alpha)| \frac{\lambda'_u(\alpha)}{\lambda_u(\alpha)} du \right] \to 0$$

d) There is a neighborhood $U(\alpha)$ such that

$$\sup\{|\lambda'''_t(\beta) - \bar{\lambda}'''_t(\beta)| : \beta \in U(\alpha)\} \to 0$$

in P_α-probability as $t \to \infty$; there exists $\delta > 0$ satisfying

$$\sup_{t \in \mathbf{R}_+} E_\alpha[\sup\{|\lambda'''_t(\beta)|^{2+\delta} : \beta \in U(\alpha)\}] < \infty$$

and

$$\sup_{t \in \mathbf{R}_+} E_\alpha[\sup\{|(\log \lambda_t)'''(\beta)|^{2+\delta} : \beta \in U(\alpha)\}] < \infty$$

Evidently, it is no small matter to verify these assumptions; Ogata (1978) does so for Poisson processes, stationary renewal processes (Exercise 1.31 and Example 9.27), and the self-exciting point process of Hawkes and Oakes (1974) (Exercise 9.22).

In view of compactness of Θ, given Assumption 9.19a) there exists for each t a maximum likelihood estimator $\hat{\alpha}$ satisfying $L_t(\hat{\alpha}) \geq L_t(\alpha)$ for all $\alpha \in \Theta$, but not necessarily fulfilling the *likelihood equation*

$$L'_t(\alpha) = 0 \tag{9.43}$$

With sufficiently many more of the assumptions above, consistency of maximum likelihood estimators can be established without recourse to (9.43), but asymptotic normality cannot (compare Theorem 5.19). We begin with three preliminary results, included because they are the principal means through which the special context of stationary point processes enters.

Lemma 9.21. Under Assumptions 9.19a), c), d), and e), for each t and α, and with $\bar{L}$ given by (9.41),

$$E_\alpha[\bar{L}'_t(\alpha)] = 0 \tag{9.44}$$

and

$$E_\alpha[(L'_t(\alpha))^2] = -E_\alpha[\bar{L}''_t(\alpha)] = tE_\alpha\left[\frac{\bar{\lambda}'_0(\alpha)^2}{\bar{\lambda}_0(\alpha)}\right] \tag{9.45}$$

Proof: Given these assumptions, derivatives, integrals, and expectations can be interchanged; consequently,

$$E_\alpha[\bar{L}_t'(\alpha)] = E_\alpha\left[-\int_0^t \bar{\lambda}_u'(\alpha)\,du + \int_0^t \frac{\bar{\lambda}_u'(\alpha)}{\bar{\lambda}_u(\alpha)}\,dN_u\right]$$

$$= E_\alpha\left[\int_0^t \frac{\bar{\lambda}_u'(\alpha)}{\bar{\lambda}_u(\alpha)}[dN_u - \bar{\lambda}_u(\alpha)\,du]\right] = 0,$$

giving (9.44). Derivation of (9.45) is similar. □

Lemma 9.22. Let α be fixed and suppose that X is $\mathcal{F}^N$-predictable and P_α-stationary with $E_\alpha[X_t^2] < \infty$. Then almost surely

$$t^{-1}\int_0^t X_u\,du \rightarrow E_\alpha[X_0] \tag{9.46}$$

and

$$t^{-1}\int_0^t \frac{X_u}{\bar{\lambda}_u(\alpha)}\,dN_u \rightarrow E_\alpha[X_0] \tag{9.47}$$

Proof: Because (9.46) is a standard Césaro limit theorem for stationary processes (see, e.g., Karlin and Taylor, 1975), in order to establish (9.47) it suffices to show that

$$t^{-1}\int_0^t \frac{X_u}{\bar{\lambda}_u(\alpha)}[dN_u - \bar{\lambda}_u(\alpha)\,du] \rightarrow 0$$

almost surely. By Theorem 2.9, the stochastic integral process is a mean zero martingale under P_α and consequently (9.47) ensues by a strong law of large numbers for martingales (see, e.g., Hall and Heyde, 1980, Theorem 2.19). □

If the definition of stochastic intensity is interpreted as "$dN_t = \bar{\lambda}_t(\alpha)\,dt$ in the small," then (9.46) and (9.47) are a statement that "$dN_t = \bar{\lambda}_t(\alpha)\,dt$ in the large."

Lemma 9.23. Given α_0, α, define

$$\Lambda_1(\alpha_0,\alpha) = \int_0^1 [\bar{\lambda}_u(\alpha) - \bar{\lambda}_u(\alpha_0)]\,du + \int_0^1 \log\frac{\bar{\lambda}_u(\alpha_0)}{\bar{\lambda}_u(\alpha)}\,dN_u$$

Then

$$E_{\alpha_0}[\Lambda_1(\alpha_0,\alpha)] \geq 0 \tag{9.48}$$

with equality if and only if $\bar{\lambda}_0(\alpha) = \bar{\lambda}_0(\alpha_0)$ almost surely with respect to P_{α_0}.

Proof: Because $\bar{\lambda}(\alpha)$ is the $(P_\alpha, \mathcal{F}^N)$-stochastic intensity of N,

$$E_{\alpha_0}[\Lambda_1(\alpha_0,\alpha)] = \int_0^1 E_{\alpha_0}[\bar{\lambda}_u(\alpha) - \bar{\lambda}_u(\alpha_0)]\,du$$

$$+ \int_0^1 E_{\alpha_0}\left[\log\left(\bar{\lambda}_u(\alpha_0)/\bar{\lambda}_u(\alpha)\right)\bar{\lambda}_u(\alpha_0)\right] du$$

$$= E_{\alpha_0}\left[\bar{\lambda}_0(\alpha_0)\left\{\frac{\bar{\lambda}_0(\alpha)}{\bar{\lambda}_0(\alpha_0)} - 1 + \log\frac{\bar{\lambda}(\alpha_0)}{\bar{\lambda}_0(\alpha)}\right\}\right],$$

from which the lemma follows by application of the inequality $\log x - 1 + 1/x \geq 0$, which holds for $x > 0$ with equality if and only if $x = 1$. □

The expression in (9.48) is analogous to the Kullback-Liebler information in classical statistics; together with Assumption 9.19b), Lemma 9.23 furnishes an identifiability criterion: $\alpha = \alpha_0$ if and only if $E_{\alpha_0}[\Lambda_1(\alpha_0, \alpha)] = 0$.

We come now to the main limit theorems of the section.

Theorem 9.24. If Assumptions 9.19a), b), and d) hold and if the "true" parameter value α satisfies Assumption 9.20a), then the maximum likelihood estimators $\hat{\alpha}$ are weakly consistent: under P_α,

$$\hat{\alpha} \to \alpha \tag{9.49}$$

in probability.

Proof: (Sketch). Given a neighborhood U_0 of α, by Lemma 9.23 and Assumption 9.19b) there is $\varepsilon > 0$ such that

$$E_\alpha[\Lambda_1(\alpha,\gamma)] \geq 3\varepsilon \tag{9.50}$$

for all $\gamma \in \Theta \backslash U_0$. By Assumptions 9.19a) and d), for each such γ there is a neighborhood $U(\gamma)$ satisfying

$$E_\alpha\left[\inf_{U(\gamma)} \bar{\lambda}_0(\beta) - \bar{\lambda}_0(\alpha) + \bar{\lambda}_0(\alpha)\log\frac{\bar{\lambda}_0(\alpha)}{\sup_{U(\gamma)}\bar{\lambda}_0(\beta)}\right] \geq E_\alpha[\Lambda_1(\alpha,\gamma)] - \varepsilon$$

and by compactness $\Theta\backslash U_0$ can be covered by finitely many such neighborhoods, say $U(\gamma_1), \ldots, U(\gamma_k)$. Predictability of stochastic intensities shows in conjunction with Lemma 9.22 that given $\delta > 0$ there is a (random) time T_0 such that for $t > T_0$ and $i = 1, \ldots, k$,

$$t^{-1}\bar{L}_t(\alpha) - \sup_{U(\gamma_i)} t^{-1}\bar{L}_t(\beta) \geq E_\alpha[\Lambda_1(\alpha,\gamma)] - 2\varepsilon \geq \varepsilon,$$

the last inequality in consequence of (9.50). Thus there exists a random time T_1 depending on U_0 and ε such that

$$\sup\{\bar{L}_t(\beta):\beta\in U_0\}\geq\sup\{\bar{L}_t(\beta):\beta\in\Theta\backslash U_0\}+\varepsilon t$$

for all $t>T_1$. Finally, Assumption 9.20a) implies that the same holds also for $L_t(\cdot)$ with probability converging to 1 as $t\to\infty$, from which we infer that $P_\alpha\{\hat{\alpha}\in U_0\}\to 1$. Since U_0 was arbitrary, (9.49) holds. □

Asymptotic normality of the estimation error $t^{1/2}[\hat{\alpha}-\alpha]$ is demonstrated by the classical argument of showing that the first derivative of the log-likelihood function (L_t in this case) is asymptotically normally distributed, then—as in Theorem 5.19—expanding the log-likelihood function in a Taylor series about a solution to the likelihood equation (9.43).

Theorem 9.25. Suppose that Assumptions 9.19a), c), and e) and 9.20c) are satisfied. Then under P_α,

$$t^{-1/2}L_t'(\alpha)\xrightarrow{d}N(0,I(\alpha)),\tag{9.51}$$

where $I(\alpha)$ is the Fisher information given by (9.42).

Proof: (Sketch). By Assumptions 9.19,

$$\bar{L}_t'(\alpha)=\int_0^t\frac{\bar{\lambda}_u'(\alpha)}{\bar{\lambda}_u(\alpha)}[dN_u-\bar{\lambda}_u(\alpha)\,du],$$

so by Theorem 2.9, $(\bar{L}_t'(\alpha))$ is a mean zero P_α-martingale. Let $dM_u(\alpha)=dN_u-\bar{\lambda}_u(\alpha)\,du$ be the $(P_\alpha,\bar{\mathcal{F}}^N)$-innovation martingale; then by stationarity of $\bar{\lambda}(\alpha)$ and (9.46),

$$t^{-1}\int_0^t\left[\frac{\bar{\lambda}_u'(\alpha)}{\bar{\lambda}_u(\alpha)}\right]^2 d\langle M(\alpha)\rangle_u=t^{-1}\int_0^t\frac{\bar{\lambda}_u(\alpha_0)^2}{\bar{\lambda}_u(\alpha_0)}\,du$$

$$\to E_\alpha\left[\frac{\bar{\lambda}_0(\alpha)^2}{\bar{\lambda}_0(\alpha)}\right]=I(\alpha)\tag{9.52}$$

In view of ergodicity of N and (9.52), it follows by appeal to Theorem B.21 that $t^{-1/2}\bar{L}_t(\alpha)\xrightarrow{d}N(0,I(\alpha))$ and Assumption 9.20c) suffices to convert this to (9.51). □

Only standard reasoning is required in order to derive the final result of the section.

Theorem 9.26. Suppose that the maximum likelihood estimators $\hat{\alpha}$ satisfy the likelihood equation (9.43) and that Assumptions 9.19a), c), e), and f), and 9.20b), c), and d) are fulfilled. Then under P_α,

$$t^{1/2}[\hat{\alpha} - \alpha] \xrightarrow{d} N(0, I(\alpha)^{-1}) \tag{9.53}$$

and

$$2[L_t(\hat{\alpha}) - \alpha] \xrightarrow{d} \chi_1^2, \tag{9.54}$$

where χ_1^2 is the chi-squared distribution with 1 degree of freedom.

Proof: (Sketch). As in the proof of Theorem 5.19, within error converging to zero in probability, there holds the Taylor expansion

$$0 = t^{-1/2} L_t'(\hat{\alpha}) \cong t^{-1/2} L_t'(\alpha) + t^{-1} L_t''(\alpha) t^{1/2} [\hat{\alpha} - \alpha]$$

The first term converges in distribution to $N(0, I(\alpha))$ by Theorem 9.25, while by arguments analogous to those employed to prove Theorem 9.24, $t^{-1} L_t''(\alpha) \to I(\alpha)$ in probability; consequently, (9.53) holds. Similarly,

$$2[L_t(\hat{\alpha}) - L_t(\alpha)] \cong 2L_t'(\alpha)[\hat{\alpha} - \alpha] + t^{-1} L_t''(\alpha)\{t^{1/2}[\hat{\alpha} - \alpha]\}^2,$$

so (9.54) follows from (9.53), Theorem 9.25, and Theorem 9.24. □

We illustrate with stationary renewal processes.

Example 9.27. (Stationary renewal processes). Let $\{F(\alpha, \cdot) : \alpha \in \Theta\}$ be a finite-dimensional family of distribution functions on $\mathbf{R}_+$; let

$$H(\alpha, t) = \frac{\int_0^t [1 - F(\alpha, x)]\, dx}{\int_0^\infty [1 - F(\alpha, x)]\, dx},$$

which for a stationary renewal process is the distribution of the forward recurrence time at t for every t (Exercise 9.3). The log-likelihood function is given by (8.41). Assumptions 9.19a) and b) are satisfied if $\alpha \to F(\alpha)$ is continuous and if $\{F(\alpha)\}$ is identifiable, while c) – f) must be imposed by fiat in the form of existence of derivatives of $\alpha \to F(\alpha, x)$ and moments of the $F(\alpha)$. Integrability in Assumptions 9.20 is ensured by stipulating the existence of moments of $F(\alpha)$ or $H(\alpha)$, but the convergence statements there are satisfied automatically. □

9.3 SPECTRAL ANALYSIS

Although we switch to frequency-domain analysis of stationary point processes, as opposed to the time/space-domain analysis of Sections 9.1 and 9.2, estimation techniques from Section 9.1 retain central roles. We begin with a spectral representation theorem for stationary point processes; related structural aspects for stationary point processes on $\mathbf{R}$, in particular a

Wold decomposition fundamental to linear prediction, are presented in Section 9.4. In the remainder of this section we deduce from the spectral representation corresponding representations for reduced second moment and covariance measures, and thereafter examine estimation of frequency-domain descriptors. Even though we deal only with second moment structure, we retain the assumption of stationarity because it permits application of limit theorems derived in Section 9.1. Most other treatments assume less but prove correspondingly less. Solely for simplicity we confine attention to stationary point processes on **R**.

Let $\mathfrak{D}$ denote the space of infinitely differentiable functions on **R** whose support is compact ("test functions") and let $\mathcal{S}$ be the set of "rapidly decreasing" functions on **R**. [For topology and functional analysis background and details, consult Rudin (1973, Chap. 7) or other standard references.] Given a function ψ in $\mathcal{S}$, denote its Fourier transform by $\tilde{\psi}$, that is,

$$\tilde{\psi}(v) = \int e^{ivx}\psi(x)\,dx \tag{9.55}$$

Here is the key result of the section.

Theorem 9.28. Let N be a stationary point process on **R** admitting a moment of order 2. Then there exists a unique complex-valued random measure Z on **R** with orthogonal increments such that for each $\psi \in \mathfrak{D}$,

$$N(\psi) = \int \tilde{\psi}(v) Z(dv) \tag{9.56}$$

We term (9.56) the *spectral representation* of N; there and elsewhere below integrals with respect to Z are in the L^2-sense. From (9.56),

$$N(B) = \int Z(dv) \int_B e^{ivx}\,dx$$

for each bounded set B and in particular for $u < t$,

$$N_t - N_u = \int (e^{itv} - e^{iuv})(iv)^{-1} Z(dv)$$

Proof of Theorem 9.28: (Sketch). Consider the centered process defined by $X(\psi) = N(\psi) - \nu\lambda(\psi)$, where ν is the intensity of N, viewed as a *random distribution* on $\mathfrak{D}$ (i.e., a random continuous linear functional). Let G be the spectral measure (Itô, 1953, Theorem 3.1) induced by X:

$$E[X(\psi)X(\varphi)] = G(\tilde{\psi}\tilde{\varphi}) \tag{9.57}$$

for ψ, $\varphi \in \mathfrak{D}$. The Fourier transform $\psi \to \tilde{\psi}$ is a homeomorphism of the space $\mathcal{S}$, endowed with the "Schwartz" topology, onto itself; moreover, $\tilde{\mathfrak{D}} = \{\tilde{\psi} : \psi \in \mathfrak{D}\}$ is dense in $\mathcal{S}$, and $\mathcal{S}$ is dense in $L^2(G)$; consequently, $\tilde{\mathfrak{D}}$ is dense in $L^2(G)$ in *its* usual topology. Given $\psi \in \mathfrak{D}$ define $Z_0(\tilde{\psi}) = X(\psi)$; as a mapping of $\tilde{\mathfrak{D}} \subset L^2(G)$ into $L^2(P)$, Z_0 is a linear (obvious from the definition) isometry, the latter since

$$\|Z_0(\psi)\|^2_{L^2(P)} = E[|X(\psi)|^2] = \|\tilde{\psi}\|^2_{L^2(G)}$$

Because $\tilde{\mathfrak{D}}$ is dense in $L^2(G)$, Z_0 can be extended to become a linear isometry from $L^2(G)$ into $L^2(P)$, and the formula $Z_0(B) = Z_0(1_B)$ defines a random measure with orthogonal increments satisfying

$$N(B) - \nu\lambda(B) = \int Z_0(dv) \int_B e^{ivx}\, dx;$$

conversion to a random measure Z fulfilling (9.56) is immediate. □

The spectral measure is the fundamental object of interest in frequency-domain analysis of stationary point processes even though (but in other senses precisely because) it depends on the distribution of the process only through the reduced second moment measure (as shown in Proposition 9.30).

Definition 9.29. Let N be a stationary point process with spectral representation (9.56). The measure

$$F(dv) = E[|Z(dv)|^2] \tag{9.58}$$

is the *spectral measure* of N. The measure G satisfying (9.57) is the *covariance spectral measure* of N. If it exists, the density f of G is termed the (covariance) *spectral density function* of N.

It is important to work with the covariance spectral density function because it exists under reasonably broad assumptions, whereas the spectral measure F is typically not absolutely continuous. Moreover, the spectral density function can be estimated within our setting.

The crucial relationship linking second moment measures and spectral measures is the following. For $\psi \in \mathfrak{D}$ let

$$\hat{\psi}(x) = (2\pi)^{-1} \int e^{-ivx} \psi(v)\, dv \tag{9.55a}$$

denote its inverse Fourier transform.

Proposition 9.30. Let N be a stationary point process on $\mathbf{R}$ admitting moment of order 2. Then the reduced second moment measure μ_*^2 and spectral measure F satisfy the Parseval relations

$$\mu_*^2(\psi) = F(\tilde{\psi}) \tag{9.59}$$

and

$$F(\psi) = \mu_*^2(\hat{\psi}) \tag{9.60}$$

for $\psi \in \mathcal{D}$.

Proof: It is enough to establish (9.59). Given φ_1, $\varphi_2 \in \mathcal{D}$, define $\psi \in \mathcal{D}$ by

$$\psi(z) = \int \varphi_1(z+x)\varphi_2(x)\,dx; \tag{9.61}$$

then by calculations similar to those in the proof of Lemma 9.8,

$$\mu_*^2(\psi) = E[N(\varphi_1)N(\varphi_2)]$$

$$= E[N(\varphi_1)\,\overline{N(\varphi_2)}]$$

(the bar denotes complex conjugation)

$$= E[Z(\tilde{\varphi}_1)\,\overline{Z(\tilde{\varphi}_2)}]$$

[Z is the spectral measure of (9.56)]

$$= E[(\int Z(dv) \int e^{ivx}\varphi_1(x)\,dx)(\int \overline{Z(du)} \int e^{-iuy}\varphi_2(y)\,dy)]$$

$$= \iint E[Z(dv)\,\overline{Z(du)}]\tilde{\varphi}_1(v)\tilde{\varphi}_2(-u)$$

$$= \int F(dv)\tilde{\varphi}_1(v)\tilde{\varphi}_2(-v) = F(\tilde{\psi})$$

Hence (9.59) holds for $\psi \in \mathcal{D}$ of the form (9.61); since the set of such ψ is dense in $\mathcal{D}$, proof of (9.59) is complete. □

For reduced covariance measure

$$\rho_*(dx) = \mu_*^2(dx) - \nu^2\,dx \tag{9.62}$$

the following version of Proposition 9.30 holds.

Proposition 9.31. For N a stationary point process on **R** admitting moment of order 2, the covariance spectral measure G and reduced covariance measure ρ_* satisfy the Parseval relations

$$\rho_*(\psi) = G(\widetilde{\psi}), \tag{9.63}$$

$$G(\psi) = \rho_*(\hat{\psi}) \tag{9.64}$$

for all $\psi \in \mathfrak{D}$. If the spectral density function f exists, then

$$f(v) = (2\pi)^{-1}E^*\left[\int (N - \nu\lambda)(dy)e^{-ivy}\right] \tag{9.65}$$

for each v, where ν is the intensity of N. □

If the Palm distribution P_0 admits a decomposition

$$E_0[N(dx)] = \varepsilon_0(dx) + \nu_0(x)\,dx, \tag{9.66}$$

with ν_0, the *Palm intensity function* (often referred to in the literature as the conditional intensity function) a positive function on $\mathbf{R}$, then (9.65) can be written as

$$f(v) = (2\pi)^{-1}\left[\nu + \int (\nu\nu_0(x) - \nu^2)e^{-ivx}\,dx\right] \tag{9.67}$$

Turning now to inference, our statistical model is the following. With the sample space and the point process N fixed, let $\mathcal{P}$ consist of all probability measures P with respect to which

1) N is stationary and ergodic;
2) N admits a moment of order 2, and hence reduced second moment measure μ_*^2 and spectral measure F. (When it exists the spectral density function is denoted by f.)

For estimation of the spectral measure we employ techniques developed in Section 9.1, specifically those described in Example 9.11. From (9.60), given a compact subset K of $\mathbf{R}$ we obtain the substitution estimator

$$\hat{F}(\psi) = \hat{\mu}_*^2(\hat{\psi}) = \lambda(K)^{-1}\int_K N(dx)\int N(dy)\,\hat{\psi}(y-x), \tag{9.68}$$

where $\hat{\psi}$ is the inverse Fourier transform of ψ and $\hat{\mu}_*^2$ is given by (9.27) with $g(x,y) = 1_K(x)/\lambda(K)$. These estimators are unbiased by Proposition 9.9 and are strongly uniformly consistent.

Theorem 9.32. Let $\mathcal{K}$ be a compact subset of $\mathfrak{D}$, all elements of which have support in a single compact subset of $\mathbf{R}$. Then with $\hat{F}$ given by (9.68), almost surely with respect to P,

$$\lim_{\delta(K)\to\infty}\ \sup_{\psi\in\mathcal{K}} |\hat{F}(\psi) - F(\psi)| = 0 \tag{9.69}$$

Proof: That with ψ fixed

$$\lim_{\delta(K)\to\infty} \hat{F}(\psi) = \lim_{\delta(K)\to\infty} \hat{\mu}_*^2(\hat{\psi}) = \mu_*^2(\hat{\psi}) = F(\psi) \tag{9.70}$$

follows directly from Theorem 9.12 (not from Theorem 9.14—$\hat{\psi}$ does not belong to $\mathfrak{D}$ in general). Let $\mathbf{M}_p(\nu)$ be the set of elements μ of $\mathbf{M}_p$ such that $\lim \mu(K)/\lambda(K) = \nu$; then according to (9.13), $P\{N \in \mathbf{M}_p(\nu)\} = 1$. The inverse Fourier transform is continuous, so given $\varepsilon > 0$ there exists a compact subset K_1 of $\mathbf{R}$ satisfying

$$\sup\left\{\int_{K_1^c} |\hat{\psi}|\, d\mu_*^2 : \psi \in \mathcal{K}\right\} < \varepsilon \tag{9.71}$$

and $\varphi_1, \ldots, \varphi_L$ in $\mathcal{K}$ such that for each $\psi \in \mathcal{K}$ there exists $\ell(\psi) \in \{1, \ldots, L\}$ for which

$$\| \hat{\psi} - \hat{\varphi}_{\ell(\psi)} \|_{K1} < \varepsilon \tag{9.72}$$

(the norm is the uniform norm on K_1). Then in the decomposition

$$\begin{aligned}\sup_{\mathcal{K}} |\hat{F}(\psi) - F(\psi)| &= \sup_{\mathcal{K}} |\hat{\mu}_*^2(\hat{\psi}) - \mu_*^2(\hat{\psi})| \\ &\le \sup_{\mathcal{K}} |\hat{\mu}_*^2(\hat{\psi}) - \hat{\mu}_*^2(\hat{\varphi}_{\ell(\psi)})| + \max_{\ell} |\hat{\mu}_*^2(\hat{\varphi}_\ell) - \mu_*^2(\hat{\varphi}_\ell)| \\ &\quad + \sup_{\mathcal{K}} |\mu_*^2(\hat{\varphi}_{\ell(\psi)}) - \mu_*^2(\hat{\psi})|\end{aligned} \tag{9.73}$$

the second term converges to zero almost surely by (9.70). In regard to the third, we have that for $\psi \in \mathcal{K}$

$$|\mu_*^2(\hat{\varphi}_{\ell(\psi)}) - \mu_*^2(\hat{\psi})|$$

$$\leqslant |\mu_*^2(\hat{\varphi}_{\ell(\psi)}; K_1) - \mu_*^2(\hat{\psi}; K_1)| + |\mu_*^2(\hat{\varphi}_{\ell(\psi)}; K_1^c) - \mu_*^2(\hat{\psi}; K_1^c)|$$

$$\leqslant \mu_*^2(K_1) \| \hat{\varphi}_{\ell(\psi)} - \hat{\psi} \|_\infty + 2\varepsilon \sup\{\|\hat{\psi}\|_\infty : \psi \in \mathcal{K}\}$$

[by (9.71)]

$$\le \mu_*^2(K_1) \|\varphi_{\ell(\psi)} - \psi\|_1 + 2\varepsilon \sup\{\|\psi\|_1 : \psi \in \mathcal{K}\}$$

(Rudin, 1973, Theorem 7.5)

$$\leqslant \mu_*^2(K_1) \sup\{\|\varphi_{\ell(\psi)} - \psi\|_\infty : \psi \in \mathcal{K}\} + 2\varepsilon \sup\{\|\psi\|_\infty : \psi \in \mathcal{K}\}$$

$$\le \text{constant} \times \varepsilon;$$

thus the third term in (9.73) converges to zero. By Theorem 9.14, $\hat{\mu}_*^2$ satisfies (9.71)—but with ε there replaced by 2ε—for all sufficiently large $\delta(K)$ almost surely, so the preceding argument applies to the first term in (9.73) as well. □

Concerning estimation of the spectral density function, for the substitution estimators

$$\hat{f}(v) = (2\pi\lambda(K))^{-1} \int_K N(dx) \int (N - v\lambda)(dy)e^{-iv(y-x)} \tag{9.74}$$

engendered by (9.67) we have not only strong uniform consistency but also asymptotic normality, the latter given following the analogous result corresponding to Theorem 9.32.

Theorem 9.33. Assume that P is ergodic and that with respect to P, N admits continuous spectral density function satisfying (9.65). Then for each compact subset K_0 of $\mathbf{R}$, almost surely with respect to P,

$$\lim_{\delta(K)\to\infty} \sup_{v\in K_0} |\hat{f}(v) - f(v)| = 0 \quad \square \tag{9.75}$$

For the proof, see Karr (1985c).

It is instructive to compare the estimator $\hat{f}$ with periodogram estimators commonly used in statistical analysis of stationary point processes (see, e.g., Brillinger, 1975b, or Cox and Lewis, 1966). Extended to our setting the periodogram is given by

$$\hat{f}^*(v) = |(2\pi\lambda(K))^{-1/2} \int_K N(dx)e^{-ivx}|^2; \tag{9.76}$$

even for $K = [0,T]$ with $T\to\infty$ the periodogram is not a consistent estimator of the spectral density function, let alone asymptotically normal (the spectral density function is related to its variance), although in fairness it must be mentioned that it does have useful properties in terms of interpretation of data. In addition, the periodogram (9.76), unlike the estimator $\hat{f}$ of (9.74), is computable from the observations $\mathcal{F}^N(K)$, and its computation is linear rather than quadratic in $N(K)$.

The central limit theorem associated with Theorem 9.32 is presented without proof.

Theorem 9.34. If the hypotheses of Theorem 9.15 are satisfied, then with $\hat{F}$ given by (9.68) there exists a centered Gaussian process $\{X(\psi): \psi \in \mathcal{D}\}$ such that for each ψ

$$\lambda(K)^{1/2}[\hat{F}(\psi) - F(\psi)] \xrightarrow{d} X(\psi) \tag{9.77}$$

as $\delta(K) \to \infty$; the covariance function of X depends only on the moment measures μ^1, μ^2, μ^3, μ^4. □

Finally, we give the central limit theorem corresponding to Theorem 9.33. In this context we fix the compact set K_0 appearing there and introduce the error processes $\alpha(v) = \lambda(K)^{1/2}[\hat{f}(v) - f(v)]$.

Theorem 9.35. Assume that the hypotheses of Theorem 9.15 are fulfilled and that the spectral density function exists, satisfies (9.65), and is continuous; suppose further that

$$E^*\left[\int |y|(N - v\lambda)(dy)\right] < \infty \tag{9.78}$$

Let K_0 be a fixed compact subset of E and define error processes $\{\alpha(v) : v \in K_0\}$ as above. Then there exists a continuous Gaussian process $\{Y(v) : v \in K_0\}$ such that as $\delta(K) \to \infty$, $\alpha(v) \xrightarrow{d} Y(v)$ for each v. The covariance function of Y depends only on the moment measures μ^1, μ^2, μ^3, μ^4. □

9.4 LINEAR STATE ESTIMATION

Because of their tractable second moment structure stationary point processes are natural candidates for linear prediction, that is, calculation of minimum mean squared error *linear* state estimators of the future of the process. Suppose that N is a stationary point process observed over a bounded set A; then we seek for each function f (assumed without loss of generality to vanish on A) a function $\hat{f}$ on A and a constant a_f such that the linear state estimator $\hat{N}(f) = N(\hat{f}) + a_f$ satisfies

$$E[(N(f) - \hat{N}(f))^2] \leq E[(N(f) - N(g) - a)^2] \tag{9.79}$$

for every function g on A and every $a \in \mathbf{R}$. Thus $\hat{N}(f)$ is the optimal linear predictor of $N(f)$ given the observations $\mathcal{F}^N(A)$. This section concentrates on linear prediction because there exist no results of generality pertaining to nonlinear state estimation for stationary point processes, even on $\mathbf{R}$. Stationarity, although exceedingly useful for some purposes, is less powerful (at least hitherto unexploited) for nonlinear state estimation because distributional characteristics other than moments are involved. We examine two approaches to linear state estimation, a time/space-domain approach applicable on general spaces and a frequency domain/spectral analysis approach for stationary point processes on $\mathbf{R}$. For the latter, observations are $\mathcal{F}_t^N$ from (9.37a), so that there are "infinitely many" data; hence it is natural to consider "moving" problems of prediction of the point process over an interval $(t, t+u]$ extending a fixed length into the future. Martingale methods do not apply because a proper history cannot be specified; however, spectral analysis provides very usable techniques. As part of their development we derive a Wold decomposition of a stationary point process on $\mathbf{R}$.

We first analyze linear prediction in the time/space domain. Let us introduce the centered process

$$M(f) = N(f) - \nu\lambda(f), \tag{9.80}$$

where ν is the P-intensity of N. Given data $\mathcal{F}^N(A)$ representing observation of N over a bounded set A with $\lambda(A) > 0$ and a function f vanishing on A, we wish to calculate that function $\hat{f}$ on A for which

$$E[(M(f) - M(\hat{f})^2] \le E[(M(f) - M(g))^2] \tag{9.81}$$

for every function g on A. Thus $M(\hat{f})$ is the optimal (i.e., minimum mean squared error) linear predictor of the unobserved random variable $M(f) = N(f) - \nu\lambda(f)$ given the observations $\mathcal{F}^N(A)$, and $M(\hat{f}) + \nu\lambda(f) = N(\hat{f}) - \nu\lambda(f - \hat{f})$ the optimal linear state estimator of $N(f)$. With P known, solution of (9.81) is straightforward; see Karr (1985c) for analysis of combined inference and linear state estimation for stationary point processes.

Proposition 9.36. Assume that N admits a moment of order 2 with respect to P and let ρ_* be the reduced covariance measure. Given f vanishing on A, the function $\hat{f}$ satisfying (9.81) is the unique function such that

$$\int_A \hat{f}(y)\rho_*(dy - x) = \int f(y)\rho_*(dy - x) \tag{9.82}$$

for almost all $x \in A$ with respect to Lebesgue measure.

Proof: By standard Hilbert space theory, $M(\hat{f})$ is the projection of $M(f)$ onto the linear space spanned by $\{M(g) : g \in L^2(A)\}$ and hence the unique solution to the normal equations

$$E[(M(\hat{f}) - M(f))M(g)] = 0, \tag{9.83}$$

where g varies over $L^2(A)$. With ρ the *un*reduced covariance measure

$$\begin{aligned} E[M(f)M(g)] &= \mathrm{Cov}(N(f), N(g)) \\ &= \int\int g(x)f(y)\rho(dx, dy) \\ &= \int_A g(x) \int f(y)\rho_*(dy - x)\, dx; \end{aligned}$$

consequently, (9.83) implies that for every g,

$$0=\int_A g(x)\left[\int_A \hat{f}(y)\rho_*(dy-x)-\int f(y)\rho_*(dy-x)\right]dx,$$

confirming (9.82). □

Still in the context of time/space-domain inference, an alternative approach undertakes to derive directly a transition kernel K from A to A^c such that for each function f on A^c

$$M(\hat{f})=\int_A N(dx)K(x,f); \tag{9.84}$$

this is done in the following result.

Proposition 9.37. If there exists a transition kernel K: $A \to A^c$ such that

$$\rho_*(dz-x)=\int_A \rho_*(dy-x)K(y,dz) \tag{9.85}$$

as measures on A^c for almost every $x \in A$, then (9.84) holds. □

The proof is left as Exercise 9.28.

We next describe the spectral approach to linear prediction for stationary point processes on **R**. It should not be dismissed as merely an application of the Parseval identity (9.59) to replace the reduced second moment measure μ_*^2 by the spectral measure F because in particular it permits rather explicit solution of "moving" linear prediction problems, in other words, derivation of optimal linear predictors of a stationary point process N over an interval $(t, t+u]$ that moves as the observations $\bar{\mathcal{F}}_t^N=\mathcal{F}^N((-\infty,t])$ unfold. As preliminary steps we state a Wold decomposition for a stationary point process on **R** and a backward moving average representation for the purely nondeterministic part.

Let N be a stationary point process on **R** and let $\bar{\mathcal{F}}_{-\infty}^N=\cap\{\bar{\mathcal{F}}_t^N: t \in \mathbf{R}\}$ be the infinitely remote past of N.

Definition 9.38. The stationary point process N is *deterministic* if $\bar{\mathcal{F}}_{-\infty}^N=\mathcal{F}^N$ and *purely nondeterministic* if $\bar{\mathcal{F}}_{-\infty}^N=\{\emptyset,\Omega\}$ (within null sets).

These are the extreme cases; in the one all knowledge about realizations of N is part of the infinitely remote past and in the other no nondeterministic information belongs to the infinitely remote past. The lattice process $\Sigma_{-\infty}^{\infty}\varepsilon_{X+n}$ of Example 1.60 is deterministic because the offset

X can be determined from $\mathcal{F}_t^N$ regardless of $t \in \mathbf{R}$; by contrast a Poisson process (more generally, any ergodic stationary point process) is purely nondeterministic (Exercise 9.29).

Stationarity suffices to eliminate intermediate cases, although of course both components may be present simultaneously. Thus we obtain the *Wold decomposition* for a stationary point process.

Theorem 9.39. Let N be a stationary point process on $\mathbf{R}$ admitting moment of order 2. Then there exists a unique decomposition $N = N^1 + N^2$ in which the point processes N^1, N^2 are stationary, N^1 is deterministic, N^2 is purely nondeterministic, and the centered processes $N^i - E[N^i]$, $i = 1,2$, are orthogonal. The deterministic part N^1 is zero if and only if there exists a spectral density function f [see (9.74)] satisfying

$$\int_{\mathbf{R}} \frac{\log f(v)}{1+v^2}\, dv > -\infty \quad \square \tag{9.86}$$

Neither Theorem 9.39 nor Theorem 9.41, the latter providing a backward moving average representation for purely nondeterministic point processes, is proved in detail. However, we do sketch a smoothing approach that permits direct application of results for mean square continuous stationary stochastic processes (see Daley, 1971, and Vere-Jones, 1974, for details).

Lemma 9.40. Given a stationary point process N on $\mathbf{R}$ with intensity v and moment of order 2, the smoothed process

$$X_t = \int_{-\infty}^{t} e^{-(t-s)}\, dN_s \tag{9.87}$$

is strictly stationary with $E[X_t] = v$ for all t. The spectral measure F_X of X is related to the spectral measure F of N via

$$F_X(dv) = (1+v^2)^{-1} F(dv) \quad \square \tag{9.88}$$

The centered process M of (9.80) and $\tilde{X}_t = X_t - v$ generate the same history and are hence together deterministic or purely nondeterministic, $\tilde{X}$ in the time series sense (see, e.g., Brillinger, 1981). The well-known Wold decomposition $\tilde{X} = \tilde{X}^1 + \tilde{X}^2$, with $\tilde{X}^1$ deterministic and $\tilde{X}^2$ purely nondeterministic, translates at once to a decomposition for M and thence to that for N.

Prediction for a deterministic stationary point process given observations $\mathcal{F}_t^N$ is completely trivial, so by Theorem 9.39 we can hereafter suppose that N is purely nondeterministic. The spectral representation of Theorem 9.28 is complemented in this case by a backward moving average representation.

Theorem 9.41. A stationary point process N on $\mathbf{R}$ is purely nondeterministic if and only if there exists a representation

$$M(\psi) = \int H * \psi(s)\widetilde{Z}(ds), \tag{9.89}$$

where H is a distribution on $\mathcal{S}$ with support contained in $(-\infty,0]$, $\widetilde{Z}$ is a random measure with orthogonal increments and $E[|\widetilde{Z}(dt)|^2] = v\,dt$ [the asterisk denotes convolution (see Rudin, 1973)]. □

For linear prediction the key points are that M and $\widetilde{Z}$ engender the same history because the distribution H in (9.89) is supported in $(-\infty,0]$, and hence the representation is invertible, and that linear prediction of $\widetilde{Z}$, by virtue of its having orthogonal, stationary increments, is essentially trivial. Thus one obtains the following solution to the "moving" linear prediction problem.

Theorem 9.42. Given the setting of Theorem 9.41 there exists a distribution K supported in $(-\infty,0]$ such that

$$\widetilde{Z}(\varphi) = \int K * \varphi\, dM, \tag{9.90}$$

for each $\varphi \in \mathcal{D}$. Hence given $u > 0$ and $\psi \in \mathcal{D}$ vanishing outside $[0,u]$, if for each t we define $\psi_t(x) = \psi(t-x)$, then the minimum mean squared error linear predictor of

$$M(\psi_t) = \int_t^{t+u} \psi(s-T)\, dN_s,$$

given the observations $\mathcal{F}_t^N$, is

$$\hat{M}(\psi_t) = \int_{-\infty}^{t} K * H(\psi_{y-t})\, dM_y \tag{9.91}$$

Proof: (Sketch). Invertibility of (9.89) implies that (9.90) holds; given this,

$$M(\psi_t) = \int_{-\infty}^{t+u} H * \psi_t\, d\widetilde{Z},$$

which together with orthogonality of the increments of $\widetilde{Z}$ and the property that M and $\widetilde{Z}$ generate the same history gives that the optimal linear predictor is

$$\hat{M}(\psi_t) = \int_{-\infty}^{t} H * \psi_t \, d\widetilde{Z}$$

Substitution of (9.90) into this expression verifies (9.91). □

In case N admits a rational covariance spectral density function more precise results are available (see Jowett and Vere-Jones, 1972). The Fourier transform of the distribution $K * H$ appearing in (9.91) has been termed the *spectral characteristic for prediction* (see, Yaglom, 1962); one can formulate a frequency-domain analog of (9.91) using it. Note that the measure (in ψ_t) defined by (9.91) may be absolutely continuous even though N is a point process; this happens, for example, if N is Poisson.

We conclude with brief analysis of a Poisson cluster process.

Example 9.43. Let N be a Neyman-Scott cluster process (see Section 7.5) in which the cluster center process is Poisson with rate η and cluster center-cluster member distances are independent and identically exponentially distributed with parameter α; we do not restrict the distribution of the number of cluster members except to the extent that it admit finite variance σ^2 and mean μ. Then the smoothed process X of (9.88) has rational spectral density function

$$f_X(v) = \frac{\eta[\alpha^2(\sigma + \mu) + \mu v^2]}{\alpha^2 + v^2} \tag{9.92}$$

(Exercise 9.32) and consequently N has rational spectral density function of the generic form

$$f(v) = \left| \frac{a(b + iv)}{(c + iv)} \right|^2 \tag{9.93}$$

In the notation of Theorem 9.42 and (9.93), it follows that for each t

$$M(\psi_t) = (b - c) \int_0^t e^{-cs} \psi(s)\, ds + \int_{-\infty}^{t} e^{b(t-s)}\, dN_s, \tag{9.94}$$

which also is left to the reader to confirm. □

EXERCISES

9.1. Let N be a stationary point process on $\mathbf{R}$ with intensity $v > 0$.

a) Prove that $P\{N(-\infty, 0) = N(0, \infty) = \infty\} + P\{N \equiv 0\} = 1$.

b) Construct an example in which both probabilities are nonzero.

9.2. Let N be a stationary renewal process on $\mathbf{R}$. Prove that if N has independent increments, then it is a Poisson process.

9.3. Let N be a stationary renewal process on $\mathbf{R}$ with interarrival distribution F and let $H(t) = m^{-1} \int_0^t [1 - F(u)]\, du$, where m is the mean of F. Prove that for every t the forward recurrence time $W_t = \sup\{s : N((t, t+s]) = 0\}$ has distribution F.

9.4. Let N be a stationary point process on $\mathbf{R}$ [with respect to (θ_s)] and let P^* be the Palm measure. (We assume the representation $N = \Sigma_{n=-\infty}^{\infty} \varepsilon_{T_n}$ with $\ldots < T_{-1} < T_0 \leq 0 < T_1 < \cdots$; let $U_i = T_i - T_{i-1}$ be the interarrival times.)

a) Prove that $P^*\{T_0 \neq 0\} = 0$.

b) Prove that for each positive random variable Z,

$$E[Z] = E^*[\int_0^{T_1} Z \circ \theta_s \, ds]$$

(In particular, $E^*[T_1] = 1$.)

c) Prove that under the Palm distribution $P_0 = \nu^{-1} P^*$ the interarrival sequence (U_i) is strictly stationary.

9.5. Verify that the random measures M defined by (9.7) satisfy (9.8).

9.6. Let (K_α) be linearly ordered bounded, convex sets in $\mathbf{R}^d$ with $\delta(K_\alpha) \to \infty$. Prove that $\lambda(\{x : d(x, \partial K_\alpha) \leq 1\})/\lambda(K_\alpha) \to 0$.

9.7. Prove Lemma 9.8.

9.8. Let N be a stationary Poisson process on $\mathbf{R}$. Calculate the second moment measure and reduced covariance measure, and interpret the results in view of the independent increments property of N.

9.9. Show that all moment and cumulant measures of a stationary point process are invariant under appropriate diagonal groups.

9.10. Let N be a stationary point process with moment of order k. Devise analogs of the estimators (9.25) and (9.27) for the reduced kth-order moment measure μ_*^k and verify that they are unbiased. [In the latter case an analog of (9.26) must be imposed.]

9.11. Let N be a stationary point process admitting moment of order 2. Devise unbiased estimators of the reduced covariance measure and establish their asymptotic properties.

9.12. Let X_n, X be random variables for which there exist cumulant sequences $(c_{n\ell})$, (c_ℓ), respectively. Prove that $X_n \xrightarrow{d} X$ if and only if $c_{n\ell} \to c_\ell$ for each ℓ.

9.13. Let a be a stationary Poisson process on $\mathbf{R}^d$ that under P_ν has intensity ν. Compare asymptotic behavior of the estimators $\hat{\nu} = N(K)/\lambda(K)$ of (9.24a) with that of the alternative estimators $\hat{\nu}^* = [\lambda(K)^{-1} \int_K N(dx) N(J + x)]^{1/2}$, where $J = (0,1]^d$. [The latter arise from the role of ν in connection with the reduced second moment measure: $\mu_*^2(J) = \nu^2$; $(\hat{\nu}^*)^2 = \hat{\mu}_*^2(J)$ as given by (9.32).]

9.14. Let N be a stationary point process on $\mathbf{R}$ with respect to an unknown probability P and suppose that observation is initiated at the time of an event of N.

a) Argue that one can take that time to be T_0 and that the law of the observations is not P but rather the Palm distribution P_0.
b) Show that the (P-)intensity of N satisfies $\nu = 1/E_0[U_i]$.
c) Imposing suitable hypotheses, describe the asymptotic behavior of the estimators $\hat{\nu}^* = [n^{-1}\Sigma_{i=1}^n U_i]^{-1}$.

9.15. Let N be a stationary point process on $\mathbf{R}$, observed asynchronously, and suppose that the parameter of interest is $m = E_0[U_i]$, the mean interarrival time under the Palm distribution.
a) Calculate the bias of the $\mathcal{F}_t^N$-estimators $\hat{m} = N_t^{-1}\Sigma_{i=1}^{N_t} U_i$.
b) Describe large-sample properties.

9.16. Suppose that under the probability P_p, $0 < p \leq 1$, N' is the p-thinning of a stationary point process N on $\mathbf{R}^d$, with known intensity ν, whose law does not depend on p. Only N' is observable.
a) Prove that N' is stationary with P_p-intensity $p\nu$.
b) Establish—given appropriate assumptions—asymptotic behavior of the $\mathcal{F}^N(K)$-estimators $\hat{p} = N'(K)/[\nu\lambda(K)]$.

9.17. (Continuation of Exercise 9.16) Continue to assume that the law of N does not depend on p, but suppose now that ν is unknown and that N is observable. Describe joint asymptotic properties of the $\mathcal{F}^N(K) \vee \mathcal{F}^{N'}(K)$-estimators $\hat{\nu} = N(K)/\lambda(K)$ and $\hat{p} = N'(K)/N(K)$.

9.18. Let $\widetilde{N}$ be a stationary Poisson process on $\mathbf{R}^2$ with intensity ν, and for each r let $B_r(x)$ be the open disk with radius r centered at x. Introduce the *interpoint distance process* (with K fixed and it is a point process on $\mathbf{R}_+$)

$$N_K(r) = \frac{1}{2}\int_K \widetilde{N}(dx)\widetilde{N}(B_r(x) \cap K)$$

(the number of interpoint distances, among points in K, not exceeding r). Assume that K is compact and convex. Prove that as $\delta(K) \to \infty$, $\{N_K(r/\lambda(K)^{1/2}: r \geq 0\} \xrightarrow{d} N$, where N is a Poisson process on $\mathbf{R}_+$ with $E[N(r)] = \nu\pi r^2/2$.

9.19. Let N be a stationary point process and introduce the *aggregation* functional $a(C) = \mu^2(C \times C)/\nu(C)$, where μ^2 is the unreduced second moment measure.
a) Prove that $a(C) = \lambda(C)^{-1}E_0[\int \lambda(C \cap (C - x))N(dx)]$. [Interpretation: $a(C)$ is large if there are many other points nearby given that there is one at the origin, and hence measures aggregation (clustering) at scale size C.]
b) Prove that if N is Poisson with intensity ν, then $a(C) = 1 + \nu\lambda(C)$.
c) Construct unbiased estimators of $a(\cdot)$ given that the intensity of N is known.

9.20. In the context and notation of Theorem 9.17, prove that the random variable $Z = \lim_{r\to\infty} \lambda(B_r)^{-1}\int_{B_r} N(dx)1(N \circ \theta_x \in \Gamma)$ is invariant.

9.21. Let N be a stationary point process on $\mathbf{R}$ with $\mathcal{F}^N$-stochastic intensity $(\bar{\lambda}_t)_{t\in\mathbf{R}_+}$, where $\mathcal{F}_t^N = \mathcal{F}^N((-\infty, t])$.

a) Prove that there exists an $\mathcal{F}^N$-stochastic intensity (λ_t) satisfying $\lambda_t = E[\bar{\lambda}_t|\mathcal{F}_t^N]$ almost surely for each t.

b) Prove (by example) that λ need not fulfill either (9.38) or (9.39).

9.22. Let N be a stationary point process on $\mathbf{R}$ that under P_α admits $\mathcal{F}^N$-stochastic intensity

$$\bar{\lambda}_t(\alpha) = \nu + \beta \int_{(-\infty,t)} e^{-\gamma(t-u)}\, dN_u,$$

where $\alpha = (\nu,\beta,\gamma)$ with $\nu > 0$ and $0 < \beta < \gamma$. (Note that the interval of integration *excludes* t.)

a) Prove that N is a Poisson cluster process.

b) Verify that Assumptions 9.19 and 9.20 are fulfilled.

c) Calculate the Fisher information (9.42).

9.23. Deduce (9.67) from (9.66).

9.24. Let N be a stationary Poisson process on $\mathbf{R}^d$. Calculate the spectral measure of N.

9.25. Let N be a stationary renewal process on $\mathbf{R}$ with interarrival distribution F. Calculate the spectral measure of N.

9.26. Let N be a stationary point process on $\mathbf{R}$ stipulated to admit (covariance) spectral density function f, and introduce the periodogram

$$J_t(\nu) = t^{-1}|(2\pi)^{-1/2} \int_0^t e^{i\nu u}\, dN_u|^2$$

a) Prove that under the hypotheses of Theorem 9.15, $J_t(\nu) \xrightarrow{d} f(\nu)^2 X$ for each ν as $t \to \infty$, where X has a χ^2-distribution with 1 degree of freedom.

b) Discuss implications concerning use of $J_t(\nu)$ as estimator of $f(\nu)$.

9.27. Let $N = \Sigma_{n=-\infty}^{\infty} \varepsilon_{(T_n,X_n)}$ be a marked point process on $\mathbf{R} \times \{1,2\}$ (i.e., events are of two types) such that the processes $N_i = \Sigma\, 1(X_n = i)\varepsilon_{T_n}$, $i = 1,2$, are stationary and admit spectral density functions f_1, f_2, respectively.

a) Show that the *(cross-)covariance measure* $\eta(A \times B) = \text{Cov}(N_1(A), N_2(B))$ admits a representation $\eta(dt \times du) = h(t-u)\,dt\,du$ for some function h.

b) Define the cross-spectral density

$$f_{12}(\nu) = (2\pi)^{-1} \int_{-\infty}^{\infty} e^{-i\nu u} h(u)\, du,$$

with h as in a), and the *coherence* between N_1 and N_2 as the function

$$C(\nu) = \frac{|f_{12}(\nu)|^2}{f_1(\nu) f_2(\nu)}$$

Show that $0 \le C(\nu) \le 1$.

c) Discuss estimation of $C(\nu)$ from observations $\mathcal{F}_t^{N_1} \vee \mathcal{F}_t^{N_2}$.

9.28. Prove Proposition 9.37.

9.29. Prove that a stationary Poisson process on **R** is purely nondeterministic.

9.30. Prove Lemma 9.40.

9.31. Let $\widetilde{N} = \Sigma_{n=-\infty}^{\infty} \varepsilon_{T_n}$ be a Poisson process on **R** with intensity α and consider the point process $N = \widetilde{N} + \Sigma_{n=-\infty}^{\infty} \varepsilon_{T_n+h}$, where $h \in \mathbf{R}$. [Interpretation: N is a process of two-point clusters, the two members of each cluster separated by h.]

a) Prove that N is stationary and calculate its intensity, reduced second moment measure, and reduced covariance measure.

b) Calculate the spectral measure of N.

c) Derive a backward moving-average representation for N.

9.32. Verify unproved assertions in Example 9.43.

NOTES

The pervasive influences on the chapter are Neveu (1977) on the theory of stationary point processes and Krickeberg (1982) on inference in a general setting. Concerning theory, Matthes et al. (1978), despite its being difficult to read, contains much on Palm distributions and some on ergodic theory, with emphasis on Poisson cluster processes; along these lines seminal papers are Matthes (1963a), Kerstan and Matthes (1964a), and Franken et al. (1965). Cox and Isham (1980) is a more elementary exposition of selected aspects of the theory.

Other expository treatments of inference—albeit only in one dimension—are Brillinger (1975b, 1978); the latter, an extended and sometimes forced analogy between point processes and time series, yields numerous insights but no grand synthesis.

Of results in the introduction, Theorem 9.12 is due to Nguyen and Zessin (1979a); we have corrected a minor error in their proof. Proposition 9.4, which generalizes the property that a translation invariant measure is a multiple of Lebesgue measure, is proved, for example, in Krickeberg (1974).

Some omissions merit mention. Stationary Poisson cluster processes are understood in many respects; Matthes et al. (1978) is the most complete reference, but omits inference. Orderliness and Khinchin/Koroljuk theorems have been investigated in some generality (see e.g., Daley, 1974; Leadbetter, 1968, 1972a). Applications as diverse as cosmology (Neyman and Scott, 1972), earthquakes (Vere-Jones, 1970), queueing systems (Franken et al., 1981, which treats in addition marked point processes that are stationary only in one coordinate (see also Krickeberg, 1982, where other versions of stationarity and isotropy are considered), stereology (Mecke and Stoyan, 1983), and others have been examined. Analysis of *real* data is the focus of Diggle (1983) and Ripley (1981).

Section 9.1

In Krickeberg (1982) the fundamental role of Palm measures in inference for stationary point processes was first recognized and explored systematically. The estimators (9.21) and (9.28) are formulated there; as described in the text, they are averages of translations of N placing each observed point at the origin.

Our development of moment measures is incomplete: factorial moment measures, which figure in the proof of Theorem 9.15 and through which, for example, one can characterize simple stationary point processes, have been omitted. They are treated in Krickeberg (1974, 1981, 1982), with additional statistical aspects in Hanisch (1983) and Osher (1983).

Theorem 9.12, the basic consistency theorem, is an immediate consequence of Theorem 9.2; the uniform versions, Theorems 9.13 and 9.14, are due to Krickeberg (1982) and Karr (1985c), respectively. Jolivet (1981) is the source of Theorems (.15 and 9.16; proofs, as in Theorem 10.12, are accomplished by time-series techniques, at whose heart are the computational rules of Leonov and Shiryayev (1959).

The "clean" moment theory for stationary point process is counterbalanced by a difficult distribution theory; typical consequences are that linear state estimation is well developed (Section 9.4) while little is known about nonlinear state estimation and that distributions of test statistics often cannot be calculated. Results that are known pertain mainly to interarrival distributions for stationary point processes on **R**; Cramér et al. (1971), Leadbetter (1969b), and Eriksson (1978) are representative.

Section 9.2

This section could equally well be titled "Intensity-Based Inference." Although obviously true, Theorem 9.18 seems not to have appeared before; there are some subtleties involved since $\mathcal{F}^N$-stochastic intensities need not be stationary. Benevensite and Jacod (1975) treats additional aspects. The remainder of the section—Assumptions 9.19 and 9.20, Lemmas 9.21 – 9.23, Theorems 9.24 – 9.26, and Example 9.27—is from Ogata (1978). Although martingale limit theorems are used, the structure of the argument is very much that from the classical case of i.i.d. data; see Ibragimov/Has'minskii (1981), the most complete source, or Wilks (1962), which is more elementary. Intensity-based estimation and model comparison for stationary point process models of precipitation is carried out in Smith and Karr (1985).

The dearth of hypothesis tests stems from lack of distribution theory. Problems that have been addressed include testing whether a stationary point process is Poisson (Brown and Silverman, 1978; Cox and Lewis, 1966; Davies, 1977) and discrimination between Poisson and particular Poisson cluster processes (Kryscio and Saunders, 1983).

Section 9.3

The brevity of this section reveals our preference for time/space-domain over frequency-domain methods. More complete presentations are Brillinger (1972, 1975b, 1978), Daley (1971), and Vere-Jones (1974); the key early works are those of Bartlett (1963, 1964). Theorem 9.28 is a special case of results in Itô (1953). Taken from Vere-Jones (1974), Proposition 9.30 is but one of several inversion formulas. Theorems 9.32, 9.33, and 9.35 are due to Karr (1985c). The estimators (9.74) are, however, of limited practical value because they are not computable from observation over a compact set. Nor is the periodogram (9.76) free from flaws: the spectral density function is related to its second moment rather than the first (Exercise 9.26). Properties of periodograms are described in Brillinger (1975b) and Cox and Lewis (1966); behavior in the Poisson case has been known since Bartlett (1963). Pham (1981) treats estimation of parametrically specified spectral density functions, while Rice (1975) considers estimation of the spectral density function in factored form.

Section 9.4

Jowett and Vere-Jones (1972) is a nice development of linear prediction for stationary point processes on **R** and from it are taken Theorems 9.39, 9.41 and 9.42, the main results of the section. Definition 9.38 introduces properties well known for time series (see, e.g., Brillinger, 1981). Ibragimov and Rozanov (1978) and, from a more elementary perspective, Karlin and Taylor (1975) discuss the analogous problem of linear prediction for stationary processes on **R**.

Exercises

9.2. See Daley (1971b).

9.4. Proofs are given in Neveu (1977). Part b), first proved in some generality by Slivnyak (1962), shows that for inference based on i.i.d. copies of a stationary point process, each copy need be observed only over one interarrival interval.

9.6. See Nguyen and Zessin (1979a); convexity systematically eliminates boundary effects.

9.10. See (9.32) and Theorems 9.13 and 9.14.

9.13. Similar effects arise, for example, depending on whether one views the parameter of a Poisson distribution as the mean or the variance.

9.14. Failure to use the Palm distribution to describe synchronous observations has led to confusion and errors in the past.

9.18. Kryscio and Saunders (1983) contains analogous features of the nearest neighbor indicator process $W_t(x) = \int_{K_1} (N(B_t(x) \cap K) = 1) N(dx)$ not only for Poisson but also for certain Poisson cluster processes. Roberts (1969) contains partial results on the unresolved problem of the distribution of the number of nearest neighbors of a point in a planar Poisson process.

9.19. This functional is proposed in Jolivet (1978); "aggregation" lacks possibly misleading connotations of "clustering."

9.21. See Exercise 2.16.

9.22. The process is described in Hawkes and Oakes (1974).

9.26. See Brillinger (1975b).

9.27. Coherence is used by Brillinger (1974) to study certain queueing systems; it can be interpreted as a measure of linear predictability of one process given observation of the other.

9.31,
9.32. See Jowett and Vere-Jones (1972) for these and other examples.

10

Inference for Stochastic Processes Based on Point Process Samples

This chapter, unlike its predecessors, focuses not on inference for point processes but rather on inference for other stochastic processes, specifically, Markov processes, stationary processes on **R**, and stationary random fields, given observations resulting from a point process sampling mechanism. It is also a sample in itself: of many specific models it examines only the three mentioned above, and only with Poisson process samples. In each, stochastic structure of the sampling process is crucial, and we overlook a substantial literature on deterministic, discrete sampling of stochastic processes. There is as yet nothing resembling a general theory of point process sampling—even martingale methods are difficult to implement effectively—and central problems are only incompletely identified. We use marked point processes to unify the problem formulation, but this produces a coherent statement, not broadly applicable techniques for solution. Nevertheless, problems that arise in point process sampling *are* of practical interest and *do* lead to challenging mathematics, and it is for these reasons that some are presented here.

To impart structure and harmony to the chapter, we concentrate on *Poisson sampling* of stochastic processes. One facet, Poisson sampling of renewal processes, has been treated in Section 8.1; in more general terms the problem, on the line, is as follows. Given a process $X = (X(t))$ on **R** with unknown probability law, but usually of fairly narrowly specified structure, and a Poisson process $N = \Sigma\, \varepsilon_{T_n}$ on **R**, often but not invariably presumed homogeneous, such that X and N are independent, the goals are to effect statistical inference for the law of X and state estimation concerning unobserved past values or not yet observed (and possibly never to be observed) future values, given as observations single realizations of the Poisson samples $X(T_n)$, that is, the values of X at the points of N, together possibly with observations of N itself. Observable aspects are encapsulated in the marked point process

$$\bar{N} = \sum \varepsilon_{(T_n, X(T_n))}, \tag{10.1}$$

in which each point of N is marked by the current value of X. As in Chapter 9, asymptotics arise given observations over $[0,t]$ as $t \to \infty$. For Poisson sampled random fields, N becomes a Poisson process on $\mathbf{R}^d$.

Before proceeding we consider the foundation issue of whether inference concerning the law $\mathcal{L}_X$ is even possible. Is $\mathcal{L}_X$ uniquely determined by the law of the Poisson sample sequence $(X(T_n))$? Under the assumption that N is homogeneous the answer is affirmative given no restriction on X other than that the $X(T_n)$ be random variables. A spatial version appears in Section 10.3.

Proposition 10.1. Let X_1, X_2 be measurable stochastic processes, each continuous in probability, and let N be a Poisson process with rate one independent of X_1 and X_2. If $(X_1(T_n)) \stackrel{d}{=} (X_2(T_n))$, then

$$X_1 \stackrel{d}{=} X_2 \tag{10.2}$$

Proof: It is important to remember that (10.2) means only equality of finite-dimensional distributions. With n and $x_1, \ldots, x_n$ fixed, define functions

$$F_i(t_1, \ldots, t_n) = P\{X_i(t_j) \le x_j, j = 1, \ldots, n\}, \qquad i = 1, 2;$$

given positive integers $k_1, \ldots, k_n$, let $K_j = k_1 + \cdots + k_j$ and let $S_j = T_{K_j}$. Then the integrals

$$\int_0^\infty \cdots \int_0^\infty F_i\left(u_1, u_1 + u_2, \ldots, \sum_{\ell=1}^n u_\ell\right) \left[\prod_{j=1}^n \frac{u_j^{k_j - 1} e^{-u_j}}{(k_j - 1)!}\right] du_1 \cdots du_n$$

$$= E[F_i(S_1, \ldots, S_n)] = P\{X_i(S_j) \le x_j, j = 1, \ldots, n\}$$

are equal by assumption. Multiplication by $\Pi_{j=1}^n (1 - \theta_j)^{k_j}$, where $|\theta_j| \le 1$, followed by summation over $k_1, \ldots, k_n$, confirms that

$$\int_0^\infty \cdots \int_0^\infty F_1\left(u_1, \ldots, \sum_{\ell=1}^n u_\ell\right) \exp\left(-\sum_{j=1}^n \theta_j u_j\right) du_1 \ldots du_n$$

$$= \int_0^\infty \cdots \int_0^\infty F_2\left(u_1, \ldots, \sum_{\ell=1}^n u_\ell\right) \exp\left(-\sum_{j=1}^n \theta_j u_j\right) du_1 \cdots du_n \tag{10.3}$$

By the assumption of continuity in probability, F_1 and F_2 are continuous; therefore, (10.3) implies that $F_1 = F_2$. Since n and the x_j were arbitrary, (10.2) holds. □

Of course, Proposition 10.1 resolves only the most basic inference question; whether, for example, $\mathcal{L}_X$ can be determined from observation of a single realization of the marked point process $\bar{N}$ remains unsettled. For the Markov process model of Section 10.1, the Poisson sample sequence $(X(T_n))$ is a Markov chain, to which, given assumptions of irreducibility and recurrence, well-known methods apply. Indeed, there is a converse, Theorem 10.3, to Proposition 10.1: if the Poisson sample sequence is Markov, then so must be X itself. For binary (0-1 valued) Markov processes consistent estimation is possible even if the rate of N is unknown and the observations are only the thinned point process $N' = \Sigma X(T_n)\varepsilon_{T_n}$; this situation has been discussed, but from the viewpoint of partially observed Poisson processes, in Sections 6.3 and 7.5. In Section 10.1 we also consider combined inference and state estimation, for which, at least in the finite-state-space case, a satisfactory solution is available.

In Section 10.2 we examine Poisson sampling of stationary stochastic processes on **R**. Although the law of X may not be recoverable from single realizations, consistent estimation of the mean and the covariance function is possible from asynchronous and synchronous observations of the Poisson sample sequence. This is a context in which Poisson sampling is superior to sampling at regularly spaced instants $t_n = n\Delta$, for because of "aliasing," estimation of even the covariance function given regular samples is impossible in general.

Finally, in Section 10.3 we consider Poisson sampling of stationary random fields, that is, stochastic processes with a multidimensional parameter set. Results mimic those in Section 10.2 but are more fragmentary.

10.1 MARKOV PROCESSES

In this section we examine Poisson sampling for Markov processes with finite state space. To this end, let X be a standard Markov process with finite state space E, let P_i be the law of X under the initial condition $X_0 = i$, let

$$P_t(i,j) = P_i\{X_t = j\} \tag{10.4}$$

be the transition function, let $A = P_0'$ be the generator matrix, and for each $\lambda > 0$ let

$$U^\lambda(i,j) = \int_0^\infty e^{-\lambda t} P_t(i,j)\, dt \tag{10.5}$$

be the λ-potential matrix. Then $U^\lambda = (\lambda I - A)^{-1}$, so that (P_t) is determined by U^λ for each $\lambda > 0$. Suppose that $N = \Sigma\, \varepsilon_{T_n}$ is a Poisson process with rate λ and that X and N are independent.

The following result is immediate.

Proposition 10.2. Under the assumptions above, the Poisson sample process $(X(T_n))$ is a Markov chain with transition matrix $Q = \lambda U^\lambda$. □

Less apparent is the converse: a Markov Poisson sample process can arise only from a Markov process.

Theorem 10.3. Let X be a measurable process with finite state space E, and assume that X is continuous in probability; let N be a Poisson process independent of X with rate 1. If the Poisson sample process $(X(T_n))$ is a (homogeneous) Markov chain, then there exists a Markov process $\tilde{X}$ such that $X \stackrel{d}{=} \tilde{X}$.

Proof: Let $R_t(i,j) = P\{X_t = j | X_0 = i\}$; these are continuous functions of t and the transition matrix Q of $(X(T_n))$ satisfies

$$Q^n(i,j) = \int_0^\infty R_u(i,j) e^{-u} u^{n-1} [(n-1)!]^{-1}\, du$$

for each n. From the Chapman-Kolmogorov equation for $(X(T_n))$ we infer that for each n and m

$$\int_0^\infty \int_0^\infty [R_{u+v}(i,j) - \sum_{k \in E} R_u(i,k) R_v(k,j)] u^{n-1} v^{m-1} e^{-(u+v)}\, du\, dv = 0 \tag{10.6}$$

and consequently (by transform or Hilbert space arguments)

$$R_{u+v}(i,j) = \sum_{k \in E} R_u(i,k) R_v(k,j) \tag{10.7}$$

for all u, v, i, j. There exists a Markov process $\tilde{X}$ with transition function (R_t), and it is immediate that $(\tilde{X}(T_n)) \stackrel{d}{=} (X(T_n))$, from which the theorem follows by appeal to Proposition 10.1. □

We now consider statistical inference, for which we impose the complication that the rate λ of the Poisson process N is unknown, but permit N to be observed. Our goal is nonparametric estimation of λ and the generator A from observations $\mathcal{H}_t$ of the marked point process $\bar{N}$ of (10.1) over sets $[0,t] \times E$, and with asymptotics as $t \to \infty$. In physical terms $\mathcal{H}_t$ corresponds to observation of N over $[0,t]$ and of the Poisson sample sequence $(X(T_n))$ over $\{1, \ldots, N_t\}$.

Maximum likelihood estimation seems hopeless in the nonparametric setting. The transition function (P_t) satisfies

$$P_t = e^{tA} = \sum_{n=0}^\infty \frac{t^n A^n}{n!} \tag{10.8}$$

for each t and hence the $\mathcal{H}_t$-likelihood function is (Exercise 10.2)

$$L(\lambda,A) = e^{-\lambda t}\lambda^{N_t}(N_t!)^{-1}\prod_{i,j\in E}[e^{U_t(i,j)A}(i,j)]^{N_t(i,j)}, \tag{10.9}$$

where

$$N_t(i,j) = \sum_{k=1}^{N_t} 1(X(T_{k-1}) = i, X(T_k) = j) \tag{10.10}$$

is the number of $i \to j$ transitions for the Markov chain $(X(T_n))$ and

$$U_t(i,j) = \sum_{k=1}^{N_t} (T_k - T_{k-1})1(X(T_{k-1}) = i, X(T_k) = j) \tag{10.11}$$

Even in the two-state case (examined in detail in Karr, 1984a) dependence of L on A is too complicated to permit explicit solution of the likelihood equations. Of course, since L factors into a function of λ alone and a function of A alone, we can easily derive the maximum likelihood estimator of λ, with the same result as in Section 3.5: $\hat{\lambda} = N_t/t$.

For estimation of A we resort to a simplification: $(X(T_n))$ has transition matrix $Q = \lambda U^\lambda = \lambda(\lambda I - A)^{-1}$; consequently,

$$A = \lambda(I - Q^{-1}) \tag{10.12}$$

If the Markov chain $(X(T_n))$ alone were observable, the nonparametric maximum likelihood estimator of Q would be (Billingsley, 1961a, b)

$$\hat{Q}(i,j) = \frac{N_t(i,j)}{\Sigma_{k=1}^{N_t} 1(X(T_{k-1}) = i)} \tag{10.13}$$

Our estimators are a combination of (10.12) and (10.13):

$$\hat{A} = \hat{\lambda}(I - \hat{Q}^{-1}); \tag{10.14}$$

while not maximum likelihood estimators, they are strongly consistent and asymptotically normal notwithstanding, and hence inferior only to the extent that they are inefficient.

Theorem 10.4. Assume that X is irreducible (hence positive recurrent) and let $\hat{A}$ be given by (10.14). Then as $t \to \infty$, $\hat{A} \to A$ almost surely.

Proof: That $\hat{\lambda} \to \lambda$ almost surely was established in Proposition 3.23, while by the strong law of large numbers for irreducible Markov chains (Doob, 1953)

$$\lim_{n\to\infty} \frac{\Sigma_{k=1}^n 1(X(T_{k-1}) = i, X(T_k) = j)}{\Sigma_{k=1}^n 1(X(T_{k-1}) = i)} = \frac{\nu(i)Q(i,j)}{\nu(i)} = Q(i,j)$$

almost surely, where ν is the unique limit distribution of $(X(T_n))$. Since $N_t \to \infty$ almost surely, it follows, with $\hat{Q}$ given by (10.13), that $\hat{Q} \to Q$ almost surely. In particular, since Q is nonsingular, $\hat{Q}$ is nonsingular for all sufficiently large t, removing any doubts that the estimators $\hat{A}$ are well defined. Since matrix inversion is continuous, consistency is confirmed. □

Theorem 10.5. Assume that X is irreducible. Then

$$t^{1/2}[\hat{A} - A] \xrightarrow{d} N(0,R) \tag{10.15}$$

on $\mathbf{R}^d$, where $d = |E|^2$ and the covariance matrix R is given by (10.21).

Proof: Consider the random elements $(\hat{\lambda},\hat{A})$ of $\mathbf{R}^{d+1}$; by Proposition 3.21,

$$n^{1/2}[n/T_n - \lambda] \xrightarrow{d} N(0,\lambda^2), \tag{10.16}$$

while by the central limit theorem for Markov chains (Doob, 1953; see also Basawa and Prakasa Rao, 1980; Billingsley, 1961a, b), with the notation

$$\hat{Q}^* = \hat{Q}_n^* = \frac{\sum_{k=1}^n 1(X(T_{k-1}) = i, X(T_k) = j)}{\sum_{k=1}^n 1(X(T_{k-1}) = i)},$$

we have

$$n^{1/2}[\hat{Q}^* - Q] \xrightarrow{d} N(0,R_0), \tag{10.17}$$

where R_0 is the $d \times d$ matrix

$$R_0((i,j),(i',j')) = \nu(i)^{-1}\{1(i = i')[1(j = j')Q(i,j) - Q(i,j)Q(i',j')]\} \tag{10.18}$$

Again ν is the unique invariant distribution for Q. Independence of X and N allows us to combine (10.16) and (10.17):

$$n^{1/2}[(n/T_n, \hat{Q}^*) - (\lambda, Q)] \xrightarrow{d} N(0,R_1) \tag{10.19}$$

on $\mathbf{R}^{1+d}$, where $R_1(0,0) = \lambda^2$, $R_1(0,(i,j)) = 0$, and $R_1((i,j),(i',j')) = R_0((i,j),(i',j'))$. By appeal to Serfozo (1975, Theorem 8.1),

$$t^{1/2}[(\hat{\lambda}, \hat{Q}) - (\lambda, Q)] \xrightarrow{d} N(0,R_2), \tag{10.20}$$

where $R_2 = \lambda^{-1}R_1$. Now let G be the open set of points (μ, V) in $\mathbf{R}^{1+d}$ such that the matrix V is nonsingular, and let $H: G \to \mathbf{R}^d$ be the function $H(\mu,V) = \mu(I - V^{-1})$. According to (10.12), $A = H(\lambda,Q)$, while Theorem 10.4 implies that almost surely $(\hat{\lambda},\hat{Q}) \in G$ for all sufficiently large t; hence (10.14) gives $\hat{A} = H(\hat{\lambda},\hat{Q})$. With $J_H(\lambda,Q)$ the Jacobian of H evaluated at (λ,Q), it follows from (10.20) and multivariate transformation theory that (10.16) holds with (T denoting transpose and)

$$R = J_H(\lambda, Q) R_2 J_H(\lambda, Q)^T \quad \square \tag{10.21}$$

State estimation for X is not difficult. Given observations $\mathcal{F}_t^{\bar{N}} = \mathcal{F}^{\bar{N}}([0,t] \times E)$, in order to calculate, for example, the conditional distribution $P\{X_u = j | \mathcal{F}_t^{\bar{N}}\}$ we notice that if $u \leq T_{N_t}$, then by independence of X and N, provided that $u \in (T_{k-1}, T_k)$,

$$P\{X_u = j | \mathcal{F}_t^{\bar{N}}\} = \frac{P_{u-T_{k-1}}(X(T_{k-1}),j) P_{T_k - u}(j, X(T_k))}{P_{U_k}(X(T_{k-1}), X(T_k))}, \tag{10.22a}$$

where (P_t) is the transition function of X and $U_k = T_k - T_{k-1}$ is the kth interarrival time of N. For $u > T_{N_t}$,

$$P\{X_u = j | \mathcal{F}_t^{\bar{N}}\} = P_{u-T_{N_t}}(Z_t, j), \tag{10.22b}$$

where $Z_t = X(T_{N_t})$ is the most recently observed value of X. In particular, choosing $u = t$ solves the filtering problem:

$$P\{X_t = j | \mathcal{F}_t^{\bar{N}}\} = P_{V_t}(Z_t, j), \tag{10.22c}$$

where V_t is the backward recurrence time of N at t (see Chapter 8). Nowhere in (10.22) have we utilized the Poisson nature of N; these formulas are valid for any sampling process independent of X. However, in the context of combined statistical inference and state estimation the Poisson assumption will be important again.

We now examine a more restricted form of observation, but under the assumption that $E = \{0,1\}$, that is, X is a *binary Markov process* with transition rates a $(0 \to 1)$ and b $(1 \to 0)$. Suppose that neither X nor the Poisson process N is entirely observable. Instead, X is observable at t if and only if $dN_t = X_t = 1$, so that the observations are the point process

$$N' = \sum_{n=0}^{\infty} 1(X(T_n) = 1)\varepsilon_{T_n}, \tag{10.23}$$

that is, $N'_t = \int_0^t X\, dN$. This process is a thinning of N, albeit not in the sense of Definition 1.38, and has already appeared several times, viewed mainly as partial observation of N forced by X, rather than vice versa. From Theorem 8.24, N' is a renewal process whose interarrival distribution F satisfies

$$\ell_{1-F}(\alpha) = \int_0^{\infty} e^{-\alpha t}[1 - F(t)]\, dt = \frac{a + b + \alpha}{\alpha^2 + \alpha(a + b + \lambda) + \lambda a} \tag{10.24}$$

From observations $\mathcal{F}_t^{N'}$ consistent estimation of a, b, and λ is possible in the following manner. It is straightforward (Exercise 10.4) to show that there is an invertible function H satisfying

$$(a,b,\lambda) = H(\ell_{1-F}(0), \ell_{1-F}(1), \ell_{1-F}(2)),$$

so that with $\hat{F}$ the $\mathcal{F}_t^{N'}$-estimator of F given by (8.20), (a,b,λ) is estimated by substitution:

$$(\hat{a},\hat{b},\hat{\lambda}) = H\left(\int [1-\hat{F}(t)]\,dt, \int e^{-t}[1-\hat{F}(t)]\,dt, \int e^{-2t}[1-\hat{F}(t)]\,dt\right) \quad (10.25)$$

These estimators (see Karr, 1984a, for proof) are strongly consistent and asymptotically normal.

Proposition 10.6. If a, b, λ are all positive, then

a) $(\hat{a},\hat{b},\hat{\lambda}) \to (a,b,\lambda)$ almost surely;

b) $t^{1/2}[(\hat{a},\hat{b},\hat{\lambda}) - (a,b,\lambda)] \xrightarrow{d} N(0,R)$, where R is a covariance matrix not calculated here. □

The principal state estimation problem, the filtering problem of calculation of $P\{X_t = 1|\mathcal{F}_t^{N'}\}$, is solved in Theorem 6.23; variants are treated in Karr (1984a).

Reverting to the Poisson sampling scheme with observations $\bar{N} = \Sigma\varepsilon_{(T_n, X(T_n))}$, we conclude the section with analysis of combined statistical inference and state estimation. Since the main state estimation issue is filtering, that is, calculation of $P\{X_t \in (\cdot)|\mathcal{F}_t^{\bar{N}}\}$, we confine attention to it (for extension to prediction, see Exercise 10.5). By (10.8) and (10.22c), for each t and j,

$$P\{X_t = j|\mathcal{F}_t^{\bar{N}}\} = e^{V_t A}(Z_t, j) \quad (10.26)$$

where V_t is the backward recurrence time at t in the sampling process N and $Z_t = X(T_{N_t})$ the most recent observation of X. (The state space E is again an arbitrary finite set.) Following our method in Sections 7.4 and 8.3, we construct pseudo-state estimators, which approximate the true state estimators of (10.26), by substituting the estimator $\hat{A}$ based on $\mathcal{F}_t^{\bar{N}}$ into (10.26):

$$\hat{P}\{X_t = j|\mathcal{F}_t^{\bar{N}}\} = e^{V_t \hat{A}}(Z_t, j) \quad (10.27)$$

The main asymptotic feature is asymptotic normality of the difference between the pseudo- and the true state estimators.

Theorem 10.7. Assume that X is irreducible, let π be the limit distribution of X_t as $t \to \infty$, and let $d = |E|$. Then as random elements of $\mathbf{R}^d$,

$$t^{1/2}[\hat{P}\{X_t \in (.)|\mathcal{F}_t^{\bar{N}}\} - P\{X_t \in (.)|\mathcal{F}_t^{\bar{N}}\}] \xrightarrow{d} Y(\cdot), \quad (10.28)$$

where $Y(\cdot) = e^{V_\infty A_\infty}(Z_\infty, \cdot)$ and

a) $A_\infty \sim N(0,R)$, where R is the covariance matrix of (10.21);

b) V_∞ is exponentially distributed with parameter λ;

c) Z_∞ is a random element of E with distribution π;

d) A_∞, V_∞, Z_∞ are mutually independent.

Proof: It suffices to show that $(\hat{A}, V_t, Z_t) \to (A_\infty, V_\infty, Z_\infty)$ with a) – d) fulfilled. That $\hat{A} \to A_\infty$ satisfying a) is established in Theorem 10.5; $\hat{A}$ is asymptotically independent of (V_t, Z_t) by the mixing character of X together with independence of X and N. Concerning (V_t, Z_t), let F be the exponential distribution with parameter λ and $R(t) = 1 + \lambda t$ the associated renewal function [see (8.6)]. Then for each j and for $y < t$

$$
\begin{aligned}
P\{V_t > y, Z_t = j\} &= \sum_{k=0}^{\infty} P\{T_k < t - y, X(T_k) = j, T_{k+1} > t\} \\
&= \sum_{k=0}^{\infty} \int_{[0,t-y)} F^k(du) P\{X_u = j\}[1 - F(t-u)] \\
&= \int_{[0,t-y)} R(du) P\{X_u = j\}[1 - F(t-u)] \\
&= P\{X_0 = j\} e^{-\lambda t} + \lambda \int_0^{t-y} P\{X_u = j\} e^{-\lambda(t-u)}\, du \\
&\sim P\{X_0 = j\} e^{-\lambda t} + P\{X_t = j\} e^{-\lambda y} \\
&\to \pi(j) e^{-\lambda y}
\end{aligned}
$$

Consequently, $(V_t, Z_t) \xrightarrow{d} (V_\infty, Z_\infty)$ such that b) – d) are fulfilled, which completes the proof. □

This argument does not depend crucially on the Poisson nature of N. Provided that N is a renewal process with nonarithmetic interarrival distribution, Theorem 10.7 can be extended (see Karr, 1984a).

10.2 STATIONARY PROCESSES ON R

The question addressed is statistical estimation for stationary stochastic processes on **R** given Poisson samples. Since "stationary" will be used in two senses, we begin by introducing them. Terminology in the literature is not entirely standard, so the reader is warned that usage here may differ from that elsewhere.

Let $(X_t)_{t \in \mathbf{R}}$ be a real-valued process. Then X is *stationary* if its probability law is invariant under translations of the time axis:

$$(X_t)_{t \in \mathbf{R}} \stackrel{d}{=} (X_{t+s})_{t \in \mathbf{R}} \tag{10.29}$$

for every $s \in \mathbf{R}$. This relatively strong form of stationarity neither requires nor entails existence of moments, but is too restrictive in some applications, not so much because it is believed to fail as because it is difficult to confirm, for example by statistical hypothesis tests. Nonetheless, it is a potent assumption in comparison with the more commonly stipulated L^2-stationarity. The process X is *L^2-stationary* if its first and second moment structure is translation invariant; that is, if $E[X_t^2] < \infty$ for each t, is the *mean*

$$m = E[X_t] \tag{10.30a}$$

is independent of t, and if the *covariance function*

$$R(t) = \mathrm{Cov}(X_u, X_{u+t}) \tag{10.30b}$$

depends only on $|t|$. In particular (Exercise 10.8), R is symmetric and positive definite with $R(0) = \mathrm{Var}(X_t)$. The law of X is not specified by m and R except in the Gaussian case, so that they do not define an identifiable statistical model, but one may perform inference for them notwithstanding. Given m and R satisfying the properties above, there exists at least one L^2-stationary process fulfilling (10.30), namely the Gaussian process with mean m and covariance function R, which then also satisfies (10.29).

For L^2-stationary processes there holds an analog of the spectral theory for stationary point processes developed in Section 9.3. Its cornerstone is a spectral representation theorem (see Brillinger, 1981, or Rozanov, 1967); there is a complex-valued random measure (in the L^2-sense) Z with mean zero and orthogonal increments such that

$$X_t = m + \int_{-\infty}^{\infty} e^{itv} Z(dv) \tag{10.31}$$

for each t. If follows that

$$R(t) = \int_{-\infty}^{\infty} e^{itv} E[|Z(dv)|^2], \tag{10.32}$$

and the measure

$$F(dv) = E[|Z(dv)|^2] \tag{10.33}$$

is the *spectral measure* of X. If the covariance function is integrable, i.e.,

$$\int_{-\infty}^{\infty} |R(t)|dt < \infty, \tag{10.34}$$

then F admits *spectral density function* f satisfying

$$f(v) = (2\pi)^{-1} \int_{-\infty}^{\infty} e^{-ivt} R(t)\, dt \tag{10.35}$$

Our concern is with estimation of the mean, the covariance function, and under the assumption that it exists, the spectral density function, given

Poisson samples of X. Throughout, X is assumed to be continuous in probability. Although we estimate only first and second moments, we do investigate behavior of estimators under both forms of stationarity. Let $N = \Sigma_{n=-\infty}^{\infty} \varepsilon_{T_n}$, a Poisson process on **R** independent of X and with *known* intensity ν, be the sampling process. We presume that observation commences at $t = 0$ and that N itself is observable, so that the observations constitute the marked point process $\bar{N} = \Sigma_{n=0}^{\infty} \varepsilon_{(T_n, X(T_n))}$. Estimation can be based on asynchronous (real-time) data

$$\mathcal{H}_t = \mathcal{F}^{\bar{N}}([0,t] \times \mathbf{R}) \tag{10.36a}$$

or synchronous data [see (2.6)]

$$\mathcal{H}_{T_n} = \sigma(T_1, \ldots, T_n; X(T_1), \ldots, X(T_n)) \tag{10.36b}$$

The former not only is more important in applications but also can generalized to higher dimensions (Section 10.3), but the latter is still interesting, especially since the two differ in mathematical details.

We begin with estimation of the mean from asynchronous data. Even though the problem is solved easily it illustrates the differing implications of the two forms of stationarity. Given that the intensity of N is known, the obvious $\mathcal{H}_t$-estimator of m is

$$\hat{m} = (\nu t)^{-1} \int_0^t X\,dN, \tag{10.37}$$

where the integral is over $[0,t]$. Elementary properties valid under L^2-stationarity are provided by the following result, whose proof is left as Exercise 10.9.

Proposition 10.8. Suppose that X is L^2-stationary and let $\hat{m}$ be given by (10.37). Then $E[\hat{m}] = m$ for each t and

$$\text{Var}(\hat{m}) = (\nu t)^{-1}[R(0) + m^2] + t^{-1} \int_{-t}^{t} \left(1 - \frac{|s|}{t}\right) R(s)\,ds \quad \square \tag{10.38}$$

Thus the estimators $\hat{m}$ are unbiased and if the covariance function is integrable in the sense of (10.34) they are mean square consistent: $\hat{m} \to m$ in L^2, and hence also weakly consistent. Stationarity in the sense of (10.29) implies strong consistency and if X satisfies an appropriate central limit theorem, then the $\hat{m}$ are asymptotically normal.

Proposition 10.9. Assume that X is stationary and that $m = E[X_t]$ exists. Then the estimators $\hat{m}$ of (10.37) are strongly consistent.

Proof: Relative to the history $\mathcal{G}_t = \mathcal{F}^N([0,t]) \vee \mathcal{F}_\infty^X$, by independence of X and N the stochastic intensity of N remains the $\mathcal{F}^N$-stochastic intensity, namely the deterministic process identically equal to ν. Lemma 9.22 applies, and by (9.47),

$$\lim \hat{m} = \lim t^{-1} \int_0^t \frac{X}{\nu} dN = E[X_0] = m \quad \square$$

Proposition 10.10. Assume that X is stationary with $E[X_t^2] < \infty$ for each t. Suppose that the covariance function R is integrable and that

$$(t^{-1/2} \int_0^t (X_u - m)\, du \xrightarrow{d} N(0, \int R(s)\, ds) \tag{10.39}$$

Then $t^{1/2}[\hat{m} - m] \xrightarrow{d} N(0,\sigma^2)$, where

$$\sigma^2 = \nu^{-1}[R(0) + m^2] + \int_{-\infty}^{\infty} R(s)\, ds$$

Proof: Conditions for validity of (10.39) are given, for example, in Rozanov (1965, Theorem 11.2); the asymptotic variance is necessarily $\int R(s)\, ds = 2\pi f(0)$, where f is the spectral density function. For $\alpha \in \mathbf{R}$ and $\widetilde{X} = X - m$,

$$\begin{aligned}
E\left[\exp(i\alpha t^{-1/2} \int_0^t \widetilde{X}\, dN\right] &= E\left[E[\exp(i\alpha t^{-1/2} \int_0^t \widetilde{X}\, dN) | X]\right] \\
&= E\left[\exp(\nu \int_0^t [e^{i\alpha t^{-1/2}\widetilde{X}} - 1]\, du)\right] \\
&\sim E\left[\exp\left(i\nu\alpha t^{-1/2} \int_0^t \widetilde{X}\, du - \frac{1}{2}\nu\alpha^2 t^{-1} \int_0^t \widetilde{X}^2\, du\right)\right] \\
&\to \exp\left[-\frac{1}{2}\nu^2\alpha^2 \int R(s)\, ds - \frac{1}{2}\nu\alpha^2(R(0) + m^2)\right],
\end{aligned}$$

where the error converges in probability to zero by Proposition 10.9, and where we have applied (10.39) as well as Proposition 10.9 to the stationary process X^2. $\square$

One can even permit the intensity of the sampling process to be unknown. The estimators $\hat{\nu} = N_t/t$ are strongly consistent and asymptotically normal by Proposition 3.23, and the estimators $\hat{m}$ can be replaced by

$$\hat{m}^* = N_t^{-1} \int_0^t X\, dN; \tag{10.37a}$$

some properties are examined in Exercise 10.10.

For estimation of the covariance function from asynchronous data we employ kernel methods. Assume for simplicity that $m=0$. Let w be a symmetric, positive, bounded, and continuous function on **R** satisfying $\int w(x)\,dx=1$ (i.e., w is a density function) and let k_t, $t \geq 0$, be positive numbers such that $k_t \to 0$, $tk_t \to \infty$ as $t \to \infty$. As $\mathcal{H}_t$-estimator of R we take

$$\hat{R}(s) = (\nu^2 t)^{-1} \int_0^t \int_0^t w_t(s - s_1 + s_2) X(s_1) X(s_2) N^{(2)}(ds_1, ds_2), \quad (10.40)$$

where (see Section 1.7)

$$N^{(2)}(ds_1, ds_2) = N(ds_1)(N - \varepsilon_{s_1})(ds_2) = 1(s_1 \neq s_2) N(ds_1) N(ds_2),$$

and where $w_t(x) = k_t^{-1} w(x/K_t)$.

The rationale is that the probability measures associated with the density functions $w_t(\cdot)$ converge weakly to ε_0 as $t \to \infty$, so that we have the approximation

$$\hat{R}(s) \sim (\nu^2 t)^{-1} \int_0^t \int_0^t 1(s_1 - s_2 = s) X(s_1) X(s_2) N^{(2)}(ds_1, ds_2), \quad (10.41)$$

whose role as estimator of $R(s)$ is apparent. The estimators $\hat{R}$, unlike the $\hat{m}$ of (10.37), are not unbiased, but are asymptotically unbiased. Moreover, weak consistency can be established under rather broad hypotheses, which is done in the following analog of Proposition 10.9. In part b) of Theorem 10.11 we employ the fourth-order cumulant function Q of X, which exists under the assumption that $E[X_t^4] < \infty$ for each t and—because $m=0$—satisfies

$$\begin{aligned} Q(s_1, s_2, s_3) = {} & E[X(t)X(t+s_1)X(t+s_2)X(t+s_3)] \\ & - R(s_1)R(s_3 - s_2) - R(s_2)R(s_3 - s_1) \\ & - R(s_3)R(s_2 - s_1) \end{aligned} \quad (10.42)$$

for all t, s_1, s_2, s_3.

Theorem 10.11. Assume that X is L^2-stationary. Then

a) For each s and t,

$$E[\hat{R}(s)] = \int_{-t}^t \left(1 - \frac{|v|}{t}\right) w_t(s - v) R(v)\,dv \quad (10.43)$$

b) If in addition R is integrable and the fourth-order cumulant function Q is bounded and continuous and satisfies

$$\sup_{s_1, s_2} \int |Q(u + s_1, u, s_2)|\,du < \infty,$$

then

$$\mathrm{Cov}(\hat{R}(s_1),\hat{R}(s_2)) = (v^2 t)^{-1}\int_{-t}^{t}\left(1-\frac{|u|}{t}\right)Q(0,u,u) + 2R(u)^2 + R(0)^2]$$

$$\times\, w_t(s_2-u)[w_t(s_1-u)+w_t(s_1+u)]\,du$$

$$+\,O(t^{-1}); \tag{10.44}$$

in particular,

$$\lim_{t\to\infty}(tk_t)\mathrm{Cov}(\hat{R}(s_1),\hat{R}(s_2))$$

$$= v^{-2}(\int w(x)^2\,dx)[Q(0,s_1,s_1)+2R(s_1)^2+R(0)^2]1(s_1=\pm s_2) \tag{10.45}$$

Proof: (Sketch). The arguments are primarily computational, so we do not present them in detail. The expression (10.43) follows essentially immediately from (10.40): since $E[N^{(2)}(ds_1,ds_2)] = v^2\,ds_1\,ds_2$,

$$E[\hat{R}(s)] = v^{-1}\int_0^t\int_0^t w_t(s-s_1+s_2)\,R(s_1-s_2)\,ds_1\,ds_2,$$

which becomes (10.43) after change of variables.

Independence of X and N implies that

$$\mathrm{Cov}(\hat{R}(s_1),\hat{R}(s_2))$$

$$= (v^2 t)^{-2}\int_0^t\int_0^t\int_0^t\int_0^t w_t(s_1-u_1+r_1)w_t(s_2-u_2+r_2)$$

$$\times\, E[X(u_1)X(r_1)X(u_2)X(r_2)]E[N^{(2)}(du_1,dr_1)N^{(2)}(du_2,dr_2)]$$

$$-\,E[\hat{R}(s_1)]\,E[\hat{R}(s_2)]$$

The latter term is calculated using (10.43). From (10.42),

$$E[X(u_1)X(r_1)X(u_2)X(r_2)] = Q(u_1-r_1,u_1-u_2,u_1-r_2) + R(u_1-u_2)R(r_1-r_2)$$

$$+\,R(u_1-r_1)R(u_2-r_2) + R(u_1-r_2)R(r_1-u_2);$$

consequently,

$$\begin{aligned}
&\operatorname{Cov}(\hat{R}(s_1),\hat{R}(s_2))\\
&\quad = (v^2t)^{-2} \int_0^t\int_0^t\int_0^t\int_0^t w_t(s_1 - u_1 + r_1)w_t(s_2 - u_2 + r_2)\\
&\qquad \times Q(u_1 - r_1, u_1 - u_2, u_1 - r_2)E[N^{(2)}(du_1,dr_1)N^{(2)}(du_2,dr_2)]\\
&\quad + (v^2t)^{-2} \int_0^t\int_0^t\int_0^t\int_0^t w_t(s_1 - u_1 + r_1)w_t(s_2 - u_2 + r_2)R(u_1 - u_2)R(r_1 - r_2)\\
&\qquad \times \{E[N^{(2)}(du_1,dr_1)N^{(2)}(du_2,dr_2)] - v^4\,du_1\,dr_1\,du_2\,dr_2\}\\
&\quad + (v^2t)^{-2} \int_0^t\int_0^t\int_0^t\int_0^t w_t(s_1 - u_1 + r_1)w_t(s_2 - u_2 + r_2)R(u_1 - r_1)R(u_2 - r_2)\\
&\qquad \times E[N^{(2)}(du_1,dr_1)N^{(2)}(du_2,dr_2)]\\
&\quad + (v^2t)^{-2} \int_0^t\int_0^t\int_0^t\int_0^t w_t(s_1 - u_1 + r_1)w_t(s_2 - u_2 + r_2)R(u_1 - r_2)R(r_1 - u_2)\\
&\qquad \times E[N^{(2)}(du_1,dr_1)N^{(2)}(du_2,dr_2)] \qquad (10.46)
\end{aligned}$$

Exercise 10.12 implies that

$$\begin{aligned}
&E[\,N^{(2)}(du_1,dr_1)\,N^{(2)}(du_2,dr_2)]\\
&\quad = v^4\,du_1\,du_2\,dr_1\,dr_2\\
&\qquad + v^3\,du_1\,dr_1\,\varepsilon_{u_1}(du_2)\,dr_2 + v^3\,du_1\,dr_1\,du_2\,\varepsilon_{u_1}(dr_2)\\
&\qquad + v^3\,du_1\,dr_1\,\varepsilon_{r_1}(du_2)\,dr_2 + v^3\,du_1\,dr_1\,du_2\,\varepsilon_{r_1}(dr_2)\\
&\qquad + v^2\,du_1\,dr_1\,\varepsilon_{u_1}(du_2)\varepsilon_{r_1}(dr_2) + v^2\,du_1\,dr_1\,\varepsilon_{r_1}(du_2)\varepsilon_{u_1}(dr_2), \qquad (10.47)
\end{aligned}$$

which one then substitutes into (10.46). The dominant components of each term arise from $v^2\,du_1\,dr_1\,\varepsilon_{u_1}(du_2)\varepsilon_{r_1}(dr_2)$ and $v^2\,du_1\,dr_1\,\varepsilon_{r_1}(du_2)\varepsilon_{u_1}(dr_2)$, and therefore (see Masry, 1983, for details; the integrability assumption on Q is used within these omitted calculations) within $O(t^{-1})$

$$\operatorname{Cov}(\hat{R}(s_1),\hat{R}(s_1))$$

$$= (v^2t)^{-1}\left[\int_{-t}^{t}\left(1-\frac{|u|}{t}\right)w_t(s_1-u)w_t(s_2-u)Q(u,0,u)\,du\right.$$

$$\left.+\int_{-t}^{t}\left(1-\frac{|u|}{t}\right)w_t(s_1+u)w_t(s_2-u)Q(0,u,u)\,du\right]$$

$$+(v^2t)^{-1}\left[\int_{-t}^{t}\left(1-\frac{|u|}{t}\right)w_t(s_1-u)w_t(s_2-u)R^2(u)\,du\right.$$

$$\left.+\int_{-t}^{t}\left(1-\frac{|u|}{t}\right)w_t(s_1-u)w_t(s_2+u)R^2(u)\,du\right]$$

$$+(v^2t)^{-1}\left[\int_{-t}^{t}\left(1-\frac{|u|}{t}\right)w_t(s_1-u)w_t(s_2-u)R^2(u)\,du\right.$$

$$\left.+R(0)^2\int_{-t}^{t}\left(1-\frac{|u|}{t}\right)w_t(s_1+u)w_t(s_2-u)\,du\right]$$

$$+R(0)^2(v^2t)^{-1}\left[\int_{-t}^{t}\left(1-\frac{|u|}{t}\right)w_t(s_1-u)w_t(s_2-u)\,du\right.$$

$$\left.+\int_{-t}^{t}\left(1-\frac{|u|}{t}\right)w_t(s_1+u)w_t(s_2-u)R^2(u)\,du\right]$$

Since $Q(u,0,u)=Q(0,u,u)$, (10.44) holds; derivation of (10.45) from it is computational. □

The asymptotic covariance between $\hat{R}(s_1)$ and $\hat{R}(s_2)$ is zero unless $s_1 = \pm s_2$, which precludes a central limit theorem for the $\hat{R}$ as stochastic processes (the limit, with independent values at distinct points, would not admit even a separable version). Nevertheless, given additional assumptions asymptotic normality can be established for single values $\hat{R}(s)$, albeit only in a rather brute-force manner. The same argument together with the Cramér-Wold device applies to other finite-dimensional distributions.

Theorem 10.12. Assume that X admits moments of all orders and that for every $k \geq 2$ the cumulant function of order k satisfies

$$\int_{\mathbf{R}^{k-1}} |Q^{(k)}(u_1,\ldots,u_{k-1})|\,du_1\cdots du_{k-1} < \infty, \tag{10.48}$$

and that in addition the covariance function is twice continuously differentiable and $\int x^2 w(x)\,dx < \infty$. Then for each s, $(tk_t)^{1/2}[\hat{R}(s)-R(s)] \xrightarrow{d} N(0,\sigma^2(s))$, where

$$\sigma^2(s) = v^{-2}\left(\int w(x)^2\,dx\right)[Q(0,s,s) + 2R(s)^2 + R(0)^2]$$

Proof: (Sketch). For discussion of cumulant functions, see Brillinger (1981). By stationarity of X and N individually and their independence, the process $M(dt) = X(t)N(dt)$ has stationary increments. The main idea is to apply the reasoning used to prove Theorem 9.15 to M, suitably centered. The kth-order cumulant measure of M (see Definition 9.6) has reduced-form representation

$$\gamma^k(dt_1, \ldots, dt_k) = \gamma_*^k(dt_1 - t_k, \ldots, dt_{k-1} - t_k)dt_k,$$

where γ_*^k, the reduced cumulant measure of order k, has finite variation by (10.48) and properties of N. As in the proof of Theorem 9.15, to establish that

$$(tk_t)^{1/2}(\hat{R}(s) - E[\hat{R}(s)]) \xrightarrow{d} N(0,\sigma^2(s)) \qquad (10.49)$$

it suffices to show that for each $\ell \geq 3$ the ℓth-order cumulant of $\hat{R}(s) - E[\hat{R}(s)]$ is of order $(tk_t)^{1-\ell}$, for then the corresponding cumulant of $(tk_t)^{1/2}(\hat{R}(s) - E[\hat{R}(s)])$ is of order $(tk_t)^{1-\ell/2}$ and hence converges to zero. Directly from (10.40),

$$\hat{R}(s)^\ell = (v^2 tk_t)^{-\ell} \prod_{j=1}^{\ell} \int_{-t}^{t}\int_{-t}^{t} w\left(\frac{s - s_{2j-1} + s_{2j}}{k_t}\right)$$

$$\times\, 1(s_{2j-1} \neq s_{2j})M(ds_{2j-1})M(ds_{2j}),$$

from which the asserted behavior of cumulants is at least intuitively clear. Some details are given in Masry (1983) (see also Jolivet, 1981). The basis of the computations is Leonov and Shiryayev (1959).

Given that (10.49) holds, in order to complete the proof it remains to show that $(tk_t)^{1/2}(E[\hat{R}(s)] - R(s)) \to 0$, which follows from (10.43) by a Taylor series expansion of R. □

Rather than exhaust all cases, we move on to estimation of the spectral density function given synchronous data. Finite sample covariances differ from those for asynchronous data; moreover, asymptotic normality of the estimators $\hat{f}$ of (10.51) has not been confirmed. We suppose that $E[X_t] = 0$ and that $E[X_t^2] < \infty$ for all t.

Recall that the observations $\mathcal{H}_{T_n}$ comprise not only the sampled values $X(T_1), \ldots, X(T_n)$ but also the observation times $T_1, \ldots, T_n$. The spectral density function, by symmetry of the covariance function, satisfies

$$f(v) = \pi^{-1}\int_0^\infty \cos(vt)R(t)\,dt, \qquad (10.50)$$

so we take as $\mathcal{H}_{T_n}$-estimator of f the function

$$\hat{f}(v) = (v\pi n)^{-1} \sum_{\ell=1}^{n-1} \sum_{k=1}^{n-\ell} X(T_k)X(T_{k+\ell})w_n(T_{k+\ell} - T_k)\cos(v(T_{k+\ell} - T_k)) \tag{10.51}$$

where the kernel w_n is constructed in the following manner. Let h be a symmetric density function on **R** (a "spectral window"), let k_n be positive numbers with $k_n \to 0$ and $nk_n \to \infty$, let $h_n(x) = k_n^{-1}h(x/k_n)$, and finally let $w_n(t) = \tilde{h}(tk_n)$, where $\tilde{h}$ is the Fourier transform of h.

One can express (10.51) in an alternative form more amenable to computation and analysis:

$$\hat{f}(v) = \int_{-\infty}^{\infty} h_n(v-u)\hat{I}(u)\,du - (2\pi vn)^{-1} \sum_{k=1}^{n} X(T_k)^2, \tag{10.51a}$$

where

$$\hat{I}(u) = (2\pi vn)^{-1} \left|\sum_{k=1}^{n} X(T_k)e^{ivT_k}\right|^2,$$

which is analogous to periodogram estimators.

We begin with a computational result.

Proposition 10.13. a) If R is integrable, then as $n \to \infty$,

$$E[\hat{f}(v)] = \int h_n(v-u)f(u)\,du + o(1) \tag{10.52}$$

uniformly in v.

b) If $\int |tR(t)|\,dt < \infty$, then, also uniformly in v,

$$E[\hat{f}(v)] = \int h_n(v-u)f(u)\,du + O(n^{-1}) \tag{10.53}$$

Proof: From (10.51a)

$$E[\hat{f}(v)] = \int h_n(v-u)E[\hat{I}(u)]\,du - \frac{R(0)}{2\pi v}, \tag{10.54}$$

while straightforward calculations reveal that

$$E[\hat{I}(u)] = \frac{R(0)}{2\pi v} + (\pi v)^{-1} \sum_{\ell=1}^{n-1} (1 - \ell/n)\varphi_\ell(u),$$

where $\varphi_\ell(u) = E[R(T_\ell)\cos(uT_\ell)]$, which, since T_ℓ has Erlang-(ℓ, v) distribution, can be written down explicitly. If R is integrable, then

$$(\pi v)^{-1} \sum_{\ell=1}^{\infty} \varphi_\ell(u) = (\pi v)^{-1} E\left[\int_0^\infty R(t)\cos(ut)\, dN\right] = f(v),$$

so that

$$E[\hat{I}(u)] = \frac{R(0)}{2\pi v} + f(v) + o(1),$$

where the Kronecker lemma is used on the last term. Substitution into (10.54) yields (10.52).

The proof of b) is similar; we omit details. □

The final results of the section describe the covariance structure of the estimators $\hat{f}$. Proofs, due to Masry (1983), can be found there. We introduce the notation

$$\hat{J}(u) = \hat{I}(u) - (2\pi v n)^{-1} \sum_{k=1}^{n} X(T_k)^2;$$

then (10.51a) becomes

$$\hat{f}(v) = \int_{-\infty}^{\infty} h_n(v-u)\hat{J}(u)\, du \tag{10.55}$$

The asymptotic covariance of $\hat{J}$ is the key to that of $\hat{f}$.

Proposition 10.14. Suppose that $\int |tR(t)|\, dt < \infty$. Then

$$\operatorname{Cov}(\hat{J}(v_1), \hat{J}(v_2))$$

$$= n^{-1}\left[f(v_1) + \frac{R(0)}{2\pi v}\right]^2 [\Gamma_n(v_1+v_2) - \Gamma_n(v_1-v_2)] + O(n^{-1}) \tag{10.56}$$

uniformly in v_1 and v_2, where

$$\Gamma_n(v) = 2\sum_{\ell=1}^{n}\left[\frac{(1-\ell}{n}\right] Re\left[\frac{v}{[v-iv]}\right]^{\ell} \tag{10.57}$$

Theorem 10.15. Suppose that the hypotheses of Proposition 10.14 hold and that in addition h is bounded and differentiable with bounded derivative. Then

$$\operatorname{Cov}(\hat{f}(\nu_1),\hat{f}(\nu_2))$$

$$= \frac{2\pi\nu}{n}\int_{-\infty}^{\infty}\left[f(u) - \frac{R(0)}{2\pi\nu}\right]^2 h_n(\nu_1 - u)[h_n(\nu_2 + u) - h_n(\nu_2 - u)]\,du$$

$$+ O((nk_n)^{-2}\log(n^{-1}k_n^{-1})), \tag{10.58}$$

and consequently,

$$\lim_{n\to\infty}(nk_n)\operatorname{Cov}(\hat{f}(\nu_1),\hat{f}(\nu_2))$$

$$= \left(2\pi\nu\left[f(\nu_1) + \frac{R(0)}{\nu^2}\right]^2\int h(x)^2\,dx\right)1(\nu_1 = \pm\nu_2) \quad \square \tag{10.59}$$

From (10.52) and (10.59) it follows that $E[(\hat{f}(\nu) - f(\nu))^2] \to 0$ for each ν [i.e., the estimators $\hat{f}(\nu)$ are mean square consistent]. In addition, they are asymptotically uncorrelated for different values of $|\nu|$. If the sampling rate (of the Poisson process N) is controllable, it can be optimized to produce minimal asymptotic variance for a given value of ν (Exercise 10.14).

The covariance structure of the estimators $\hat{f}$ differs from that of the asynchronous data estimators [note the analogy to (10.51a)]

$$\hat{f}^*(\nu) = \int_{-\infty}^{\infty} k_t^{-1} h\left(\frac{\nu - u}{k_t}\right)\hat{H}(u)\,du - (2\pi\nu t)^{-1}\int_0^t X^2\,dN, \tag{10.60}$$

where $k_t \to 0$, $tk_t \to \infty$ and $\hat{H}$ is the *periodogram*

$$\hat{H}(u) = (2\pi\nu t)^{-1}\left|\int_0^t e^{-ius}X(s)N(ds)\right|^2 \tag{10.61}$$

In particular, if $\int|tR(t)|\,dt < \infty$, then

$$E[\hat{H}(u)] = \nu^2 f(u) + \frac{\nu R(0)}{2\pi} + O(t^{-1}) \tag{10.62}$$

and

$$\operatorname{Cov}(\hat{H}(\nu_1),\hat{H}(\nu_2))$$

$$= t^{-1}\left[\nu^2 f(\nu_1) + \frac{\nu R(0)}{2\pi}\right][\Delta_t(\nu_1 + \nu_2) + \Delta_t(\nu_1 - \nu_2)] + O(t^{-1}), \tag{10.63}$$

where Δ_t is the Féjer kernel $\Delta_t(\nu) = t^{-1}[2\nu^{-1}\sin(\nu t/2)]^2$, which also arises in time series analysis (see Masry, 1978a, or Exercise 10.15). However, despite finite sample differences, asymptotics are the same.

10.3 STATIONARY RANDOM FIELDS

This section is a spatial version of Section 10.2, but rather than a grand synthesis it is incomplete, and ends with problems, not solutions. Before proceeding to the main results, we introduce random fields and present a multidimensional analog of Proposition 10.1.

A random field is a stochastic process with a multidimensional (Euclidean) parameter set, and stationarity is translation invariance in the obvious sense.

Definition 10.16. a) A *random field* on $\mathbf{R}^d$ is a measurable stochastic process $(Y(x))_{x\in\mathbf{R}^d}$ taking values in $\mathbf{R}$.

b) The random field Y is *stationary* if for each $y \in \mathbf{R}^d$,

$$(Y(x+y))_{x\in\mathbf{R}^d} \stackrel{d}{=} (Y(x))_{x\in\mathbf{R}^d} \tag{10.64}$$

As in the one-dimensional case, if Y is stationary and if $E[|Y(x)|^2]<\infty$ for each x, then Y is L^2-*stationary* and there exist a *covariance function*

$$R(y) = \mathrm{Cov}(Y(x), Y(x+y)) \tag{10.65}$$

and *spectral measure* F satisfying

$$R(y) = \int_{\mathbf{R}^d} e^{i<y,v>} F(dv) \tag{10.66}$$

for each y, where $<\cdot,\cdot>$ denotes the inner product on $\mathbf{R}^d$. However, whereas in one dimension the covariance function is symmetric and hence a function of t only through $|t|$, in the multidimensional case the stronger condition of isotropy is not implied by stationarity.

Definition 10.17. An L^2-stationary random field Y is *isotropic* if for every rotation γ of $\mathbf{R}^d$, $\mathrm{Cov}(Y\gamma x), Y(\gamma y)) = \mathrm{Cov}(Y(x), Y(y))$ for all x, $y \in \mathbf{R}^d$.

For an isotropic random field the covariance function reduces to a function R on $\mathbf{R}_+$: $\mathrm{Cov}(Y(x), Y(y)) = R(|x-y|)$ and the spectral measure satisfies $F\gamma^{-1} = F$ for each γ.

For simplicity we often assume that $E[Y(x)] = 0$, in which case there exists a *spectral representation* analogous to (9.56):

$$Y(x) = \int_{\mathbf{R}^d} e^{i<v,x>} Z(dv), \tag{10.67}$$

where Z is a mean zero random measure with orthogonal increments and $E[Z(A_1)\overline{Z(A_2)}] = F(A_1 \cap A_2)$, and where F is the spectral measure. For isotropic random fields more detailed spectral representations can be derived; these and additional material can be found in Yadrenko (1983).

Our main topic in the section is estimation of means and covariance functions for stationary random fields given Poisson samples. That is, let Y be a L^2-stationary random field on $\mathbf{R}^d$ and let $N = \Sigma\, \varepsilon_{X_n}$ be a stationary Poisson process on $\mathbf{R}^d$, independent of Y, with known intensity $\nu > 0$. The data are observations of the marked point process

$$\bar{N} = \Sigma\, \varepsilon_{(X_n, Y(X_n))} \tag{10.68}$$

as in Sections 10.1 and 10.2, with each point X_n of N marked by the value of Y at X_n. We continue to assume that the sampling process is observable. Without the linear order structure of $\mathbf{R}$ we can consider only "asynchronous" data

$$\mathcal{H}(K) = \mathcal{F}^{\bar{N}}(K \times \mathbf{R}), \tag{10.69}$$

where K is a compact—and ordinarily, convex—subset of $\mathbf{R}^d$.

Before examining estimation of the mean and covariance function we present a generalization of Proposition 10.1 to the effect that the law of the marked point process determines that of the random field. Indeed, in the following result neither process need be stationary; however, minimal regularity conditions are required: Y must be continuous in probability and the mean measure of N must place positive mass in every open set.

Theorem 10.18. Let Y be a random field on $\mathbf{R}^d$ that is continuous in probability, let N be a Poisson process on $\mathbf{R}^d$ with diffuse mean measure μ satisfying $\mu(G) > 0$ for every open set G, assume that Y and N are independent, and let $\bar{N}$ be the marked point process of (10.68). Then the law of $\bar{N}$ determines that of Y.

Proof: Let Y_1, Y_2 be random fields fulfilling the hypotheses of the theorem, with associated marked point processes $\bar{N}_1$, $\bar{N}_2$, respectively, and suppose without loss of generality that $|Y_1| \le 1$ and $|Y_2| \le 1$. Then for h a function on $\mathbf{R}^d$ with $0 \le h < 1$ and $f(x,u) = -\log(1 - uh(x))$, $|u| \le 1$, we have

$$E[\exp(-\int h(x)Y_1(x)\mu(dx))] = E[\exp(-\int [1 - e^{-f(x,Y_1(x))}]\,\mu(dx))]$$

$$= E[E[\exp(-\int f(x,Y_1(x))N(dx))|Y_1]]$$

$$= E[\exp(-\int f\, d\bar{N}_1)]$$

$$= E[\exp(-\int h(x)Y_2(x)\mu(dx))];$$

the last equality is a reversal of the first three. Consequently, by Theorem 1.12 the random measures $M_1(dx) = Y_1(x)\mu(dx)$ and $M_2(dx) = Y_2(x)\mu(dx)$ are identically distributed. Given $x_1, \ldots, x_k$ in $\mathbf{R}^d$, for each i choose open sets G_{in} such that $G_{in} \downarrow \{x_i\}$. Then $\mu(G_{in}) > 0$ for each i and n, while $\mu(G_{in}) \to 0$ for each i. Thus by continuity in probability of Y_1 and Y_2,

$$(Y_1(x_1), \ldots, Y_1(x_k)) = \lim\left(\frac{M_1(G_{1n})}{\mu(G_{1n})}, \ldots, \frac{M_1(G_{kn})}{\mu(G_{kn})}\right)$$

$$\stackrel{d}{=} \lim\left(\frac{M_2(G_{1n})}{\mu(G_{1n})}, \ldots, \frac{M_2(G_{kn})}{\mu(G_{kn})}\right)$$

$$= (Y_2(x_1), \ldots, Y_2(x_k));$$

the limits are in the sense of convergence in distribution. Hence $Y_1 \stackrel{d}{=} Y_2$. □

Suppose now that Y is an L^2-stationary random field on $\mathbf{R}^d$ with unknown mean and covariance function to be estimated from observations $\mathcal{H}(K)$ of the Poisson sample process $\bar{N}$. (Once again N is a stationary Poisson process with known intensity ν.) By analogy with (10.37) we introduce $\mathcal{H}(K)$-estimators

$$\hat{m} = [\nu\lambda(K)]^{-1}\int_K Y\,dN; \tag{10.70}$$

as customary, the "sample size" K is suppressed. [Were ν unknown, one could use instead the estimators $\hat{m}^* = N(K)^{-1}\int_K Y\,dN$, whose properties are in some ways superior even when ν is known (see Karr, 1986a).]

Quadratic mean consistency of the $\hat{m}$ is one consequence of the following analog of Proposition 10.8, whose computational proof is omitted.

Proposition 10.19. Let $\hat{m}$ be given by (10.70). Then for each K, $E[\hat{m}] = m$ and

$$\operatorname{Var}(\hat{m}) = [\nu\lambda(K)]^{-1}[R(0) + m^2]$$

$$+ \lambda(K)^{-1}\int R(y)\left[\frac{\lambda(K \cap (K - y))}{\lambda(K)}\right]dy, \tag{10.71}$$

where $K - y = \{x - y : x \in K\}$. □

To establish consistency not only of the $\hat{m}$ but also of estimators of the covariance function, we need the following. Recall from Chapter 9 that for a compact, convex set K, $\delta(K)$ denotes the supremum of radii of Euclidean balls contained in K.

Lemma 10.20 For each $y \in \mathbf{R}^d$, $\lim_{\delta(K)\to\infty} [\lambda(K \cap (K - y))/\lambda(K)] = 1$. □

We can now show consistency of the $\hat{m}$.

Proposition 10.21. Assume that the covariance function is integrable:

$$\int_{\mathbf{R}^d} |R(y)|\, dy < \infty \tag{10.72}$$

and let $\hat{m}$ be given by (10.70). Then $\lim_{\delta(K)\to\infty} \operatorname{Var}(\hat{m}) = 0$, and hence the estimators are mean square consistent.

Proof: Evidently, the first term in (10.71) converges to zero since $\delta(K)\to\infty$ implies that $\lambda(K)\to\infty$. By Lemma 10.20, (10.72), and the dominated convergence theorem,

$$\lim \int R(y)\left[\frac{\lambda(K\cap(K-y))}{\lambda(K)}\right] dy = \int_{\mathbf{R}^d} R(y)\, dy;$$

therefore, the second term converges to zero as well. □

For estimation of the covariance function we suppose for simplicity that $m = 0$. By analogy to the one-dimensional case, given observations $\mathcal{H}(K)$ the estimator is

$$\hat{R}(x) = [\nu^2\lambda(K)]^{-1} \int_K \int_K w_K(x - x_1 + x_2) Y(x_1) Y(x_2) N^{(2)}(dx_1, dx_2), \tag{10.73}$$

where $N^{(2)}(dx_1, dx_2) = N(dx_1)(N - \varepsilon_{x_i})(dx_2)$, where the kernel w is a positive, bounded, isotropic density function on $\mathbf{R}^d$, and where $w_K(x) = \alpha_K^{-d} w(x/\alpha_K)$, with $\alpha_K \to 0$ and $\alpha_K^d \lambda(K) \to \infty$ as $\delta(K) \to \infty$. Motivation is the same as for the one-dimensional version (10.40) and properties are substantially similar.

Theorem 10.22. Assume that Y is L^2-stationary, that the covariance function is continuous and integrable, and that the fourth-order cumulant function Q [see (10.42)] exists and satisfies

$$\sup_{x_1,x_2} \int_{\mathbf{R}^d} |Q(x+x_1,x,x_2)|\,dx < \infty$$

Then

a) For each x,

$$E[\hat{R}(x)] = \int_{K-K} w(y)R(x-\alpha_K y)\frac{\lambda(K\cap(K-x+\alpha_K y))}{\lambda(K)}\,dy, \quad (10.74)$$

and hence $E[\hat{R}(x)] \to R(x)$ as $\delta(K)\to\infty$.

b) For each x_1 and x_2,

$$\lim_{\delta(K)\to\infty} \lambda(K)\alpha_K^d \operatorname{Cov}(\hat{R}(x_1),\hat{R}(x_2))$$

$$= (\nu^{-2}\int_{\mathbf{R}^d} w(x)^2\,dx)[Q(0,x_1,x_1)+2R(x_1)^2+R(0)^2]1(x_1=\pm x_2) \quad (10.75)$$

Note the exact correspondence between (10.75) and (10.45).

Proof: (Sketch). The calculations are nearly the same as those used to prove Theorem 10.11. As when $d=1$, $E[N^{(2)}(dx_1,dx_2)] = \nu^2\,dx_1\,dx_2$, so that

$$E[\hat{R}(x)] = \lambda(K)^{-1}\int_K\int_K w_K(x-x_1+x_2)R(x_1-x_2)\,dx_1\,dx_2$$

$$= \int_K w(y)R(x-\alpha_K y)\frac{\lambda(K\cap(K-x+\alpha_K y))}{\lambda(K)}\,dy,$$

which is the first part of (10.74). As $\delta(K)\to\infty$, $\alpha_K\to 0$ by assumption while $\lambda(K\cap(K-x+\alpha_K y))/\lambda(K) \sim \lambda(K\cap(K-x))/\lambda(K)$, which converges to 1 by Lemma 10.20, giving the second part of (10.74).

Concerning b),

$$E[\hat{R}(x)\hat{R}(y)] = [\nu^2\lambda(K)]^{-2}\int_K\int_K w_K(x-x_1+x_2)w_K(y-y_1+y_2)$$
$$\times E[Y(x_1)Y(x_2)Y(y_1)Y(y_2)]$$
$$\times E[N^{(2)}(dx_1,dx_2)N^{(2)}(dy_1,dy_2)],$$

which is expanded as in the proof of Theorem 10.11. First of all,

$$E[Y(x_1)Y(x_2)Y(y_1)Y(y_2)]$$
$$= Q(x_2-x_1,y_1-x_1,y_2-x_1)+R(x_2-x_1)R(y_2-y_1)$$
$$+R(y_1-x_1)R(y_2-x_2)+R(y_2-x_1)R(x_2-y_1)$$

In addition, (10.47) remains valid with only notational changes. The dominant contributions to $\text{Cov}(\hat{R}(x),\hat{R}(y))$ are analogous to those in the one-dimensional case, with the result (Karr, 1986a) that within $O(\lambda(K)^{-1})$,

$$\begin{aligned}
&\text{Cov}(\hat{R}(x),\hat{R}(y))\\
&= [v^2\lambda(K)]^{-1}\int_K w_K(x-z)[w_K(y-z)+w_K(y+z)]\\
&\quad\times[Q(0,z,z)+2R(z)^2+R(0)^2]\,\frac{\lambda(K\cap(K-z))}{\lambda(K)}dz,\\
&= [\alpha_K^d v^2\lambda(K)]^{-1}\int_K w(z)[w(z+\alpha_K^{-1}(y-x))+w(z+\alpha_K^{-1}(y+x))]\\
&\quad\times[Q(0,x-\alpha_K z,x-\alpha_K z)+2R(x-\alpha_K z)^2+R(0)^2]\\
&\quad\times\frac{\lambda(K\cap(K-x+\alpha_K z))}{\lambda(K)}dz,
\end{aligned} \tag{10.76}$$

from which (10.75) follows by another application of Lemma 10.20 in conjunction with the remaining assumptions. (It is also necessary to observe that $\int_{K-K} w(z)^2\,dz \to \int w(z)^2\,dz$, but this is shown easily.) □

Given sufficiently strong assumptions on Y, asymptotic normality of $\hat{m}$ and $\hat{R}$ can be shown by application of Theorem 9.15 to the process $M(dx) = Y(x)N(dx)$ (see also Theorem 10.12). For details and variations, see Karr (1986a).

State estimation for random fields given point process samples has a significant applications component. For example, mineral and oil reserves must be estimated and mapped from a small number of test drillings, and areal precipitation must be inferred from rain gage measurements. We consider only linear state estimation. Let Y be an L^2-stationary random field on $\mathbf{R}^d$ with known mean $m = 0$ and known covariance function R. We retain the assumption that the sampling process N is Poisson, independent of X and observable. Given observations of the marked point process $\bar{N}$ over a bounded set A in $\mathbf{R}^d$, we wish to calculate for each x the optimal (in the MMSE sense) linear state estimator $\hat{Y}(x)$ of $Y(x)$. "Linear" we take to mean that

$$\hat{Y}(x) = \int_A h(x,z)Y(z)N(dz), \tag{10.77}$$

with h a function on $\mathbf{R}^d \times A$ that is to be determined.

The optimal function h^* can be characterized as the solution to an integral equation.

Proposition 10.23. With A fixed, the optimal linear state estimator corresponds to the unique solution h^* to the equation

$$R(x-y) = \nu \int_A h^*(x,z)R(y-z)\,dz + h^*(x,y)R(0), \qquad (10.78)$$

for $x \in \mathbf{R}^d$ and $y \in A$. Moreover, with $\hat{Y}^*(x) = \int_A h^*(x,z)Y(z)N(dz)$,

$$E[(\hat{Y}^*(x) - Y(x))^2] = h^*(x,x)R(0) \qquad (10.79)$$

Proof: The following argument, although heuristic, illustrates the key ideas. For fixed A and x, the optimal linear state estimator $\hat{Y}(x)$, by Hilbert space theory, fulfills the normal equations

$$E[(\hat{Y}(x) - Y(x))Y(y)N(dy)] = 0, \qquad y \in A;$$

therefore, h^* must satisfy

$$\nu R(x-y)\,dy = E[\hat{Y}(x)Y(y)N(dy)]$$

$$= E[(\int_A h^*(x,z)Y(z)N(dz))Y(y)N(dy)]$$

$$= \int_A h^*(x,z)R(y-z)E[N(dz)N(dy)]$$

$$= \int_A h^*(x,z)R(y-z)[\nu^2\,dz\,dy + \nu\varepsilon_y(dz)\,dy],$$

which is the same as (10.78). Derivation of (10.79) from (10.78) is a straightforward calculation. □

In a sense (10.78) is a renewal equation (see Chapter 8) whose formal solution is

$$h^*(x,y) = \sum_{k=0}^{\infty} (-\nu)^k R(0)^{-(k+1)} R_A^k(x,y), \qquad (10.80)$$

where $R_A^0(x,y) = R(x-y)$ and

$$R_A^{k+1}(x,y) = \int_A R_A^k(x,y)R(z-y)\,dz$$

The restriction to linear state estimators of the form (10.77) engenders some undesirable behavior: for a given realization of (Y,N) the estimator $\hat{Y}$ does not reproduce Y at the points of N even though the values $Y(X_n)$ are known exactly. This defect, which arises because the points of N are random in location and number, can be rectified by redefining $\hat{Y}(X_n)$ to be $Y(X_n)$, but optimality properties become unclear. Alternatively, one could treat the X_n as fixed and apply a multivariate function approximation technique to the data $Y(X_n)$ to produce an estimator $\hat{Y}(\cdot)$, but again optimality is unclear.

Yet another approach is to allow estimators that are linear functionals of the marked point process $\bar{N}$, that is,

$$\hat{Y}(x) = \int_{A \times \mathbf{R}} h(x,y,u)\bar{N}(dy,du), \tag{10.81}$$

although this fails to alleviate the difficulty. An argument analogous to the proof of Proposition 10.23 identifies the optimal solution h^*.

Proposition 10.24. The MMSE linear state estimator of the form (10.81) corresponds to the function h^* satisfying

$$\frac{dE[Y(x)1(Y(y) \in (\cdot))]}{dP\{Y(y) \in (\cdot)\}}(u)$$

$$= h^*(x,y,u) + \int_{A \times \mathbf{R}} h^*(x,y',u') \frac{dP\{Y(y') \in du', Y(y) \in (\cdot)\}}{dp\{Y(y) \in (\cdot)\}}(u)\,dy' \quad \square \tag{10.82}$$

There is some resemblance between this procedure and the "disjunctive kriging" of Matheron (1976). Solution of (10.82) requires more than just the covariance function of Y, but not full knowledge of the distribution: only bivariate distributions appear in (10.82), which is usable if these can be calculated.

Combined inference and state estimation remains to be treated. Given Poisson samples of Y observed over K, it is clear what one *might* do. First, estimate R using (10.73), then—assuming that $m = 0$ and that linear state estimators of the form (10.77) are sought—substitute $\hat{R}$ for R in (10.80) to obtain an estimator $\hat{h}^*$, and finally form pseudo-state estimators

$$\hat{\hat{Y}}(x) = \int_K \hat{h}^*(x,z)Y(z)N(dz) \tag{10.83}$$

A myriad of complications ensues, including whether there is convergence in (10.80) when $\hat{R}$ replaces R, the dual role of the observations in (10.83) and, for asymptotics, the dependence of the domain of integration (as well as the estimators) on A.

As a mirror of reality we end here, with unresolved questions and vagueness rather than answers and certainty.

EXERCISES

10.1. Prove Proposition 10.2.

10.2. Derive the likelihood function $L(\lambda, A)$ of (10.9).

10.3. Verify that (10.22a) and (10.22b) obtain for any observable sampling process N independent of X.

10.4. Let N' be the renewal process given by (10.23), where X is a binary Markov process with (positive) transition rates a $(0 \to 1)$ and b $(1 \to 0)$ and $N = \Sigma\, \varepsilon_{T_n}$ is Poisson with rate λ and independent of X.
a) Prove that the interarrival distribution F of N' satisfies (10.24).
b) Prove that there is a function H such that

$$(a,b,\lambda) = H(\ell_{1-F}(0), \ell_{1-F}(1), \ell_{1-F}(2))$$

10.5. After defining suitable pseudo-state estimators, extend Theorem 10.7 to state estimators $P\{X_{t+s} = j \mid \mathcal{F}_t^N\}$ (where $s > 0$ is fixed), used for prediction of X.

10.6. Let X be a binary Markov process with transition rates a $(0 \to 1)$ and b $(1 \to 0)$ observed at the "jittered" random times $T_n = n\,\Delta + \delta_n$, where $\Delta > 0$ and the δ_n are i.i.d. random variables independent of X with $P\{|\delta_n| < \Delta/2\} = 1$. (The latter condition prevents "crossed" observations.) Assume that the distribution G of the δ_n is known and that observations are the marked point process $N = \Sigma\, \varepsilon_{(T_n, X(T_n))}$.
a) Prove that $(X(T_n))$ is a Markov chain and calculate its transition matrix.
b) Devise $\mathcal{F}_t^N$-estimators of a and b that are strongly consistent and asymptotically normal; verify these properties.

10.7. Let X, N be as in Section 10.1, let (L_t) be the *left*-continuous backward recurrence time process of N [see (8.25)], and let $Z_t = X(T_n)$ be the most recently observed value of X.
a) Show that the $(P_A, \mathcal{F}^N)$-stochastic intensity of the point process $N_t(i,j)$ of (10.10) is $\lambda_t(A,i,j) = P_{Lt}(Z_{t-}, i)\, A(i,j)$, where (P_t) and A are the transition function and generator of X.
b) Describe large sample (i.e., as $t \to \infty$) behavior of the martingale estimators

$$\hat{A}(i,j) = \int_0^t \hat{P}_{L_s}(Z_{s-}, i)\, dN_s(i,j),$$

where $\hat{P}_u(\cdot\,,\cdot) = \exp(u\hat{A})(\cdot\,,\cdot)$.

10.8. Let R be the covariance function of an L^2-stationary process X on $\mathbf{R}$. Prove that R is symmetric and positive definite and that $R(0) = \mathrm{Var}(X_t)$ for each t.

10.9. Prove Proposition 10.8.

10.10. Show that under the hypotheses of Propositions 10.9 and 10.10 the estimators $\hat{m}^*$ of (10.37a) are strongly consistent and asymptotically normal.

10.11. (Continuation of Exercise 10.10) Calculate the asymptotic efficiency of the estimators $\hat{m}$ of (10.37) relative to that of the estimators $\hat{m}^*$ of (10.37a).

10.12. Verify (10.47).

10.13. Prove that under the assumptions and notation of Theorem 10.12, for each k the reduced kth-order cumulant measure γ_*^k of the process $M(dt) = X(t)N(dt)$ has finite total variation.

10.14. Let $\hat{f}(v)$ be the $\mathcal{H}_{T_n}$-estimators of the spectral density function given by (10.51) and let v be fixed. Determine the value of the sampling rate ν that minimizes the asymptotic variance in (10.59).

10.15. Confirm that the $\mathcal{H}_t$-estimators $\hat{f}^*(v)$ of the spectral density function given by (10.60) fulfill (10.62) and (10.63) provided that $\int |tR(t)|\, dt < \infty$.

10.16. Prove Lemma 10.20.

10.17. a) Formulate analogs of the estimators (10.73) under the stipulation that Y is a stationary, isotropic random field on $\mathbf{R}^d$ with mean zero and $\mathrm{Cov}(Y(x), Y(y)) = R(|x-y|)$.

b) Establish a corresponding analog of Theorem 10.22.

c) Determine the efficiency that would be lost if instead of the estimators in a) one were to ignore isotropy of Y and use instead the estimators $\hat{R}$ in (10.40).

NOTES

Even in one dimension there is as yet no semblance of a general theory of point process sampling of stochastic processes. This chapter is somewhat unified but hardly complete: it omits more specific models than it presents. Of couse, when the "field" consists only of more or less isolated special cases, little else is possible. Two forms of point process sampling have been investigated in some breadth: deterministic—usually regular—and Poisson sampling; both assume independence of the sampling process and the process being sampled. Until Karr (1986a), nearly all statistical results dealt with one-dimensional parameter sets.

Regular sampling has been studied for birth/death processes (Keiding, 1975), binary Markov processes (Brown et al., 1977, 1979), and stationary processes (time series). For time series there arises (see Masry, 1978b) the phenomenon of "aliasing": from regular observations spaced Δ apart one cannot recover the spectral density function at frequencies above the Nyquist frequency $2\pi/\Delta$.

By contrast, Poisson sampling has been known at least since Shapiro and Silverman (1960) to be alias-free for a very broad class of stationary processes. (Proposition 10.1 and Theorem 10.18 have similar implications.) Masry (1978b, 1983) examines other alias-free sampling methods.

Deterministic but not necessarily regular sampling is considered for diffusion processes in Le Breton (1975) and for time series by Robinson (1977).

The sole result in the introduction, Proposition 10.1, is due to Kingman (1963); it is generalized to random fields in Theorem 10.18.

Section 10.1

Theorem 10.3, the converse to the elementary Proposition 10.2, comes from Kingman (1963). The substitution estimators $\hat{A}$ of (10.14), and Theorems 10.4 and 10.5 concerning them, are new here but based on ideas in Karr (1984a), the source as well of Proposition 10.6 and the estimators to which it applies. More primitive versions of the latter appear in Karr (1982). Although formally new, Theorem 10.7 is a fairly straightforward extension of Karr (1984a, Theorem 10.1), which pertains only to binary Markov processes; the proof here is less intricate.

Statistical inference for Markov chains and processes with finite state space given complete observations is developed in Billingsley (1961a,b) and less completely in Basawa and Prakasa Rao (1980); identification of the estimator $\hat{Q}$ of (10.13) as nonparametric maximum likelihood estimator is noted in all. Doob (1953) remains perhaps the most lucid presentation of limit theorems underlying the asymptotics.

Section 10.2

The main sources are Masry (1978a, 1983) on synchronous and asynchronous Poisson sampling, respectively, of stationary processes on **R**. Thus the latter provides the estimators $\hat{R}$ of (10.40) and Theorems 10.11 and 10.12 concerning their properties, while from the former come the estimators $\hat{f}$ in (10.51), Propositions 10.13 and 10.14, and Theorem 10.15. Some details of proofs are given in these two papers; Masry and Lui (1976) contains related material. Central limit theorems once again rest ultimately on Leonov and Shiryayev (1959) (see also Brillinger, 1972).

Background on stationary processes can be found in Cramér and Leadbetter (1967) and Rozanov (1965); Karlin and Taylor (1975) is a more elementary, less complete exposition.

Section 10.3

All the results are taken from Karr (1986a). Theorem 10.18 generalizes, even in one dimension, Proposition 10.1, while each remaining aspect—estimators and properties alike—has an apparent analog in Section 10.2.

The whole topic of random fields has developed only rather recently. Problems on which research has centered are local behavior, extreme values, Markov properties, asymptotics and linear extrapolation; intricacies and subtleties are legion [central limit theorems are especially difficult (see, e.g., Deo, 1975)]. General sources are Adler (1981), Rozanov (1982), and Yadrenko (1983).

Exercises

10.4. See Karr (1982, 1984a), Theorem 8.24, and Exercise 3.15.

10.6. See Karr (1984a).

10.7. See also Section 8.4.

10.13. See also Theorem 9.15.

10.14, 10.15. Proofs are given in Masry (1978a).

10.17. See Karr (1986a).

Appendix A
Spaces of Measures

This appendix should not be regarded as more than a very selective compilation of definitions and results; motivation and proofs are equally absent. Cohn (1980), Dunford and Schwartz (1958), Matthes et al. (1978), and Rudin (1975) are representative treatments of a more substantial nature; Kallenberg (1983) contains an appendix at an intermediate level of detail.

Our setting is a locally compact Hausdorff space E with Borel σ-algebra $\mathscr{E}$. Let $\mathscr{B}$ be the family of bounded (relatively compact) sets in $\mathscr{E}$. By C, C_0, C_K we denote the set of bounded, continuous functions on E, the set of bounded, continuous functions vanishing at infinity ($f \in C_0$ if and only if f is continuous and for each $\varepsilon > 0$ there is a compact set K such that $|f(x)| < \varepsilon$ for all $x \notin K$), and the set of continuous functions on E with compact support. Both C and C_0 are Banach spaces under the uniform norm, the latter is separable, and C_K is dense in it. The Baire σ-algebra $\sigma(C_K)$ coincides with $\mathscr{E}$.

Except at the end of the ensuing discussion we restrict attention to positive measures.

Definition A.1. A measure μ on E is a *Radon measure* if $\mu(K) < \infty$ for each compact set K.

Evidently, μ is a Radon measure if and only if $|\mu(f)| = |\int f \, d\mu| < \infty$ for every $f \in C_K$. Some other basic properties merit mention.

Proposition A.2. If μ is a Radon measure, then

a) μ is σ-finite;

b) μ is regular: for each $A \in \mathscr{E}$,

$$\mu(A) = \sup\{\mu(K) : K \subset A,\ K \text{ is compact}\}$$

$$= \inf\{\mu(G) : A \subset G,\ G \text{ is open}\} \quad \square$$

Among Radon measures we distinguish several key subclasses.

Definition A.3. A Radon measure μ is

a) A *point* (or *counting*) *measure* if $\mu(A) \in \mathbf{N}$ for every $A \in \mathcal{B}$;

b) A *simple point measure* if μ is a point measure and if $\mu(\{x\}) \leq 1$ for each $x \in E$;

c) *Diffuse* if $\mu(\{x\}) = 0$ for every $x \in E$.

The fundamental point measures are the *point masses* ε_x defined by

$$\begin{aligned} \varepsilon_x(A) = 1(x \in A) &= 1 \qquad \text{if } x \in A \\ &= 0 \qquad \text{if } x \notin A \end{aligned}$$

(The alternative notation δ_x and terminology *Dirac measure* are widespread, especially for $E = \mathbf{R}$.) Theorem A.4 exhibits a point measure as a countable sum of point masses.

A point x is an *atom* of μ if $\mu(\{x\}) > 0$; the set of atoms is countable. A point measure consists solely of atoms of mass 1. At the opposite extreme, diffuse measures have no atoms at all and hence are analogous to continuous distribution functions on $\mathbf{R}$; corresponding to the familiar decomposition of a distribution function into discrete and continuous (but not necessarily absolutely continuous) parts is the following representation.

Theorem A.4. A Radon measure μ can be decomposed as

$$\mu = \mu_d + \sum_{i=1}^{K} a_i \varepsilon_{x_i}, \tag{A.1}$$

where μ_d is diffuse, $0 \leq K \leq \infty$, $a_i > 0$ for each i and the x_i are distinct points in E. The decomposition is unique within reordering of the (x_i, a_i). $\square$

The Radon measure μ is termed *purely atomic* if the diffuse component is zero. A purely atomic measure is a point measure if and only if $a_i \in \mathbf{N}$ for each i and in this case (x_i) can have no accumulation points in E. For a simple point measure, $a_i = 1$ for each i, giving $\mu(A) = \Sigma_{i=1}^{K} \varepsilon_{x_i}(A) = \Sigma_{i=1}^{K} 1(x_i \in A)$ as the number of points in A.

We denote by $\mathbf{M}$ the set of Radon measures on E and by $\mathbf{M}_p$, $\mathbf{M}_s$, $\mathbf{M}_a$, $\mathbf{M}_d$ the subsets of point measures, simple point measures, purely atomic measures, and diffuse measures. Once measurability structure is established, a *random measure* (purely atomic random measure, diffuse random measure) is by definition a random element of $\mathbf{M}$ ($\mathbf{M}_a$,$\mathbf{M}_d$) and a *point process* (simple point process) a random element of $\mathbf{M}_p$ ($\mathbf{M}_s$).

Unlike function spaces, the set $\mathbf{M}$ admits refreshingly simple measurability and topological structure. Essentially all functionals of interest are measurable with respect to the σ-algebra we are about to introduce, and many are continuous with respect to a Polish topology whose Borel σ-algebra coincides with it.

On $\mathbf{M}$ we define the σ-algebra

$$\mathcal{M} = \sigma(\mu \to \mu(f) : f \in C_K); \tag{A.2}$$

that is, $\mathcal{M}$ is the smallest σ-algebra on $\mathbf{M}$ rendering measurable each mapping $\mu \to \mu(f), f \in C_K$. Characterization in terms of sets is also possible.

Proposition A.5. The σ-algebra $\mathcal{M}$ is generated by the mappings $\mu \to \mu(A)$, $A \in \mathcal{B}$. □

In particular, $\mathcal{M}$ is countably generated. Each of the sets $\mathbf{M}_p$, $\mathbf{M}_s$, $\mathbf{M}_a$, $\mathbf{M}_d$ belongs to $\mathcal{M}$, and we define $\mathcal{M}_p = \mathcal{M} \cap \mathbf{M}_p, \ldots$.

For our purposes only the vague topology on $\mathbf{M}$ is required, but for completeness we mention also the weak topology as well as the strong topology engendered by the total variation norm.

Definition A.6. Suppose that μ_n, $n \geq 1$, and μ are elements of $\mathbf{M}$; then (μ_n) converges *vaguely* to μ if $\mu_n(f) \to \mu(f)$ for every $f \in C_K$.

We write vague convergence simply as $\mu_n \to \mu$. In many situations the definition is adequate for direct verification, but alternative forms are available.

Theorem A.7. For μ_n, $\mu \in \mathbf{M}$, the following assertions are equivalent:

a) $\mu_n \to \mu$;

b) For every compact set K, $\overline{\lim}\ \mu_n(K) \leq \mu(K)$ and for every bounded open set G, $\underline{\lim}\ \mu_n(G) \geq \mu(G)$;

c) $\lim \mu_n(A) = \mu(A)$ for every $A \in \mathcal{B}$ with $\mu(\partial A) = 0$;

d) $\lim \mu_n(f) = \mu(f)$ for every bounded, measurable function f on E vanishing outside some compact set and satisfying $\mu(D_f) = 0$, where D_f is the discontinuity set of f. □

In fact, if $\mu_n \to \mu$ and f_n, f are such that (f_n) is uniformly bounded, if the f_n are all zero outside a common compact set, and if $f_n(x_n) \to f(x)$ whenever $x_n \to x$ in E, then $\mu_n(f_n) \to \mu(f)$.

The mapping $x \to \varepsilon_x$ of E into $\mathbf{M}_p$ is one-to-one and vaguely continuous; thus we sometimes regard E as a subset of $\mathbf{M}_p$.

Measurability structure and the vague topology are intimately related.

Theorem A.8. The σ-algebra $\mathcal{M}$ is the Borel (and Baire) σ-algebra relative to the vague topology. □

While $\mathbf{M}_p$ is closed, none of $\mathbf{M}_s$, $\mathbf{M}_a$, $\mathbf{M}_d$ is.

For random measures and point processes, defined (Definition 1.1) as random elements of $(\mathbf{M},\mathcal{M})$ and $(\mathbf{M}_p,\mathcal{M}_p)$, respectively, convergence in distribution can be investigated with standard methods (set forth, e.g., in Billingsley, 1968) by virtue of the next result.

Theorem A.9. The spaces $\mathbf{M}$ and $\mathbf{M}_p$ are Polish (i.e., metrizable as complete, separable metric spaces) in the vague topology. □

In establishing convergence in distribution the key step is usually to show tightness, which is ordinarily done by direct or indirect appeal to Prohorov's theorem (see Billingsley, 1968), necessitating, in turn, a tractable characterization of relatively compact subsets, in this case of $\mathbf{M}$ or $\mathbf{M}_p$. For random measures and point processes this step is only rarely difficult because the following result implies (see Lemma 1.20) that tightness reduces to that of one-dimensional distributions.

Theorem A.10. For a subset Γ of $\mathbf{M}$ the following assertions are equivalent:

a) Γ is relatively compact in the vague topology;
b) $\sup\{\mu(A) : \mu \in \Gamma\} < \infty$ for each $A \in \mathcal{B}$;
c) $\sup\{|\mu(f)| : \mu \in \Gamma\} < \infty$ for every $f \in C_K$. □

The set of *finite* elements of $\mathbf{M}$ can in addition be endowed with the *weak* topology (which despite the name is stronger than the vague topology): (μ_n) converges weakly to μ, written $\mu_n \xrightarrow{w} \mu$, if $\mu_n(f) \to \mu(f)$ for every $f \in C_K$ (i.e., $\mu_n \to \mu$ vaguely) and in addition $\mu_n(E) \to \mu(E)$ or, equivalently, if $\mu_n(f) \to \mu(f)$ for every $f \in C$. Analogs of the results above hold (see Kallenberg, 1983). Duality considerations also lead to the weak topology; they stem from the Riesz representation theorem.

Theorem A.11. Given a positive linear functional L on C_K, there is a unique element μ of $\mathbf{M}$ such that $L(f) = \mu(f)$ for all $f \in C_K$. □

The vector space V of differences $\mu - \lambda$, where μ, λ are *finite* elements of $\mathbf{M}$, is normed by the *total variation norm*

$$\|\mu\| = \sup\{\textstyle\sum|\mu(A_i)| : (A_i) \text{ is a finite partition of } E\}; \qquad \text{(A.3)}$$

this norm—applied to probability measures—is utilized in Sections 1.6 and 2.4. Its properties on V include the following.

Theorem A.12. a) V is a Banach space under the total variation norm;

b) V is the Banach space dual of C_0;

c) The weak*-topology induced on V by C_0 is the weak topology as defined above. □

Appendix B
Continuous-Time Martingales

Even more than Appendix A, this appendix is selective and incomplete. All the subtlety and much of the beauty of the theory are missing from our presentation. With one exception no proofs are given, but some heuristic arguments germane to point processes are sketched.

The setting, save in Theorem B.21, where the history and martingale have particular form, is a probability space $(\Omega,\mathcal{F},P)$ together with a history $\mathcal{H} = (\mathcal{H}_t)_{t\geq 0}$ satisfying the "conditions habituelles" ($\mathcal{H}$ is right-continuous and $\mathcal{H}_0$ contains all null sets in $\mathcal{H}_\infty = \vee_{t\geq 0} \mathcal{H}_t$). Important concepts: adapted processes, stopping times, predictable processes, ... are mentioned—albeit briefly—in Section 2.1. The qualification "with respect to $\mathcal{H}$" is omitted throughout.

We recall that a process X is càdlàg (*continue à droite, limité à gauche*) if almost surely the path $t \rightarrow X_t(\omega)$ is right-continuous on $[0,\infty)$ and admits left-hand limits on $(0,\infty)$. For martingales this can be proved rather than stipulated, but since in this book "càdlàguity" holds by construction (point processes are right-continuous), we incorporate it in the definition. The reader is warned that several of the results below are not valid without right continuity.

Definition B.1. a) An adapted càdlàg process M is a *martingale* if $E[|M_t|] < \infty$ for each t and if whenever $s < t$,

$$E[M_t|\mathcal{H}_s] = M_s \tag{B.1}$$

b) An adapted càdlàg process Z is a *submartingale* if $E[|Z_t|] < \infty$ for each t and if $s < t$ implies that

$$E[Z_t|\mathcal{H}_s] \geq Z_s \tag{B.2}$$

c) An adapted càdlàg process Z is a *supermartingale* if the process $-Z = (-Z_t)$ is a submartingale.

In connection not only with point processes but also Wiener processes (see, e.g., Kallianpur, 1980, or Liptser and Shiryayev, 1978) square integrable martingales are particularly important.

Definition B.2. A martingale M is *square integrable* if

$$\sup_{t \geq 0} E[M_t^2] < \infty \tag{B.3}$$

If M is a square integrable martingale, then by Jensen's inequality for conditional expectations, the process M^2 is a submartingale.

The important concept of the predictable variation of a square integrable martingale M stems from a particular decomposition of the submartingale M^2. Before presenting it, we introduce further terminology.

Definition B.3. An adapted process A is an *increasing process* if $A_0 = 0$ and if each sample path $t \to A_t$ is nondecreasing and right-continuous. If, in addition, $\sup_t E[A_t] < \infty$, then A is *integrable*.

Predictable increasing processes, one component of the Doob-Meyer decomposition (B.5), can be characterized via martingales.

Theorem B.4. An integrable increasing process A is predictable if and only if for every positive martingale M and each $t \geq 0$,

$$E[\int_0^t M\,dA] = E[\int_0^t M_-\,dA], \tag{B.4}$$

where M_- is the process of left-hand limits of M. □

We cannot pursue here the rather profound analogy between supermartingales and superharmonic functions (see, e.g., Neveu, 1975), but within this context one distinguishes supermartingales analogous to potentials (i.e., superharmonic functions vanishing "at the boundary").

Definition B.5. A supermartingale Z is a *potential* if $Z \geq 0$ and if $\lim_{t\to\infty} E[Z_t] = 0$.

We come now to the fundamental decomposition theorem, due to Doob for discrete-time supermartingales and to Meyer (1965) in the continuous-time case.

Theorem B.6. Given a potential Z, the following statements are equivalent:

a) There exists a predictable increasing process A (necessarily integrable) such that for all t,

$$Z_t = E[A_\infty|\mathcal{H}_t] - A_t; \tag{B.5}$$

b) Z is a class (D): the family of random variables $\{Z_T$: T is a finite stopping time$\}$ is uniformly integrable. □

When it exists, the *Doob-Meyer decomposition* (B.5) is unique within an evanescent process (satisfying $P\{X_t = 0$ identically in $t\} = 1$). Note that the process $M_t = E[A_\infty|\mathcal{H}_t]$ is a unformly integrable martingale; thus one interpretation of (B.5) is that the nonmartingale part of a supermartingale is predictable.

A point process N is evidently a submartingale and hence given sufficient integrability admits a Doob-Meyer decomposition $N = M + A$, with M a martingale and A, the *compensator* of N, a predictable increasing process. (In Chapter 2, the compensator was derived using dual predictable projections, an approach at once more general and more amenable to computation.) The fundmental tool for analysis of the innovation martingale M is the predictable variation process $<M>$, which results from the Doob-Meyer decomposition of the submartingale M^2.

Theorem B.7. Let M be a square integrable martingale. Then there exists a unique predictable increasing process $<M>$ such that $<M>_0 = M_0^2$ and such that $M^2 - <M>$ is a martingale. □

Definition B.8. The process $<M>$ is the *predictable variation* of M.

Since a square integrable martingale has orthogonal increments, for $s < t$,

$$E[(M_t - M_s)^2|\mathcal{H}_s] = E[<M>_t - <M>_s|\mathcal{H}_s]$$

By analogy with Theorem 2.17, if t is fixed, then almost surely

$$<M>_t = M_0^2 + \lim_{n\to\infty} \sum_{k=1}^{2^n} E[(M_{kt/2^n} - M_{(k-1)t/2^n})^2|\mathcal{H}_{(k-1)t/2^n}] \tag{B.6}$$

For example, if N is a point process with stochastic intensity λ (Definition 2.28), then

$$<M>_t = \lim \sum_{k=1}^{2^n} E\left[\left(N_{kt/2^n} - N_{(k-1)t/2^n} - \int_{(k-1)t/2^n}^{kt/2^n} \lambda_s\, ds\right)^2 \Big| \mathcal{H}_{(k-1)t/2^n}\right]$$

$$= \lim \sum_{k=1}^{2^n} E[(N_{kt/2^n} - N_{(k-1)t/2^n})^2|\mathcal{H}_{(k-1)t/2^n}]$$

$$- 2\lim \sum_{k=1}^{2^n} \left(\int_{(k-1)t/2^n}^{kt/2^n} \lambda_s \, ds \right) E[N_{kt/2^n} - N_{(k-1)t/2^n} | \mathcal{H}_{(k-1)t/2^n}]$$

$$+ \lim \sum_{k=1}^{2^n} \left(\int_{(k-1)t/2^n}^{kt/2^n} \lambda_s \, ds \right)^2$$

(by predictability of λ)

$$= \lim \sum_{k=1}^{2^n} E[N_{kt/2^n} - N_{(k-1)t/2^n} | \mathcal{H}_{(k-1)t/2^n}] = \int_0^t \lambda_s \, ds,$$

where the last equality is by (2.23).

The space of square integrable martingales becomes a Hilbert space under the norm $\|M\| = E[<M>_\infty]$; the inner product is constructed under the assumption that the polar identity holds, leading to predictable covariation processes.

Definition B.9. For square integrable martingales M and N,

$$<M,N> = \frac{1}{4}[<M+N> - <M-N>] \tag{B.7}$$

is the *predictable covariation* of M and N.

Orthogonality is then defined in the obvious manner.

Definition B.10. Square integrable martingales M and N are *orthogonal*, written $M \perp N$, if $<M,N> \equiv 0$.

Evidently, $<M,M> = <M>$. The characterization that $M^2 - <M>$ is a martingale extends: $<M,N>$ is the unique predictable càdlàg process such that $<M,N>_0 = M_0 N_0$ and such that $MN - <M,N>$ is a martingale. Orthogonality is then characterized easily.

Theorem B.11. Square integrable martingales M and N are orthogonal if and only if MN is a martingale. □

Stochastic integration with respect to martingales is treated in depth and generality in (among others) Elliott (1982), Jacod (1979), Kallianpur (1980), and Metivier and Pellaumial (1980). We do not require such generality: integrals with respect to innovation martingales for point processes exist pathwise as Lebesgue-Stieltjes integrals. Given a process C and martingale M for which the *stochastic integral* process $\int_0^t C \, dM$ exists—whether by general theory or by specialized construction—we denote it by $C*M$. Here as throughout the book integrals $\int_0^t$ are over the closed interval $[0,t]$. The central property underlying martingale inference for point processes is that the stochastic integral $C*M$ of a predictable process with respect to a martingale is itself a martingale; heuristically,

$$E[(C*M)_{t+\Delta t}-(C*M)_t|\mathcal{H}_t]\cong E[\,C_t(M_{t+\Delta t}-M_t)|\mathcal{H}_t]$$
$$=C_tE[M_{t+\Delta t}-M_t|\mathcal{H}_t]=0$$

by predictability of C followed by the property that M is a martingale. Theorem 2.9 contains a formal proof based on elementary predictable processes and the monotone class theorem. A similarly heuristic argument to the effect that

$$\Delta<C*M>_t \cong E[\{(C*M)_{t+\Delta t}-(C*M)_t\}^2|\mathcal{H}_t]$$
$$\cong C_t^2E[M_{t+\Delta t}-M_t)^2|\mathcal{H}_t]\cong C_t^2\Delta<M>_t$$

provides motivation for the second part of the next result.

Theorem B.12. If M is a square integrable martingale and C is a bounded, predictable process, then the stochastic integral $C*M$ is a square integrable martingale with

$$<C*M>_t=\int_0^t C^2\,d<M> \quad \square \tag{B.8}$$

More generally we have the following property.

Theorem B.13. Suppose that $M(1)$, $M(2)$ are square integrable martingales, that $C(1)$, $C(2)$ are bounded, predictable processes, and that $\widetilde{M}_t(i)=\int_0^t C(i)\,dM(i)$, $i=1,2$. Then

$$<\widetilde{M}(1),\widetilde{M}(2)>_t=\int_0^t C_u(1)C_u(2)\,d<M(1),M(2)>_u \quad \square \tag{B.9}$$

Especially, if $M(1)$ and $M(2)$ are orthogonal, then so are $\widetilde{M}(1)$, $\widetilde{M}(2)$; this property is used, for example, in the proof of Proposition 5.4.

Only a minor role is played in this book by the quadratic variation of a square integrable martingale, but for completeness we mention it briefly. The formal definition is that the quadratic variation of M is the process

$$[M]_t=<M^c>_t+\sum_{s\leqslant t}(\Delta M_s)^2, \tag{B.10}$$

where M^c is the continuous martingale part of M [realized from an orthogonal decomposition $M=M^c+M^d$, where M^c is a continuous martingale, M^d is a "compensated sum of jumps," and $M^c \perp M^d$ (see Jacod, 1979, for formalities)]. The more easily understood approach is via a counterpart of (B.6):

$$[M]_t = M_0^2 + \lim_{n\to\infty} \sum_{k=1}^{2^n} (M_{kt/2^n} - M_{(k-1)t/2^n})^2; \qquad \text{(B.11)}$$

thus if $M_t = N_t - \int_0^t \lambda\, ds$ is the innovation martingale of a point process N with stochastic intensity λ, then $[M]_t = N_t$ for each t. The dual predictable projection of the quadratic variation is the predictable variation.

None of the three grand classes of martingale results—inequalities, convergence theorems, and optional sampling theorems—is central to this book, but no presentation can ignore them. We present parallel versions of each, for supermartingales and for martingales.

Martingale inequalities are important because they lead to the convergence theorems, but are also the most important link between martingale theory and classical analysis.

Theorem B.14. If Z is a supermartingale, then for each t and $a > 0$,

$$P\{\sup_{s\le t} Z_t \ge a\} \le a^{-1}\{E[Z_0] - E[Z_t^-]\} \quad \square \qquad \text{(B.12)}$$

Theorem B.15. For each $p > 1$ there exist universal constants c_p, C_p such that for each martingale M and each $t \ge 0$,

$$c_p E[[M]_t^{p/2}] \le E[(\sup_{s\le t} |M_s|)^p] \le C_p E[[M]_t^{p/2}] \quad \square \qquad \text{(B.13)}$$

Of the sometimes bewildering variety of martingale convergence thoerems, the following are two of the most useful.

Theorem B.16. If Z is a supermartingale with $\sup_t E[Z_t^-] < \infty$, then there exists a random variable $Z_\infty \in L^1(P)$ such that $Z_t \to Z_\infty$ almost surely as $t \to \infty$. $\square$

Theorem B.16 applies in particular to nonnegative martingales and supermartingales, among them likelihood ratios; in general, the convergence $Z_t \to Z_\infty$ is not realized in $L^1(P)$. For martingales L^1-convergence is equivalent to uniform integrability and in this case the martingale has very simple structure.

Theorem B.17. For a martingale M the following assertions are equivalent:

a) M is uniformly integrable;

b) There is a random variable $M_\infty \in L^1(P)$ such that $M_t \to M_\infty$ almost surely and in $L^1(P)$;

c) There is a random variable $M_\infty \in L^1(P) \cap \mathcal{H}_\infty$ such that $M_t = E[M_\infty | \mathcal{H}_t]$ for each t. □

Optional sampling theorems pertain to preservation of martingale or supermartingale properties when the constant times in Definition B.1 are replaced by stopping times.

Theorem B.18. Let Z be a supermartingale for which there exists a random variable $Y \in L^1(P)$ satisfying $Z_t \geq E[Y|\mathcal{H}_t]$ for each t (i.e., Z is bounded below by a uniformly integrable martingale, and hence fulfills the hypotheses of Theorem B.16). Then for all stopping times $T_1 \leq T_2 \leq \infty$, Z_{T_1} and Z_{T_2} belong to $L^1(P)$ and $E[Z_{T_2}|\mathcal{H}_{T_1}] \leq Z_{T_1}$. □

Theorem B.19. Let $M_t = E[M_\infty|\mathcal{H}_t]$ be a uniformly integrable martingale. Then for each stopping time $T \leq \infty$, $M_T = E[M_\infty|\mathcal{H}_T]$. Consequently, for stopping times $T_1 \leq T_2 \leq \infty$, $E[M_{T_2}|\mathcal{H}_{T_1}] = M_{T_1}$. □

Often a process fails to be, for example, a martingale only because integrability or square integrability requirements fail, yet the property in question holds for certain "stopped" processes. "Localization" makes this idea precise and usable.

Definition B.20. A process X is said to possess a property (P) (e.g., that of being a martingale) *locally* if there exist stopping times $T_1 \leq T_2 \leq \cdots \to \infty$ almost surely, such that for each n the stopped process $(X_{t \wedge T_n})$ has property (P). The T_n are *localizing times*.

For example, M is a *local martingale* if there are localizing times T_n such that each process $(M_{t \wedge T_n})$ is a martingale. In general—the most obvious exception being the convergence theorems, which degenerate—results above carry over if both hypotheses *and* conclusions are localized correspondingly.

Ordinarily, the localizing times depend on the property (P) but for a simple point process $N = \Sigma\, \varepsilon_{T_n}$ and $\mathcal{H} = \mathcal{F}^N$ the points T_n serve universally. For example, if $E[N_t] < \infty$ for each t, then the compensator exists, the innovation martingale M is locally square integrable, and hence there exists a locally integrable, predictable, increasing process $<M>$ such that $M^2 - <M>$ is a local martingale. For clarity and ease of exposition, in the text we have imposed hypotheses sufficiently strong to avoid localization; most of the results extend, however. When the underlying space is [0,1], as in Chapter 5, no harm ensues, but for processes on $[0,\infty)$, localization is mandatory in all interesting cases.

We conclude with a central limit theorem specific to point process martingales. Among its applications is Theorem 9.25.

Theorem B.21. Let N be a point process with continuous compensator A and $\mathcal{F}^N$-innovation martingale M, and suppose that Y is a predictable process satisfying

$$t^{-1}\int_0^t Y^2\,dA \to \sigma^2 \tag{B.14}$$

in the sense of convergence in probability, where $0<\sigma^2<\infty$. Then

$$t^{-1/2}\int_0^t Y\,dM \xrightarrow{d} N(0,\sigma^2) \tag{B.15}$$

Proof: (Sketch). The argument is substantially that employed by Kutoyants (1975) to establish a similar result for integrals with respect to diffusion processes.

For each predictable process B, the process

$$M_t(B) = \exp\left[-\int_0^t B\,dN; -\int_0^t (e^B-1)\,dA\right]$$

is a local square integrable martingale [it is the likelihood ratio associated with the probability measure under which N has compensator $d\tilde{A} = e^B\,dA$ (see Theorem 2.31)]. Glossing over technicalities we infer via the optional sampling theorem that

$$E[M_T(B)] = E[M_0(B)] = 1 \tag{B.16}$$

for each stopping time T.

By appeal to (B.14), define a random time change (T_t) by $\int_0^{T_t} Y^2\,dA = t\sigma^2$ [strictly speaking, $t\sigma^2$ must be replaced by $(1-\varepsilon)t\sigma^2$, but these kinds of details can be dealt with using the methods of Kutoyants, 1975, and we do not dwell on them]. With t fixed and (B.16) applied to $B_s = iuY_s/t^{1/2}$, where $u\in\mathbf{R}$ and $i^2=-1$,

$$1 = E[M_{T_t}(iut^{-1/2}Y)]$$

$$= E[\exp\{iut^{-1/2}\int_0^{T_t} Y\,dN - \int_0^{T_t} (e^{iut^{-1/2}Y}-1)\,dA\}]$$

$$= E\left[\exp\left\{iut^{-1/2}\int_0^{T_t} Y\,dN - iut^{-1/2}\int_0^{T_t} Y\,dA + \frac{u^2}{2t}\int_0^{T_t} Y^2\,dA\right\}\right.$$

$$\times \exp\left\{ -\int_0^{T_t} \sum_{k=3}^{\infty} \frac{(iut^{-1/2}Y)^k}{k!} dA \right\} \Bigg]$$

$$= e^{u^2\sigma^2/2} E[\exp\{ iut^{-1/2} \int_0^{T_t} Y\, dM \} R_t],$$

where in view of (B.14), $R_t \to 1$ in probability. Consequently,

$$t^{-1/2} \int_0^{T_t} Y\, dM \xrightarrow{d} N(0,\sigma^2)$$

To complete the proof it remains to show that

$$t^{-1/2} [\int_0^t Y\, dM - \int_0^{T_t} Y\, dM] \to 0$$

in probability, to which once more the reasoning of Kutoyants (1975) applies with only minor modification. □

References

Aalen, O. O. (1975). *Statistical Inference for a Family of Counting Processes*. Ph.D. dissertation, The University of California, Berkeley.

Aalen, O. O. (1976). Nonparametric inference in connection with multiple decrement models. *Scand. J. Statist.* 3, 15 – 27.

Aalen, O. O. (1977). Weak convergence of stochastic integrals related to counting processes. *Z. Wahrscheinlichkeitstheorie und verw. Geb.* 38, 261 — 277.

Aalen, O. O. (1978). Nonparametric inference for a family of counting processes. *Ann. Statist.* 6, 701 – 726.

Aalen, O. O. (1980). A model for nonparametric regression analysis of counting processes. *Lect. Notes Statist.* 2, 1 – 25.

Aalen, O. O., and Hoem, J. M. (1978). Random time changes for multivariate counting processes. *Skand. Akt.* 1978, 81 – 101.

Aalen, O. O., and Johansen, S. (1978). An empirical transition matrix for nonhomogeneous Markov chains based on censored observations. *Scand. J. Statist.* 5, 141 – 150.

Adler, R. J. (1981). *The Geometry of Random Fields*. Wiley, New York.

Akritas, M. G., and Johnson, R. A. (1981). Asymptotic inference in Lévy processes of discontinuous type. *Ann. Statist.* 9, 604 – 614.

Albrecht, P. (1981). Über einige Eigenschaften des gemischten Poissonprozesse. *Mitt. Verein. Schweiz. Versicherungsmath.* 1981, 241 – 250.

Albrecht, P. (1982a). Testing the goodness of fit of a mixed Poisson process. *Insurance Math. Econ.* 1, 27 – 33.

Albrecht, P. (1982b). On some statistical methods connected with the mixed Poisson process. *Scand. Actuarial J.* 1982, 1 — 14.

Aldous, D. (1978). Stopping times and tightness. *Ann. Probability 6*, 335 – 340.

Ambartzumian, R. V. (1965). Two inverse problems concerning the superposition of recurrent point processes. *J. Appl. Prob.* 2, 449 – 454.

Ambartzumian, R. V. (1972). Palm distributions and superpositions of independent point processes in $\mathbf{R}^n$. In *Stochastic Point Processes,* Lewis, ed. Wiley, New York.

Ambartzumian, R. V., and Nahapetian, B. S. (1980). The Palm distribution and limit theorems for random point processes. *Akad. Nauk. Armjan. SSR Dokl.* 71, 87 – 90.

Ammann, L. P., and Thall, P. F. (1979). Count distributions, orderliness and invariance of Poisson cluster processes. *J. Appl. Prob.* 16, 261 – 273.

Andersen, P. K. (1982). Testing goodness of fit of Cox's regression and life model. *Biometrics* 38, 67 – 77.

Andersen, P. K. (1983). Comparing survival distributions via hazard ratio estimates. *Scand. J. Statist.* 10, 77 – 86.

Andersen, P.K., and Borgan, O. (1985). Counting process models for life history data: a review. *Scand. J. Statist.* 12, 97 — 158.

Andersen, P. K., and Gill, R. D. (1982). Cox's regression model for counting processes: a large sample study. *Ann. Statist.* 10, 1100 – 1120.

Andersen, P. K., Borgan, O., Gill, R. D., and Keiding, N. (1982). Linear nonparametric tests for comparison of counting processes, with application to censored survival data. *Internat. Statist. Rev.* 50, 219 – 244.

Anderson, T. W. (1958). *An Introduction to Multivariate Statistical Analysis*. Wiley, New York.

Anderson, T. W., and Darling, D. A. (1952). Asymptotic theory of certain 'goodness of fit' criteria based on stochastic processes. *Ann. Math. Statist.* 23, 193 – 212.

Araujo, A., and Giné, E. (1980). *The Central Limit Theorem for Real and Banach-Valued Random Variables*. Wiley, New York.

Asher, R. B., and Lainiotis, D. G. (1977). Adaptive estimation of doubly stochastic Poisson processes. *Inform. Sci.* 12, 245 – 261.

Athreya, K. B., Tweedie, R. L., and Vere-Jones, D. (1980). Asymptotic behaviour of point processes with Markov dependent intervals. *Math. Nachr.* 99, 301 – 313.

Baba, Y. (1982). Maximum likelihood estimation of parameters in birth and death processes by Poisson sampling. *J. Oper. Res. Soc. Japan* 25, 99 – 112.

Baddeley. A, (1980). A limit theorem for statistics of spatial data. *Adv. Appl. Prob.* 12, 447 – 461.

Bailey, K. R. (1983). The asymptotic joint distribution of regression and survival parameter estimates in the Cox regression model. *Ann. Statist.* 11, 39 – 48.

Balakrishnan, A. V. (1972). A martingale approach to linear recursive state estimation. *SIAM J. Control* 10, 754 — 766.

Banys, R. (1975). The convergence of sums of dependent point processes to Poisson processes. *Litovsk. Mat. Sb.* 15, 11 – 23.

Banys, R. (1979a). The convergence of superpositions of integer-valued random measures. *Lithuanian Math. J.* 19, 1 – 15.

Banys, R. (1979b). Limit theorems for superpositions of multidimensional integer-valued random processes. *Lithuanian Math. J.* 19, 15 – 24.

Banys, R. (1980). On superposition of random measures and point processes. *Lect. Notes Statist.* 2, 26 – 37.

Banys, R. (1982). On the convergence of superpositions of point processes in the space $D[0,1]^2$. *Litovsk. Mat. Sb.* 22, 3 – 7.

Barlow, R. E. (1968). Likelihood ratio tests for restricted families of probability distributions. *Ann. Math. Statist.* 39, 547 – 560.

Barlow, R. E., and Proschan, F. (1969). A note on tests for monotone failure rate based on incomplete data. *Ann. Math. Statist.* 40, 595 – 600.

Barlow, R. E., and Proschan, F. (1975). *Statistical Theory of Reliability and Life Testing: Probability Models.* Silver Spring, Md.

Barlow, R. E., Bartholomew, D. J., Bremner, J. M., and Brunk, H. D. (1972). *Statistical Inference under Order Restrictions.* Wiley, New York.

Barndorff-Nielson, O. (1965). Identifiability of mixtures of exponential families. *J. Math. Anal. Appl.* 12, 115 – 121.

Barndorff-Nielson, O., and Yeo, G. F. (1969). Negative binomial processes. *J. Appl. Prob.* 6, 633 — 647.

Bártfai, P., and Tomkó, J. (1981). *Point Processes and Queueing Problems.* North-Holland, Amsterdam.

Bartholomew, D. J. (1956). Tests for randomness in a series of events when the alternative is trend. *J. Roy. Statist. Soc. B* 18, 234 – 239.

Bartlett, M. S. (1954). Processus stochastiques ponctuels. *Ann. Inst. H. Poincaré* 14, 35 – 60.

Bartlett, M. S. (1963). The spectral analysis of point processes. *J. Roy. Statist. Soc. B* 25, 264 – 296.

Bartlett, M. S. (1964). The spectral analysis of two-dimensional point processes. *Biometrika* 51, 299 – 311.

Bartlett, M. S. (1967). The spectral analysis of line processes. *Proc. Fifth Berkeley Symp. Math. Statist. Prob.* III, 135 – 153.

Bartlett, M. S. (1974). The statistical analysis of spatial pattern. *Adv. Appl. Prob.* 6, 336 – 358.

Bartlett, M. S. (1975). *The Statistical Analysis of Spatial Pattern.* Chapman & Hall, London.

Bartoszyński, R., Brown, B. W., McBride, C., and Thompson, J. R. (1981). Some nonparametric techniques for estimating the intensity function of a cancer related nonstationary Poisson process. *Ann. Statist.* 9, 1050 – 1060.

Basawa, I. V., and Prakasa Rao, B. L. S. (1980). *Statistical Inference for Stochastic Processes*. Academic Press, New York.

Baudin, M. (1981). Likelihood and nearest-neighbor distance properties of multidimensional Poisson cluster processes. *J. Appl. Prob.* 18, 879 - 888.

Begun, J. M., Hall, W. J., Huang, W.-M., and Wellner, J. A. (1983). Information and asymptotic efficiency in parametric-nonparametric models. *Ann. Statist.* 11, 432 - 452.

Belkin, B. (1971). First passage to a general threshold for a process corresponding to sampling at Poisson times. *J. Appl. Prob.* 8, 573 - 588.

Beneveniste, A., and Jacod, J. (1975). Intensité stochastique du processus ponctuel stationnaire, reconstruction du processus ponctuel à partir de son intensité stochastique. *C. R. Acad. Sci. Paris* 280, A821 - A824.

Berman, M. (1977). Some multivariate generalizations of results in univariate stationary point processes. *J. Appl. Prob.* 14, 748 - 757.

Berman, M. (1978). Regenerative multivariate point processes. *Adv. Appl. Prob.* 10, 411 - 430.

Berman, M. (1981). Inhomogeneous and modulated gamma processes. *Biometrika* 68, 143 - 152.

Beutler, F. J., and Dolivo, F. B. (1976). Recursive integral equations for the detection of counting processes. *J. Appl. Math. and Opt.* 3, 65 - 72.

Beutler, F. J., and Leneman, O. A. Z. (1966a). The theory of stationary point processes. *Acta Math.* 116, 159 - 197.

Beutler, F. J., and Leneman, O. A. Z. (1966b). Random sampling of random processes: stationary point processes. *Inform. and Control* 9, 325 — 346.

Bickel, P. J. (1969). Tests for monotone failure rate, II. *Ann. Math. Statist.* 40, 1250 - 1260.

Bickel, P. J., and Doksum, K. A. (1968). Tests for monotone failure rate based on normalized spacings. *Ann. Math. Statist.* 40, 1216 - 1235.

Bickel, P. J., and Doksum, K. A. (1977). *Mathematical Statistics: Basic Ideas and Selected Topics*. Holden-Day, San Francisco.

Billingsley, P. (1961a). *Statistical Inference for Markov Processes*. University of Chicago Press, Chicago.

Billingsley, P. (1961b). Statistical methods in Markov chains. *Ann. Math. Statist.* 32, 12 - 40.

Billingsley, P. (1968). *Convergence of Probability Measures*. Wiley, New York.

Billingsley, P., and Tøpsøe, F. (1967). Uniformity in weak convergence. *Z. Wahrscheinlichkeitstheorie und verw. Geb.* 7, 1 - 16.

Birnbaum, A. (1954). Statistical methods for Poisson processes and exponential populations. *J. Amer. Statist. Assoc.* 49, 254 – 266.

Blackwell, D. (1948). A renewal theorem. *Duke Math. J.* 15, 145 – 150.

Blanc-Pierre, A., and Sultan, R. (1969). Sur deux modèles de fonctions aléatoires dérivées d'un processus de Poisson. *C. R. Acad. Sci. Paris* 269, A326 – A328.

Blumenthal, R. M., and Getoor, R. K. (1968). *Markov Processes and Potential Theory*. Academic Press, New York.

Blumenthal, S., Greenwood, J. A., and Herbach, L. (1971). Superimposed nonstationary renewal processes. *J. Appl. Prob.* 8, 184 – 192.

Boel, R. K., and Beňes, V. (1980). Recursive nonlinear estimation of a diffusion acting as the rate of an observed Poisson process. *IEEE Trans. Inform. Theory* IT-26, 561 – 575.

Boel, R., Varaiya, P., and Wong, E. (1975). Martingales on jump processes, I: Representation results, and II: Applications. *SIAM J. Control* 13, 999 – 1061.

Böker, F., and Serfozo, R. F. (1983). Ordered thinnings of point processes and random measures. *Stochastic Process. Appl.* 15, 113 – 132.

Borgan, O. (1984). Maximum likelihood estimation in a parametric counting process model, with application to censored failure time data and multiplicative models. *Scand. J. Statist.* 11, 1 – 16.

Borgan, O., and Ramlau-Hansen, H. (1985). Demographic incidence rates and estimation of intensities with incomplete information *Ann. Statist.* 13, 564 – 582.

Boswell, M. T. (1966). Estimating and testing trend in a stochastic process of Poisson type. *Ann. Math. Statist.* 37, 1564 – 1573.

Boswell, M. T., and Brunk, H. D. (1969). Distribution of likelihood ratio in testing against trend. *Ann. Math. Statist.* 40, 371 – 380.

Breiman, L. (1963). The Poisson tendency in traffic distribution. *Ann. Math. Statist.* 34, 308 – 311.

Brémaud, P. (1972). *A Martingale Approach to Point Processes*. Ph.D. dissertation, The University of California, Berkeley.

Brémaud, P. (1975a). An extension of Watanabe's characterization theorem. *J. Appl. Prob.* 8, 396 – 399.

Brémaud, P. (1975b). La méthode des semi-martingales en filtrage lorsque l'observation est un processus ponctuel marqué. *Lect. Notes Math.* 511, 1 – 18.

Brémaud, P. (1975c). On the information carried by a stochastic point process. *Cahiers du CETHEDEC* 43, 43 – 70.

Brémaud, P. (1976). Bang-bang controls of point processes. *Adv. Appl. Prob.* 8, 385 – 394.

Brémaud, P. (1979). Optimal thinning of a point process. *SIAM J. Control and Optimization* 17, 222 – 230.

Brémaud, P. (1981). *Point Processes and Queues: Martingale Dynamics*. Springer-Verlag, Berlin.

Brémaud, P., and Jacod, J. (1977). Processus ponctuels et martingales: résultats récents sur la modèlisation et le filtrage. *Adv. Appl. Prob.* 9, 362 – 416.

Breslow, N. (1975). Analysis of survival data under the proportional hazards model. *Internat. Statist. Rev.* 43, 45 – 58.

Breslow, N., and Crowley, J. (1974). A large sample study of the life table and product limit estimators under random censorship. *Ann. Statist.* 2, 437 — 453.

Brillinger, D. R. (1969). The asymptotic representation of the sample distribution function. *Bull. Amer. Math. Soc.* 75, 545 – 547.

Brillinger, D. R. (1972). The spectral analysis of stationary interval functions. *Proc. Sixth Berkeley Symp. Math. Statist. Prob.* 1, 483 – 513.

Brillinger, D. R. (1974). Cross-spectral analysis of processes with stationary increments, including the $G/G/\infty$ queue. *Ann. Probability* 2, 815 – 827.

Brillinger, D. R. (1975a). Identification of point process systems. *Ann. Probability* 3, 909 – 929.

Brillinger, D. R. (1975b). Statistical inference for stationary point processes. In *Stochastic Processes and Related Topics,* Puri, ed. Academic Press, New York.

Brillinger, D. R. (1978). Comparative aspects of the study of ordinary time series and point processes. In *Developments in Statistics,* Vol. 1, Krishnaiah, ed. Academic Press, New York.

Brillinger, D. R. (1979). Analyzing point processes subjected to random deletions. *Canad. J. Statist.* 7, 21 – 27.

Brillinger, D. R. (1981). *Time Series: Data Analysis and Theory.* Holden-Day, San Francisco.

Brown, M. (1969). An invariance property of Poisson processes. *J. Appl. Prob.* 6, 453 – 458.

Brown, M. (1970). A property of Poisson processes and its application to equilibrium of particle systems. *Ann. Math. Statist.* 41, 1935 – 1941.

Brown, M. (1971). Discrimination of Poisson processes. *Ann. Math. Statist.* 42, 773 – 776.

Brown, M. (1972). Statistical analysis of nonhomogeneous Poisson processes. In *Stochastic Point Processes*, Lewis, ed. Wiley, New York.

Brown, M. (1980). Bounds, inequalities and monotonicity properties for some specialized renewal processes. *Ann. Probability* 8, 227 – 240.

Brown, M., and Solomon, H. (1974). Some results for secondary processes generated by a Poisson process. *Stochastic Process. Appl.* 2, 337 – 348.

Brown, M., Solomon, H., and Stephens, M. A., (1977) Estimation of parameters of zero-one processes by interval sampling. *Oper. Res.* 25, 493 — 505.

Brown, M., Solomon, H., Stephens, M. A. (1979). Estimation of parameters of zero-one processes by Interval sampling: an adaptive approach. *Oper. Res.* 27, 606 — 615.

Brown, T. C. (1978). A martingale approach to the Poisson convergence of simple point processes. *Ann. Probability* 6, 615 – 628.

Brown, T. C. (1979). Position dependent and stochastic thinnings of point processes. *Stochastic Process. Appl.* 9, 189 – 193.

Brown, T. C. (1981). Compensators and Cox convergence. *Proc. Cambridge Philos. Soc.* 90, 305 – 319.

Brown, T. C. (1983). Some Poisson approximations using compensators. *Ann. Probability* 11, 726 – 744.

Brown, T. C., and Kupka, J. (1983). Ramsey's theorem and Poisson random measures. *Ann. Probability* 11, 904 – 908.

Brown, T. C., and Silverman, B. W. (1978). Short distances, flat triangles, and Poisson limits. *J. Appl. Prob.* 15, 815 — 825.

Brown, T. C., Silverman, B. W., and Milne, R. K. (1981). A class of 2-type point processes. *Z. Wahrscheinlichkeitstheorie und verw. Geb.* 58, 299 – 308.

Brown, W. M. (1963). Sampling with random jitter. *J. SIAM* 11, 460 – 473.

Bucy, R. S., and Kalman, R. (1961). New results in linear filtering and prediction theory. *J. Basic Engg.* 83, 95 – 108.

Burton, R. M., and Waymire, E. (1984). Scaling limits for point random fields. *J. Multivariate Anal.* 15, 237 – 251.

Byth, K. (1981). θ-stationary point processes and their second-order analysis. *J. Appl. Prob.* 18, 864 – 878.

Cane, V. R. (1977). A class of nonidentifiable stochastic models. *J. Appl. Prob.* 14, 475 – 482.

Carter, D. S., and Prenter, P. M. (1972). Exponential spaces and counting processes. *Z. Wahrscheinlichkeitstheorie und verw. Geb.* 21, 1 – 19.

Chen, Y. Y., Hollander, M., and Langberg, N. A. (1983). Testing whether new is better than used with randomly censored data. *Ann. Statist.* 11, 267 – 274.

Chong, F. S. (1981). A point process with second order Markov dependent intervals. *Math. Nachr.* 103, 155 – 163.

Chung, K. L. (1972). The Poisson process as renewal process. *Period. Math. Hungar.* 2, 41 – 48.

Chung, K. L. (1974). *A Course in Probability Theory*. Academic Press, New

York.
Chung, K. L., and Doob, J. L. (1965). Fields, optionality and measurability. *Amer. J. Math.* 87, 397 – 424.
Çinlar, E. (1968). On the superposition of m-dimensional point processes. *J. Appl. Prob.* 5, 169 – 176.
Çinlar, E. (1972). Superposition of point processes. In *Stochastic Point Processes*, Lewis, ed. Wiley, New York.
Çinlar, E. (1975a). *Introduction to Stochastic Processes.* Prentice-Hall, Englewood Cliffs, N.J.
Çinlar, E. (1975b). Markov renewal theory: a survey. *Management Sci.* 21, 727 – 752.
Çinlar, E., and Agnew, R. A. (1968). On the superposition of point processes. *J. Roy. Statist. Soc. B* 30, 576 – 581.
Çinlar, E., and Jacod, J. (1981). Representation of semimartingale Markov processes in terms of Wiener processes and Poisson random measures. In *Seminar on Stochastic Processes 1981*, Çinlar, Chung, and Getoor, eds. Birkhäuser, Boston.
Çinlar, E., and Jagers, P. (1973). Two mean values which characterize the Poisson process. *J. Appl. Prob.* 10, 678 – 681.
Clarkson, D. B., and Wolfson, D. B. (1983). An application of a displaced Poisson process. *Statist. Neerl.* 37, 21 – 28.
Clevenson, M. L., and Zidek, J. W. (1977). Bayes linear estimators of the intensity function of the nonstationary Poisson process. *J. Amer. Statist. Assoc.* 72, 112 – 120.
Cliff, A. D., and Ord, J. K. (1975). Model building and the analysis of spatial pattern in human geography. *J. Roy. Statist. Soc. B* 37, 297 – 348.
Cohn, D. L. (1980). *Measure Theory.* Birkhäuser, Boston.
Cowan, R. (1980). Further results on single-lane traffic flow. *J. Appl. Prob.* 16, 523 – 531.
Cox, D. R. (1955). Some statistical models connected with series of events. *J. Roy. Statist. Soc. B* 17, 129 – 164.
Cox, D. R. (1962). *Renewal Theory.* Methuen, London.
Cox, D. R. (1965). On the estimation of the intensity function of a stationary point process. *J. Roy. Statist. Soc. B* 27, 332 – 337.
Cox, D. R. (1972a). The statistical analysis of dependencies in point processes. In *Stochastic Point Processes*, Lewis, ed. Wiley, New York.
Cox, D. R. (1972b). Regression models and life tables. *J. Roy. Statist. Soc. B* 34, 187 – 220.
Cox, D. R. (1975). Partial likelihood. *Biometrika* 62, 269 – 276.
Cox, D. R., and Hinkley, D. V. (1974). *Theoretical Statistics.* Chapman & Hall, London.

Cox, D. R., and Isham, V. (1980). *Point Processes.* Chapman & Hall, London.

Cox, D. R., and Lewis, P. A. W. (1966). *The Statistical Analysis of Series of Events.* Chapman & Hall, London.

Cox, D. R., and Lewis, P. A. W. (1972). Multivariate point processes. *Proc. Sixth Berkeley Symp. Math. Statist. Prob.* 3, 401 — 448.

Cramér, H., and Leadbetter, M. R. (1967). *Stationary and Related Stochastic Processes.* Wiley, New York.

Cramér, H., Leadbetter, M. R., and Serfling, R. J. (1971). On distribution function-moment relationships in a stationary point process. *Z. Wahrscheinlichkeitstheorie und verw. Geb.* 18, 1 – 8.

Crump, K. S. (1975). On point processes having an order statistic structure. *Sankhya A* 37, 396 – 404.

Csörgö, M., and Révész, P. (1981). *Strong Approximation in Probability and Statistics.* Academic Press, New York.

Daley, D. J. (1971). Weakly stationary point processes and random measures. *J. Roy. Statist. Soc. B* 33, 406 – 428.

Daley, D. J. (1972). Asymptotic properties of stationary point processes with generalized clusters. *Z. Wahrscheinlichkeitstheorie und verw. Geb.* 21, 65 – 76.

Daley, D. J. (1973). Poisson and alternating renewal processes with superposition a renewal process. *Math. Nachr.* 57, 359 – 369.

Daley, D. J. (1974). Various concepts of orderliness for point processes. In *Stochastic Geometry*, Harding and Kendall, eds. Wiley, New York.

Daley, D. J. (1976). Queueing output processes. *Adv. Appl. Prob.* 8, 395 – 415.

Daley, D. J. (1982a). Stationary point processes with Markov-dependent intervals and infinite intensity. *J. Appl. Prob.* 19A, 313 – 320.

Daley, D. J. (1982b). Infinite intensity mixtures of point processes. *Math. Proc. Cambridge Philos. Soc.* 92, 109 – 114.

Daley, D. J., and Milne, R. K. (1973). Theory of point processes: a bibliography. *Internat. Statist. Rev.* 41, 183 – 201.

Daley, D. J., and Milne, R. K. (1975). Orderliness, intensities and Palm-Khinchin equations for multivariate point processes. *J. Appl. Prob.* 12, 383 – 389.

Daley, D. J., and Oakes, D. (1974). Random walk point processes. *Z. Wahrscheinlichkeitstheorie und verw. Geb.* 30, 1 – 16.

Daley, D. J., and Vere-Jones, D. (1972). A summary of the theory of point processes. In *Stochastic Point Processes*, Lewis, ed. Wiley, New York.

Daniels, H. E. (1963). The Poisson process with a curved absorbing boundary. *Bull. Internat. Statist. Inst.* 40, 994 – 1008.

Darling, D. A. (1953). The Kolmogorov-Smirnov, Cramér-von Mises tests.

Ann. Math. Statist. 28, 823 – 838.
Davies, R. B. (1977). Testing the hypothesis that a point process is Poisson. *Adv. Appl. Prob.* 9, 724 – 746.
Davis, D. J. (1952). An analysis of some failure data. *J. Amer. Statist. Assoc.* 47, 113 – 150.
Davis, M. H. A. (1976). The representation of martingales of jump processes. *SIAM J. Control* 14, 623 – 638.
Davis, M. H. A., Kailath, T., and Segall, A. (1975). Nonlinear filtering with counting observations. *IEEE Trans. Inform. Theory* IT-21, 143 – 150.
Debes, H., Kerstan, J., Liemant, A., and Matthes, K. (1970). Verallgemeinerungen eines Satzes von Dobruschin I. *Math. Nachr.* 47, 183 – 244.
Debes, H., Kerstan, J., Liemant, A., and Matthes, K. (1971). Verallgemeinerungen eines Satzes von Dobruschin III. *Math. Nachr.* 50, 299 – 383.
Deheuvels, P. (1983). Point processes and multivariate extreme values. *J. Multivariate Anal.* 13, 257 – 272.
Dellacherie, C. (1972). *Capacités et Processus Stochastiques.* Springer-Verlag, Berlin.
Dellacherie, C., and Meyer, P.-A. (1980). *Potentiel et Probabilités.* North-Holland, Amsterdam.
Dempster, A. P., Laird, N. M., and Rubin, D. B. (1977). Maximum likelihood from incomplete data via the EM algorithm. *J. Roy. Statist. Soc. B* 39, 1 – 38.
Deo, C. M. (1975). A functional central limit theorem for stationary random fields. *Ann. Probability* 3, 708 – 715.
Diggle, P. J. (1979). On parameter estimation and goodness-of-fit testing for spatial point patterns. *Biometrics* 35, 87 – 101.
Diggle, P. J. (1983). *Statistical Analysis of Spatial Point Patterns.* Academic Press, New York.
Dobrushin, R. L. (1956). On the Poisson law for distributions of particles in space. *Ukrain. Mat. Z.* 8, 127 – 134.
Dobrushin, R. L. (1968). The description of a random field by means of its conditional distributions. *Theor. Prob. Appl.* 13, 197 – 224.
Doléans-Dade, C. (1970). Quelques applications de la formule de changement de variables pour les semimartingales. *Z. Wahrscheinlichkeitstheorie und verw. Geb.* 16, 181 – 194.
Dollard, J. D., and Friedman, C. N. (1979). *Product Integration.* Addison-Wesley, Reading, Mass.
Donsker, M. (1952). Justification and extension of Doob's heuristic approach to the Kolmogorov-Smirnov theorems. *Ann. Math. Statist.* 23, 277 – 283.

Doob, J. L. (1949). Heuristic approach to the Kolmogorov-Smirnov theorems. *Ann. Math. Statist.* 20, 393 – 403.

Doob, J. L. (1953). *Stochastic Processes*. Wiley, New York.

Driscoll, M. F., and Weiss, N. A. (1974). Random translations of stationary point processes. *J. Math. Anal. Appl.* 48, 423 – 433.

Dubins, L. E., and Freedman, D. A. (1967). Random distribution functions. *Proc. Fifth Berkeley Symp. Math. Statist. Prob.* 2, 183 – 214.

Dudley, R. M. (1968). Distances of probability measures and random variables. *Ann. Math. Statist.* 39, 1563 – 1572.

Dudley, R. M. (1972). Speeds of metric probability convergence. *Z. Wahrscheinlichkeitstheorie und verw. Geb.* 22, 323 – 332.

Dudley, R. M. (1973). Sample functions of Gaussian processes. *Ann. Probability* 1, 66 – 103.

Dudley, R. M. (1974). Metric entropy and the central limit theorem in $C(S)$. *Ann. Inst. Fourier* 24(2), 49 – 60.

Dudley, R. M. (1978). Central limit theorems for empirical measures. *Ann. Probability* 6, 899 – 929.

Dunford, N., and Schwartz, J. T. (1958). *Linear Operators I: General Theory*. Interscience, New York.

Durbin, J. (1961). Some methods for constructing exact tests. *Biometrika* 48, 41 – 55.

Durbin, J. (1971). Boundary-crossing problems for the Brownian motion and Poisson processes and techniques for computing the power of the Kolmogorov-Smirnov test. *J. Appl. Prob.* 8, 431 – 453.

Durbin, J. (1973a). *Distribution Theory for Tests Based on the Sample Distribution Function*. SIAM, Philadelphia.

Durbin, J. (1973b). Weak convergence of the sample distribution function when parameters are estimated. *Ann. Statist.* 1, 279 – 290.

Dvoretsky, A., Kiefer, J., and Wolfowitz, J. (1953a). Sequential decision problems for processes with continuous time parameter: testing hypotheses. *Ann. Math. Statist.* 24, 254 – 264.

Dvoretsky, A., Kiefer, J., and Wolfowitz, J. (1953b). Sequential decision problems for processes with continuous time parameter: problems of estimation. *Ann. Math. Statist.* 24, 403 – 415.

Dvoretsky, A., Kiefer, J., and Wolfowitz, J. (1956). Asymptotic minimax character of the sample distribution function and of the classical multinomial estimator. *Ann. Math. Statist.* 27, 642 – 669.

Ebe, M. (1969). On sets of motion of point processes. *J. Oper. Res. Soc. Japan* 11, 97 – 113.

Efron, B. (1977). Efficiency of Cox's likelihood function for censored data. *J. Amer. Statist. Assoc.* 72, 557 – 565.

El-Karoui, N., and Lepeltier, J.-P. (1977). Représentation des processus ponctuels multivariés à l'aide d'un processus de Poisson. *Z. Wahrscheinlichkeitstheorie und verw. Geb.* 39, 111 — 133.

Elliot, R. J. (1976). Stochastic integrals for martingales of a jump process with partially accessible jump times. *Z. Wahrscheinlichkeitstheorie und verw. Geb.* 36, 213 - 226.

Elliott, R. J. (1982). *Stochastic Calculus and Applications*. Springer-Verlag, New York.

Epstein, B. (1960). Tests for the validity of the assumption that the underlying distribution of life is exponential. *Technometrics* 2, 83 - 101.

Eriksson, B. (1978). An approximation of the variance of counts for a stationary point process. *Scand. J. Statist.* 5, 111 - 115.

Fedorseev, V. I., and Shirokov, F. P. (1976). Spatial-temporal nonlinear filtration for Poisson random fields. *Problems Inform. Transmission* 12, 20 - 28.

Feigin, P. D. (1976). Maximum likelihood estimation for continuous time stochastic processes. *Adv. Appl. Prob.* 8, 712 - 736.

Feigin, P. D. (1979). On the characterization of point processes with the order statistic property. *J. Appl. Prob.* 16, 297 - 304.

Feigin, P. D. (1981). Conditional exponential families and a representation theorem for asymptotic inference. *Ann. Statist.* 9, 597 - 603.

Feller, W. (1971). *An Introduction to Probability Theory and Its Applications*, Vol. 2, 2nd ed. Wiley, New York.

Fellows, A., and Granara, J. (1977). Une caractérisation de processus de Poisson. *Z. Wahrscheinlichkeitstheorie und verw. Geb.* 39, 71 - 79.

Ferguson, T. S. (1967). *Mathematical Statistics: A Decision-Theoretic Approach*. Academic Press, New York.

Feuerverger, A., and Mureika, R. A. (1977). The empirical characteristic function and its applications. *Ann. Statist.* 5, 88 - 97.

Fichtner, K. H. (1975). Charakterisierung Poissonscher zufälliger Punktfolgen und infinitesimale Verdunnungsschemata. *Math. Nachr.* 68, 93 — 104.

Fisher, L. (1972). A survey of the mathematical theory of multidimensional point processes. In *Stochastic Point Processes,* Lewis, ed. Wiley, New York.

Fishman, P. H., and Snyder, D. L. (1976). The statistical analysis of space-time point processes. *IEEE Trans. Inform. Theory* IT-22, 257 - 274.

Fleischmann, K. (1981). Ergodicity properties of infinitely divisible stochastic point processes. *Math. Nachr.* 102, 127 - 135.

Fleming, T. R. (1978a). Asymptotic distribution results in competing risks estimation. *Ann. Statist.* 6, 1071 – 1079.

Fleming, T. R. (1978b). Nonparametric estimation for nonhomogeneous Markov processes in the problem of competing risks. *Ann. Statist.* 6, 1057 – 1070.

Foldes, A., and Rejto, L. (1981a). Strong uniform consistency for nonparametric survival curve estimators from randomly censored data. *Ann. Statist.* 9, 122 – 129.

Foldes, A., and Rejto, L. (1981b). A LIL type result for the product limit estimator. *Z. Wahrscheinlichkeitstheorie und verw. Geb.* 56, 75 – 86.

Fortet, R. (1951). Random functions from a Poisson process. *Proc. Second Berkeley Symp. Math. Statist. Prob.* 373 – 385.

Franken, P. (1963). Approximation durch Poissonische Prozesse. *Math. Nachr.* 26, 101 – 114.

Franken, P. (1964). Approximation der Verteilungen von Summen unabhangiger nichtnegativer ganzzahliger Zufällsgrossen durch Poissonische Verteilungen. *Math. Nachr.* 27, 303 – 340.

Franken, P., and Streller, A. (1980). Reliability analysis of complex repairable systems by means of marked point processes. *J. Appl. Prob.* 17, 154 – 167.

Franken, P., Liemant, A., and Matthes, K. (1965). Stationäre zufällige Punktfolgen III. *Jber. Deutsch. Math.-Verein.* 67, 183 – 202.

Franken, P., König, D., Arndt, U., and Schmidt, V. (1981). *Queues and Point Processes*. Akademie-Verlag, Berlin.

Freed, D. S., and Shepp, L. A. (1982). A Poisson process whose rate is a hidden Markov process. *Adv. Appl. Prob.* 14, 21 – 36.

Freedman, D. (1962). Poisson processes with a random arrival rate. *Ann. Math. Statist.* 33, 924 – 929.

Fritz, J. (1969). Entropy of point processes. *Studia Sci. Math. Hungar.* 4, 389 – 400.

Fritz, J. (1970). Generalization of McMillan's theorem to random set functions. *Studia Sci. Math. Hungar* 5, 369 – 394.

Fujisaki, M., Kallianpur, G., and Kunita, H. (1972). Stochastic differential equations for nonlinear filtering problems. *Osaka J. Math.* 9, 19 – 40.

Gacs, P., and Szasz, D. (1975). On a problem of Cox concerning point processes in $\mathbf{R}^k$ of "controlled variability." *Ann. Probability* 3, 597 – 607.

Gaenssler, P. (1984). *Empirical Processes: On some Basic Results from the Probabilistic Point of View*. IMS Lecture Notes-Monograph Series 3, 1 – 179.

Gaenssler, P., and Stute, W. (1976). On uniform convergence of measures with application to uniform convergence of empirical distributions. *Lect. Notes Math.* 566, 45 – 56.

Gaenssler, P., and Stute, W. (1979). Empirical processes: a survey of results for independent and identically distributed random variables. *Ann. Probability* 7, 193 – 243.

Galtchouk, L., and Rozovskii, B. (1971). The disruption problem for a Poisson process. *Theor. Prob. Appl.* 16, 712 – 716.

Galun, S. A., and Trifonov, A. P. (1982). Detection and estimation of the instant of change of intensity of a Poisson flow. *Automat. Remote Control* 43, 782 – 790.

Gaver, D. P. (1963). Random hazard in reliability problems. *Technometrics* 5, 211 – 226.

Gaver, D. P., and Lewis, P. A. W. (1980). First order autoregressive gamma sequences and point processes. *Adv. Appl. Prob.* 12, 727 – 745.

Gehan, E. A. (1965). A generalized Wilcoxon test for comparing arbitrarily singly-censored data. *Biometrika* 52, 203 — 233.

Geman, D., and Horowitz, J. (1973). Remarks on Palm measures. *Ann. Inst. H. Poincaré* 9, 215 – 232.

Geman, S., and Hwang, C.-R. (1982). Nonparametric maximum likelihood estimation by the method of sieves. *Ann. Statist.* 10, 401 — 414.

Gill, R. D. (1980a). Nonparametric estimation based on censored observations of a Markov renewal process. *Z. Wahrscheinlichkeitstheorie und verw. Geb.* 53, 97 – 116.

Gill, R. D. (1980b). *Censoring and Stochastic Integrals*. Math. Centre Tracts 124. Mathematisch Centrum, Amsterdam.

Gill, R. D. (1983). Large sample behaviour of the product-limit estimator on the whole line. *Ann. Statist.* 11, 49 – 58.

Gill, R. D. (1984). Understanding Cox's regression model: a martingale approach. *J. Amer. Statist. Assoc.* 79, 441 – 447.

Girsanov, I. (1960). On transforming a certain class of stochastic processes by absolutely continuous substitution of measure. *Theor. Prob. Appl.* 5, 285 – 301.

Glotzl, E. (1979). On the singularity of Σ point processes. *Math. Nachr.* 92, 211 – 213.

Goldman, J. R. (1967a). Infinitely divisible point processes in $\mathbf{R}^n$. *J. Math. Anal. Appl.* 17, 133 – 146.

Goldman, J. R. (1967b). Stochastic point processes: limit theorems. *Ann. Math. Statist.* 38, 771 – 779.

Govindarajulu, Z. (1975). *Sequential Statistical Procedures*. Academic Press, New York.

Grandell, J. (1971a). On stochastic processes generated by a stochastic intensity function. *Skand. Aktuar. Tidskrift* 54, 204 – 240.

Grandell, J. (1971b). A note on the linear estimation of the intensity in a doubly stochastic Poisson field. *J. Appl. Prob.* 8, 612 – 614.

Grandell, J. (1972a). On the estimation of intensities in a stochastic process generated by a stochastic intensity sequence. *J. Appl. Prob.* 9, 542 – 556.

Grandell, J. (1972b). Statistical inference for doubly stochastic Poisson processes. In *Stochastic Point Processes,* Lewis, ed. Wiley, New York.

Grandell, J. (1976). Doubly stochastic Poisson processes. *Lect. Notes Math.,* 529.

Grandell, J. (1977). Point processes and random measures. *Adv. Appl. Prob.* 9, 502 – 526.

Greenwood, P. E., and Shiryayev, A. N. (1985). *Contiguity and the Statistical Invariance Principle.* Gordon and Breach, London.

Gregoire, G. (1984). Negative binomial point processes. *Stochastic Process. Appl.* 16, 179 – 188.

Grenander, U. (1950). Stochastic processes and statistical inference. *Ark. Math.* 1, 195 – 277.

Grenander, U. (1951). On empirical spectral analysis of stochastic processes. *Ark. Math.* 1, 503 – 531.

Grenander, U. (1956). On the theory of mortality measurement, II. *Skand. Akt.* 39, 125 – 153.

Grenander, U. (1981). *Abstract Inference.* Wiley, New York.

Grigelionis, B. (1963). On the convergence of random step processes to a Poisson process. *Theor. Prob. Appl.* 8, 177 – 182.

Grigelionis, B. (1973). On the nonlinear filtering theory and absolute continuity of measures corresponding to stochastic processes. *Lect. Notes Math.* 330, 80 – 94.

Grigelionis, B. (1975). Random point processes and martingales. *Litovsk. Mat. Sb.* 15, 101 – 114.

Grigelionis, B. (1980). A martingale approach to the statistical problems of point processes. *Scand. J. Statist.* 7, 190 – 196.

Gyorfi, L. (1972). Poisson processes defined on an abstract space. *Studia Sci. Math. Hungar.* 7, 243 – 248.

Haberland, E. (1975). Infinitely divisible recurrent point processes. *Math. Nachr.* 70, 259 – 264.

Hadjiev, D. I. (1978). On the filtering of semimartingales in case of observation of point processes. *Theor. Prob. Appl.* 23, 169 – 178.

Haight, F. (1965). Counting distributions for renewal processes. *Biometrika* 52, 395 – 403.

Hajek, J., and Sidak, Z. (1967). *Theory of Rank Tests.* Academic Press, New York.

Hall, P. (1982). On Starr and Vardi's estimates of the number of transmission sources. *J. Appl. Prob.* 19, 52 – 63.

Hall, P., and Heyde, C. C. (1980). *Martingale Limit Theory and Its*

Applications. Academic Press, New York.

Hall, W. J., and Wellner, J. A. (1980). Confidence bands for a survival curve from censored data. *Biometrika* 67, 133 – 143.

Hanen, A. (1973). Mesures aléatoires stationnaires et mesure de Palm. *Ann. Inst. H. Poincaré B* 9, 311 – 325.

Hanisch, K.-H. (1982). On inversion formulae for *n*-fold Palm distributions of point processes on LCS-spaces. *Math. Nachr.* 106, 171 – 179.

Hanisch, K.-H. (1983). Reduction of the nth moment measures and the special case of the third moment measure of stationary and isotropic planar point processes. *Math. Operationsforsch. Statist.* 14, 421 – 435.

Hanisch, K.-H., and Stoyan, D. (1979). Formulas for the second-order analysis of marked point processes. *Math. Operationsforsch. Statist.* 10, 555 – 560.

Harris, T. E. (1963). *The Theory of Branching Processes*. Springer-Verlag, Berlin.

Harris, T. E. (1968). Counting measures, monotone random set functions. *Z. Wahrscheinlichkeitstheorie und verw. Geb.* 10, 102 – 119.

Harris, T. E. (1971). Random measures and motions of point processes. *Z. Wahrscheinlichkeitstheorie und verw. Geb.* 18, 85 – 115.

Hawkes, A. G. (1971). Spectra of some self-exciting and mutually exciting point processes. *Biometrika* 58, 83 – 104.

Hawkes, A. G., and Oakes, D. (1974). A cluster process representation of a self-exciting point process. *J. Appl. Prob.* 11, 493 — 503.

Helland, I. S. (1982). Central limit theorems for martingales with discrete or continuous time. *Scand. J. Statist.* 9, 79 – 94.

Hida, T. (1953). On some asymptotic properties of Poisson processes. *Nagoya Math. J.* 6, 29 – 36.

Holgate, P. (1965). Tests of randomness based on distance methods. *Biometrika* 52, 345 – 353.

Holgate, P. (1972). The use of distance methods for the analysis of spatial distribution of points. In *Stochastic Point Processes*, Lewis, ed. Wiley, New York.

Hollander, M., and Proschan, F. (1972). Testing whether new is better than used. *Ann. Math. Statist.* 43, 1136 – 1146.

Hollander, M., and Proschan, F. (1979). Testing to determine the underlying distribution using randomly censored data. *Biometrics* 35, 393 – 401.

Hollander, M., and Wolfe, D. A. (1973). *Nonparametric Statistical Methods*. Wiley, New York.

Horváth, L. (1984). Strong approximation of renewal processes. *Stochastic Process. Appl.* 18, 127 – 138.

Huber, P. (1981). *Robust Statistics*. Wiley, New York.

Ibragimov, I. A., and Has'minskii, R. Z. (1981). *Statistical Estimation: Asymptotic Theory*. Springer-Verlag, New York.

Ibragimov, I. A., and Rozanov, Yu. A. (1978). *Gaussian Random Processes*. Springer-Verlag, New York.

Isham, V. (1980). Dependent thinning of point processes. *J. Appl. Prob.* 17, 987 – 995.

Isham, V., and Westcott, M. (1979). A self-correcting point process. *Stochastic Process. Appl.* 8, 335 – 347.

Isham, V., Shanbhag, D. N., and Westcott, M. (1975). A characterization of the Poisson process using forward recurrence times. *Math. Proc. Cambridge Philos. Soc.* 78, 513 – 516.

Itô, K. (1953). Stationary random distributions. *Mem. Coll. Sci. Univ. Kyoto, Ser. A.* 28, 209 — 223.

Itô, K. (1972). Poisson point processes attached to Markov processes. *Proc. Sixth Berkeley Symp. Math. Statist. Prob.* 3, 225 – 239.

Ito, Y. (1977). Superposition of distinguishable point processes. *J. Appl. Prob.* 14, 200 – 204.

Ito, Y. (1978). Superposition and decomposition of stationary point processes. *J. Appl. Prob.* 15, 481 – 493.

Ito, Y. (1980). On renewal processes decomposable into IID components. *Adv. Appl. Prob.* 12, 672 – 688.

Ivanoff, G. (1982). Central limit theorems for point processes. *Stochastic Process. Appl.* 12, 171 – 186.

Jacobsen, M. (1982). Statistical analysis of counting processes. *Lect. Notes Statist.* 12.

Jacobsen, M. (1984). Maximum likelihood estimation in the multiplicative intensity model: a survey. *Internat. Statist. Rev.* 52, 193 – 207.

Jacod, J. (1975a). Two dependent Poisson processes whose sum is still a Poisson process. *J. Appl. Prob.* 12, 170 – 172.

Jacod, J. (1975b). Multivariate point processes: predictable projection, Radon-Nikodym derivatives, representation of martingales. *Z. Wahrscheinlichkeitstheorie und verw. Geb.* 31, 235 – 253.

Jacod, J. (1979). Calcul stochastique et problèmes de martingales. *Lect. Notes Math.* 714.

Jagers, P. (1972). On the weak convergence of superpositions of point processes. *Z. Wahrscheinlichkeitstheorie und verw. Geb.* 22, 1 – 7.

Jagers, P. (1973). On Palm probabilities. *Z. Wahrscheinlichkeitstheorie und verw. Geb.* 26, 17 – 32.

Jagers, P. and Lindvall, T. (1973). Three theorems on the thinning of point processes. *Adv. Appl. Prob.* 5, 14 – 15.

Jagers, P., and Lindvall, T. (1974). Thinning and rare events in point processes. *Z. Wahrscheinlichkeitstheorie und verw. Geb.* 28, 89 – 98.

Jain, N., and Marcus, M. B. (1975). Central limit theorems for $C(S)$-valued random variables. *J. Funct. Anal.* 19, 216 – 231.

Jewell, W. S. (1960). Properties of recurrent event processes. *Oper. Res.* 8, 446 – 472.

Johansen, S. (1978). The product limit estimator as maximum likelihood estimator. *Scand. J. Statist.* 5, 195 – 199.

Johansen, S. (1983). An extension of Cox's regression model. *Inter. Statist. Rev.* 51, 165 – 174.

Johnson, N. L. and Kotz, S. (1975). A vector multivariate hazard rate. *J. Multivariate Anal.* 5, 53 – 66.

Jolivet, E. (1978). Caractérisation et test du caractère agrégatif des processus ponctuels stationnaires sur $\mathbf{R}^2$. *Lect. Notes Math.* 636, 1 – 25.

Jolivet, E. (1981). Central limit theorem and convergence of empirical processes for stationary point processes. In *Point Processes and Queueing Problems,* Bártfai and Tomkó, eds. North-Holland, Amsterdam.

Jorgensen, D. W. (1961). Multiple regression analysis of a Poisson process. *J. Amer. Statist. Assoc.* 56, 235 – 245.

Jowett, J. H., and Vere-Jones, D. (1972). The prediction of stationary point processes. In *Stochastic Point Processes,* Lewis, ed. Wiley, New York.

Kabanov, Yu. A. (1979). The capacity of a channel of Poisson type. *Theor. Prob. Appl.* 23, 143 – 147.

Kabanov, Yu. M., and Liptser, R. S. (1983). On convergence in variation of the distribution of multivariate point processes. *Z. Wahrscheinlichkeitstheorie und verw. Geb.* 63, 475 – 485.

Kabanov, Yu. M., Liptser, R. S., and Shiryayev, A. N. (1980a). Some limit theorems for simple point processes (martingale approach). *Stochastics* 3, 203 – 216.

Kabanov, Yu. M., Liptser, R. S., and Shiryayev, A. N. (1980b). Representation of integer-valued random measures and local martingales by means of random measures with deterministic compensators. *Mat. Sb.* 111, 293 – 307.

Kabanov, Yu. M., Liptser, R. S., and Shiryayev, A. N. (1983). Weak and strong convergence of the distributions of counting processes. *Theor. Prob. Appl.* 28, 303 – 336.

Kailath, T. (1970). The innovations approach to detection and estimation theory. *Proc. IEEE* 58, 680 – 695.

Kailath, T., and Segall, A. (1975). Radon-Nikodym derivatives with respect to measures induced by discontinuous independent increment processes. *Ann. Probability* 3, 449 – 464.

Kalbfleisch, J. D., and Prentice, R. L. (1980). *The Statistical Analysis of*

Failure Time Data. Wiley, New York.
Kalbfleisch, J. D., and Prentice, R. L. (1981). Estimation of the average hazard rate. *Biometrika* 68, 105 – 112.
Kallenberg, O. (1973). Characterization and convergence of random measures and point processes. *Z. Wahrscheinlichkeitstheorie und verw. Geb.* 27, 9 – 21.
Kallenberg, O. (1974). Extremality of Poisson and sample processes. *Stochastic Process. Appl.* 2, 73 – 83.
Kallenberg, O. (1975a). Limits of compound and thinned point processes. *J. Appl. Prob.* 12, 269 – 278.
Kallenberg, O. (1975b). On symmetrically distributed random measures. *Trans. Amer. Math. Soc.* 202, 105 – 121.
Kallenberg, O. (1977). Stability of critical cluster fields. *Math. Nachr.* 77, 7 – 43.
Kallenberg, O. (1979). L^p intensities of random measures. *Stochastic Process. Appl.* 9, 155 — 161.
Kallenberg, O. (1980). On conditional intensities of point processes. *Z. Wahrscheinlichkeitstheorie und verw. Geb.* 41, 205 – 220.
Kallenberg, O. (1982). Conditioning in point processes. Technical report 82-7, Department of Mathematics, Chalmers Tekniska Högskola and University of Göteborg, Göteborg, Sweden.
Kallenberg, O. (1983). *Random Measures*, 3rd ed. Akademie-Verlag, Berlin/Academic Press, New York.
Kallenberg, O. (1984). An informal guide to the theory of conditioning in point processes. *Internat. Statist. Rev.* 52, 151 – 164.
Kallianpur, G. (1980). *Stochastic Filtering Theory*. Springer-Verlag, New York.
Kallianpur, G., and Striebel, C. (1968). Estimation of stochastic systems: arbitrary system process with additive white noise observation errors. *Ann. Math. Statist.* 39, 785 – 801.
Kalman, R. E. (1960). A new approach to linear filtering and prediction problems. *J. Basic Engg.* 82, 35 – 45.
Kaplan, E. L., and Meier, P. (1958). Nonparametric estimation from incomplete observations. *J. Amer. Statist. Assoc.* 53, 457 – 481.
Karlin, S., and Studden, W. J. (1966). *Tchebycheff Systems: With Applications in Analysis and Statistics*. Interscience, New York.
Karlin, S., and Taylor, H. M. (1975). *A First Course in Stochastic Processes,* 2nd ed. Academic Press, New York.
Karp, S., and Clark, J. R. (1970). Photon counting: a problem in classical noise theory. *IEEE Trans. Inform. Theory* IT-16, 672 – 680.
Karr, A. F. (1976a). Two extreme value processes arising in hydrology. *J. Appl. Prob.* 13, 190 – 194.

Karr, A. F. (1976b). A conditional expectation for random measures. Technical report, Department of Mathematical Sciences, The Johns Hopkins University, Baltimore, Md.

Karr, A. F. (1978). Derived random measures. *Stocastic Process. Appl.* 8, 159 – 169.

Karr, A. F. (1979). Classical limit theorems for measure-valued Markov processes. *J. Multivariate Anal.* 9, 234 – 247.

Karr, A. F. (1982). A partially observed Poisson process. *Stochastic Process. Appl.* 12, 249 – 269.

Karr, A. F. (1983). State estimation for Cox processes on general spaces. *Stochastic Process. Appl.* 14, 209 – 232.

Karr, A. F. (1984a). Estimation and reconstruction for zero-one Markov processes. *Stochastic Process. Appl.* 16, 219 – 255.

Karr, A. F. (1984b). Combined nonparametric inference and state estimation for mixed Poisson processes. *Z. Wahrscheinlichkeitstheorie und verw. Geb.* 66, 81 – 96.

Karr, A. F. (1984c). The martingale method: introductory sketch and access to the literature. *Oper. Res. Lett.* 3, 59 – 63.

Karr, A. F. (1985a). State estimation for Cox processes with unknown probability law. *Stochastic Process. Appl.* 20, 115 — 131.

Karr, A. F. (1985b). Inference for thinned point processes, with application to Cox processes. *J. Multivariate Anal.* 16, 368 — 392.

Karr, A. F. (1985c). Estimation of Palm measures of stationary point processes. Technical report, Department of Mathematical Sciences, The Johns Hopkins University, Baltimore, Md.

Karr, A. F. (1985d). Inference and state estimation for Poisson processes with one observable component. Technical report, Department of Mathematical Sciences, The Johns Hopkins University, Baltimore, Md.

Karr, A. F. (1985e). Inference for Poisson processes based on integral data Technical report, Department of Mathematical Sciences, The Johns Hopkins University, Baltimore, Md.

Karr, A. F. (1985d). Inference for stationary random fields given Poisson samples. *Adv. Appl. Prob.* (to appear).

Karr, A. F. (1986*b*). Maximum likelihood estimation in the multiplicative intensity model. *Ann. Statist.* (to appear).

Kavvas, M. L., and Delleur, J. (1981). A stochastic cluster model of daily rainfall occurrences. *Water Resources Res.* 17, 1151 – 1160.

Keiding, N. (1974). Estimation in the birth process. *Biometrika* 61, 71 – 80.

Keiding, N. (1975). Maximum likelihood estimation in the birth and death process. *Ann. Statist.* 3, 363 – 372.

Kerstan, J. (1962). Teilprozesse Poissonscher Prozesse. *Trans. Third*

Prague Conf. 377 – 403.

Kerstan, J., and Matthes, K. (1964a). Stationäre zufällige Punktfolgen II. *Jber. Deutsch. Math.-Verein.* 66, 106 – 118.

Kerstan, J., and Matthes, K. (1964b). Verallgemeinerung eines Satzes von Sliwnjak. *Rev. Roumaine Math. Pures Appl.* 9, 811 – 829.

Kerstan, J., and Matthes, K. (1965). A generalization of the Palm-Khinchin theorem. (In Russian.) *Ukrain. Mat. Z.* 17, 29 – 36.

Kerstan, J., Matthes, K., and Prehn, U. (1971). Verallgemeinerungen eines Satzes von Dobruschin II. *Math. Nachr.* 51, 149 – 188.

Khinchin, A. Ya. (1956a). Streams of events without aftereffects. *Theor. Prob. Appl.* 1, 1 – 15.

Khinchin, A. Ya. (1956b). On Poisson streams of events. *Theor. Prob. Appl.* 1, 248 – 255.

Khinchin, A. Ya. (1960). *Mathematical Methods in the Theory of Queueing.* Griffin, London.

Kiefer, J. (1972). Skorohod embedding of multivariate random variables and the sample distribution function. *Z. Wahrscheinlichkeitstheorie und verw. Geb.* 24, 1 – 35.

Kiefer, J., and Wolfowitz, J. (1956). Consistency of the maximum likelihood estimator in the presence of infinitely many incidental parameters. *Ann. Math. Statist.* 27, 887 – 906.

Kiefer, J., and Wolfowitz, J. (1957). Sequential tests of hypotheses about the mean occurrence time of a continuous parameter Poisson process. *Naval Res. Logist. Quart.* 3, 205 – 219.

Kingman, J. F. C. (1963). Poisson counts for random sequences. *Ann. Math. Statist.* 34, 1217 – 1232.

Kingman, J. F. C. (1964a). Random discrete distributions. *J. Roy. Statist. Soc. B* 37, 1 – 22.

Kingman, J. F. C. (1964b). On doubly stochastic Poisson processes. *Proc. Cambridge Philos. Soc.* 60, 923 – 930.

Kingman, J. F. C. (1967). Completely random measures. *Pacific J. Math.* 21, 59 – 78.

Kolmogorov, A. N., and Tihomirov, V. M. (1961). ε-entropy and ε-capacity of sets in functional spaces. *Amer. Math. Soc. Trans.* 17, 277 – 364.

Komatsu, T. (1976). Statistics of processes with jumps. *Lect. Notes Math.* 550, 276 – 289.

Komlós, J., Major, P., and Tusnády, G. (1975). An approximation of partial sums of independent random variables and the sample distribution function, I. *Z. Wahrscheinlichkeitstheorie und verw. Geb.* 32, 111 — 131.

Komlós, J., Major, P., and Tusnády, G. (1976). An approximation of partial sums of independent random variables and the sample distribution function, II. *Z. Wahrscheinlichkeitstheorie und verw. Geb.* 34, 33 — 58.

Konecny, F. (1984). Parameter estimation for point processes with partial observations: a filtering approach. *Systems Control Lett.* 4, 281 – 286.

König, D., and Schmidt, V. (1980). Imbedded and nonimbedded stationary characteristics of queueing systems with varying rates and point processes. *J. Appl. Prob.* 17, 753 – 767.

Koziol, J. A., and Green, S. B. (1976). A Cramér-von Mises statistic for randomly censored data. *Biometrika* 63, 465 – 474.

Kozlov, O. K. (1976). Gibbsian description of point random fields. *Theor. Prob. Appl.* 21, 339 – 355.

Krickeberg, K. (1972). The Cox process. *Symp. Math.* 9, 151 – 167.

Krickeberg, K. (1974). Moments of point processes. In *Stochastic Geometry*, Harding and Kendall, eds. Wiley, New York.

Krickeberg, K. (1976). An alternative approach to the Glivenko-Cantelli theorems. *Lect. Notes Math.* 566, 57 – 67.

Krickeberg, K. (1980). Statistical problems on point processes. *Banach Center Publ.* 6, 197 — 223.

Krickeberg, K. (1981). Moment analysis of stationary point processes in $\mathbf{R}^d$. In *Point Processes and Queueing Problems,* Bártfai and Tomkó, eds. North-Holland, Amsterdam.

Krickeberg, K. (1982). Processus ponctuels en statistique. *Lect. Notes Math.* 929, 205 – 313.

Kryscio, R. J., and Saunders, R. (1983). On interpoint distances for planar Poisson processes. *J. Appl. Prob.* 20, 513 — 528.

Kunita, H. (1971). Asymptotic distribution of the non-linear filtering errors of Markov processes. *J. Multivariate Anal.* 1, 365 – 393.

Kunita, H., and Watanabe, S. (1967). On square integrable martingales. *Nagoya Math. J.* 30, 209 – 245.

Kurtz, T. G. (1974). Point processes and completely monotone set functions. *Z. Wahrscheinlichkeitstheorie und verw. Geb.* 31, 57 – 67.

Kutoyants, Yu. A. (1975). Local asymptotic normality for the diffusion type processes. *Izv. Akad. Nauk. Armen. SSR* 10, 103 – 112.

Kutoyants, Yu. A. (1979). Intensity parameter estimation of an inhomogeneous Poisson process. *Prob. Control Inform. Theory* 8, 137 – 149.

Kutoyants, Yu. A. (1980). *Estimation of Parameters of Random Processes.* Akad. Nauk. Armyan. SSR, Erevan.

Kutoyants, Yu. A. (1982). Multidimensional parameter estimation of the intensity function of inhomogeneous Poisson processes. *Prob. Control Inform. Theory* 11, 325 – 334.

Lawless, J. F. (1982). *Statistical Models and Methods for Lifetime Data*. Wiley, New York.

Lawrance, A. J. (1973). Dependency of intervals between events in superposition processes. *J. Roy. Statist. Soc. B* 35, 306 – 315.

Leadbetter, M. R. (1966). On streams of events and mixtures of streams. *J. Roy. Statist. Soc. B* 28, 218 – 227.

Leadbetter, M. R. (1968). On three basic results in the theory of stationary point processes. *Proc. Amer. Math. Soc.* 19, 115 – 117.

Leadbetter, M. R. (1969a). On certain results for stationary point processes and their application. *Bull. Internat. Statist. Inst.* 43, 309 — 320.

Leadbetter, M. R. (1969b). On the distributions of the times between events in a stationary stream of events. *J. Roy. Statist. Soc. B* 31, 295 – 302.

Leadbetter, M. R. (1972a). On basic results of point process theory. *Proc. Sixth Berkeley Symp. Math. Statist. Prob.* 3, 449 — 462.

Leadbetter, M. R. (1972b). Point processes generated by level crossings. In *Stochastic Point Processes,* Lewis, ed. Wiley, New York.

Le Breton, A. (1975). On continuous and discrete sampling for parameter estimation in diffusion type processes. *Math. Programming Studies* 5, 124 – 144.

LeCam, L. (1960a). Locally asymptotically normal families of distributions. *Univ. Calif. Publ. Statist.* 3, 37 – 98.

LeCam, L. (1960b). An approximation theorem for the Poisson binomial distribution. *Pacific J. Math.* 10, 1181 – 1197.

Lee, L. (1982). Distribution-free tests for comparing trends in Poisson series. *Stochastic Process. Appl.* 12, 107 – 113.

Lee, P. M. (1967). Infinitely divisible stochastic processes. *Z. Wahrscheinlichkeitstheorie und verw. Geb.* 7, 147 – 160.

Lee, P. M. (1968). Some examples of infinitely divisible point processes. *Studia Sci. Math. Hungar.* 3, 219 – 224.

Lehmann, E. (1959). *Testing Statistical Hypotheses.* Wiley, New York.

Lehmann, E. (1983). *Theory of Point Estimation.* Wiley, New York.

Lenglart, E. (1977). Relation de domination entre deux processus. *Ann. Inst. H. Poincaré* 13, 171 – 179.

Leonov, V. P., and Shiryayev, A. N. (1959). On a method of calculation of semi-invariants. *Theor. Prob. Appl.* 4, 319 – 329.

Lévy, P. (1954). Processus semi-Markoviens. *Proc. Internat. Cong. Math.* (Amsterdam) 3, 416 – 426.

Lewis, P. A. W. (1964). A branching Poisson process model for the analysis of computer failure patterns. *J. Roy. Statist. Soc. B* 26, 398 – 456.

Lewis, P. A. W. (1965). Some results on tests for Poisson processes. *Biometrika* 52, 67 – 78.

Lewis, P. A. W. (1966). A computer program for the statistical analysis of series of events. *IBM Systems J.* 5, 202 — 225.

Lewis, P. A. W. (1967). Non-homogeneous branching Poisson processes. *J. Roy. Statist. Soc. B* 29, 343 - 354.

Lewis, P. A. W. (1969). Asymptotic properties and equilibrium conditions for branching Poisson processes. *J. Appl. Prob.* 6, 355 - 371.

Lewis, P. A. W. (1972a). Recent results in the statistical analysis of univariate point processes. In *Stochastic Point Processes,* Lewis, ed. Wiley, New York.

Lewis, P. A. W. (1972b). *Stochastic Point Processes: Statistical Analysis, Theory, and Applications*. Wiley, New York.

Lewis, P. A. W., and Shedler, G. S. (1977). Analysis and modelling of point processes in computer systems. *Bull. Internat. Statist. Inst.* 47(2), 193 - 210.

Lewis, T., and Govier, L. J. (1964). Some properties of counts for certain types of point processes. *J. Roy. Statist. Soc. B* 26, 325 - 337.

Liemant, A. (1969). Invariante zufällige Punktfolgen. *Wiss. Z. Friedrich-Schiller-Universität Jena Math.-Naturwiss.-Reihe* 18, 361 — 372.

Liemant, A. (1975). Verallgemeinerungen eines Satzes von Dobruschin V. *Math. Nachr.* 70, 387 - 390.

Liemant, A., and Matthes, K. (1974). Verallgemeinerungen eines Satzes von Dobruschin IV. *Math. Nachr.* 59, 311 — 317.

Liemant, A., and Matthes, K. (1977). Verallgemeinerungen eines Satzes von Dobruschin VI. *Math. Nachr.* 80, 7 - 18.

Liese, F. (1980). Uberlagerung verdunnter und schwach abhangiger Punktprozesse. *Math. Nachr.* 95, 177 - 186.

Lindgren, G. (1980). Point processes of exits by bivariate Gaussian processes and extremal theory of the chi-squared process and its concomitants. *J. Multivariate Anal.* 10, 181 - 206.

Lindsay, B. G. (1983a). The geometry of mixture likelihoods. I. A general theory. *Ann. Statist.* 11, 86 - 94.

Lindsay, B. G. (1983b). The geometry of mixture likelihoods. II. The exponential family. *Ann. Statist.* 11, 783 - 792.

Lindvall, T. (1976). An invariance principle for thinned random measures. *Studia Sci. Math. Hungar.* 11, 269 - 275.

Lindvall, T. (1981). On coupling of discrete renewal processes. *Z. Wahrscheinlichkeitstheorie und verw. Geb.* 48, 57 - 70.

Lindvall, T. (1982). On coupling of continuous time renewal processes. *J. Appl. Prob.* 19, 82 - 89.

Linkov, Yu. N. (1981). Asymptotic power of statistical criteria for counting processes. *Prob. Inform. Transmission* 17, 196 - 205.

Liptser, R. S., and Shiryayev, A. N. (1978). *Statistics of Random Processes,* Vols. I and II. Springer-Verlag, Berlin.

Liptser, R. S., and Shiryayev, A. N. (1980). A functional central limit theorem for semimartingales. *Theor. Prob. Appl.* 25, 667 – 688.

Lo, A. Y. (1982). Bayesian nonparametric statistical inference for Poisson point processes. *Z. Wahrscheinlichkeitstheorie und verw. Geb.* 59, 55 – 66.

Loève, M. (1977). *Probability Theory,* Vol. I. Springer-Verlag, New York.

Lorentz, G. G. (1966). *Approximation of Functions.* Holt, Reinhart and Winston, New York.

Lundberg, O. (1940). *On Random Processes and Their Application to Sickness and Accident Statistics.* Almqvist & Wiksells, Uppsala.

Macchi, O. (1971). Stochastic point processes and multicoincidences. *IEEE Trans. Inform. Theory* IT-17, 1 – 7.

Macchi, O. (1975). The coincidence approach to point processes. *Adv. Appl. Prob.* 7, 83 – 122.

Macchi, O. (1977). Stochastic point processes in pure and applied physics. *Bull. Internat. Statist. Inst.* 47(2), 211 – 241.

Macchi, O., and Picinbono, B. (1972). Estimation and detection of weak optical signals. *IEEE Trans. Inform. Theory* IT-18, 562 — 573.

Mann, N. R., Schafer, R. E., and Singpurwalla, N. D. (1974). *Methods for Statistical Analysis of Reliability and Life Data.* Wiley, New York.

Marshall, A. W., and Proschan, F. (1965). Maximum likelihood estimation for distributions with monotone failure rate. *Ann. Math. Statist.* 36, 69 – 77.

Masry, E. (1978a). Poisson sampling and spectral estimation of continuous-time processes. *IEEE Trans. Inform. Theory* IT-24, 173 – 183.

Masry, E. (1978b). Alias-free sampling: an alternative conceptualization and its applications. *IEEE Trans. Inform. Theory* IT-24, 317 – 324.

Masry, E. (1983). Nonparametric covariance estimation from irregularly spaced data. *Adv. Appl. Prob.* 15, 113 – 132.

Masry, E. and Lui, M.-C. C. (1976). Discrete spectral estimation of continuous-parameter processes—a new consistent estimate. *IEEE Trans. Inform. Theory* IT-22, 289 – 312.

Matheron, G. (1976). A simple substitute for conditional expectation: the disjunctive kriging. In *Advanced Geostatistics in the Mining Industry*, Guariscio, David, and Huijbregts, eds. Reidel, Boston.

Matthes, K. (1963a). Stationäre Zufällige Punktfolgen I. *Jber. Deutsch. Math.-Verein.* 66, 66 – 79.

Matthes, K. (1963b). Unbeschrankt teilbare Verteilungsgesetze stationärer zufälliger Punktfolgen. *Wiss. Z. Hochsch. Elektrotech. Ilm* 9, 235 – 238.

Matthes, K., Kerstan, J., and Mecke, J. (1978). *Infinitely Divisible Point Processes.* Wiley, New York.

Matthes, K. Warmuth, W., and Mecke, J. (1979). Bemerkungen zu einer Arbeit von Nguyen Xuan Xanh und Hans Zessin. *Math. Nachr.* 88, 117 – 127.

Maziotto, G., and Szpirglas, J. (1980). Équations du filtrage pour un processus de Poisson mélangé à deux indices. *Stochastics* 4, 89 – 119.

McFadden, J. A. (1962). On the lengths of intervals in a stationary point process. *J. Roy. Statist. Soc. B* 24, 364 – 382.

McFadden, J. A. (1965). The entropy of a point process. *SIAM J. Appl. Math.* 13, 988 – 994.

McFadden, J. A., and Weissblum, W. (1963). Higher order properties of a stationary point process. *J. Roy. Statist. Soc. B* 25, 413 – 431.

Mecke, J. (1967a). Zum Problem der Zeiligbarkeit stationärer rekurrenter zufälliger Punktfolgen. *Math. Nachr.* 35, 311 – 321.

Mecke, J. (1967b). Stationäre zufällige Masse auf lokalkompakten abelschen Gruppen. *Z. Wahrscheinlichkeitstheorie und verw. Geb.* 9, 36 – 58.

Mecke, J. (1968). Eine charakteristische Eigenschaft der doppelt stochastischen Poissonschen Prozesse. *Z. Wahrscheinlichkeitstheorie und verw. Geb.* 11, 74 – 81.

Mecke, J. (1972). Zufällige Masse auf lokalkompakten Hausdorffschen Raumen. *Beiträge Anal.* 3, 7 – 30.

Mecke, J. (1976). A characterization of the mixed Poisson processes. *Rev. Roumaine Math. Pures Appl.* 21, 1355 – 1360.

Metivier, M., and Pellaumail, J. (1980). *Stochastic Integration.* Academic Press, New York.

Meyer, P.-A. (1965). *Probability and Potentials.* Ginn-Blaisdell, Waltham, Mass.

Meyer, P.-A. (1969). Guide détaillé de la théorie "générale" des processus. *Lect. Notes Math.* 51, 140 – 165.

Meyer, P.-A. (1971). Processus de Poisson Ponctuels, d'après K. Itô. *Lect. Notes Math.* 191, 177 — 180.

Meyer, P.-A. (1976). Un cours sur les intégrales stochastiques. *Lect. Notes Math.* 511, 245 – 400.

Miles, R. E. (1970). On the homogeneous planar Poisson process. *Math. Biosci.* 6, 85 – 127.

Millar, P. W. (1979). Asymptotic minimax theorems for the sample distribution function. *Z. Wahrscheinlichkeitstheorie und verw. Geb.* 48, 233 – 252.

Milne, C. (1982). Transient behaviour of the interrupted Poisson process. *J. Roy. Statist. Soc. B* 44, 398 – 405.

Milne, R. K. (1970). Identifiability for random translations of Poisson processes. *Z. Wahrscheinlichkeitstheorie und verw. Geb.* 15, 195 –

201.
Milne, R. K. (1971). Simple proofs of some theorems on point processes. *Ann. Math. Statist.* 42, 368 – 372.
Mirasol, N. M. (1963). The output of an M/G/∞ queueing system is Poisson. *Oper. Res.* 11, 282 – 284.
Mogyorodi, J. (1971). Some remarks on the rarefaction of the renewal processes. *Lit. Mat. Sb.* 11, 303 – 315.
Mogyorodi, J. (1972). Rarefaction of renewal processes, I and II. *Studia Sci. Math. Hungar.* 7, 285 – 305.
Mogyorodi, J. (1973a). Rarefaction of renewal processes, III and IV. *Studia Sci. Math. Hungar.* 8, 21 – 38.
Mogyorodi, J. (1973b). Rarefaction of renewal processes, V and VI. *Studia Sci. Math. Hungar.* 8, 193 – 209.
Moore, E., and Pyke, R. (1968). Estimation of transition distributions of a Markov renewal process. *Ann. Inst. Statist. Math.* 20, 411 – 424.
Moran, P. A. P. (1970). A note on the serial coefficients. *Biometrika* 57, 670 – 673.
Mori, T. (1971). On random translations of point processes. *Yokohama Math. J.* 19, 119 – 139.
Mori, T. (1975). Stationary random measures and renewal theory. *Yokohama Math. J.* 23, 31 – 54.
Mori, T. (1980). A generalization of Poisson point processes with application to a classical limit theorem. *Z. Wahrscheinlichkeitstheorie und verw. Geb.* 54, 331 – 340.
Moyal, J. E. (1962). The general theory of stochastic population processes. *Acta Math.* 108, 1 – 31.
Naes, T. (1982). The asymptotic distribution of the estimator of the regression parameter in Cox's regression model. *Scand. J. Statist.* 9, 107 – 115.
Nawrotzki, K. (1962). Ein Grenzwertsatz fur homogene zufällige Punktfolgen. *Math. Nachr.* 24, 201 – 217.
Nelson, W. B. (1970). Hazard plotting methods for analysis of life data with different failure modes. *J. Qual. Technol.* 2, 126 – 141.
Nelson, W. B. (1972). Theory and applications of hazard plotting for censored failure data. *Technometrics* 14, 945 – 966.
Nelson, W. B. (1982). *Applied Life Data Analysis*. Wiley, New York.
Neveu, J. (1965). *Mathematical Foundations of the Calculus of Probability*. Holden-Day, San Francisco.
Neveu, J. (1968). Sur la structure des processus ponctuels stationnaires. *C. R. Acad. Sci. Paris* A267, 561 – 564.
Neveu, J. (1975). *Discrete-Parameter Martingales.* North-Holland, Amsterdam.

Neveu, J. (1976). Sur les mesures de Palm de deux processus ponctuels stationnaires. *Z. Wahrscheinlichkeitstheorie und verw. Geb.* 34, 199 – 203.

Neveu, J. (1977). Processus ponctuels. *Lect. Notes Math.* 598, 249 – 447.

Newman, D. S. (1970). A new family of point processes which are characterized by their second moment properties. *J. Appl. Prob.* 7, 338 – 358.

Neyman, J., and Scott, E. L. (1958). Statistical approach to problems of cosmology. *J. Roy. Statist. Soc. B* 20, 1 – 43.

Neyman, J., and Scott, E. L. (1972). Processes of clustering and applications. In *Stochastic Point Processes,* Lewis, ed. Wiley, New York.

Neyman, J., Scott, E. L., and Shane, C. D. (1956). Statistics of images of galaxies with particular reference to clustering. *Proc. Third Berkeley Symp. Math. Statist. Prob.* 3, 75 – 111.

Nguyen, X. X. (1979). Ergodic theorems for subadditive spatial processes. *Z. Wahrscheinlichkeitstheorie und verw. Geb.* 48, 159 – 176.

Nguyen, X. X., and Zessin, H. (1976a). Punktprozesse mit Wechselwirkung. *Z. Wahrscheinlichkeitstheorie und verw. Geb.* 37, 91 – 126.

Nguyen, X. X., and Zessin, H. (1976b). Martin-Dynkin boundary of mixed Poisson processes. *Z. Wahrscheinlichkeitstheorie und verw. Geb.* 37, 191 – 200.

Nguyen, X. X., and Zessin, H. (1979a). Ergodic theorems for spatial processes. *Z. Wahrscheinlichkeitstheorie und verw. Geb.* 48, 133 – 158.

Nguyen, X. X., and Zessin, H. (1979b). Integral and differential characterizations of the Gibbs process. *Math. Nachr.* 88, 105 – 115.

Nguyen-van-Huu. (1979). A limit theorem for functionals of a Poisson process. *Comment. Math. Univ. Carolin.* 20, 547-564.

Oakes, D. (1981). Survival times: aspects of partial likelihood. *Internat. Statist. Rev.* 49, 235 – 264.

Ogata, Y. (1978). Asymptotic behavior of maximum likelihood estimators for stationary point processes. *Ann. Inst. Statist. Math.* 30, 243 – 251.

Ogata, Y., and Vere-Jones, D. (1984). Inference for earthquake models: a self-correcting model. *Stochastic Process. Appl.* 17, 337 – 347.

Osher, J. (1983). On estimators for the reduced second moment measure of point processes. *Math. Operationsforsch. Statist.* 14, 63 – 71.

Osher, J., and Stoyan, D. (1981). On the second-order and orientation analysis of planar stationary point processes. *Biometrics J.* 23, 523 – 533.

Ososkov, G. A. (1956). A limit theorem for flows of similar events. *Theor. Prob. Appl.* 1, 248 – 255.

Palm, C. (1943). Intensitätsschwankungen in Fernsprechverkehr. *Ericsson Technics* 44, 1 – 189.

Papangelou, F. (1970). The Ambrose-Kakutani theorem and the Poisson process. *Lect. Notes Math.* 160, 234 – 240.

Papangelou, F. (1972). Integrability of expected increments of point processes and a related change of scale. *Trans. Amer. Math. Soc.* 165, 483 – 506.

Papangelou, F. (1974a). On the Palm probabilities of processes of points and processes of lines. In *Stochastic Geometry*, Harding and Kendall eds. Wiley, New York.

Papangelou, F. (1974b). The conditional intensity of general point processes and an application to line processes. *Z. Wahrscheinlichkeitstheorie und verw. Geb.* 28, 207 — 226.

Papangelou, F. (1976). Point processes on spaces of flats and other homogeneous spaces. *Proc. Cambridge Philos. Soc.* 80, 297 – 314.

Pham, T. D. (1981). Estimation of the spectral parameters of a stationary point process. *Ann. Statist.* 9, 615 – 627.

Polak, M., and Szynd, D. (1977). An approximation for Poisson processes defined on an abstract space. *Bull. Acad. Polon. Sci.* 25, 1037 – 1043.

Pollard, D., and Strobel, J. (1979). On the construction of random measures. *Bull. Soc. Math. Grèce* 20, 67 – 80.

Pons, O. (1981). Test sur la loi d'un processus ponctuel. *C. R. Acad. Sci. Paris. Ser. I Math.* 292, 91 – 94.

Port, S. C., and Stone, C. J. (1973). Infinite particle systems. *Trans. Amer. Math. Soc.* 178, 307 – 340.

Prékopa, A. (1956a). On stochastic set functions, I. *Acta Math. Hungar.* 8, 215 — 256.

Prékopa, A. (1956b). On stochastic set functions, II and III. *Acta Math. Hungar.* 8, 337 – 400.

Prékopa, A. (1957). On Poisson and composed Poisson stochastic set functions. *Studia Math.* 16, 142 – 155.

Prékopa, A. (1958). On secondary processes generated by a random point distribution of Poisson type. *Ann. Univ. Sci. Budapest Sect. Math.* 1, 153 – 170.

Prentice, R. L., and Kalbfleisch, J. D. (1979). Hazard rate models with covariates, *Biometrics* 35, 25 – 39.

Prentice, R. L., and Self, S. G. (1983). Asymptotic distribution theory of Cox-type regression models with general relative risk form. *Ann. Statist.* 11, 803 – 813.

Proschan, F. and Pyke, R. (1967). Tests for monotone failure rate. *Proc. Fifth Berkeley Symp. Math. Statist. Prob.* 3, 293 – 312.

Przyborowski, J., and Wilenski, H. (1940). Homogeneity of results in testing samples from a Poisson series. *Biometrika* 31, 313 – 323.

Puri, M. L., and Sen, P. K. (1971). *Nonparametric Methods in Multivariate Analysis*. Wiley, New York.

Puri, P. S. (1982). On the characterization of point processes with the order statistic structure without the moment condition. *J. Appl. Prob.* 19, 39 – 51.

Pyke, R. (1959). The supremum and infimum of the Poisson process. *Ann. Math. Statist.* 30, 568 – 576.

Pyke, R. (1961a). Markov renewal processes: definition and preliminary properties. *Ann. Math. Statist.* 32, 1231 – 1242.

Pyke, R. (1961b). Markov renewal processes with finitely many states. *Ann. Math. Statist.* 32, 1243 – 1259.

Pyke, R. (1965). Spacings. *J. Roy. Statist. Soc. B* 27, 395 – 449.

Pyke, R. (1968). The weak convergence of the empirical process with random sample size. *Proc. Cambridge Philos. Soc.* 64, 155 – 160.

Pyke, R., and Schaufele, R. (1964). Limit theorems for Markov renewal processes. *Ann. Math. Statist.* 35, 1746 – 1764.

Quenouille, M. H. (1949). Problems in plane sampling. *Ann. Math. Statist.* 20, 355 – 375.

Qureishi, A. S. (1964). The discrimination between two Weibull processes. *Technometrics* 6, 57 – 76.

Racicot, R. L. (1971). Fitting a filtered Poisson process. *J. Inst. Appl. Math.* 7, 260 – 272.

Räde, L. (1972a). *Thinning of Renewal Processes: A Flow Graph Study*. Göteborg, Sweden.

Räde, L. (1972b). Limit theorems for thinning of renewal processes. *J. Appl. Prob.* 9, 847 – 851.

Radecke, R. M. (1977). Entwicklungsgesetze stationärer markierter Punktprozesse. *Math. Nachr.* 81, 83 – 167.

Radecke, R. M. (1979). Uber eine spezielle Klasse abhangiger Verschiebungen Poissonscher Punktprozesse. *Math. Nachr.* 92, 129 – 138.

Ramlau-Hansen, H. (1983). Smoothing counting process intensities by means of kernel functions. *Ann. Statist.* 11, 453 – 466.

Randles, D. H., and Wolfe, D. A. (1971). *Introduction to the Theory of Nonparametric Statistics*. Wiley, New York.

Rao, M. and Wedel, R. (1968). Poisson processes as renewal processes invariant under translations. *Ark. Mat.* 8, 539 –– 541.

Rebolledo, R. (1977a). Remarques sur la convergence en loi des martingales vers martingales continues. *C. R. Acad. Sci. Paris* 285, 465 – 468.

Rebolledo, R. (1977b). Remarques sur la convergence en loi des martingales vers martingales continues. *C. R. Acad. Sci. Paris* 285,

517 – 520.

Rebolledo, R. (1978). Sur les applications de la théorie des martingales à l'étude statistique d'une famille de processus ponctuels. *Lect. Notes Math.* 636, 27 – 70.

Rebolledo, R. (1980). Central limit theorem for local martingales. *Z. Wahrscheinlichkeitstheorie und verw. Geb.* 51, 269 – 286.

Redner, R. A., and Walker, H. F. (1984). Mixture densities, maximum likelihood and the EM algorithm. *SIAM Rev.* 26, 195 – 239.

Rényi, A. (1964a). On an extremal property of the Poisson process. *Ann. Inst. Stat. Math.* 16, 129 – 133.

Rényi, A. (1964b). On two mathematical models of the traffic on a divided highway. *J. Appl. Prob.* 1, 311 – 320.

Rényi, A. (1967). Remarks on the Poisson process. *Studia Sci. Math. Hungar.* 2, 119 – 123.

Rice, J. (1975). Estimated factorization of the spectral density of a stationary point process. *Adv. Appl. Prob.* 7, 801 – 817.

Ripley, B. D. (1976a). Locally finite random sets: foundations for point process theory. *Ann. Probability* 4, 983 – 994.

Ripley, B. D. (1976b). The foundations of stochastic geometry. *Ann. Probability* 4, 995 – 998.

Ripley, B. D. (1976c). On stationarity and superposition of point processes. *Ann. Probability* 4, 999 – 1005.

Ripley, B. D. (1976d). The second-order analysis of stationary point processes. *J. Appl. Prob.* 13, 255 — 266.

Ripley, B. D. (1977). Modelling spatial patterns. *J. Roy. Statist. Soc. B* 39, 172 – 212.

Ripley, B. D. (1981). *Spatial Statistics*. Wiley, New York.

Ripley, B. D., and Kelly, F. P. (1977). Markov point processes. *J. London Math. Soc. II Ser.* 15, 188 – 192.

Roberts, F. D. K. (1969). Nearest neighbors in a Poisson ensemble. *Biometrics* 56, 401 – 406.

Robinson, P. M. (1977). Estimation of a time series from irregularly spaced data. *Stochastic Process. Appl.* 6, 9 – 24.

Rohde, H., and Grandell, J. (1972). On the removal time of aerosol particles from the atmosphere by precipitation scavenging. *Tellus* 24, 443 – 454.

Rohde, H., and Grandell, J. (1981). Estimates of characteristic times for precipitation scavenging. *J. Atmospheric Sci.* 38, 370 – 386.

Rolski, T. (1981). Stationary random processes associated with point processes. *Lect. Notes Statist.* 5.

Rootzén, H. (1976). A note on the central limit theorem for doubly stochastic Poisson processes. *J. Appl. Prob.* 13, 809 – 813.

Ross, S. M. (1969). Optimal dispatching of a Poisson process. *J. Appl. Prob.* 6, 692 – 699.

Ross, S. M. (1982). *Stochastic Processes.* Wiley, New York.

Roussas, G. (1969). Nonparametric estimation of the transition distribution function of a Markov process. *Ann. Math. Statist.* 40, 1386 – 1400.

Roussas, G. G. (1972). *Contiguity of Probability Measures: Some Applications in Statistics*. Cambridge University Press, Cambridge.

Rozanov, Yu. A. (1967). *Stationary Random Processes.* Holden-Day, San Francisco.

Rozanov, Yu. A. (1982). *Markov Random Fields*. Springer-Verlag, New York.

Rubin, I. (1972). Regular point processes and their detection. *IEEE Trans. Inform. Theory* IT-18, 547 – 557.

Rubin, I. (1974). Regular jump processes and information processing. *IEEE Trans. Inform. Theory* IT-20, 617 – 624.

Rudemo, M. (1972). Doubly stochastic Poisson processes and process control. *Adv. Appl. Prob.* 4, 318 – 338.

Rudemo, M. (1973a). State estimation for partially observed Markov chains. *J. Math. Anal. Appl.* 44, 581 – 611.

Rudemo, M. (1973b). Point processes generated by transitions of Markov chains. *Adv. Appl. Prob.* 5, 262 – 286.

Rudemo, M. (1973c). On a random transformation of a point process to a Poisson process. In *Mathematics and Statistics,* Chalmers Tekniska Högskola, Göteborg, Sweden.

Rudemo, M. (1975). Prediction and smoothing for partially observed Markov chains. *J. Math. Anal. Appl.* 49, 1 – 23.

Rudin, W. (1973). *Functional Analysis.* McGraw-Hill, New York.

Rudin, W. (1975). *Real and Complex Analysis*. McGraw-Hill, New York.

Ryll-Nardzewski, C. (1961). Remarks on processes of calls. *Proc. Fourth Berkeley Symp. Math. Statist. Prob.* 2, 455 – 465.

Savage, I. R. (1956). Contributions to the theory of rank order statistics—the two sample case. *Ann. Math. Statist.* 27, 590 – 615.

Saw, J. G. (1975). Tests on the intensity of a Poisson process. *Comm. in Statist.* 4, 777 – 782.

Sellke, T., and Siegmund, D. (1983). Sequential analysis of the proportional hazards model. *Biometrika* 70, 315 – 326.

Sen, P. K. (1981). The Cox regression model, invariance principle for some induced quantile processes and some repeated significance tests. *Ann. Statist.* 9, 109 – 121.

Sendov, B. (1969). Some questions of the theory of approximation of functions and sets in the Hausdorff metric. *Russian Math. Surveys* 24, 143 – 183.

Serfling, R. J. (1978). Some elementary results on Poisson approximation in a sequence of Bernoulli trials. *SIAM Rev*. 20, 567 – 579.

Serfling, R. J. (1980). *Approximation Theorems of Mathematical Statistics*. Wiley, New York.

Serfozo, R. F. (1972a). Conditional Poisson processes. *J. Appl. Prob*. 9, 288 – 302.

Serfozo, R. F. (1972b). Processes with conditional stationary independent increments. *J. Appl. Prob*. 9, 303 – 315.

Serfozo, R. F. (1975). Functional limit theorems for stochastic processes based on embedded processes. *Adv. Appl. Prob*. 7, 123 – 139.

Serfozo, R. F. (1977). Compositions, inverses and thinnings of random measures. *Z. Wahrscheinlichkeitstheorie und verw. Geb*. 37, 253 – 265.

Shanbhag, D. N., and Westcott, M. (1977). A note on infinitely divisible point processes. *J. Roy. Statist. Soc. B* 39, 331 – 332.

Shapiro, H. S., and Silverman, R. A. (1960). Alias-free sampling of random noise. *J. SIAM* 8, 225 – 248.

Shiryayev, A. N. (1981). Martingales: recent developments, results and applications. *Internat. Statist. Rev*. 49, 199 – 233.

Shorrock, R. W. (1975). Extremal processes and random measures. *J. Appl. Prob*. 12, 316 – 323.

Simar, L. (1976). Maximum likelihood estimation of a compound Poisson process. *Ann. Statist*. 4, 1200 – 1209.

Slivnjak, I. M. (1962). Some properties of stationary flows of homogeneous random events. *Theor. Prob. Appl*. 7, 347 – 352.

Slud, E. V. (1984). Sequential linear rank tests for 2-sample censored survival data. *Ann. Statist*. 12, 551 — 571.

Smith, J. A. (1981). *Point Process Models of Rainfall*. Ph.D. dissertation, The Johns Hopkins University.

Smith, J. A., and Karr, A. F. (1983). A point process model of summer season rainfall occurrences. *Water Resources Res*. 19, 95 – 103.

Smith, J. A., and Karr, A. F. (1985). Statistical inference for point process models of rainfall. *Water Resources Res*. 21, 73 – 79.

Smith, W. L. (1955). Regenerative stochastic processes. *Proc. Roy. Soc. A* 232, 6 – 31.

Smith, W. L. (1958). Renewal theory and its ramifications. *J. Roy. Statist. Soc. B* 20, 243 – 284.

Smythe, R. T. (1975). Ergodic properties of marked point processes in $\mathbf{R}^\nu$, *Ann. Inst. H. Poincaré*, 11, 109 – 125.

Snyder, D. L. (1972a). Filtering and detection for doubly stochastic Poisson processes. *IEEE Trans. Inform. Theory* IT-18, 91 – 102.

Snyder, D. L. (1972b). Smoothing for doubly stochastic Poisson processes.

IEEE Trans. Inform. Theory IT-18, 558 – 562.

Snyder, D. L. (1973). Information processing for observed jump processes. *Inform. and Control.* 22, 69 – 78.

Snyder, D. L. (1975). *Random Point Processes.* Wiley-Interscience, New York.

Solomon, H., and Wang, P. C. C. (1972). Nonhomogeneous Poisson fields of random lines with applications to traffic flow. *Proc. Sixth Berkeley Symp. Math. Statist. Prob.* 3, 383 – 400.

Starr, N. (1974). Optimal and adaptive stopping based on capture rates. *J. Appl. Prob.* 11, 294 – 301.

Stone, C. J. (1968a). On a theorem of Dobrushin. *Ann. Math. Statist.* 39, 1391 – 1401.

Stone, C. J. (1968b). Infinite particle systems and multidimensional renewal theory. *J. Math. Mech.* 18, 201 – 227.

Störmer, H. (1969). Zur Uberlagerung von Erneurrungsprozessen. *Z. Wahrscheinlichkeitstheorie und verw. Geb.* 13, 9 – 24.

Stoyan, D. (1983). *Comparison Methods for Queues and Other Stochastic Processes.* (English edition edited by D. J. Daley.) Wiley-Interscience, New York.

Stoyan, D., and Mecke, J. (1983). *Stochastische Geometrie.* Akademie-Verlag, Berlin.

Stoyanov, J. M., and Vladeva, D. I. (1982). Estimation of unknown parameters of continuous time stochastic processes by observations at random moments. (In Russian.) *C. R. Acad. Bulgare Sci.* 35, 153 – 156.

Strassen, V. (1964). An invariance principle for the LIL. *Z. Wahrscheinlichkeitstheorie und verw. Geb.* 3, 211 – 226.

Strassen, V. (1965). The existence of probability measures with given marginals. *Ann. Math. Statist.* 36, 423 – 439.

Strauss, D. J. (1972). A model for clustering. *Biometrika* 62, 467 – 475.

Streit, F. (1978). Optimality properties of statistical tests for Poisson processes. *Adv. Appl. Prob.* 10, 330 – 331.

Stulajter, F. (1982). Some nonlinear statistical problems of a Poisson process. *Kybernetika* (Prague) 18, 397 – 407.

Sundt, B. (1982). On the problem of testing whether a mixed Poisson process is homogeneous. *Insurance Math. Econ.* 1, 253 – 254.

Susarla, V., and Van Ryzin, J. (1980). Large sample theory for an estimator of the true survival time from censored samples. *Ann. Statist.* 8, 1002 – 1016.

Takacs, L. (1957). On secondary stochastic processes generated by a multidimensional Poisson process. *Publ. Math. Inst. Hungar. Acad. Sci.* 2, 71 – 79.

Tanner, M. A., and Wong, W. H. (1983). The estimation of the hazard function from randomly censored data by the kernel method. *Ann. Statist.* 11, 989 – 993.

Tapia, R. A., and Thompson, J. R. (1978). *Nonparametric Probability Density Estimation.* The Johns Hopkins University Press, Baltimore, Md.

Teicher, H. (1961). Identifiability of mixtures. *Ann. Math. Statist.* 32, 244 – 248.

Thédeen, T. (1964). A note on the Poisson tendency in traffic distribution. *Ann. Math. Statist.* 35, 1823 – 1824.

Thédeen, T. (1967a). Convergence and invariance questions for point systems in $\mathbf{R}^1$ under random motion. *Ark. Mat.* 7, 211 – 239.

Thédeen, T. (1967b). On stochastic stationarity of renewal processes. *Ark. Math.* 7, 249 – 263.

Thédeen, T. (1969). On road traffic with free overtaking. *J. Appl. Prob.* 6, 524 – 549.

Thompson, H. R. (1955). Spatial point processes, with application to ecology. *Biometrika* 42, 102 – 115.

Thomson, P. J. (1982). Signal estimation using an array of recorders. *Stochastic Process. Appl.* 13, 201 – 214.

Tocher, K. D. (1950). Extension of the Neyman-Pearson theory to the discrete case. *Biometrika* 37, 130 – 144.

Tøpsøe, F. (1977). Uniformity in convergence of measures. *Z. Wahrscheinlichkeitstheorie und verw. Geb.* 39, 1 – 30.

Tsiatis, A. A. (1978). An example of nonidentifiability in competing risks. *Scand. Actuar. J.* 1978, 235 – 239.

Tsiatis, A. A. (1981). A large sample study of Cox's regression model. *Ann. Statist.* 9, 93 – 108.

Tucker, H. G. (1964). An estimate of the compounding distribution of a compound Poisson distribution. *Theor. Prob. Appl.* 8, 195 – 200.

Tulya-Muhika, S. (1971). A characterization of *E*-processes and Poisson processes in $\mathbf{R}^n$. *Z. Wahrscheinlichkeitstheorie und verw. Geb.* 20, 199 – 216.

Valkeila, E. (1982). A general Poisson approximation theorem. *Stochastics* 7, 159 – 171.

van der Hoeven, P. C. T. (1982). Une projection de processus ponctuels. *Z. Wahrscheinlichkeitstheorie und verw. Geb.* 61, 483 – 499.

van der Hoeven, P. C. T. (1983). *On Point Processes.* Math. Centre Tracts 165. Mathematisch Centrum, Amsterdam.

van der Plas, A. P. (1983). On the estimation of the parameters of Markov probability models using macro data. *Ann. Statist.* 11, 78 – 85.

van Schuppen, J. (1977). Filtering, prediction and smoothing for counting process observations: a martingale approach. *SIAM J. Appl. Math.* 32, 552 – 570.

van Zuijlen, M. C. A. (1978). Properties of the empirical distribution function for independent nonidentically distributed random variables. *Ann. Probability* 6, 250 – 266.

Vapnik, V. N., and Chervonenkis, A. Ya. (1971). On the uniform convergence of relative frequencies of events to their probabilities. *Thoer. Prob. Appl.* 16, 264 – 280.

Varaiya, P. (1975). The martingale theory of jump processes. *IEEE Trans. Automat. Control* AC-20, 34 – 42.

Vardi, Y. (1979). Asymptotically optimal sequential estimation: the Poisson case. *Ann. Statist.* 7, 1040 – 1051.

Vardi, Y. (1980). On a stopping time of Starr and its use in estimating the number of transmission sources. *J. Appl. Prob.* 17, 235 – 242.

Vardi, Y. (1982a). Nonparametric estimation in renewal processes. *Ann. Statist.* 10, 772 – 785.

Vardi, Y. (1982b). Nonparametric estimation in the presence of length bias. *Ann. Statist.* 10, 616 – 620.

Vardi, Y., Shepp, L. A., and Kaufman, L. (1975). A statistical model for positron emission tomography. *J. Amer. Statist. Assoc.* 80, 8 — 20.

Vere-Jones, D. (1968). Some applications of probability generating functionals to the study of input-output systems. *J. Roy. Statist. Soc. B* 30, 321 — 333.

Vere-Jones, D. (1970). Stochastic models for earthquake occurrence. *J. Royal Statist. Soc. B* 32, 1 — 62.

Vere-Jones, D. (1974). An elementary approach to the spectral theory of stationary random measures. In *Stochastic Geometry,* Harding and Kendall, eds. Wiley, New York.

Vere-Jones, D. (1975a). A renewal equation for point processes with Markov-dependent intervals. *Math. Nachr.* 68, 133 – 139.

Vere-Jones, D. (1975b). On updating algorithms for inference for stochastic processes. In *Perspectives in Probability and Statistics,* Gani, ed. Academic Press, New York.

Vere-Jones, D. (1978). Space-time correlations for micro-earthquakes—a pilot study. *Suppl. Adv. Appl. Prob.* 10, 73 – 87.

Vere-Jones, D. (1982). On the estimation of frequency in point process data. *J. Appl. Prob.* 19A, 383 — 394.

Vervaat, V. (1967). Thinned and stretched Poisson processes. (Dutch) *Stat. Neerl.* 21, 245 – 268.

Voekel, J. G., and Crowley, J. (1984). Nonparametric inference for a class of semi-Markov processes with censored observations. *Ann. Statist.* 12, 142 – 160.

Volkonskii, V. A. (1960). An ergodic theorem for the distribution of the duration of fades. *Theor. Prob. Appl.* 5, 323 — 326.

von Waldenfels, W. (1968). Charakteristische Funktionale zufälliger Masse. *Z. Wahrscheinlichkeitstheorie und verw. Geb.* 10, 279 – 283.

Wald, A. (1947). *Sequential analysis*. Wiley, New York.

Wan, C. B., and Davis, M. H. A. (1977). The general point process disorder problem. *IEEE Trans. Inform. Theory* IT-23, 538 – 540.

Watanabe, S. (1964). On discontinuous additive functionals and Lévy measures of a Markov process. *Japan. J. Math.* 34, 53 – 70.

Watanabe, S. (1977). Poisson point process of Brownian excursions and its applications to diffusion processes. *Proc. Symp. Pure Math.* 31, 153 – 164.

Watson, G. S., and Leadbetter, M. R. (1964a). Hazard analysis I. *Biometrika* 51, 175 – 184.

Watson, G. S., and Leadbetter, M. R. (1964b). Hazard analysis II. *Sankhya* 26A, 110 – 116.

Waymire, E., and Gupta, V. K. (1981). The mathematical structure of rainfall representations, 2: A review of the theory of point processes. *Water Resources Res.* 17, 1273 – 1286.

Waymire, E., and Gupta, V. K. (1983). An analysis of the Pólya point process. *Adv. Appl. Prob.* 15, 39 – 53.

Wegmann, H. (1977). Characterization of Palm distributions and infinitely divisible random measures. *Z. Wahrscheinlichkeitstheorie und verw. Geb.* 39, 257 – 262.

Weiss, P. (1979). On the singularity of Poisson processes. *Math. Nachr.* 92, 111 – 115.

Wellner, J. A. (1977). A martingale inequality for the empirical process. *Ann. Probability* 5, 303 – 308.

Wellner, J. A. (1978). Limit theorems for the ratio of the empirical distribution function to the true distribution function. *Z. Wahrscheinlichkeitstheorie und verw. Geb.* 45, 73 – 88.

Wellner, J. A. (1982). Asymptotic optimality of the product limit estimator. *Ann. Statist.* 10, 595 – 602.

West, S. A. (1977). Upper and lower probability inferences pertaining to Poisson processes. *J. Amer. Statist. Assoc.* 72, 448 – 452.

Westcott, M. (1971). On existence and mixing results for cluster point processes. *J. Roy. Statist. Soc. B* 33, 290 – 300.

Westcott, M. (1972). The probability generating functional. *J. Austral. Math. Soc.* 14, 448 – 466.

Whitt, W. (1972). Limits for the superposition of m-dimensional point processes. *J. Appl. Prob.* 9, 462 – 465.

Whitt, W. (1973). On the quality of Poisson approximations. *Z.*

Wahrscheinlichkeitstheorie und verw. Geb. 28, 23 – 36.
Whitt, W. (1982). Approximating a point process by a renewal process. I. Two basic methods. *Oper. Res.* 30, 125 – 147.
Wichura, M. J. (1970). On the construction of almost uniformly convergent random variables with given weakly convergent laws. *Ann. Math. Statist.* 41, 284 – 291.
Wickwire, K. H. (1981). Optimal stopping problems for differential equations perturbed by Poisson processes. *Naval Logist. Res. Quart.* 28, 423 – 429.
Wilks, S. S. (1962). *Mathematical Statistics*. Wiley, New York.
Willie, J. S. (1982a). Measuring the association of a time series and a point process *J. Appl. Prob.* 19, 597 — 608.
Willie, J. S. (1982b). Covariation of a time series and a point process. *J. Appl. Prob.* 19, 609 — 618.
Wisniewski, T. K. M. (1972). Bivariate stationary point processes: fundamental relations and first recurrence times. *Adv. Appl. Prob.* 4, 296 – 317.
Wold, H. (1948). On stationary point processes and Markov chains. *Skand. Aktuar.* 31, 229 – 240.
Wu, C. F. J. (1983). On the convergence properties of the EM algorithm. *Ann. Statist.* 11, 95 – 103.
Yadrenko, M. I. (1983). *Spectral Theory of Random Fields*. Optimization Software, Inc., New York.
Yaglom, A. M. (1962). *An Introduction to the Theory of Stationary Random Functions*. Prentice-Hall, Englewood Cliffs, N.J.
Yandell, B. S. (1983). Nonparametric inference for rates and densities with censored serial data. *Ann. Statist.* 11, 1119 – 1135.
Yor, M. (1976). Représentation des martingales de carré intégrable relative aux processus de Wiener et de Poisson à *n* paramètres. *Z. Wahrscheinlichkeitstheorie und verw. Geb.* 39, 121 – 129.
Zaat, J. C. A. (1964). Interrupted Poisson processes, a model for the mechanical thinning of beet plants. *Stat. Neerl.* 18, 311 – 324.
Zacks, S. (1971). *Theory of Statistical Inference*. Wiley, New York.
Zakai, M. (1969). On the optimal filtering of diffusion processes. *Z. Wahrscheinlichkeitstheorie und verw. Geb.* 11, 230 – 243.
Zeephongsekul, P. (1981). Laplace functional approach to point processes occurring in a traffic model. *Ann. Probability* 9, 1034 – 1040.

Index